Bear and Human

The Archaeology of Northern Europe
Volume 3

General Editors
Paul S. Johnson, *University of Nottingham*
Sam Turner, *Newcastle University*

Editorial Board
Ian Armit, *University of York*
Manuel Fernández-Götz, *University of Edinburgh*
Julie Lund, *Universitetet i Oslo*
Karen Milek, *Durham University*
Neil Price, *Uppsala universitet*
Magali Watteaux, *Université Rennes 2*
Mark White, *Durham University*

Advanced Studies on the Archaeology and History of Hunting, edited by the Centre for Baltic and Scandinavian Archaeology (ZBSA), Volume 3

Bear and Human

Facets of a Multi-Layered Relationship from Past to Recent Times, with Emphasis on Northern Europe

Edited by Oliver Grimm,
in cooperation with Daniel Groß, Alexandra Pesch, Olof Sundqvist, and Andreas Zedrosser

A volume based on papers presented at a conference at Orsa Predator Park, Dalarna, Sweden, Oct. 16th to 18th, 2019

Volume 3.3

BREPOLS

British Library Cataloguing in Publication Data

A catalogue record for this book is available from the British Library.

Copy editor: Gundula Lidke

Layout, typesetting, and image editing: Matthias Bolte and Cornelia Lux-Kannenberg. Cover design by TopicA.

The editor has made every possible effort to identify all copyright owners. In the case that copyrights have not been cleared, please contact the editor.

© 2023, Brepols Publishers n.v., Turnhout, Belgium.

This is an open access publication made available under a CC BY-NC 4.0 International License: https://creativecommons.org/licenses/by-nc/4.0/. No part of this publication may be reproduced, stored in a retrieval system, or transmitted, in any form or by any means, for commercial purposes, without the prior permission of the publisher, or as expressly permitted by law, by licence or under terms agreed with the appropriate reprographics rights organization.

D/2023/0095/164

ISBN: 978-2-503-60611-8 (3 vols)

e-ISBN: 978-2-503-60613-2 (3 vols)

Volume DOI: 10.1484.M.TANE-EB.5.134321

Three-volume set : DOI: 10.1484/M.TANE-EB.5.133678

Printed in the EU on acid-free paper.

Conference sponsors

Vetenskapsrådet (Swedish Research Council)

Kungl. Gustav Adolfs Akademien för svensk folkkultur (The Royal Gustavus Adolphus Academy for Swedish Folk Culture)

Stockholms universitet (Stockholm University)

Zentrum für Baltische und Skandinavische Archäologie (Centre for Baltic and Scandinavian Archaeology)

Book sponsors

Kungl. Gustav Adolfs Akademien för svensk folkkultur (The Royal Gustavus Adolphus Academy for Swedish Folk Culture)

Länsstyrelsen Västerbotten (County Administration of Västerbotten, Sweden)

International Council for Game and Wildlife Conservation (CIC)

List of contents

Book 1

List of contents 7

Foreword by Oliver Grimm 13

Chapter 1 – Bear and human: Facets of a multi-layered relationship – introduction, discussion and synthesis 17

"Bear and human" – introduction, discussion and synthesis
Oliver Grimm 19

Chapter 2 – Bears in biology (Europe) 49

Conservation status and distribution of the brown bear in Europe
Andreas Zedrosser and Jon E. Swenson 51

The history of the Scandinavian Brown Bear Research Project – a formidable success story
Jon E. Swenson and Sven Brunberg 63

The management of brown bears in Sweden, Norway and Finland
Michael Schneider, Andreas Zedrosser, Ilpo Kojola and Jon E. Swenson 77

Genetics of brown bears in northern Europe
Alexander Kopatz 99

Hibernation ecology of brown bears in Sweden
Andrea Friebe, Jon E. Swenson and Andreas Zedrosser 109

The social system of a "nonsocial" species, the brown bear
Andreas Zedrosser, Shane C. Frank, Jennifer E. Hansen, Sam M. J. G. Steyaert, J. E. Swenson . . 121

Sexually selected infanticide as a mating strategy in brown bears
Andreas Zedrosser, Sam M. J. G. Steyaert and Jon E. Swenson 129

Bears – fact and fiction about bear hunting and intelligence
Oliver Grimm, Andreas Zedrosser and Jon E. Swenson 137

Chapter 3 – Bear hunting (Europe) 147

Bear hunting in the later Middle Ages and early modern period, viewed from the perspective of art history and contemporary textual sources
Richard Almond 149

Chapter 4 – Animal agency (northern Europe) 173

Posthuman bears: Sight, agency, and baiting in Early Modern England
Liam Lewis 175

Chapter 5 – Bears in long-term archaeo(zoo)logical studies (northern Europe) 185

Brown bears in burials and entertainment in later prehistoric to modern Britain (*c.* 2400 BC – AD 1900s)
Hannah J. O'Regan 187

Bears and humans in Sweden – 10,000 years of interactions from the Mesolithic to the Middle Ages
Ola Magnell 209

Zooarchaeological brown bear (*Ursus arctos*) finds in eastern Fennoscandia
Kristiina Mannermaa, Tuija Kirkinen and Suvi Viranta-Kovanen 235

The history of the brown bear (*Ursus arctos* L.) in the northern German lowlands
Ulrich Schmölcke 265

In the company of bears: The role and significance of the bear from the perspective of the Holocene hunter-gatherer-fishers of the East European Plain forest zone (10th–3rd millennium BC)
Ekaterina A. Kashina and Anastasia A. Khramtsova 291

Chapter 6 – Bears in archaeo(zoo)logical, focused analysis (northern Europe) 315

The White One: How to frame the narrative of the world's oldest intact polar bear skeleton, specimen S10673 from Finnøy, southwestern Norway, in a museum display
Kristin Armstrong Oma and Elna Siv Kristoffersen 317

The bear minimum. Reconsidering ursine remains and depictions at Pitted Ware culture (*c.* 3200–2300 BC) sites in Sweden
Tobias Lindström 335

The Kainsbakke bears and changing patterns in the human-bear relationship through the Danish Mesolithic and Neolithic
Lutz Klassen and Kristian Murphy Gregersen 351

Bears and the Viking Age transition in Sweden
John Ljungkvist and Karl-Johan Lindholm. 387

Book 2

The occurrence of *Ursus arctos* in relation to other faunal remains in burials during the
Late Iron Age (560/70–1050 CE) in Uppland, Sweden
Hannah Strehlau. 417

Bear bones from the Viking Age cult place at Frösö church – the unifying factor in
bear-human relationships in Viking Age Jämtland, northern Sweden
Ola Magnell. 429

Bear claws in Iron Age burials on Gotland, Sweden – a first survey
Jane Jordahl, John Ljungkvist and Sabine Sten. 453

Claws in Late Iron Age graves (*c*. 550–1100 CE) and bones in a castle (post 1500) –
Ursus arctos in the Åland archipelago
Rudolf Gustavsson and John Ljungkvist. 469

The power of the paw. Multi-species perspectives on the bear claw burial tradition in
a long-time perspective in South Norway
Anja Mansrud. 489

Bear skin burials revisited: Norway and Sweden, mainly Migration Period
Oliver Grimm. 533

Sámi bear graves – results from archaeological and zooarchaeological excavations and
analyses in the Swedish part of Sápmi
Elisabeth Iregren. 547

Sámi bear graves in Norway – hidden sites and rituals
Ingrid Sommerseth. 587

Bear bones at Saami offering sites
Marte Spangen, Anna-Kaisa Salmi, Tiina Äikäs and Markus Fjellström. 603

Bear skin trade in the late 1st/early 2nd millennium AD – what do we know from Russian sources?
Andrei V. Zinoviev. 619

The bear cult in medieval Novgorod, based on archaeological finds
Elena A. Tianina. 629

Evidence of bear remains in a cremation burial in the Moscow region (Burial 5, Kremenye burial
ground on the upper river Oka, 12th century)
Alexander S. Syrovatko, Natalia Svirkina and Liudmila Plekhanova 639

Chapter 7 – Bears in the history of religion (northern Europe) 653

Bears in Old Norse religion with specific references to the *berserkir*
Olof Sundqvist 655

"The Bear Ceremonial" and bear rituals among the Khanty and the Sami
Håkan Rydving 677

The songs and rituals of the Finno-Karelian bear hunt: Gifts, seduction and mimesis in the forest
Vesa Matteo Piludu 693

The Finno-Karelian bear feast and wedding: The bruin as a guest of honour of the village
Vesa Matteo Piludu 723

The Finno-Karelian bear skull rituals: Bringing the bruin home to ensure its regeneration
Vesa Matteo Piludu 745

The human-bear relationship among swidden cultivators and forest peasants in Savonia, Finland, and central Scandinavia
Marja-Liisa Keinänen 761

Karhurokka – traditional bear meat soup and other bear meat recipes from Finland
Tuija Kirkinen 777

Bear skins as a church offering
Teppo Korhonen 781

Bears in churches: Skins, paws, and claws from Norway
Jahn Børe Jahnsen 795

Book 3

Chapter 8 – Bears in literary studies and the history of ideas (northern Europe) 825

Bears, kennings and skaldic poetry
Maria Cristina Lombardi 827

The role of bears in Old Norse literature – a bestiary concept?
Agneta Ney 839

The bear in popular belief, legend and fairy tale
Klaus Böldl 851

Killer bears and bear killers in 19th-century Sweden
Karin Dirke 863

From monster to endangered animal: Three bear stories by Selma Lagerlöf
Claudia Lindén 875

Bears as *pares*: Some notes on bear stories in Zapinejie (Arkhangelskaya oblast, northern part of the Russian Federation) and the tendency to equality in human-bear relations
Andrey V. Tutorski 887

Chapter 9 – Bears in philology (northern, central and eastern Europe) . . 901

Bjørnestad, *Bjørnbåsen*, and *Godfardalen*: Bear/human relations as referred to in place names from southwestern Norway
Inge Særheim 903

Germanic "bear" and Germanic personal names before *c.* AD 1000 with elements referring to "bear"
Robert Nedoma 921

The Slavic word for "bear"
Jürgen Udolph 933

Chapter 10 – Bears in image science (northern Europe) 941

Stone Age amber bear figurines from the Baltic Sea area
Daniel Groß and Peter Vang Petersen 943

The bear in Late Iron Age and Viking Period Scandinavian art – a survey
Sigmund Oehrl 961

Bears in Swedish imagery, AD 1000–2000
Åsa Ahrland and Gert Magnusson 991

Chapter 11 – Bears in Classical Antiquity 1017

Bear und human in Greco-Roman antiquity
Florian Hurka 1019

Bears in Early and Middle Byzantine art (330–1204)
Martina Horn 1031

Chapter 12 – Further reading: Bears in a broader perspective 1063

The role of bears in the Late Bronze and Early Iron Ages in southern Germany, with a focus on the Hallstatt period
Melanie Augstein 1065

Tracking former royal dignity: The bear in medieval German literature
Sabine Obermaier 1091

"The Bear's Son Tale": Traces of an ursine genealogy and bear ceremonialism in a pan-European oral tradition
Roslyn M. Frank 1107

The bear in European folktales – with a special focus on Scandinavian variants
Angelika Hirsch 1121

The role of the bear in the Russian folk tale: Personage, plot type, and behavioural scenarios
Inna Veselova 1133

Bears bring spring: An anthropological view on the role of the bear in middle European winter feasts
Jet Bakels and Anne Marie Boer 1147

What are those bears doing there? On a painting from early Italian art
Henk van Os 1159

Bear-human interactions: Archaeological and ethnographic investigations in North American indigenous cultures
Kerry Hull 1163

Bears in the starry sky
Ernst Künzl 1185

Chapter 8

Bears in literary studies and the history of ideas (northern Europe)

The bear that illustrates H. Falk's instructions (H. Falk, Underrättelser om björnskall, *Stockholm 1828, here the title page) on how to kill such an animal looks weary and suspicious, rather than aggressive and bold. Surprisingly, Falk, a hunter in the period when a state bear extermination policy was in force, expresses how the majestic bear evokes fear and awe and hopes for its survival (see Dirke, this volume).*

Bears, kennings and skaldic poetry

By Maria Cristina Lombardi

Keywords: Skaldic poetry, kenning, bear

Abstract: "Metamorphosis and identity are the two limits of human existence, incompatible with one another, but complementary in that human life exists in a movement between these two limits" (LICHTENSTEINS 1963). This is particularly true in the poetic language, where the use of metaphors makes crossing between the two possible and easy. In skaldic poetry the close connection among different domains (animal, human, material) and the ontological border instability between them is so relevant that it has originated in a vast and rich amount of literary and linguistic expressions through which contacts and interlacements have led to transformations and metamorphosis. Skaldic kennings (rethorical tropes which are notably creative and frequent in this genre) are far more numerous and complex than those in Eddic lays. Skalds, Viking court poets who composed their texts for the drótt *(the king and his followers), had to show their skills and inventiveness through a massive use of these complicated linguistic ornaments, which they embedded in quite short texts where metaphors based on animals often occurred. In my paper I will investigate some skaldic stanzas with kennings containing animal metaphors, and then I will discuss those presenting* heiti *for "bear" mentioned in the* Þulur – *lists of poetic terms at the end of* Skáldskaparmál *in Snorri's Edda – where under the name "bear" we find epithets and periphrases to be used as alternatives to the common substantive "bear", by this way satisfying the quest of synonyms typical of Old Norse skaldic verse. My hypothesis is that Scandinavian poetry – with its mythological lore essentially based on war and battle – contains kennings with animals traditionally connected with the battlefield (wolves, eagles, ravens, etc.) rather than with bears; the latter animals being associated with hunting and belonging to non-Germanic traditions.*

"Metamorphosis and identity are the two limits of human existence, incompatible with one another, but complementary in that human life exists in a movement between these two limits" (LICHTENSTEINS 1963, 214). This is particularly true in the poetic language, where the massive use of metaphors makes crossing between the two possible and easy. This peculiarity of poetic language represents the real power of poetry; since ancient times poetry has been capable of creating new landscapes – phantastic realities, where creativity and artistic discourse have been expressed without hindrances.

The close connection among different domains (animal, human, material) and the ontological border instability between them are so relevant that they have originated in a vast and rich amount of literary and linguistic expressions through which contacts and interlacements have led to transformations and metamorphosis. The history of literature is full of animal metaphors which connect poetry to the animal world, and writing to game (BLEAKLEY 2000, 26–27). Ovid's *Metamorphoses* represent perhaps the most famous example of this kind, narrating myths and legends already spread and known among Greeks and Romans. In Old Norse sagas *berserksgangr* can be compared to a kind of metamorphosis; special warriors consecrated to Odin became sorts of bears in battle, transforming into bear-like beings in their ferocity, aggressivity, loosing humanity (see SUNDQUIST, this volume). They usually preserved this evil nature even in their normal lives as some *Íslendingasǫgur* show (cf. *Eyrbyggja saga* 1957, chap. 25).

If we view Old Norse culture as more or less directly deriving from an Indo-European population which is supposed to have settled down in Scandinavia around the 2nd millennium BC (HEDEAGER 2004, 225–228; JENNBERT 2004, 206–213), adapting to the ecology and in contact with other significant cultures, in such a context Odin's shamanic nature is particularly relevant especially with regard to bear traditions. Germanic cultures had taboos against uttering the bear's name, which led to the replacement of the Indo-European word for "bear" by noa words stemming from the colour brown (bear, bjǫrn, etc.; cf. NEDOMA, this volume; FROG 2008, 12–13). We have an example of this in one of the oldest lays of the poetic Edda, *Vǫlundarkviða*, where "bear" is referred to directly as *brunni*, "the brown one", a noa word for "bear". Moreover, hunters with a freshly killed bear are a topic of folktales in northeastern Scandinavia, and stories about bear-men are spread in those areas: we have many "memorates" collected by Bengt af Klinterberg describing experiences of men transforming into bears and *vice versa* (e.g. *Lappen som gick björn* "The Sami who transformed into a bear", cf. AF KLINTERBERG 1986, 305–306). But it must be said that Old Norse sources and folktales are very late and therefore hardly reliable concerning ancient rituals and beliefs in general, and of course this also applies to those about bears.

Since the 12th century, written sources have been preserved from medieval Iceland, many of them in Old Norse language. Different works, both in prose and poetry, and various literary genres – such as sagas or the collection of heroic and mythologic stories called the poetic Edda – appeared and were copied in Scandinavia. As far as poetry is concerned, skaldic poems, a very peculiar genre presenting a quite sophisticated form, hosted special rhethorical devices which will be analysed here.

I will briefly mention an Icelandic saga which has been taken as an example of text preserving a bear ritual, *Hrólfs saga kraka*, as well as some stanzas in the poetic Edda. Then I will concentrate on skaldic poetry because skalds were absolutely innovative and revolutionary in linguistic inventiveness, through their exceptionally complicated kennings where personifications of natural elements – trees acting as warriors, ships and swords acting as animals (horses, serpents, bears) – constantly occur. By applying cognitive linguistic theory by George Lakoff and Mark Johnson (LAKOFF/JOHNSON 2003) which studies the relationships between language and thought to skaldic poetry, we find recurring models expressing the perception and the conception of the world which concretise what Peter Gärdenfors maintains: if we want to suggest a particular quality (mobility, strength, power, etc.) of an object, for example a sword, a ship (i.e. rapidity, aggressivity, or strength), we can do that by using an adjective or, more effectively, through calling it directly by the name of a being (an animal) having these qualities (GÄRDENFORS 2000).

Among sagas, *Hrólfs saga kraka*, analysed by SCHJØDT (2003, 261–278) as a narrative based on initiation rites, has been taken as an example of initiation into the status of *berserkr* warriors. Unfortunately the saga is very late, and it is difficult to take for granted what such a late narrative sequence tells about initiation as if it would mirror old rituals (for example those relating women to bear rituals, always providing women with rings through which they must look at the ceremony).[1] The Scandinavian Germanic society had undergone many crucial changes since the pre-Christian period (into which the *berserkr* rites must presumably be placed). Moreover, there is no external evidence for such rites in Germanic communities while there is much for Sami rites.

---

1   Clive Tolley has recently treated the accounts of coupling between a bear-man and a woman in *Hrólfs saga kraka* and *Gesta Danorum* Book X, arguing for the transmission and adaptation of these narrative complexes from Sami traditions (TOLLEY 2007).

Similarly, according to Frog and other scholars, sources related to the Sami are quite recent, and it cannot be demonstrated with certainty that those rituals (as described in *Hrólfs saga kraka*) really existed in the past (FROG 2008, 1–50).

Nevertheless the Sami, although influenced from outside themselves – i.e. by the Scandinavians – preserved stories and practices connected to the bear (as the above quoted "memorates" collected by af Klinterberg show).[2] But *Hrólfs saga kraka* is a completely different type of source – it is Icelandic and thus very far from the areas which Sami culture had influenced (Norway, Sweden).

The poetic Edda presents almost exclusively negative images of bears as destroying and killing creatures; they appear as symbols of bad luck in Gunnar's premonitory speech, full of scary images, which he makes before leaving for Atli's kingdom in *Atlakviða*, 11 (one of the Sigurðr-cycle poems; cf. *Edda* 1962, 242; *Edda [transl.]* 1923, 486):

> *Ulfr mun ráða*
> *arfi Niflunga,*
> *gamlir, granverðir,*
> *ef Gunnars missir,*
> *birnir blacfiallir*
> *bíta þreftǫnnom,*
> *gamna greystóði,*
> *ef Gunnarr né kømrað.*

> The wolves then shall rule / the wealth of the Niflungs,
> wolves aged and grey-hued, / if Gunnar is lost,
> and black-coated bears / with rending teeth bite,
> and make glad the dogs, / if Gunnar returns not.

They appear also in Kostbera's dream about the terrible outcome of her husband's (Högni, Gunnar's brother) journey, when she tries to dissuade him from leaving, in *Atlamál*, 17 (*Edda* 1962, 250; *Edda [transl.]* 1923):

> *Bjǫrn hugða ec hér inn kominn, bryti up stocca,*
> *hristi svá hramma, at vér hrædd yrðim;*
> *munni oss mǫrg hefði, svá at vér mættim ecci;*
> *þar var oc þrǫmmon þeygi svá lítil.*

> A bear saw I enter, / the pillars he broke,
> and he brandished his claws / so that craven we were;
> with his mouth seized he many, / and nought was our might,
> and loud was the tumult, / not little it was.

Unlike skaldic poetry, which is a typical product of the Viking age, Eddic lays have their roots in old Germanic lore (FROG 2008, 6). *Vǫlundarkviða* ("The song of the marvellous craftsman Vǫlundr"), for instance, has correspondences in other Germanic traditions, i.e. in Anglo-Saxon elegies (the so-called *Deor's Lament* mentioning Weland, the Old English name of Vǫlundr) and the so-called Franks

---

2   The bear rites recorded from northern Siberia are remarkably homogeneous (ie. bear rites where the bear had become an object of cult, a totem animal for some tribes, their divine progenitor, among Ob Hungarians); see HONKO et al. 1994, 120–135; PENTIKÄINEN 2007, 45–47 ; see also RYDVING, this volume.

Casket, an early 8th-century decorated ivory casket, partly preserved at the British Museum in London and partly at the Bargello Museum in Florence. Furthermore, *Vǫlundarkviða* shows some relationships with Sami culture.[3] It starts with a prose passage where three brothers (one of them is Vǫlundr), said to be the sons of the Finns' king, are described as skilful hunters. They marry three swan-maidens (who are in fact valkyries) who after seven years fly away from them. Vǫlundr's brothers leave in search for their wives, while Vǫlundr stays at home and waits for his wife, forging rings in order to attract her. Here the presence of rings (always appearing in traditional after-hunting rituals attended by women) forged for a woman is particularly interesting (*Edda* 1962, 118; *Edda [transl.]* 1923, 257):

Kom þar af veiði
veðreygr scyti,
Vǫlundr, líðandi
um langan veg.

Vǫlundr home
from his hunting came,
from a weary way,
the weather-wise bowman.

In stanzas 9 and 10 we learn that he sits on a bear skin while cooking bear steaks, at which point the animal is referred to directly as *brúnni*, "the brown one", the above mentioned noa word for "bear" (*Edda* 1962, 118; *Edda [transl.]* 1923, 258–259):

9.
Gecc brúnni
bero hold steikia, […].

A brown bear's flesh / would he roast with fire, […].

10.
Sat á berfialli,
bauga talði, […].

On the bear skin he rested, / and counted the rings […].

In stanza 17 Vǫlundr bares his teeth like an animal, and he is referred to as a *dyrr*, an "animal", when in captivity (*Edda* 1962, 119; *Edda [transl.]* 1923, 261):

17.
Ámun eru augu
ormi þeim inum frána,
tenn hánum teygjask.

The glow of his eyes / is like gleaming snakes,
His teeth he gnashes.

---

3   Hatto (1980, 327) shows that the folktale has sub-Arctic roots related to totemic ancestry.

Nevertheless we must say that these kinds of allusions to bear-myths are quite rare in Old Norse sources. Gods and myths of the Vikings have drawn attention away from ancient customs among Scandinavians, in particular those living in the northeastern area. From archaeological evidence (DuBois 2014, 47–48), we know that hunter-gathers, closely dependent on seasonal patterns, replicated their concern for survival in art and ritual (totemic figures, representing mythical ancestors for the clan or the community; cf. Zachrisson 2008, 32–39). In this regard Thomas DuBois observes that already then "the relative paucity of depictions of the bear – held as supernaturally powerful throughout ancient North Eurasia – may also indicate a totemic status of that animal, this time through taboos against representation" (DuBois 2014, 48). Generally also pronouncing the totemic entity's name was forbidden.

Coming to skaldic poetry, we must keep in mind that the first skaldic texts were composed orally and that they were written down only after the 12th century. The genre survived even after the end of the Viking age, until the 14th century, with a flourishing of Christian skaldic compositions. Skalds were particularly concerned with linguistic experiments; their kennings (very peculiar "half-metaphors") were highly creative and far more numerous and complex than those in Eddic lays.[4] Skalds were Viking court poets who composed their texts for the *drótt* (the king or jarl, and his followers). They usually described their king's or chieftain's warlike enterprises, but also their own love sufferings or other topics, and they had to show their skills and inventiveness through a massive use of kennings embedded in quite short texts (where kennings represented a relevant portion). Skalds elaborated endlessly on the form of their verses and the poetic components used for the battlefield, its protagonists and actions: texts were full of metaphors for "warrior", "sword", or "ship", many of which contained animal names.

Gaston Bachelard compares metamorphosis to metaphor, by arguing that the true "animalizing imagination" tends to deform metaphors towards metamorphosis (Bachelard 1965, 51). The passage of the animal into the language implies a double movement – that of the animal into the language (the animalisation of language) and that of the language into the animal, making the animal loose its nature and assume the value of a sign-symbol (ibid.).

However, kennings – even the metaphorical ones (which have a metaphor as a "Base word") – do not permit a complete metamorphosis, because they bind and limit metaphors to their referents by "Determinants" (in the genitive), always containing terms belonging to the Referent's habitat, as in *sævar bjǫrn* = "the bear" (Base word) "of the sea" (Determinant)" = the ship (Referent).

Here the ship is called *bjǫrn* "bear" (a metaphor), while the term in the genitive *sævar* "of the sea" connects the bear to the sea, thus suggesting the referent's domain and habitat. An element of the original domain is therefore preserved. Here an animal metaphor is used to indicate an unanimate object (ship), thus giving the object the aesthetic and intrinsic qualities of the animal; the bear (imposing build, strength, energy, violence) transforms the ship into a hybrid (a bear-ship being) – a strong and dangerous war-ship.

We have many variations of this kenning for "ship" (mentioning other animals); the ship may be called a "horse" (the horse of the sail), a metaphor indicating a speedy means of transport, while the

---

[4] A kenning is a stylistic figure that speaks to knowledge and reason and involves word-play and logical relationships. The fact that kennings occur in other poetic corpora suggests that they are an inheritance from Germanic poetry. The skaldic kennings differ in two characteristic features from the older ones: in their variability, and in their building of complex clusters. A *heiti* "name, appellation, designation" is a synonym in place of the normal word for something. *Heiti* are distinguished from kennings in that a *heiti* is a simple word, whereas a kenning is a circumlocution in the form of a phrase or a compound word (cf. Marold 2012, lxxv–lxxxv).

term in the genitive "of the sail" (a metonym, since the sail is a part of the ship) connects the horse to the ship, thus revealing the Referent's domain.[5]

Other variations of this kenning may also contain synonyms both for "bear" and for "sea" (quoted in Snorri's *Edda* [*Skáldskaparmál*], which I will soon analyse), i.e. *iugtanni hvalranns* "the greedy-toothed one" (an epithet for "bear") "of the whale-house" (a kenning for "the sea"), or *sævar vetrliði* "the winter-survivor of the sea".

DuBois (2014) claims that when an animal appears capable of signalling important events such as seasonal shifts – as we can infer from the kenning "winter survivor" for "bear", it probably indicates some kind of sacral relevance of it. Thus, calling the bear "winter survivor" in Viking poems points out the animal's hibernation and its awakening as a type of re-birth in spring. We might view the presence of epithets like this in Old Norse kennings as a precious piece of stratified knowledge belonging to old memories and traditions which had remained active in the Viking age, when they were combined with new deities and other rituals (ibid.).

The concentrated language usually characterising kennings makes it a difficult cognitive process for present day readers to decode them, but this was not the case for the original audience of skaldic poetry. They knew the complicated lists of synonymic names for the different referents, i.e. "the bear of the sail", or "the bear of the keel", both for "ship", or "the horse of the whale-house", also for "ship". The surprise effect was rather produced by how these were combined, their sound effects, and the alternation of the strong and weak accents in metrics. Personificated trees with spears and swords instead of bows and leaves (referring to warriors), and bears with keels and sails going on the sea instead of walking in the forest: such images of hybrids, strange creatures halfway between human and animal, or monsters (a kind of science fiction *ante litteram*) are suggested by kenning combination mechanisms.

But let's return to the bear. In my research on Viking poetry I have found that bear-mentions are few and generally not based on myths but rather on natural observation. So, how does the bear appear and act in skaldic kennings and *heiti*?

In this paper I will consider some kennings in a skaldic stanza (*lausavísa*) by Markús Skeggjason, a famous Icelandic skald who was also law-speaker in Iceland from 1084 until his death in 1107.[6] In Ari Þorgilsson's *Íslendingabók* (*Íslendingabók/Landnámabók* 1968), Markús is mentioned as an important informant for Ari about the lives of the earlier law-speakers in Iceland.

This *lausavísa*, quoted in Snorri's *Edda* (*Skáldskaparmál* 157) and in Óláfr Þórðarson's *Third Grammatical Treatise* (cf. Ólsen 1884, 76) presents several kennings, with synonyms for "bear" as metaphorical Base words.[7]

I will investigate and compare them with other kennings and *heiti* for "bear" mentioned in the (Younger) Edda, which was compiled and written down by Snorri Sturluson in the 1220s. In particular, I will analyse its part devoted to poetic diction, called *Skáldskaparmál*, and the *Þulur*-lists of poetic terms. Here, under the name "bear", we find epithets and periphrases to be used as alternatives to the common substantive "bear", this way satisfying the quest of synonyms typical of Old Norse skaldic verse. They refer to the animal's physical appearance and to its behaviour, as shown in folklore and popular traditions, rather than to mythological stories.

---

5   Already in the 13th century, Óláfr Þórðarson's *Third Grammatical Treatise* describes and analyses kennings and poetic compounds; he explains that a ship can be paraphrased as "animal of the sea" and illustrates how metaphors move from an animate being (bear) to an inanimate object (ship; cf. Ólsen 1884, 76).
6   Markús was the son of Skeggi Bjarnason and possibly a brother of the poet Þórarinn Skeggjason. He appears to have had close ties to the Church; during his time as a lawspeaker, and with his guidance, Gizurr Ísleifsson, bishop of Skálholt (1081–1118), established the Icelandic tithe laws (*Íslendingabók/Landnámabók* 1968, I, 22).
7   The stanza is in *dróttkvætt* metre (the metre of the chieftain's followers).

In *Skáldskaparmál* the kennings for "bear" follow those for "wolf" – which are far more numerous –, and skaldic stanzas where synonyms for "wolf" are quoted come after them. Here we find frequent allusions to mythological stories involving wolves, i.e. Fenrir (the wolf who will swallow Odin in the Ragnarök, the final fight between gods and giants and all evil powers), or Geri and Freki (the two wolves – whose names both mean "the ravenous" or "the greedy one" – who are said to accompany the god Odin).[8] By the way, the name Freki (Old Norse *frekr* "greedy", Old English *frec* "desirous, greedy", and Old High German *freh* "greedy") occurs also among "bear" names.

After the *heiti* referring to "wolf", a short list of synonyms for "bear" follows. These describe the bear by cultural and natural images associated to the animal (*Skáldskaparmál* 322; *Snorri Sturluson* 1998, 88; *Edda [transl.]* 1997, 136):

Bjǫrn:
fetviðnir húnn
vetrliði bersi
fress íugtanni,
ifjungr glúmr
jölfuðr vilskarpr
bera jórekr
riti frekr
blómr ysjungr

Bear:
forest-walker (also a name for outlaws: *skoggangr*) cub
winter-survivor, grizzly
snarler greedy-tooth,
goshawk or the hooded one, the dark one
the yellow-bum (also one of Odin's names deriving from *jálnir* "one who makes noise"), shrivelled-gut,
she-bear, *Iorekr* (proper name, meaning "rich of horses", maybe a sea king)
scratcher the greedy one
the mighty one bustler.

Later in the same text, in the above mentioned *Þúlur* at the end of *Skáldskaparmál*, we have another list of bear synonyms (*Þúlur, Skáldskaparmál* 510; *Snorri Sturluson* 1998, 132; *Edda [transl.]* 1997):

bersi blómr
bera elgviðnir
blájaxl ísólfr
ok breiðvegi
bestingr bassi
balti hlébarðr
úlfr frekr vilnir
jórekr mǫsni

---

8   Both names can be interpreted as nominalised adjectives, according to Lindow 2001, 38.

grizzly the mighty one
she-bear elk-destroyer
blue (dark) molar ice-wolf
and broadway
haltered (therefore: a prisoner, an outlaw) she-bear
growler leopard (*hlébarðr*: loanword, according to the dictionary of Cleasby/Vigfússon 1957, 270, used indiscriminately for wolves and bears)
wolf greedy robber
jórekr (rich of horses) *mǫsni* (untranslated term).

The list goes on (*Þulur*, *Skáldskaparmál* 511; Snorri Sturluson 1998, 132; *Edda [transl.]* 1997, 164):

Fetviðnir húnn
fress vetrliði
íugtanni jálfuðr
ifjungr vilskarpr.

forest-walker, cub,
snarler, winter-survivor,
greedy-tooth, yellow-bum,
hooded one, shrivelled-gut.

Most denominations coincide in the two different catalogues, but some are specific for each list, e.g. nine ones mentioned in the *Þulur*: *elgviðnir*, dark-tooth, ice-wolf and broadway, haltered, *bassi*, growler, rough, robber, *mǫsni*, as well as three mentions in the other list: the dark one, scratcher, bustler.

Five of them are *nomina agentis*, describing the bear as an active subject, and are based on how the human eye saw the bear: forest-walker, scratcher, bustler, robber, growler, greedy, dark-tooth, winter-survivor.

Some coincide with those for "wolf", probably being confused by skalds who used them indifferently (sometimes the bear is even called a wolf). Unlike those for "wolf", no bear name is connected to mythology. The great majority is based on natural observation, some are based on folklore, as can be inferred from folk tales and popular legends (especially those spread in Sami areas in northeastern Scandinavia, i.e. "the hooded one"). Only "ice-wolf", clearly referring to "polar bear", shows a typical kenning mechanism based on an environment exchange pattern, like the following one: The ship is sea-travelling, the bear is forest-(walker)travelling; they exchange their space. Thus we have this proportion – the bear: the forest = the ship: the sea; consequently, we have the following logical conclusion that if the bear of the sea is the ship = the bear is the ship of the forest, according to Aristotle's analogical concept of metaphor (*Aristotle, Poetica,* 22).

Kennings gave the *dróttkvætt* poet a rich and transforming diction, and Markús Skeggjason, whose stanza quoted by Snorri in *Skáldskaparmál* 157 shows synonyms for bear occupying the basical position in all kennings, was extremely skilful in using them. In fact Markús was able to juxtapose unlike kennings in a series of metamorphoses, making a ship turn into a bear (or into a horse, or a serpent). And he goes on with verbs expressing the chosen animal's typical aggressive and rapid movements (*Skáldskaparmál*; Snorri Sturluson 1998, 75; *Edda [transl.]* 1995, 125):

Fjarðlinna óð fannir
fast vetrliði rastar;
hljóp of húna gnípur

hvalranns íugtanni;
bjǫrn gekk framm á fornar
flóðs hafskíða slóðir;
skúrǫrðigr braut skorðu
skers glymfjǫtur bersi.

Of the fjord snake waded through the snowdrifts
firmly the bear of the current;
jumped over the peaks of the mastheads
of the whale-house the bear;
the bear went forward on the old
on the flood sea skis' track;
the storm-breasting broke through of the prop
the skerry's clashing fetter the bear.

The bear is called by its common name only in the 5th and 8th lines (*bjǫrn* and *bersi*). Otherways it is called by *heiti*. We imagine an acrobatic bear which *óð* (waded), *hljóp* (jumped), *gekk* (went), *braut* (broke through) in his typical environment: the mountains. Although the stanza describes a ship sailing on the sea, the "metaphor bear" (which is the Base word of the four kennings) guides our imagination. The ship-bear waded through snowdrifts (of the fjord snake: the ship), jumped over the peaks (of the whale-house: the sea), went forward on the old tracks (of the sea skis: the ships), broke through the clashing fetter of the skerry. There is a conceptual "sea-mountain-exchange" according to a well-known skaldic tradition of the oxymoronic kenning, based on the association of contrary elements.

In the first kenning the bear is called *vetrliði*: one who has passed a winter = a winter-old bear, a winter-survivor: *vetrliði rastar* "the winter-survivor <bear> of the current": ship. The verb *vaða* "to wade, to go through shallow water or snow", reinforces the suggestion of coastal waters. It is spring, and the ship springs up again like a bear who wakes after sleeping all winter long. In the second couplet the ship leaps over the peaks of the whale-house (the waves) on the old tracks (on the old commercial sea lanes). The last couplet depicts the "bear of the prop", towering high on a sea-storm, while the waves clash through the rocky islands. In the last kenning the word *skorða* "the prop" points out that the ship is in the dock. It means that the journey is over, the cub has grown into an adult bear. The double image by which the poet plays with and combines different natural elements and linguistic domains results in an effective synthesis of ambiguities and multiple associations. Snowdrifts, peaks, or tracks suggest a mountainous area. We have therefore a double exposition of a bear moving and jumping in its natural world and at the same time of a bear-like ship behaving like an animal in the sea; all these environmental elements are connected to kennings for "sea".

At first Markús does not call the animal by its real name, he does it only in the second *helmingr* (half-stanza) of the *lausavísa*, at first he chooses "winter-survivor", "greedy tooth", and then he mentions "bear": "storm-battling little bear". The skaldic aesthetic of not openly naming things was a hiding one. It described objects and beings by epithets and periphrases. Obscurity could provide a secret code as well as create a powerful language known and understood only by a selected group of individuals bound by particular relationships. This sophisticated method was very much appreciated by kings and their followers, the *drótt*, from which the main and more solemn metre *dróttkvætt* takes its name.

Besides Markús's kennings, we have other skaldic "bear- metaphors" designating "ship":
- in a stanza in *hrynhenda* (flowing metre) in *Magnússdrápa* (Drápa about Magnús) by Arnórr jarlaskáld Þórðarson, 11th century: *hlébarðs hanka* – "the bear (leopard) of the cleat": ship,

- in some stanza fragments called *Ferðavísur* (Stanzas about journeys) by Hofgarða-Refr Gestsson, 11th century: *bjǫrn undinna festa* – "the bear of twisted moorings": ship, *vetrliði skíða* – "the bear (one who passed the winter) of planks": ship, *skautjalfaðar* – "*skyldir* (the requisitioner) of the sail-bear": ship,
- in a stanza of the anonymous *Plácitusdrápa*, 54, 12th century: *unnfress* – "the wave-bear": ship, and
- in a late Christian skaldic poem *Kátrínardrápa* (Drápa about St Cathrine), 16, by Kálfr Hallson, 14th century: *fress ǫldu* – "the bear of the wave": ship.

I want to point out two more interesting examples where *bjǫrn* appears in kennings with an unusual referent: poetry. In *Friðþjófsrímur* poetry is called both *ljóða bjǫrn* (II,59,3) and *ljóða fress* (I,59,4) "bear of songs" (LARSSON 1893, 134). These seem to be variations of Markús Skeggjason's above-quoted kenning: *flóðs bjǫrn Skáld*, 157, for "ship". Here the bear becomes a metaphor for "poetry". This interesting transformation of the bear into poetry is not easy to explain. One possibility may derive from an original metaphor poetry = ship, since in the Old Norse myth of poetry, the mead of poetry is carried on a ship by two dwarfs. They had stolen it from a wise man called Kvasir. The dwarfs slew Kvasir and brewed the mead of poetry with his blood. *ljóða bjǫrn* and *ljóða fress* might thus be rare mythological kennings with "bear" as a Base word. With "bear" being a famous metaphor for "ship", they could be immediately identified as "the ship of songs", a well-known kenning for poetry.

We have seen that the animals occurring in kennings express different connotations for the ship. In a way they contribute to form the concept of a ship, by suggesting its manifold qualities – as they do in male proper names such as *-bjǫrn-*, *-ulfr-*, etc. (*Þorbjǫrn*, *Geirbjǫrn*, etc.), evoking particular characteristics associated to them.[9]

Summarising, the bear mentions in skaldic poetry are significantly fewer compared with those of other animals. That may suggest that this animal would belong to an older type of symbols associated with an ancient society of hunters rather than of warriors: a world dominated by other rituals and values than those in force in the Viking era, of which skaldic poetry is a typical expression.

Therefore my hypothesis is that the Viking cultural heritage – with its mythological lore essentially based on war and battle values – hosts descriptions with animals traditionally connected to the battle field: wolves, eagles, ravens, etc. The bear is rather connected with hunting and its rituals. This is evident also in the above quoted saga *Hrólfs saga kraka*, and in some east Scandinavian folktales, where remnants of ceremonies before and after bear-hunting are preserved.

BIBLIOGRAPHY

Primary sources:
Aristotle, Poetica: Aristotle, Poetica (Bari 1988).
Atlakviða: see *Edda* 1962.
Atlamál: see *Edda* 1962.

Edda 1962: *Edda*. Die Lieder des Codex Regius. Ed. G. Nekkel, red. H. Kuhn (Heidelberg 1962).
Edda (transl.) 1923: A. H. Bellows, The poetic Edda (New York 1923).
Edda (transl.) 1995: Edda. Ed./transl. A. Faulkes (London 1995).

Edda (transl.) 1997: Edda. Transl. A. Faulkes (London 1997).
Eyrbyggja saga 1957: Eyrbyggja saga. Ed. Einar Ólafur Sveinsson. Íslenzk fornrit IV (Reykjavík 1957).

Íslendingabók/Landnámabók 1968: Íslendingabók/Landnámabók. Íslenzk Fornrit I,22 (Reykjavík 1968).

Óláfr Þórðarson Third Grammatical Treatise: see ÓLSEN 1884.

---

9 In Swedish folktales, Thor is sometimes connected to the Ursus constellation (the Great Bear); cf. ROSÉN 1913, 213–244; STRÖM 1967, 81–82.

Skáldskaparmál: see *Snorri Sturluson* 1998.
Snorri Sturluson 1998: *Snorri Sturluson, Edda: Skáldskaparmál*. 2 volumes. Ed. A. Faulkes (Exeter 1998).

Vǫlundarkviða: see *Edda* 1962.

Secondary sources:

Bachelard 1965: G. Bachelard, Lautréamont (Paris 1965).

Bleakley 2000: A. Bleakley, The Animalizing Imagination. Totemism, Textuality and Ecocriticism (London 2000).

Cleasby/Vigfússon 1957: R. Cleasby/G. Vigfússon, An Old Icelandic-English Dictionary. 2nd ed. with suppl. by W. Craigie (Oxford 1957).

DuBois 2014: T. A. DuBois, Underneath the Self-Same Sky: Comparative Perspectives, in Sami Finnish and Medieval Scandinavian Astral Lore. In: T. R. Tangherlini (ed.), Nordic Mythology. Interpretations, Intersections, and Institutions (Berkeley, Los Angeles 2014).

Frog 2008: Frog, Vǫlundr and the bear in Norse Tradition. In: A. Zanchi (ed.), Skáldamjöðurinn: Selected Proceedings of the UCL Graduate Symposia in Old Norse Literature and Philology, 2005–2006 (London 2008) 1–50.

Gärdenfors 2000: P. Gärdenfors, Conceptual Spaces: The Geometry of Thought (Cambridge [Mass.] 2000).

Hatto 1980: A. Hatto, The Swan Maiden: A Folktale of North Eurasian Origin? In: A. Hatto, Essays on Medieval German and Other Poetry. Anglica Germanica Series 2 (Cambridge 1980) 267–297.

Hedeager 2004: L. Hedeager, Dyr og andre mennesker – mennesker og andre dyr. Dyreornamentikkens trascendentale realitet. In: A. Andrén/K. Jennbert/C. Raudvere (eds.), Vägar till Midgård (Lund 2004) 219–252.

Honko et al. 1994: L. Honko/S. Timonen/M. Branch/ K. Bosley (eds.), The Great Bear. A Thematic Anthology of Oral Poetry in the Finno-Ugrian Languages (Oxford 1994).

Jennbert 2004: K. Jennbert, Människor och djur. In: A. Andrén/K. Jennbert/C. Raudvere (eds.), Vägar till Midgård (Lund 2004) 183–218.

Af Klinterberg 1986: B. Af Klinterberg, Svenska folksägner (Stockholm 1986).

Lakoff/Johnson 2003: G. Lakoff/M. Johnson, Metaphors We Live By (Chicago 2003).

Larsson 1893: L. Larsson, Sagan och rimorna om Friðþiófr hinn Frœkni (Lund 1893).

Lichtensteins 1963: H. Lichtensteins, The Dilemma of Human Identity. Notes on Self-Transformation, Self-Objectivation, and Metamorphosis. Journal of the American Psychoanalytic Association 11, 1963, 173–223.

Lindow 2001: J. Lindow, Norse Mythology: A Guide to Gods, Heroes, Rituals, and Beliefs (Oxford 2001).

Marold 2012: E. Marold, The aesthetic function of the kenning. In: D. Whaley (ed.), Poetry from the Kings' Sagas 1: From Mythical Times to c. 1035. Skaldic Poetry of the Scandinavian Middle Ages 1 (Turnhout 2012) lxxv–lxxxv.

Ólsen 1884: B. M. Ólsen, Den tredje og fjærde grammatiske afhandling i Snorres Edda, tilligemed de grammatiske afhandlingers prolog og to andre tillæg (København 1884).

Pentikäinen 2007: J. Pentikäinen, Golden King of the Forest. The Lore of the Northern Bear (Helsinki 2007).

Rosén 1913: H. Rosén, Freykult och djurkult. Fornvännen 8, 1913, 213–244.

Schjødt 2003: J. P. Schjødt, Initiation, liminalitet og tilegnelse af numinøs viden. En undersøgelse af strukutr og symbolik i førkristen nordisk religion. Unpubl. PhD thesis, Det Teologiske Fakultet, Aarhus University (Aarhus 2003).

Ström 1967: F. Ström, Nordisk hedendom (Göteborg, Lund 1967).

Tolley 2007: C. Tolley, Hrólfs saga kraka and Sami Bear Rites (London 2007).

Zachrisson 2008: I. Zachrisson. The Sámi and their Interaction with the Nordic Peoples. In: S. Brink/N. Price (eds.), The Viking World (London 2008) 32–39.

Prof. Maria Cristina Lombardi
Naples University, L'Orientale
Naples
Italy
mclombardi@unior.it

The role of bears in Old Norse literature – a bestiary concept?

By Agneta Ney

Keywords: Heroic, sagas, myth, shape-shifting, she-bear

Abstract: The purpose of this article is to discuss the role of bears in Old Norse literature. The study is mainly based on a comparison between the Legendary Sagas and The Poetic Edda as well as legal, onomastic, and runic sources. Themes such as shapeshifting, bears in dreams, and bears in mythical wisdom are analysed. Gender aspects are taken into consideration, for example the role of she-bears in the narrative, and also the perspective of the contemporary medieval audience. Regarding the audience's expectations, the personal name Bera, meaning she-bear, may have functioned as a signal for what was to come, namely, a story featuring a female with a strong connection to the wild. In a social hierachical perspective, the bear seems to be connected to the political elite as a symbol of power and strength.

Introduction

Studying Old Norse literature makes you aware not only of its greatness, but also of its many-sided genres, for example, *Eddukvæði* (The Poetic Edda) and the whole corpus of the Icelandic Sagas. Although these sources were actually written down in Iceland, mostly in the 13th–14th centuries, they were preceded by a widespread oral tradition going back to the Viking Age, and in some cases even to the Migration Era. Even the medieval Icelandic and Norwegian written law-codes are, to a certain extent, influenced by orally performed legislative formula. The law-codes, as a part of the Old Norse literature, will, together with a number of Eddic lays and sagas, be taken into account in the present study (Vésteinn Ólason 1998, 19–20; Gísli Sigurðsson 2004, 56–60; 2008, 19–28).

The background of saga-writing is a wide-ranging subject, but important aspects of it are, among other things, the settlement in a new country and the gradual process of social and political changes. According to *Íslendingasögur* (The Sagas of Icelanders), the most well-known group of sagas, social ties, feud and honour were the main topics in Icelandic society during the Viking Age. In contrast to the above-mentioned literature, the *Fornaldarsögur* (The Legendary Sagas) take place outside of Iceland, mostly in other Scandinavian countries, and have a mythical-heroic character. Main themes in several Legendary Sagas are war and love affairs, but they also contain elements of folktales and the supernatural. The stories are supposed to have occurred before the settlement of Iceland. The function of this genre was not to tell "how it really was", but to entertain, and, if orally presented, the audience certainly loved to listen to these kinds of episodes, possibly performed with dramatic gestures and voices (cf. Gunnell 1995, 80–92; Lassen et al. 2003, 7–16).

The Poetic Edda is divided into mythological and heroic lays, the former telling of the heathen gods and goddesses, and other myths, and the latter of male and female heroes, passion, betrayal, and revenge. The poems were collected and written down in the 13th century. They were, however, composed much earlier, though not before the 9th century (Jónas Kristjánsson/Vésteinn Ólason 2014, 19).

After this brief introduction to the land and the literature – where do the bears fit in? In the following, I would like to present the themes of my study, as well as emphasising the need to cross disciplinary borders, especially when studying cultural history. Such a perspective is often mirrored in legal sources. Provided that source criticism is taken into account (cf. Ney [Breisch] 1994, 48–53), law-codes may thus give detailed information on the society's structure and thoughts, for example on the imagined boundaries between the social and the wild, as well as on laws for handling bears in real life. Therefore, I would like to begin with some comments on the occurence of bears in the medieval Icelandic law *Grágás* ("Grey Goose"). This leads further to my main purpose, to discuss the role of bears in the Icelandic sagas and in the Eddic lays, and if there is a bestiary concept that is a cultural, commonly shared one, and how it may be defined (cf. Hastrup 1985, 136–154). In order to throw some light on this, I have gathered examples from the narratives that concern themes of shapeshifting, the meaning of the bear in personal names, on the bear as a symbol in dreams, and its role in mythical wisdom and knowledge. In analysing this concept, gender and social aspects are taken into consideration.

The bear in legal sources

When it comes to bears in "real life", the Icelandic fauna did not (and does not) include bears. However, *Grágás* does mention polar bears (*hvítabjörn*) and brown bears (*viðbjörn*). Polar bears did, apparently, sometimes reach the island on ice floes from Greenland, and the law regulates how men should handle them. It was a legal right to hunt polar bears and use their skin and meat (*Grágás* 29, 32). The law also regulates how brown bears should be handled, if they were brought to Iceland. Some brown bears may thus have been shipped over. However, according to the law, this was a serious crime; the owner of the bear and the captain of the ship got exiled, and the steersman had to pay a fee. If the bear got loose and harmed a person or some property, the man who had brought the bear was responsible. There are specific rules for domesticated bears (*alibjörn*) to prevent damages. According to the law, the owner of a domesticated bear should handle the animal as if it was a dog, and if someone injured this bear he should pay a fee for it, and if there was a more serious injury, the punishment was exile. The law does not say for how long, but in most cases the duration of exile was three years. If the bear-owner set his animal on someone and the person got hurt or killed, the owner was outlawed, and the bear had no right to live (*Grágás* 259, 261, 349).

Grágás was written down in the middle of the 13th century, but was replaced by *Járnsíða* ("Ironside") in 1271, after the Icelandic Freestate fell under Norwegian sovereignty. Under the new law, polar bears and brown bears were still mentioned in connection with legal hunting. However, domesticated bears were no longer a legal matter, most likely due to internal or external changes. Nevertheless, a long time after, in the early modern centuries, they were allowed in Scandinavia as, for example, valuable king's gifts and for entertainment (*Járnsíða*, 176, 184; Berg 1965, 93–112; Hastrup 1985, 207–208, 232–237; Gunnar Karlsson 1992, ix–xxxiii; Haraldur Bernharðsson et al. 2005, 13–25).

Transformation into a bear

Bears have indeed over time been hunted for skin and meat, but were also seen as a symbol of savagery and strength. The strength of a bear is, among other things, a frequent motif in sayings and folklore tales (see BÖLDL, this volume). In the latter, people encounter bears in the woods, sometimes in the shape of a man-bear, the offspring of bear and human (BYOCK 1998, xxvi–xxviii; KUUSELA 2019, 96–107).

The transformation of a man into a bear is related in The Legendary Saga *Hrólfs saga kraka* (*The Saga of King Hrolf Kraki*), which was composed in its present form around 1400 (cf. LANZING 2009, 405–430). However, it is estimated that a king named *Hrólfr* lived in Scandinavia in the 6th century. In this saga, a man called *Björn* ("bear"), the son of a king up in the north of Norway, and a young girl, *Bera* ("she-bear"), love each other. *Björn* has a stepmother, Queen *Hvít* ("White"). She was, according to the saga, the daughter of the King of the Sámi and lived, before marrying *Björn*'s father, in the Finnmark region in northern Norway. This may connect her with the Sámi religion and shamanism. The queen tries to seduce *Björn*, but in vain. When he refuses to be her lover, she is hurt by his rejection and wants to punish him with witchcraft (*Hrólfs saga kraka*, 47; *The Saga of King Hrolf Kraki*, 34–40):

Hún lýstr nú til hans með úlfhanzka ok segir, at hann skyldi verða at einum hídbirni ólmum ok grimmum, [...].

She then struck him with her wolfskin gloves, telling him to become a cave bear, grim and savage, [...].

He will never be released from the spell (cf. JÓNAS KRISTJÁNSSON 1988, 352–353; BYOCK 1998, vii–xxxii; STRÖMBÄCK 2000, 24–25; AALTO 2010, 123–124).

Björn disappears, and no one knows where to or why. But one day, *Bera* encounters a bear. The bear is approaching, but not threatening her. She recognises his eyes and follows him into his cave. Since the spell has only transformed *Björn* into a bear by day, he is "retransformed" at night. *Bera* stays with him in the cave, but one night *Björn* tells her that he will be hunted the next day and killed, so this will be their last night. He continues as follows: Firstly *Bera* must go to his father, the king, and ask him to give her "whatever is under the beast's left shoulder" (it is his ring, and the ring will show the bear's real identity). Secondly, she must be aware of Queen *Hvít* – she will be suspicious and try to make her eat of the bear's meat, but this she must not do as she is pregnant and will give birth to three boys, and if she tastes the meat it will affect the children. Thirdly, *Björn* tells her to go home to her father and give birth, and he adds (*Hrólfs saga kraka*, 49; *The Saga of King Hrolf Kraki*, 38):

"Ok ef þú mátt eigi annast þá heima fyrir ósköpum þeira ok ófyrirleitni, leiddu þá í burtu, ok farðu hingat til hellisins með þá. Hér muntu sjá kistu með þremr gólfum. Þat munu segja rúnir þær, sem þar eru hjá, hvat hverr þeira skal eignast. Vápn þrjú eru í berginu, ok skal þat hverr hafa, sem honum er ætlat. Sá, sem fyrst kemr til, sonr okkarr, skal Þórir heita, annarr Elg-Fróði, þriði Böðvarr, ok þykki mér þess líkara, at þeir verði eigi litlir fyrir sér ok þeira nöfn munu lengi uppi vera."

"If you are not able to raise them [their sons] at home, because of their strange and uncontrollable natures, bring them here to the cave. You will find here a chest with three bottoms. Runes are carved on it, and they will tell what each of the boys should receive as his inheritance. Three weapons are imbedded in the rock, and each of our sons shall have the one intended for him. Our firstborn will be called Elg-Frodi, the second son, Thorir, and the third, Bodvar. It seems to me most likely that they will not be weaklings and that their names will long be remembered."[1]

Björn then foretells *Bera* many other things, but soon the bear-shape comes over him again. He leaves the cave and *Bera* follows. Hunters are waiting outside the cave, "circling the side of the mountain". A large pack of hounds are in front of the men, and the bear tries to escape. In the end, the men form a ring around the bear. After slaying all the dogs and one man, he gets exhausted, and he is finally killed by the king's men (*Hrólfs saga kraka*, 49–50; *The Saga of King Hrolf Kraki*, 38–39; cf. OEHRL 2013, 299).

As instructed, *Bera* takes the ring from the bear's left shoulder. Queen *Hvít* prepares a great feast with the bear's meat for the men – and at this feast she forces *Bera* to taste the meat (*Hrólfs saga kraka*, 51; *The Saga of King Hrolf Kraki*, 38–39):

Hún [Hvít] bitar fyrir hana [Bera], ok þat verðr af leiknum, at hún etr þann bita. Drottning skerr þá annan ok lætr í munn henni, ok þar kemr lítit korn niðr af þeim bita, ok sló þá úr munni sér ok kveðst eigi meir mundu eta, þó at hún píndi hana eða deyddi.
Drottning mælti: "Vera kann, at nú dugi nokkut," ok hló at.

The queen cut a small piece of the meat for Bera, and in the end Bera ate it. The queen then cut another piece and put it into Bera's mouth. Bera swallowed a small morsel of it, then spat the rest out of her mouth. She declared that she would not eat any more, even if she were to be tortured or killed.
"It may be," said the queen, "that this bit will be enough," and she burst out laughing.

When *Bera* gives birth to the triplets, the first boy is half man, half elk. He is an elk below the navel, and is called *Elg-Fróði*. The second boy has dog's feet and is called *Þórir Hundsfótr*. However, the third boy, *Böðvarr*, is without any blemish. *Fróði* has an "unyielding nature", and thus he is brought into the bear's cave, gets the short sword that was intended for him and moves up into the mountains. *Þórir* is also brought into the cave, and takes an axe from the rock. He then goes to see his brother *Fróði*, and later becomes king over the Gauts. *Böðvarr* takes revenge for his father's fate and kills the queen, and after the king's death he takes over the kingdom. I will leave the rest of the story untold; however, *Böðvarr*, according to the saga is, like his father, connected with a transformation into a bear-shape, and is a warrior (*Hrólfs saga kraka*, 50–57; *The Saga of King Hrolf Kraki*, 39–44; cf. BYOCK 1998, xxix).[2]

The bear motif in *Hrólfs saga kraka* leads to political power and prosperity, at least for the sons with bear ancestry. The male protagonist, *Böðvarr*, also shows some similarities with a special folk tale called "Bear's Son Tale" (see FRANK, this volume; cf. HARRIS 1973, 25–53; MITCHELL 1991, 58–59; BYOCK 1998, xxvi–xxviii; STRÖMBÄCK 2000, 79–87, 187). Characteristic elements of this tale are, among others, the hero's animal ancestry and a special sword, and these fit in with *Böðvarr*. As in fairy tales, the author of *Hrólfs saga kraka* is using the epic number three – three sons, three weapons, three

---

1 In the Icelandic edition used here, the first born is said to be *Þórir*, but in the English translation it is *Elg-Fróði*. The difference probably depends on textual variations in different manuscripts/editions. In a following episode in the Icelandic text, it is said that *Elg-Fróði* was actually the first born; see below in the present article; cf. GUÐNI JÓNSSON 1965, xxv–xxvi; BYOCK 1998, xxxvi.
2 *Böðvarr* got the nickname *Biarki*, a diminutive of *biari* (bear), that is "little bear", cf. LIND 1920/1921, column 22. He may be defined as a berserk; cf. SUNDQVIST, this volume.

attempts to get the best weapon out of the rock, all in order to increase the attention of the audience. The significance of animal ancestry for male virtues as strength and courage is obvious in *Hrólfs saga kraka*, but there are also some other notable gendered aspects that consider the female. Firstly, Queen *Hvít* knows how to use witchcraft for transforming a man into a bear; secondly, *Bera* is defined with strong connections to the wild and the knowledge of runes – and she is the bear's match and the mother of his children.

The bear in onomastics

The name *Bera* leads me to onomastics. Was it a common name, as the male counterpart *Björn* was? In comparison with the "Scandinavian Dictionary of Runes" (*Nordiskt runnamnslexicon*) by the Swedish runologist Lena Peterson, this female name is not represented in her huge collection of runic material from the Viking Age (from Sweden, Norway, Denmark, and the British Isles). While the name *Bera* is not represented at all, not even as an element, the male name *Bjǫrn* is the second most frequent name (with 118 examples, after the leading *Sveinn* with 147 examples), and as a second element -*bjǫrn* is the second most common one (in 51 compounds), after another wild animal, -*ulfr* (in 52 compounds).[3]

However, *Bera* is mentioned in Old Norse literature, but only in a few places. It could be of interest to take a closer look at the contexts in which the name appears. As a consequence of its lexical meaning and the eventual specific contexts, hypothetically this female bear name may function as a signal to the audience of a coming theme with a close connection to the wild (cf. SMITH 2016, 307–308).

Occurrences of the she-bear (Bera)

In the heroic Eddic lay *Atlamál in grœnlenzku* (*The Greenlandic poem of Atli*), the name *Bera* (along with the compound *Kostbera*) is found, and is used for the same person, but there is no *Björn* whatsoever. *Bera* is the wife of *Högni*, a brother-in-law of *Sigurðr Fáfnisbani* (Sigurd the Dragonslayer). She has seen a bear in her dreams, and tells her husband of its frightful appearance (*Atlamál in grœnlenzku*, 386:17; *The Greenlandic Poem of Atli*, 220:17):

"*Bjǫrn hugða ek hér inn kominn,*
brytti upp stokka,
hristi svá hramma
at vér hrædd yrðim;
munni oss mǫrg hefði,
svá at vér mættim ekki,
þar var ok þrǫmmun
þeygi svá lítil."

"I thought I saw a bear come in here, *smash up the panelling,*
swing with his paws so that we were afraid;
many of us he held in his mouth so that we were helpless;
his lumbering made no small amount of noise."

---

3 However, the names *Bersa*/*Birsa* are found on runic stones in Uppland, Sweden. There are some variations but the same meaning: *Bersa*: [**birsa**] U 189, **bersu** U79B, **birsu** U 79A, and *Bersi*, meaning "small bear"; cf. PETERSON 2004, 6, 46; 2007, 41, and passim; cf. also JANZÉN 1947, 35, 44. The element *kost*- in *Kostbera* is probably derived from the Old Icelandic noun "*kost*", m., (with the meaning "choice"), cf. LIND 1910, column 715. The element -*bera* has also been interpreted as "to carry", cf. VON SEE et al. 2009, 590–591, with references.

The context is that *Bera*'s husband and his brother *Gunnarr* have been invited to go to their brother-in-law *Atli*. The invitation is followed by a message from *Atli*'s wife *Guðrún*. It consists of both a written runic message and a gold ring, encircled with wolf's hair. *Bera* is the only one who can interpret the runic message; it is a warning for her not to go. She therefore tries to convince *Högni* to stay. We do not know if she has actually had the dream, or just says that she has. However, *Högni* interprets his wife's dream as meaning that terrible weather is to follow, and that the bear is a symbol of a storm (*Atlamál in grœnlenzku*, 386:18; *The Greenlandic Poem of Atli*, 220:18):

"Veðr mun þar vaxa,
verða ótt snemma,
hvítabjǫrn hugðir,
þar mun hregg austan."

"That must mean a storm coming, it'll soon be dawn;
thinking of a *white bear*, that's a blizzard from the east."

It has been said that the polar bear mentioned in the stanza above may contribute to the poem's Greenlandic connection (JÓNAS KRISTJÁNSSON/VÉSTEINN ÓLASON 2014, 134, 147–148). However, it is *Högni* who defines the bear as a polar bear, not *Bera*. Another plausible explanation is that *Atlamál in grœnlenzku* was preserved among the Greenlanders and came from a Greenlandic narrator when the Eddic poems were being compiled (JÓNAS KRISTJÁNSSON 1988, 70; LÖNNROTH 2016, 408).

In *Völsunga saga* (*The Saga of the Volsungs*), a Legendary Saga written down in the middle of the 13th century, with influences from the heroic lays about *Sigurðr Fáfnisbani*, *Bera* bursts out in the corresponding episode (*Völsunga saga*, 202; *The Saga of the Volsungs*, 98, 13–26; cf. GLAUSER 2007, 13–26):

"Björn hugða ek hér inn koma", segir hún, "ok braut upp konungs hásæti ok hristi svá hrammana, at vér urðum öll hrædd, ok hafði oss öll senn sér í munni, svá at ekki máttum vér, ok stóð þar af mikil ógn."

"I thought a bear entered here", she said. "He *destroyed the king's throne* and waved his paws about so much that we all grew afraid. He had us all in his mouth together, so that we could do nothing and great terror arose."

Högni answers (*Völsunga saga*, 202; *The Saga of the Volsungs*, 98, 13–26):

*"Þar mun koma veðr m*ikit, er þú ætlaðir hvítabjörn.*"

"A strong tempest will come, where you thought it a white bear."

One interesting difference between the Eddic lay and the saga is the destruction done by the bear; in the former he "smashed up the panelling", and in the latter he "destroyed the king's throne" (cf. my italics in the quoted stanzas). Thus, historical changes are reflected by the contemporary saga author's choice of words. The outrageous bear is threatening a symbol of power and thereby the political structure of the medieval kingdom.

The name *Bera* is used in only two Legendary Sagas, and there in connection with bears. In comparison, there are a few examples from The Sagas of the Icelanders, for instance in *Egils saga Skalla-grímssonar*. In this saga, there are three women named *Bera*. One is *Bera Yngvarsdóttir*, the wife of *Skalla-Grim Kveld-Úlfsson*, a suitable match for a man whose father was *Kveld-Úlfr*, depicted as a werewolf and berserk, especially when considering their names, the connection of which to the wild is evident. Their grand-daughter, the daughter of *Egil*, is another one, and a third one is *Bera Ormsdóttir*, the wife of *Egil*'s grandson *Skúli* (*Egils saga Skallagrímssonar*, ch. 1, 20, 31, 35; cf. *Hauksbók*, 127; *Landnámabók* ch. 25; cf. HEIMIR PÁLSSON 2007, 96–121). There is no obvious connection between these *Bera*-named women and the bear, but the semantic meaning and the character associated with the bear have truly effected the name-giving. Besides, *Skalla-Grím*'s *Bera*, as well as *Bera* in *Hrólfs saga kraka*, came from Norway, something that may lead to further questions on the tradition of name-giving. That the *Bera*-name recurs in *Egils saga Skallagrímssonar* is confirmed through *Hallbera Úlfsdóttir*, the mother of *Kveld-Úlfr*, and *Hallbera*s father, *Úlfr* ("wolf"), who gave her a she-bear name and her brother *Hallbjǫrn* a bear-name, which is indeed a genealogical statement at the beginning of the saga.

THE BEAR IN DREAMS

Compared to the examples from Eddic poetry and *Völsunga saga*, bears in dreams forecast future events and also comment on something that has already happened, as in the following three Legendary Sagas. In *Örvar-Odds saga*, a bear-*fylgja* in a dream reacts to how a man named *Gudmundr* acts against his brother *Oddr*. The former is on his ship waiting for good weather, and his nephew *Sigurdr* is with him. *Gudmundr* tells his nephew of his dream. He dreamt that a polar bear was lying in the bay. The bear's head and back met over the ship, and it was a frightful sight. *Sigurdr* interprets his dream to mean that the bear must be the *fylgja* of *Oddr*. A *fylgja* ("one who follows") is connected to the belief that all individuals had one or more *fylgja*, in female or animal shape. The animal *fylgja* was thought of as a symbol of a man's character. Therefore, the *fylgja* of a chieftain or a king was often in a bear's shape. *Gudmundr* and *Sigurdr* are convinced of the serious nature of this dream; because they did not let *Oddr* sail away with them, this bear-*fylgja* is obviously angry. They therefore decide to ask him to go with them (*Örvar-Odds saga*, 212–213; HASTRUP 1988, 152–153; NÄSSTRÖM 2002, 313–314).

The role of the bear as a *fylgja* in dreams also occurs in *Þorsteins saga Víkingssonar*. The Swedish King *Njörfi*'s son, *Þorsteinn*, is dreaming of 30 wolves and eight bears. One of the bears had red cheeks. *Þorsteinn* asks his younger brother to interpret this dream, and he assumes that the big bear with red cheeks must be the *fylgja* of their enemy *Jökull*, the son of the Norwegian Viking king, and the other bears are his seven brothers, and the wolves are the men coming to fight with them (*Þorsteins saga Víkingssonar*, 29).

A third example is from *Hrólfs saga Gautrekssonar*. Queen *Ingigerdr* is dreaming of wolves, polar bears, one with red cheeks, and a frightful beast (*it óarga dýr*). This dream is, with variations, repeated twice. On the first night, the queen wakes up and tells her husband, the Swedish king, *Eirekr*, of the dream. She herself interprets the wolves as men's *fylgjur* (*manna fylgjur*), the beast as King *Hrólf*'s *fylgja*, and the polar bear as the *fylgja* of the king's fosterbrother. In her own interpretation, she underlines the strength of the bear. In her second dream, there are wolves running from a ship, a frightful beast, two polar bears, and even a boar. She says, as before, that the beast must be the *fylgja* of King *Hrólfr*, but this time the animal seems bigger and worse. The boar is the *fylgja* of the king's brother and, since there were two polar bears, the king must have the good company of another king or king's son. *Ingigerdr* and *Eirekr* are in Uppsala, and, according to her dreams, an aggressive attack from King *Hrólf* is imminent (*Hrólfs saga Gautrekssonar*, 71–72, 89).

The bear in mythical wisdom and knowledge

Through the ability to foresee the future, a bear in a dream is in a way connected to mythical wisdom, as is the knowledge of runes. *Bera* in *Völsunga saga* and *Atlamál in grœnlenzku*, and *Bera* in *Hrólfs saga kraka* know how to read runes and, in another heroic poem, *Sigrdrífumál* (The Lay of Sigrdrifa), a former valkyrie, *Sigrdrífa*, transmits indispensable knowledge to *Sigurðr Fáfnisbani*, especially runic knowledge, for all kinds of troublesome situations in life.[4] The context is that *Sigrdrífa* has given the wrong king victory in battle, and has been punished (*Sigrdrífumál*, prose 314; *Lay of Sigrdrifa*, prose 167):

> *En Óðinn stakk hana svefnþorni í hefnd þess ok kvað aldri skyldu síðan sigr vega í orrostu ok kvað hana giptask skyldu.*

> And Odin pricked her with a sleep-thorn in revenge for this and said that she would never again fight victoriously in battle and said that she should be married.

Sigurðr awakens the sleeping valkyrie, and she brings him a horn full of mead. It is a memory-drink for remembering the deep knowledge transmitted to him. *Sigrdrífa* is referring to *Óðinn*, who, among other things, says that if a man wants to be "wiser in spirit than every other man", he should cut runes on a shield, on glass, gold, amulets, and on the seat of honour, and also on the bear's paw and on the wolf's claw (*Sigrdrífumál*, 317:17; *Lay of Sigrdrifa*, 168:13, 169:16):[5]

[...] á bjarnar hrammi *ok á Braga tungu,* *á úlfs klóm* *ok á arnar nefi,* *[...].*	[...] on the bear's paw and on Bragi's tounge, on the wolf's claw, and the eagle's beak, [...].

Conclusion

The transformation into a bear shape, the she-bear motif including traditions in name-giving, and bears in dreams are in my opinion the most important themes for analysing the role of bears in Old Norse literature. However, there are also some other aspects, as for example "bear fights" (cf. OEHRL 2013, 297, 299, 306–307, 312–313). Generally, bears in The Sagas of Icelanders appear as dangerous animals, a threat both to cattle and to man, and thus must be fought, and as *Grettir* comments about this kind of fight, it is the hardest trial of manhood (*Grettis saga Ásmundarsonar*, ch. 21).[6]

Since a bear was also a valuable animal, another theme could have been "the bear as a king's gift". One example derives from the tale *Auðunar þáttr Vestfirzka*, where a man named *Auðunn* is spending everything he has on buying a polar bear in Greenland. He then ships the bear to Norway, where he

---

4   The Valkyries lived in Valhall. They were Odin's maids, serving mead to the fallen warriors, and they took part in battle, choosing who should fall or survive. Some of them interacted in the human world, as for example *Sigrún*, who supported *Helgi Hundingsbani* in battle and then married him, cf. *Helgakviða Hundingsbani* I–II, 247–258, 270–283; *The First Poem of Helgi Hundingsbani*, 114–122; *A Second Poem of Helgi Hundingsbani*, 132–141.
5   Bragi was the god of poetry, cf. *Sigrdrífumál*, 317, note 17.
6   Bear fights also occur in *Víga-Glúms saga*, ch. 3, and in *Finnboga saga ramma*, ch. 11. The bear in *Finnboga saga ramma* did great damage to the cattle, and therefore was outlawed like a man, and a price was set on his head. This bear was actually made "human" in crossing the border between nature and culture, cf. TOLLEY 2009, 580; OEHRL 2013, 301.

goes ashore keeping the bear on a lead, searching for lodging. The Norwegian king wants to buy the bear, but *Auðunn* instead intends to sail to Denmark and bring the bear as a gift for the Danish king (*Auðunar þáttr Vestfirzka*, ch. 1; cf. *Vatsdæla saga*, ch. 16).

My starting point was the imagined boundary between the social and the wild, which led me to further ask if there was a bestiary concept in the poetry and sagas – and there is. Bears in sagas are quite frequent, but evidently there is a difference in role and motif depending on whether the bear occurs in a more realistic saga or in a supernatural one. In Eddic poetry, however, bears are underrepresented, and their role is in general quite modest, with the exception of the previously mentioned *Bera* character and the importance of carving runes on a bear's paw in order to achieve wisdom. The main and constantly present animal symbol in the Eddic lays is the wolf. According to the folklorist Clive Tolley, the wolf belongs entirely to the wild, while the bear, since it shares some features with humans, may be defined as a liminal creature, and liminality means the crossing of boundaries, and this in its turn has been seen as a mark of the bear's holiness and power (TOLLEY 2009, 580; NEY 2012, 483–489).

In my examples, above all those from The Legendary Sagas, the she-bear motif dominates in the shapeshifting episodes (without taking the berserkers into account). Regarding bears in dreams in this saga category, I would like to point out that even if queens are certainly bear-dreamers, this kind of dreaming is not after all restricted to a certain gender. However, the bear as a symbol of strength and courage is connected with masculinity in a social and political sense, and above all in a hierarchical perspective, since the occurence of bears is recurrently associated with the king and the king's men.

With this social hierarchical perspective in mind, I would like to comment on The Legendary Sagas as a genre. These sagas were highly appreciated at the king's court or at a chieftain's house. Stories of people "in between", in this case both bear and human, did fascinate (and still do). In connection with the audience's expectations, even if more research needs to be done before drawing further conclusions, in my opinion it seems that the she-bear has a special role. For the audience, the name *Bera* may actually have functioned as a signal for what to expect: a female with some connection with a bear or to the wild in another sense.

I will finally once again point out the importance of legal sources for our knowledge on how people thought and what they believed in, and will end with a practical example. According to the law, it was important to announce a homicide as soon as possible and to make it known to the nearest farmstead (unless the killed man had family there – in that case the risk of blood-vengeance was immediate). When announcing a homicide, the law forbade you to blame it on a bear, that is, unless the killer was actually called *Björn*, you were not to say that it was "*björn*" who did it (*Den ældre Gulathings-Lov*, ch. 156).

BIBLIOGRAPHY

Primary sources

A Second Poem of Helgi Hundingsbani: A Second Poem of Helgi Hundingsbani. In: C. Larrington (ed./transl.), The Poetic Edda (Oxford 2008) 132–141.

Atlamál in grænlenzku: Atlamál in grænlenzku. In: Jónas Kristjánsson/Vésteinn Ólason (eds.), Eddukvæði 2. Hetjukvæði (Reykjavík 2014) 383–401.

Auðunar þáttr Vestfirzka: Auðunar þáttr Vestfirzka: https://heimskringla.no.

Den ældre Gulathings-Lov: Den ældre Gulathings-Lov. In: R. Keyser/P. A. Munch (eds.), Norges Gamle Love indtil 1387, vol. 1 (Cristiania 1846): https://heimskringla.no.

Egils saga Skallagrímssonar: Egils saga Skallagrímssonar. Sveinbjörn Þórdarson (ed.). Icelandic Saga Database: https://sagadb.org.

Finnboga saga ramma: Finnboga saga ramma. Sveinbjörn Þórdarson (ed.). Icelandic Saga Database: https://sagadb.org.

Grágás: Grágás. In: Gunnar Karlsson/Kristján Sveinsson/ Mörður Árnason (eds.), Lagasafn íslenska þjóðveldisins (Reykjavík 1992).
Grettis saga Ásmundarsonar: Grettis saga Ásmundarsonar. Sveinbjörn Þordarson (ed.). Icelandic Saga Database: https://sagadb.org.
Hauksbók: Hauksbók. Udgiven efter de Arnamagnæanske håndskrifter No. 371, 544 og 675, 4o samt forskellige papirshåndskrifter; af Finnur Jónsson for det Kongelige Nordiske Oldskrift-Selskab (København 1892–1896).
Helgakviða Hundingsbani I: Helgakviða Hundingsbani I. In: Jónas Kristjánsson/Vésteinn Ólason (eds.), Eddukvæði 2. Hetjukvæði (Reykjavík 2014) 247–258.
Helgakviða Hundingsbani II: Helgakviða Hundingsbani II. In: Jónas Kristjánsson/Vésteinn Ólason (eds.), Eddukvæði 2. Hetjukvæði (Reykjavík 2014) 270–283.
Hrólfs saga Gautrekssonar: Hrólfs saga Gautrekssonar. In: Guðni Jónsson (ed.), Fornaldar sögur Norðurlanda 4 (Reykjavík 1965) 51–176.
Hrólfs saga kraka: Hrólfs saga kraka ok kappa hans. In: Guðni Jónsson (ed.), Fornaldar sögur Norðurlanda 4 (Reykjavík 1965) 1–105.
Járnsíða: Járnsíða. In: Haraldur Bernharðsson/Magnús Lyngdal Magnússon/Már Jónsson (eds.), Járnsíða og Kristinréttur Árna Þorlákssonar (Reykjavík 2005).
Landnámabók: Landnámabók. Sturlubók: https://www.snerpa.is/net/snorri/landnama.htm.

Lay of Sigrdrifa: Lay of Sigrdrifa. In: C. Larrington (ed./transl.), The Poetic Edda (Oxford 2008) 168–169.
Sigrdrífumál: Sigrdrífumál. In: Jónas Kristjánsson/Vésteinn Ólason (eds.), Eddukvæði 2. Hetjukvæði (Reykjavík 2014) 313–321.
The First Poem of Helgi Hundingsbani: The First Poem of Helgi Hundingsbani. In: C. Larrington (ed./transl.), The Poetic Edda (Oxford 2008) 114–122.
The Greenlandic Poem of Atli: The Greenlandic Poem of Atli. In: C. Larrington (ed./transl.), The Poetic Edda (Oxford 2008) 217–233.
The Saga of King Hrolf Kraki: The Saga of King Hrolf Kraki. J. L. Byock (transl.) (London et al. 1998).
The Saga of the Volsungs: The Saga of the Volsungs: The Norse Epic of Sigurd the Dragonslayer. J. L. Byock (transl.) (Berkeley et al. 1990).
Þorsteins saga Víkingssonar: Þorsteins saga Víkingssonar. In: Guðni Jónsson (ed.), Fornaldar sögur Norðurlanda 3 (Reykjavík 1965) 1–73.
Vatsdæla saga: Vatsdæla saga. Sveinbjörn Þórdarson (ed.). Icelandic Saga Database: https://sagadb.org.
Víga-Glúms saga: Víga-Glúms saga. Sveinbjörn Þórdarson (ed.). Icelandic Saga Database: https://sagadb.org.
Völsunga saga: Völsunga saga. In: Guðni Jónsson (ed.), Fornaldar sögur Norðurlanda 4 (Reykjavík 1965) 107–218.
Örvar-Odds saga: Örvar-Odds saga. In: Guðni Jónsson (ed.), Fornaldar sögur Norðurlanda 4 (Reykjavík 1965) 199–363.

Secondary sources

AALTO 2010: S. AALTO, Categorizing Otherness in the Kings' Sagas. Dissertations in Social Sciences and Business Studies 10 (Joensuu 2010).
BERG 1965: G. BERG, Tama björnar, dansande björnar och björnförare. Fataburen. Nordiska museets och Skansens årsbok 1965, 93–112: diva-portal.org/smash/get/diva2:1294369/fulltext01.pdf.
BYOCK 1998: J. L. BYOCK, Introduction. In: The Saga of King Hrolf Kraki: Translated with an Introduction by Jesse L. Byock (London et al. 1998) vii–xxxii.
GÍSLI SIGURÐSSON 2004: GÍSLI SIGURÐSSON, The Medieval Icelandic Saga and the Oral Tradition: A Discourse on Method. Publications of the Milman Parry Collection of Oral Literature 2. Transl. N. Jones (Cambridge [Mass.] 2004).
GÍSLI SIGURÐSSON 2008: GÍSLI SIGURÐSSON, Orality Harnessed: How to Read Written Sagas from an Oral Culture. In: E. Mundal/J. Wellendorf (eds.), Oral Art Forms and their Passage into Writing (Copenhagen 2008) 19–28.
GLAUSER 2007: J. GLAUSER, The Speaking Bodies of Saga Texts. In: J. Quinn/K. Heslop/T. Wills (eds.), Learning and Understanding the Old Norse: Essays in Honour of Margaret Clunies Ross. Medieval Texts and Cultures of Northern Europe 18 (Turnhout 2007) 13–26.

GUÐNI JÓNSSON 1965: GUÐNI JÓNSSON, Formáli. In: Guðni Jónsson (ed.), Fornaldar sögur Norðurlanda 4 (Reykjavík 1965) 51–176.
GUNNAR KARLSSON 1992: GUNNAR KARLSSON, Inngangur. In: Gunnar Karlsson/Kristján Sveinsson/ Mörður Árnason (eds.), Grágás. Lagasafn íslenska þjóðveldisins (Reykjavík 1992) ix–xxxiii.
GUNNELL 1995: T. GUNNELL, The Origins of Drama in Scandinavia (Martlesham, Rochester [New York] 1995).
HARALDUR BERNHARÐSSON et al. 2005: HARALDUR BERNHARÐSSON/MAGNÚS LYNGDAL MAGNÚSSON/MÁR JÓNSSON, Formali. In: Haraldur Bernharðsson/Magnús Lyngdal Magnússon/Már Jónsson (eds.), Járnsíða og Kristinréttur Árna Þorlákssonar (Reykjavík 2005) 13–25.
HARRIS 1973: R. HARRIS, The Death of Grettir and Grendel: A New Parallel. In: L. Elmevik (ed.), Scripta Islandica 24 (Uppsala 1973) 25–53.
HASTRUP 1985: K. HASTRUP, Culture and History in Medieval Iceland. An Anthropological Analysis of Structure and Change (Oxford 1985).
HEIMIR PÁLSSON 2007: HEIMIR PÁLSSON, Óðinn, Þór og Egill. Skírnir 181, 2007, 96–121.
JANZÉN 1947: A. JANZÉN, De fornvästnordiska personnamnen. In: A. Janzén (ed.), Nordisk kultur 7: Personnamn (Stockholm 1947) 22–186.

Jónas Kristjánsson 1988: Jonas Kristjánsson, Eddas and Sagas: Iceland's Medieval Literature. Transl. P. Foote (Reykjavík 1988).

Jónas Kristjánsson/Vésteinn Ólason 2014: Jónas Kristjánsson/Vesteinn Ólason, Formáli. In: Eddukvæði 2. Hetjukvæði (Reykjavík 2014) 7–204.

Kuusela 2019: T. Kuusela, Skogens ludne drott. Folktro om björnen i Jämtland och Härjedalen. Jämten: Årsbok för Jamtli, Heimbygda och Jämtlands läns konstförening 112, 2019, 96–107.

Lanzing 2009: T. Lanzing, Einn fagur aldingardur – The manuscript transmission of Hrólfs saga Kraka. In: A. Lassen/A. Ney/Ármann Jakobsson (eds.), The Legendary Sagas. Origins and Development (Reykjavík 2012) 405–430.

Lassen et al. 2003: A. Lassen/A. Ney/Ármann Jakobsson, Inledning. In: Ármann Jakobsson/A. Lassen/A. Ney (eds.), Fornaldarsagornas struktur och ideologi. Handlingar från ett symposium i Uppsala 31.8.–2.9 2001. Nordiska texter och undersökningar 28 (Uppsala 2003) 7–19.

Lind 1910: E. H. Lind, Norsk-isländska dopnamn och fingerade namn från medeltiden: Samlade och utgivna av E. H. Lind, vol. 5 (Uppsala 1910).

Lind 1920/1921: E. H. Lind, Norsk-Isländska personbinamn från medeltiden: Samlade och utgivna med förklaringar av E. H. Lind (Uppsala 1920/1921).

Lönnroth 2016: L. Lönnroth, Kommentar. In: Den poetiska Eddan. Gudadikter och hjältedikter efter Codex Regius och andra handskrifter. Översättning med inledning och kommentar av Lars Lönnroth (Stockholm 2016) 408.

Mitchell 1991: S. A. Mitchell, Heroic Sagas and Ballads: Myth and Poetics (Ithaka [New York] 1991).

Ney [Breisch] 1994: A. Ney [Breisch], Frid och fredlöshet: Sociala band och utanförskap på Island under äldre medeltid. Diss. Acta Universitatis Upsaliensis Studia Historica Upsaliensia 174 (Uppsala 1994).

Ney 2012: A. Ney, 'Uselt är att vara varg': Om vargterminologi i västnordisk litteratur och rättsuppfattning". Historisk Tidskrift, Svenska Historiska Föreningen 2012(3), 2012, 483–489.

Näsström 2002: B.-M. Näsström, Fornskandinavisk religion: En grundbok (Lund ²2002).

Oehrl 2013: S. Oehrl, Svá beitum vér björnuna á mörkinni norðr – Bear hunting and its ideological context (as a background for the interpretation of bear claws and other remains of bears in Germanic graves of the 1st millennium AD). In: O. Grimm/U. Schmölcke (eds.), Hunting in northern Europe until 1500 AD. Old traditions and regional developments, continental sources and continental influences. Papers presented at a workshop organized by the Centre for Baltic and Scandinavian Archaeology (ZBSA) Schleswig, June 16th and 17th, 2011. Schriften des Archäologischen Landesmuseums, Ergänzungsreihe 7 (Neumünster 2013) 297–332.

Peterson 2004: L. Peterson. Lexikon över urnordiska personnamn (Uppsala 2004): http://isof.se/sprak/namn/personnamn/lexikon-over-urnordiska-personnamn.html.

Peterson 2007: L. Peterson. Nordiskt runnamnslexicon (Uppsala ⁵2007).

Smith 2016: G. W. Smith, Theoretical Foundations of Literary Onomastics. In: C. Hough/D. Izdebska (eds.), The Oxford Handbook of Names and Naming (Oxford 2016) 295–309.

Strömbäck 2000: D. Strömbäck, Sejd och andra studier i nordisk själsuppfattning. Med bidrag av B. Almqvist, G. Gidlund & H. Mebius. Acta Academiæ Regiæ Gustav Adolphi 72 (Uppsala 2000).

Tolley 2009: C. Tolley, Shamanism in Norse Myth and Magic 1–2. FF Communications / Folklore Fellows 296/297 (Helsinki 2009).

Vésteinn Ólason 1998: Vésteinn Ólason, Dialogues with the Viking Age: Narration and Reprentation in the Sagas of the Icelanders. Transl. A. Wawn (Reykjavik 1998).

Von See et al. 2009: K. von See/B. La Farge/E. Picard/K. Schulz/M. Teichert, Kommentar zu den Liedern der Edda 6: Heldenlieder (Heidelberg 2009).

Docent Agneta Ney
Uppsala
Sweden
agneta.ney@telia.com

The bear in popular belief, legend and fairy tale

By Klaus Böldl

Keywords: Fairy tale, legend, popular belief, hunter's jargon, bear's son

Abstract: The bear plays a prominent role in popular belief and folkloristic traditions such as legends, fables and fairy tales. A closer look at these traditions shows that the ideas about bears are not at all uniform. Rather, numerous characteristics are attributed to the bear that can also be partly contrary: Thus it can on the one hand appear as an animal of wisdom, on the other hand it can be associated also with stupidity, for example in funny fairy tales. A certain similarity of the build of the bear to that of a human as well as the exceptional size and strength of this animal form the starting point of this varying attributions. With particular regard to Olaus Magnus's Historia de gentibus septentrionalibus *(1555), which represents the most fruitful Scandinavian source of premodern concepts of the bear, this contribution provides an overview of the most important bear motifs in fairy tale and legend as well as in popular beliefs.*

Introduction

A casual glance into folkloric, popular medical and iconographic handbooks or encyclopedias (cf. Peuckert 1927; 1930/1933; Stauch 1937; Zenker 1950; Wehrhahn-Stauch 1968; Reichstein/Ranke 1976; Paproth 1977) already provides an impression of the abundance of ideas that have been associated with the bear since time immemorial.[1] The bear plays a prominent role in many religions and mythologies, especially in the sphere of hunting. It is well represented not just in legends and fairy tales but for example also in medieval proverbs (cf. Thesaurus proverbiorum medii aevi 1995, 337–342). The importance of the bear in religious and symbolic thinking, going back as far as prehistory, is shown for one thing in the constellations of Ursa Major and Ursa Minor (cf. Künzl, this volume) and also in the fact that the Indo-European terms for the bear were substituted in the Germanic languages by a noa-name that refers to the brown colour of the animal (cf. Reichstein/Ranke 1976, 46; Björklöv 2010, 27; cf. also contributions by Nedoma, Særheim, or Udolph, this volume). Even according to surviving popular belief, anger is provoked in the bear, which understands human language and can even read thoughts, when it is called by its correct name (cf. Schön 2006, 23). The Nordic languages contain numerous names to describe the bear, of which *Nalle* and *Bamse* are without doubt the most popular, but terms like *Bestefar* (grandfather), *Gubbe* (old man), *Sötfot* (sweet foot), *Svarten* (black one), and many others have been documented and are often used in folk tales (cf. Tillhagen 1985, 131).

---

1 We are talking here, apart from the importance of the polar bear for the arctic peoples, almost exclusively about the brown bear (*Ursus arctos*); cf. Paproth 1977, col. 1194.

A major reason for the eminent importance of the bear in cultural history lies of course in its physical strength, which is superior to that of all of Europe's other animals. It is not uncommon for this to appear exaggerated into the demonic sphere in legendary traditions. In this connection, there is a remarkable belief that the growth of a bear never finishes and even the flesh of a killed bear continues to grow (cf. PEUCKERT 1927, col. 884). The observation already made by classical authors that the savagery of the bear increases extraordinarily when it is hungry or when its young are threatened is absolutely correct (cf. ZENKER 1950, col. 1143). However, to the special aura of the bear also belong – to a larger extent than with other animals – ideas of specific intellectual and magical strengths, which make the bear on the one hand into a sinister and, from a Christian perspective, frequently fiendish creature, but through which at the same time special relationships between man and bear can be created. These relationships arise above all from the frequently noted similarity between the body of man and of bear, which is supposedly demonstrated by skinned bears in particular (cf. PEUCKERT 1927, col. 884; TILLHAGEN 1985, 132). Johann Heinrich Zedler's *Universallexikon* from the era of the Enlightenment also mentions as "something wondrous" in this connection that the she-bear suckles her young not by facing them "backwards" but "forwards towards her chest with two teats, like a woman" (ZEDLER 1731, vol. 3, col. 115: "nicht 'hinterwärts', sondern 'vorwärts nach dem Brust-Kern zu mit zweyen Gesäugen, gleich einem Weibs-Bilde'"). In the *Deutsches Wörterbuch* of the Brothers Grimm, it is stated that "huntsman-like the hand, finger and gait of a human are ascribed to it, because it can rear up, straighten up and walk upright" (GRIMM/GRIMM 1854, col. 224: "weidmännisch wird ihm gleich dem menschen hand, finger und gang zugeschrieben, weil er sich aufbäumen, emporrichten und aufrecht gehen kann"). These similarities invite anthropomorphising considerations in particular; indeed, they make the bear almost into a crossover between the human and the animal sphere, which will be expanded on in more detail below. The numerous fairy tales and legends about humans turned into bears reflect such ideas, too (cf. PEUCKERT 1927, 162–163). Although many sources refer to its cleverness and also, for example, to its exceptional skill in killing its prey, the bear can due to its apparently clumsy-funny movements also end up as the representative of gormlessness and gullibility, especially in funny fairy tales and fables.

On the basis of its specific physical qualities, the bear moreover plays a prominent role in popular medicine. Healing, regenerative, fertility-supporting and strengthening effects are ascribed to the various components of the bear, its blood, fat, fur or testicles; furthermore, teeth and claws, which were also found as grave goods (cf. various contributions, this volume), could apparently fulfil amulet functions still in recent times (cf. PEUCKERT 1927, col. 901–905). A bear's tooth carried in the hunter's pocket in northern Swedish Ångermanland fulfils an apotropaic function – it is supposed to protect him against being bitten by the bear (cf. TILLHAGEN 1985, 148). A bear's tooth is also considered helpful against toothache (cf. BJÖRKLÖV 2010, 276). The right eye of a bear, ground and dried and hung in a bag around the neck, protects small children against nightmares (cf. BJÖRKLÖV 2010, 277). In many areas of Scandinavia, strengthening effects were attributed especially to the contents of the bear's gall bladder, which were regarded as early as antiquity as a remedy against the most diverse illnesses (cf. HÖFLER 1912 204–205); mixed with brandy, they were supposed to bestow strength and courage on hunters and sooth eye, kidney and other ailments. In Norrbotten (northern Sweden), this potion was administered as a tonic to women in childbirth (cf. TILLHAGEN 1985, 146). The idea that a particularly foul-tasting medicine develops a special healing property probably plays the decisive role in the case of the gall bladder contents. However, most of these popular medical concepts are based on a sympathetic logic according to which the special physical or organic qualities of the bear are transferable to humans. Thus, the paws of bears, according to popular belief charged with special nutritional potency, also play a role in folk medicine; they could, for example, be applied as a palliative during labour pains (cf. BJÖRKLÖV 2010, 278). In her *Liber subtilitatum diversarum naturarum creaturarum* (Book about the Secrets of the Various Natures of Creatures) from around the middle of the 12th century, Hildegard von Bingen warns against the consumption of bear's flesh, because this arouses sexual desire, whereas the bear's fat represents a proven hair restorer (cf. Björklöv 2010, 47).

It will have become clear already that the generally accepted premodern ideas about the bear can in no way be reduced to a uniform structure of meaning. One characteristic of the bear just seems to be that, regardless of the current context, the most diverse, often contrary features and behaviour can be ascribed to it. While the traditional material, which is attributable in the broadest sense to popular legend, is based predominantly on concrete observations and experience which demand authenticity and are being worked on with the help of various "mythical" patterns of interpretation, the bear of the popular fairy tale is a creature of pure fantasy, which can indeed be occasionally provided with legendary features but has no connection with the audience's everyday life and is rather frequently described as a simple-minded, likeable creature. Thus, fairy tales about bears can turn up in regions where bears never occurred, e.g. on the Faroe Islands. Often the bear has become a mere animal sign here that, like the cunning fox or the brutal-malicious wolf, embodies a human characteristic.

Christianity as the decisive framework of meaning for the Middle Ages and the early modern period also belongs of course to the traditions which have played a part in forming the popular image of the bear. The bear is already encountered in several places in the Old Testament; David boasts to Saul that with the help of God he has defeated a lion and a bear (1 Sam. 17:34–37),[2] and special respect for the superiority of the bear seems to be expressed in the proverbial expression of the prophet Amos: "As if a man did flee from a lion, and a bear met him" (Amos 5:19; cf. also Thesaurus proverbiorum medii aevi 1995, 340). The bear appears downright as "king of the animals" in the Old Testament, which is perhaps reflected in Master Amund's murals in the church of Södra Råda (Värmland; 1496), where the animal depicted in the centre of the creation of quadrupeds can likely be identified as a bear.[3] Overall, bear attributes such as strength, anger and a threatening nature predominate in the Bible (according to the Old Testament, the bear also falls into blind rage particularly when it is robbed of its young; cf. 2 Sam. 17:8 and Prov. 17:12), which is why the bear is already in early Christianity regarded as a demon and as a theriomorphic symbol of, among other things, the vice of anger and cruelty as well as hedonism (because of its affinity to honey). This demonising perspective can be observed in the popular perception of the bear up to the 19th century, when it can, for example in Norwegian folk tales, be dubbed the horse of the Devil.[4] Thus the frequent fights between man and bear in Romanesque sculpture can be interpreted mostly as a struggle with evil (cf. Stauch 1937, col. 1443–1444). However, the bear, just as the lion, is a thoroughly equivocal symbol in Jewish-Christian tradition. Its imposing strength can also mean God's omnipotence, and it appears, through the restraint of its wild nature, as a symbol of the converted heathen (cf. Zenker 1950, col. 1146). The aforementioned ambiguity of the bear is thus reflected in the Christian tradition, too.

Ideas expressed in classical zoology and medicine, which go back to Aristotle and Pliny in particular (cf. Zenker 1950, col. 1143–1144), and the symbolisations of bears in the Christian tradition as well as diverse experiences with the bear, which are articulated with the help of "popular-religious" patterns of interpretation and folkloristic narrative schemata – all these traditions and forms of knowledge are frequently present in the popular image of the bear at least up to the time of the Industrial Revolution. The boundaries between elements of classical general education and modern hunter's cock-and-bull-stories can thereby occasionally merge. In the case of written sources, the – conscious or subconscious – "better knowledge" of an author as opposed to his illiterate informants

---

2 A whole series of late medieval chalk paintings in Swedish country churches corroborates the popularity of this biblical passage, for example in Almunge and in Ekeby churches (Uppland).
3 Cf. medeltidbild.historiska.se/medeltidbild/visa/foto.asp?imageId=9416614.
4 Cf. Stauch 1937, col. 1446; Wehrhahn-Stauch 1968, col. 243–244. For the modern continued existence of these ideas cf. Peuckert 1927, col. 890.

must also be reckoned with in many cases. It is thus not possible to gain a purely Nordic conception of the bear, going back to autochthonous experiences and traditions.

Nevertheless, this contribution will attempt to outline the most important aspects of the spheres of popular beliefs, legends and fairy tales, which constitute the idea of the bear in premodern Scandinavia, even though – with the considerations already mentioned in mind – there cannot be a consistent demarcation of scholarly and literary conceptions. The bear chapters in Olaus Magnus's *Historia de gentibus septentrionalibus* (1555) should as far as possible serve as guidelines, since this work brings together various bodies of knowledge about the bear in an illustrative and contemporary way. A brief look at the bear in fairy tales at the end aims to round off the picture.

Olaus Magnus's *Historia de gentibus septentrionalibus*

This work from 1555 represents by far the richest Nordic source for premodern perspectives on the bear, and at the same time it is a prime example of the amalgamation of classical and medieval scholarly knowledge, mixed with experiences and tales from the indigenous sphere (Olaus Magnus 1972).[5] With his voluminous work, characterised by amazing erudition and references to the country's geography, history and culture, Olaus Magnus (1490–1557), Sweden's last Catholic archbishop, living in exile in Rome, wanted to direct the attention of the European public to the richness and special features of the north and thus stimulate the recatholisation of Sweden. For his portrayal of the animal world, which takes up approximately 300 pages of his *Historia*, he certainly not only exhausted the whole of the literature on the natural world up to Pliny, but was also able to fall back on the experiences he had gained himself as a young man in the service of the Church during extensive travels in the north of Sweden and to Norway in 1518/1519 (cf. Grape 1970, 30–46).

No less than eleven chapters of the 18th book, which concern the forest animals of the north *(De animalibus sylvestribus)*, are devoted to the bear.[6] On the same level as the zoological and economic-utilitarian perspectives (right at the beginning of the section on the bear in Chapter 24 of the 18th book, the significance of polar bear skins as altar decorations in cathedrals is referred to), ideas are assimilated in the *Historia* that appear rather incredible or unreal to today's reader. Thus Olaus Magnus also deals with one of the most popular bear fantasies, namely the belief already mentioned in Pliny's natural history, that bear cubs only win their form and shape by being licked thoroughly by their mother. The ambiguity with which Zedler refutes this assumption in his encyclopedia mentioned at the beginning allows the conclusion that 200 years later a matter of widespread zoological knowledge was still at issue here, although this fable had already been rejected in the 13th century by Albertus Magnus.[7] Olaus Magnus explains the tininess and alleged shapelessness of the bear cubs, in whose case only the claws are developed from the outset, with the she-bear's short pregnancy of supposedly only 30 days (XVIII, 26).[8] According to him, the she-bear gives birth in extreme seclusion – so that a human can hardly ever witness this process – to her five cubs (actually, only one to three) after a seven-day deep sleep, namely either in a cave or under a self-made canopy. During her deep sleep, the she-bear puts on large amounts of fat while she sucks at her front right paw. "Paw sucking",

---

5   The *Historia Animalium* by the Swiss philologist and naturalist Konrad Gessner (1516–1565), in which the bear also attracts a lot of attention, stems from the same period.
6   Hereafter, the work is referred to by book and chapter numbers, respectively, so that the places in the various editions can be found more readily and make translations easier.
7   Cf. Zedler 1731, col. 114; on the sources about the she-bear and her cubs cf. Foote 1998, vol. 3, 939; in addition: Björklöv 2010, 40–44.
8   A mistake already coming from Aristotle; actually, the female brown bear carries her young for about six months.

through which the bear is supposed to satisfy its hunger during hibernation as well, correlates with a widespread idea that has become associated with the observation that the bear's soles are renewed in February (cf. Foote 1998, 939). Special care is due in the forest on the day on which the bear changes the paw and therefore briefly awakes; this can happen around, for example, Candlemas or on St Canute's Day (13 January; cf. Tillhagen 1985, 142). This is where, in legendary traditions, the motif of the "hunger paws" (sucking on them means rescue from starvation and from dying of thirst for errant hunters) stems from (cf. Paproth 1977, col. 1195–1196). Olaus Magnus's reproductions of popular or scholarly ideas are by no means exhaustive. On the other hand, he often embellishes his sources, which go back as far as classical natural history and are occasionally misunderstood, with details that lend the portrayals a character of authenticity. In this way, he vividly substantiated the topos of the aggressive mother bear. Just as wildly as skilfully, the she-bear kills stags and even bulls which come too close to her cubs, whereupon she shoulders her prey and, standing upright, takes it into the forest to devour it (XVIII, 26).

Of all animals, only the wild boar knows how to resist an attack by a bear with its tusks, and only the hedgehog takes the risk of seeking out a bear's den and is even in the position to kill a bear. Olaus Magnus presumes, on the other hand, that the ferocity of a pack of wolves is one of the reasons why bears hibernate despite their warm fur – their reluctance to leave traces in the snow as well as the harmfulness of ice to the bear's sensitive paws are also considered, but not the obvious fact that the bear, as an animal with a predominantly vegetarian diet, finds no nutrition in winter (cf. XVIII, 27). The phenomenon of hibernation has in fact always fascinated people: In the Nordic imagination, there are often forest maidens and other sprites who provide the bear with food in its cave. The advantage of being able to sleep the cold winter away has been granted to bears by the Virgin Mary, according to a legend documented in Norway and Sweden: The bear carried her on his back over a river whereas the wolf and the horse had refused because they did not have the time (Frøstrup 1989, 36). According to another tradition, it was the Lord himself who rewarded the bear in this way, after the goat and the horse had balked at performing this service because of their constant need to eat (Tillhagen 1985, 150).

The special attention which the mother bear shows her cubs[9] can extend to children, as a later folk memory shows: In Sollefteå (northern Sweden), there was a report of a boy who had got lost in the forest in winter and had survived because he was kept warm by a bear in its cave. Every morning, a forest maiden brought a bowl of milk to the cave. On the day the bear was finally slain and the boy was rescued, there had been blood instead of milk in the bowl (cf. Tillhagen 1985, 143–144). On the other hand, however, an encounter with a bear can prove to be fatal, especially for pregnant women, since the belief – widespread in the north and also noted by Olaus Magnus – that wolves, or rather werewolves, were after the foetuses of women in particular had extended to bears in many areas. Such legends are in a strange, charged relationship with the prevailing ideas about the relations between bears and children; they do not occur in Olaus Magnus's work.

Horses might according to Olaus Magnus escape a bear by fleeing, but would never recover from this fright and were henceforth incapable of joining battle. So it was a popular stratagem to rush against a mounted enemy while wrapped in a bear skin (XVIII, 27). Bears are introduced early by their mother to the pleasure of licking honey. Suffering a lot of bee stings in the process has the positive effect of freeing them from their profusion of blood (XVIII, 28). As bears, as Aristotle knew already, are omnivores, young bears can also try to consume mandrake, which is life-threatening for them (but which admittedly does not occur in the north). They eat ants afterwards as a remedy

---

9 Actually, infanticides among brown bears are quite common; since bears are polygamous, the male tends to kill cubs when it is not their father; cf. Björklöv 2010, 85. Cf. also Zedrosser et al., this volume, on infanticide among bears.

against the poisoning (XVIII, 28). The protein-rich ants really represent a popular food, especially for she-bears (BJÖRKLÖV 2010, 91). One of the bear's weak points is its nervousness. If it is hit by the arrow from a hunter's crossbow while it is eating autumn berries, it experiences such a fright that the berries it has just eaten shoot out of its bottom like a hail shower (XVIII, 25; Fig. 1); a bear that is disconcerted in this way is then a relatively easy kill. The bear's most sensitive part of the body is its skull. If it is, while stealing honey, struck by a wooden club fitted with iron spikes, which has been fixed above the source of the honey for this purpose, this means death for the bear. Konrad Gessner, Olaus Magnus's Swiss contemporary, also notes in the first volume of his *Historia animalium* (1551) that even a slight blow to its skull could kill a bear, because the bone there is very thin and brittle (cf. BJÖRKLÖV 2010, 55).

A sensitive side of the bear which otherwise rarely appears in the literature is demonstrated by its love of music. Shepherds are at times dragged off by bears so that they can play something to them on their bagpipes (Fig. 2). Only when the bear, driven by hunger, sets off in search of food can the shepherd flee and reach safety (XVIII, 31). When Olaus Magnus subsequently shifts to writing about tamed bears from Lithuania and Russia that dance to trumpet music and then demand their reward while holding their hats out to the public (XVIII, 32; Fig. 3), the episode with the shepherd makes it clear that the musicality of the bear is a human trait that the bear possesses from the outset but which is utilised simply for human needs through training.[10] In the north, as in many regions of Europe, the taming of bears has a long history. From the Middle Ages, they provided entertainment at fairs. Moreover, they have been used as guard animals at times (BERG 1965, 93–96).

Olaus Magnus seizes on a large number of bear topoi that the zoological and medical scholarship has had at the ready since antiquity. Skilfully, he links these with information that might stem from Swedish informants he encountered on his inspection journeys. To what extent the literary construct that arose in this way actually reflects the ideas and experiences of broader rural layers of the population about the bear can scarcely be established. Very much in the sense of Renaissance philosophy, Olaus Magnus conceives the natural world and therefore the world of animals, too, as a semiotic system. Animals do not just appear here as zoological entities; as part of the order of creation, they are rather integrated into a complex structure of correspondences, similarities and representations. The apparently naïve anthropomorphisation of the fauna therefore forms no unreflected-popular layer of the *Historia*, as might be believed at first sight, even if it makes use of the wealth of the "popular superstition"; the detection of the "human" in animals is rather an integrating part of the Catholic archbishop's interpretation of the world. In this thinking, animal and human worlds reflect each other well; against this background it becomes understandable why Olaus Magnus devotes special attention to the tradition of the animal bridegroom or rather the bear's son.

The motif of the bear's son

In Chapter 30 of the 18th book, Olaus Magnus relates the story of a farmer's daughter who is dragged off by a bear to its cave. Whereas the bear initially wants to eat her, he is transformed by the sight of her from robber into lover (*ex raptore amator effectus*). In trying to nourish his lover in a befitting manner, the bear causes heavy damage among the farmer's cattle, which is why it is ultimately killed. At this point, the bear has already made the girl pregnant; the fruit of this union proves to be human in form because nature had determined to moderate the eerie that had grown from this "double

---

[10] In the same way, according to Olaus Magnus, the small Öland horse also instinctively starts to dance *humano more* as soon as it hears a string instrument (XVII, 16; cf. BÖLDL 2007, 28).

material" (*duplica materia*). When the son nonetheless behaves like a bear, this is explicable to Olaus Magnus through the fact that basically the father's characteristics had been passed on, just like all humans were descended from Adam. The son will later avenge his father's death; the Danish ruling dynasty will emerge from his descendants. Olaus Magnus took this story, which portrays a narrative complementary to reports about bears being tamed and therefore incorporated into the human world (cf. BÖLDL 2007, 132, 134), from the *Historia de omnibus Gothorum Sveonumque regibus* (1554) by his brother Johannes Magnus, but the original source is the *Gesta Danorum* by Saxo Grammaticus, completed around 1215 (cf. SAXONIS GESTA DANORUM 1931, 287–288).

It is not hard to show in this episode the motif of the bear's son that is widespread throughout the northern hemisphere (cf. STITT 1992) and which introduces mostly the fairy tale of the three abducted princesses (Aarne/Thompson Index [AaTh] 301, 301A and 301B); this tale also occurs as an independent narrative (cf. GOLTHER 1930/1933b). An English historical source from the Early Middle Ages already mentions a Beorn Beresun whose bear ears are a clearly visible sign of his animal origin. This in turn is substantiated by a story quite similar to one in Olaus Magnus (cf. EDSMAN 1956, col. 672–673). The motif of the bear's son in storytelling also occurs in legend literature, in most detail in *Hrólfs saga kraka* (cf. NEY, this volume). The bear cub is frequently characterised by hideousness or foulness in the supraregional stock of legends and fairy tales (cf. PANZER 1910, 19). In the Icelandic version, *Bjarndreingur*, the father bear moves out and returns only every four years to remind the mother that the son, becoming ever stronger, is not allowed to leave the house. Finally, however, the bear's son leaves it before the appointed time and unwittingly kills his father, the bear, with a magic axe it had received from him (subsequently the story then follows AaTh 301; cf. RITTERSHAUS 1902, 102–106). The wide geographical dissemination of the motif can be rated probably as an indication of a relatively great age; at any rate, the oldest evidence in northern Europe goes back to the 11th century. A legend recorded in 1755 is supposed to explain the origin of the Sámi "Bear Feast", a ceremonial following a successful bear hunt; since it emphasises – along with other motif-related parallels – a mixed marriage between a bear and a female human, the origin of the bear's son motif has been recognised in archaic hunting rituals (cf. EDSMAN 1956, col. 672; WARD 1977, col. 1232–1234; BJÖRKLÖV 2010, 267–272, especially 270–271; cf. also various contributions on the bear ceremonial and the bear's son, this volume).

THE BEAR IN THE SCANDINAVIAN WORLD OF FAIRY TALES AND LEGENDS

Along with narratives about the bear's son, there is also a series of fairy tales in which the similarity between human and bear finds expression in transformation phantasies. One of the best-known Norwegian fairy tales of all, *Østenfor sol og vestenfor måne* (East of the Sun and West of the Moon, AaTh 425), deals with the liaison between a girl and a bear which, however, turns out to be a king's son whose stepmother has cast a spell on him. Unrecognised, the king's son, who strolls as a bear during the day, lies every night next to the girl, until she, in her curiosity, lights a candle and lets three drops of tallow drip onto his robe, which leads to the prince having to return to his stepmother and to marry a troll woman. But the heroine knows in the end how to prevent this, since only she, as a Christian woman, is able to wash the spots of tallow out. The fairy tale shows many alignments with that about *King Valemon* (AaTh 425 A), who also has to adopt the form of a bear during the day due to a witch's spell and would be released only if he is not seen by anybody in human form for seven years. Again the drops of tallow play a role: The heroine sees the prince in his real form, and the escape of the couple from the witch's castle is of course successful this time, too.

The fairy tale of the bear-skin man (*Bärenhäuter*; AaTh 361) also needs to be mentioned in this respect; it is evidenced throughout Scandinavia (cf. UTHER I, 2004, 227). The Devil promises wealth to an impoverished soldier who has been dismissed from service, if he neither washes himself nor cuts his nails for seven years. The hero uses the money to help another man overcome adversity, who consequently promises him the hand of one of his daughters in marriage. Only the youngest of the sisters sees the good heart of the hero behind his repulsive external experience and chooses him as her bridegroom, after which the two others kill themselves out of grief in the end. The connection between this fairy tale and the bear symbolism is vague in many versions or does not exist at all. Thus the hero's hideous external appearance is explained by his work at the Devil's fireplace in the Swedish version by Liungman, representing the motif of "hell's stoker" (cf. SCHERF 1995, 48); there is no talk of a bear skin. In the versions in which the hero really slips into a bear's robe at the behest of the Devil, a bear appears – usually during the conversation between the soldier and the Devil – and is immediately shot dead by the hero. This is how it is already told in 1670 by Grimmelshausen, whose story for the first time links the fairy tale type AaTh 361 with a bear motif. This turns up again in the north in, among other things, the Danish version, *Bjørnmanden*, as recorded by Ewald Tang Christensen in 1902. Here it is also present that the Devil sows the hero into the bear's robe, and that he at first spreads fear and terror among the people until they recognise his good-heartedness and helpfulness (cf. BØDKER 1960, 52–57). Whether the bear skin motif can be interpreted as a later reflex of Old Norse berserker concepts must remain open in light of the relatively recent development history of this type (cf. GOLTHER 1930/1933a, col. 168–170; cf. SUNDQVIST, this volume, from the viewpoint of a historian of religion; cf. also HIRSCH, this volume). The high incidence of ideas of bears as dressed up or transformed humans does not necessarily allow this connection to be seen as compelling.

As already indicated, the "simple forms" in Scandinavia have not shaped a specifically Nordic bear motif; in most cases, they represent an international narrative. Often the bear is only a cipher for a quarry of especially high-quality. In the legends recorded under E 11–17 of *Types of the Swedish Folk Legend*, a mostly feminine forest spirit promises a man a bear (or an elk) as quarry for the following day, if he provides the spirit with food or tobacco, gives her a shirt to rock her infant with, or calls her by a certain name (cf. AF KLINTBERG 2010, 100–101). Even if a hunter is willing to move his campfire because it lies on an elven path, he will be rewarded in this way (cf. AF KLINTBERG 2010, 199 K 181). However, bears indeed appear very often in popular legends, but in no way exclusively in hunting contexts. The special size of giants can for instance be emphasised when a giant child uses a bear as a toy (cf. AF KLINTBERG 2010, 157). Bears are in a position to drive away the trolls which have settled down on a farm; alternatively, the nuisance guests might be robbers (cf. RÖHRICH 1977; UTHER I, 2004, 79–80; UTHER II, 2004, 54–55). The bear proves itself grateful to the human who removes a thorn from its paw (AaTh 156; a motif also known from hagiography, although it is, as for example with St Jerome, associated mostly with a lion, which becomes a faithful companion).

In the funny fairy tale about the Dalecarlians on a bear hunt (*Dalkarlarna på björnjakt*, AaTh 1225), the one of the three companions who dares to enter the bear cave has his head bitten off by its occupant. The punch line of the story comes when the Dalecarlians can no longer remember whether their companion ever possessed a head. This story, handed down in the north and also in Norway and among the Sámi, probably stems from the Orient and is well substantiated in eastern Europe in particular (cf. UTHER II, 2004, 80). However, more frequent are the fairy tales in which the bear appears as stupid and gullible, to be usually duped by the fox. In the collection of Norwegian fairy tales by Asbjørnsen and Moe, the stories of bear and fox are grouped in a series of trenchant funny stories (cf. ASBJØRNSEN/MOE 1982, 479–483). The best known of these fairy tales relates how the bear comes by its stubby tale (AaTh 2): The fox, who arouses envy in the bear with a stolen bundle of fish, recommends that the bear sticks its tail in a hole in the ice and uses it as fishing bait; thereupon the tail freezes solid and is bitten off by the fish. This anecdote is documented for the first time in the *Roman de Renart* (1178) and recorded remarkably frequently throughout Europe (cf. UTHER I, 2004, 17–18). In a Swedish version, it is the well of an old women in which the bear tries to fish. The fox betrays the bear to the woman, who beats the frozen bear while the fox is able to drink up the woman's cream undisturbed (cf. Liungman 1950, 11). In other fairy tales of this funny type, in which the wolf can from time to time take the place of the tricked bear, the fox is cheated of its proportion of butter, and the bear catches its paws in a trap; the bear is made to eat its own entrails (AaTh 21) or carry the fox. In AaTh 36, the fox rapes the she-bear; in AaTh 37, disguised as a nursemaid, it eats her children (cf. Paproth 1977, col. 1200–1201). Also very popular in the north is the motif according to which the bear gets to grab the fox by the leg after a further atrocity but lets go of it again when the fox convinces it that it is just a root (AaTh 5; cf. Paproth 1977, col. 1200; examples in Asbjørnsen/Moe 1982, 479–483; Liungman 1950, 13). In all these fairy tales, the otherwise so central physical strength of the bear is suppressed or neutralised by its stupidity. Beyond all concrete experience of the bear and related mythical-religious concepts, it is reduced here to the representation of a certain (inferior) mental disposition which finds its expression in the dealings with a certainly physically weaker but intellectually superior animal. However, a fable in the *Dialogus Creaturarum Moralizatus*, the first book printed in Sweden (1483), certainly forms a remarkable exception to this. Here it is for once the bear that cheats the wolf. It suggests to the wolf to look after him in its cave during the winter, if the wolf in return feeds it throughout the summer. Thus the bear spends a peaceful summer, but in the winter cave, the bear has nothing to offer the wolf but sucking the bear's paws (cf. Björklöv 2010, 369–370). The moral of this story is not absolutely evident. Apparently, it is about being careful in choosing your friends.

In summary, it can be recorded that the bear takes a special position above all other animals in the perception of the inhabitants of the northern hemisphere, as can be seen in popular beliefs as well as in legend and fairy tale literature in extraordinarily complex and varied bear semantics. Two factors seem to be decisive for the not unfrequently also contradictory popular image of the bear: On the one hand, the frightening size and strength of this "king" of the animals, which led to various taboo ideas and – in earlier times – also to a religious exaltation, and on the other hand its "human-like" form which makes it into a projection surface for the most varied and also all too human characteristics, which span from knowledge of healing, love of children and musicality to credulity and clumsiness. The long journey from the demonic and life-threatening ruler of the forests to the teddy bear (certainly also one of the popular manifestations of the bear) is undoubtedly of great interest for the history of mentality.

Bibliography

Asbjørnsen/Moe 1982: P. C. Asbjørnsen/J. E. Moe, Samlede eventyr 1 (Stabekk 1982).

Berg 1965: G. Berg: Tama björnar, dansande björnar och björnförare. Fataburen 1965, 93–112.

Björklöv 2010: S. Björklöv, Björnen i markerna och kulturen (Hedemora 2010).

Böldl 2007: K. Böldl, Die Frömmigkeit der Fische. Zur Zoologie der *Historia de gentibus septentrionalibus* von Olaus Magnus. In: A. Heitmann/W. Heizmann/O. Rehm (eds.), Tiere in skandinavischer Literatur und Kulturgeschichte. Repräsentationsformen und Zeichenfunktionen (Freiburg i. Br. 2007) 115–141.

Bødker 1960: L. Bødker, Danske Folkeeventyr (København 1960).

Edsman 1956: C.-M. Edsman, Björnfest. In: Kulturhistorisk leksikon for nordisk middelalder 1 (København 1956) col. 671–674.

Foote 1998: P. Foote (ed.), Olaus Magnus, Historia de Gentibus Septentrionalibus. Description of the Northern Peoples. Transl. P. Fisher/H. Higgens (London 1998).

Frøstrup 1989: J. Chr. Frøstrup, Dyr i tro og overtro (Oslo 1989).

Golther 1930/1933a: W. Golther, Bärenhäuter. In: Handwörterbuch des deutschen Märchens 1 (Berlin, Leipzig 1930/1933) 169–172.

Golther 1930/1933b: W. Golther, Bärensohn. In: Handwörterbuch des deutschen Märchens 1 (Berlin, Leipzig 1930/1933) 172–174.

Grape 1970: H. Grape, Olaus Magnus. Forskare, moralist, konstnär (Stockholm 1970).

Grimm/Grimm 1854: J. Grimm/W. Grimm, Bär. In: Deutsches Wörterbuch 1 (Leipzig 1854) col. 1122–1125.

Höfler 1912: M. Höfler, Die volksmedizinische Organotherapie und ihr Verhältnis zum Kultopfer (Stuttgart, Berlin, Leipzig 1912).

Af Klintberg 2010: B. af Klintberg, The Types of the Swedish Folk Legend (Helsinki 2010).

Liungman 1950: W. Liungman, Sveriges samtliga Folksagor i ord och bild 2 (Stockholm 1950).

Olaus Magnus 1972: Olaus Magnus, Historia de gentibus septentrionalibus, Romae 1555. Introd. J. Granlund [Reprint of Latin original edition] (Copenhagen 1972).

Panzer 1910: F. Panzer, Beowulf. Studien zur germanischen Sagengeschichte 1 (München 1910).

Paproth 1977: H.-J. Paproth, Bär, Bären. In: Enzyklopädie des Märchens 1 (Berlin, New York 1977) col. 1194–1203.

Peuckert 1927: W.-E. Peuckert, Bär. In: Handwörterbuch des deutschen Aberglaubens 1 (Berlin, Leipzig 1927) col. 881–905.

Peuckert 1930/1933: W.-E. Peuckert, Bär. In: Handwörterbuch des deutschen Märchens 1 (Berlin, Leipzig 1930/1933) 157–169.

Reichstein/Ranke 1976: H. Reichstein/K. Ranke, Bär. In: Reallexikon der Germanischen Altertumskunde 1 (Berlin, New York 1976) 45–48.

Rittershaus 1902: A. Rittershaus, Die neuisländischen Volksmärchen. Ein Beitrag zur vergleichenden Märchenforschung (Halle a. d. Saale 1902).

Röhrich 1977: L. Röhrich, Bärenführer. In: Enzyklopädie des Märchens 1 (Berlin, New York 1977) 1217–1225.

Saxonis Gesta Danorum 1931: Saxonis Gesta Danorum. Eds. J. Olrik, H. Ræder (Hauniæ 1931).

Scherf 1995: W. Scherf, Das Märchenlexikon, vols. 1–2 (München 1995).

Schön 2006: E. Schön, Folktro på fäbodvall (Ljusdal 2006).

Stauch 1937: L. Stauch, Bär. In: Reallexikon zur deutschen Kunstgeschichte 1 (Stuttgart 1937) col. 1442–1449.

Stitt 1992: M. Stitt, Beowulf and the Bear's Son: Epic Saga, and Fairytale in Northern Germanic Tradition (New York 1992).

Thesaurus proverbiorum medii aevi 1995: Thesaurus proverbiorum medii aevi, Bär. In: Lexikon der Sprichwörter des romanisch-germanischen Mittelalters, begründet von Samuel Singer, vol. 1 (Berlin, New York 1995) 337–342.

Tillhagen 1985: C.-H. Tillhagen, Jaktskrock (Stockholm 1985).

Uther I + II, 2004: H.-J. Uther, The Types of International Folktales. A Classification and Bibliography. Based on the System of Antti Aarne and Stith Thompson. Folklore Fellows Communications 284, Vols. I + II (Helsinki 2004).

Ward 1977: D. Ward, Bärensohn. In: Enzyklopädie des Märchens 1 (Berlin, New York 1977) col. 1232–1235.

Wehrhahn-Stauch 1968: L. Wehrhahn-Stauch, Bär. In: LCI. Lexikon der christlichen Ikonographie 1 (Rom, Freiburg, Basel, Wien 1968) col. 242–244.

Zedler 1731: J. H. Zedler, Bär. In: Grosses vollständiges Universal-Lexikon aller Wissenschafften und Künste, welche bishero durch menschlichen Verstand und Witz erfunden und verbessert worden 1 (Leipzig 1731) col. 114–116.

Zenker 1950: S. Zenker, Bär. In: Reallexikon für Antike und Christentum 1 (Stuttgart 1950) col. 1143–1147.

Prof. Klaus Böldl
Institut für Skandinavistik, Frisistik und Allgemeine Sprachwissenschaft (ISFAS)
– Skandinavistik –
Christian-Albrechts-Universität zu Kiel
Kiel
Germany
k.boeldl@isfas.uni-kiel.de

Fig. 1. If a bear is hit by the arrow from a hunter's crossbow while it is eating autumn berries, it experiences such a fright that the berries it has just eaten shoot out of its bottom. A bear that is disconcerted this way is then a relatively easy kill (after Olaus Magnus, Historia de gentibus septentrionalibus *[Antverpiae 1558], XVIII, 25. © Augsburg, Staats- und Stadtbibliothek Gs 6098, digitalised by the Bayerische Staatsbibliothek / Bavarian State Library. MDZ Digitale Sammlungen).*

Fig. 2. The bear's love of music leads to it dragging shepherds off, who play something on the bagpipes to it. Not until the bear goes looking for food can the shepherd flee and reach safety (after Olaus Magnus, Historia de gentibus septentrionalibus *[Antverpiae 1558], XVIII, 31. © Augsburg, Staats- und Stadtbibliothek Gs 6098, digitalised by the Bayerische Staatsbibliothek / Bavarian State Library. MDZ Digitale Sammlungen).*

Fig. 3. Tamed bears from Lithuania and Russia dance to trumpet music and then demand their reward while holding their hats out to the public. The natural musicality of the bear is utilised merely for human needs through training (after Olaus Magnus, Historia de gentibus septentrionalibus *[Antverpiae 1558], XVIII, 32. © Augsburg, Staats- und Stadtbibliothek Gs 6098, digitalised by the Bayerische Staatsbibliothek / Bavarian State Library. MDZ Digitale Sammlungen).*

Killer bears and bear killers in 19th-century Sweden

By Karin Dirke

Keywords: Bear hunting, northern Europe, Herman Falk, Llewellyn Lloyd, human-animal studies, animal agency

Abstract: The chapter discusses encounters between bears and the two 19th-century hunters Herman Falk and Llewellyn Lloyd. By taking both the hunter's and the bear's perspective into account, the 19th-century ambivalence towards hunting in general, and large carnivores in particular, is highlighted. Different forms of hunting have been categorised by anthropologist Garry Marvin, and this chapter utilises, as well as questions, these categories. From the bear's perspective, hunting was probably experienced as traumatic and destructive. On the other hand, hunters also built relationships with individual animals, thus gaining knowledge about bears. A hunt sometimes went on for hours or days. The particular animal hunted may have been hunted before, and was known to the hunters. This allowed the hunters to gain a nuanced picture of the animals and to understand their lives. In this way, the perception of the bear grew from experiences that were shared by hunters and bears. Eventually, this alternative view of the bear entailed the relative protection of the animal in Sweden.

Introduction

What happens if we think about bears and humans as two equal parts in the construction of Scandinavian hunting narratives and practices? As an effect of Cartesian dualism, animals have been viewed as objects in the western world, as lacking both agency and intentionality. This has resulted in an ignorance of the animal's perspective. This paper instead aims to discuss how ideas about bears in hunting literature and zoology grew from a shared experience of encounters between humans and bears, how they are constructed as entangled with each other, and how they influence one another in narratives about hunting.

What did the hunter know about the bear? Folklore and experience seem to point towards a tacit knowledge to which the bear itself contributes. What does this mean for ideas about bear agency, and how has it led to the protection of the bear in Sweden? The story circles around two bear hunters in 19th-century Sweden, and the bears they encountered and wrote about.

Bear hunters

Two prominent bear hunters, Herman Falk (1785–1865) and Llewellyn Lloyd (1792–1876), published works on bear hunting in the first half of the 19th century. Both shared their knowledge with the Swedish zoologist, Sven Nilsson (1787–1883). The three men were interconnected and frequently referred to each other's work. Falk was a professional bear hunter, and, as a royal forester, it was his obligation to organise bear battues (Fig. 1). The task was to arrange large numbers of people in a search party, driving the bears towards the hunters. The operation required knowledge and skill, concerning both the organisation of large numbers of people and experience in the behaviour of bears. As a former military man and a skilled hunter, Falk had both. In 1819 he published instructions on how to organise large bear hunting parties, and in 1828 this booklet was revised and republished.

Llewellyn Lloyd was an English nobleman, who during the 1820s and 1830s travelled in Sweden and Norway, hunting, fishing, and publishing travel accounts and hunting stories about his experiences (Fig. 2). His independence of means made this possible, and the threat of being charged of accidently shooting a forester while poaching in his home country encouraged him to travel. Lloyd participated in Falk's hunting parties, but also hunted bears in different ways, individually on skis, and with groups of local people. Falk instructed Lloyd on how to conduct effective bear hunting, and Lloyd frequently referred to Falk's knowledge about bears and hunting. Lloyd published several books about his travelling adventures. Mainly, they consisted of hunting and fishing stories, but they also included accounts of ethnographic observations. The books were published in several editions in English and were also translated into Swedish. The writings of these two hunters are the main source material of this study.[1] Thus, the behaviour of the bears is interpreted through an analysis of the words of their killers.

Killing animals

The relationship between the hunter and the bear revolves around the death of the animal. It is the explicit aim of the hunt, and the fear of death or injury is what primarily drives the bear's behaviour. To understand these hunters and their relationships with the bears, we thus need to understand the meaning of hunting in the 19th century. Anthropologist Garry Marvin has discussed how hunting can be defined and understood as a cultural and social pursuit. Hunting, of course, involves the killing of animals, but not all animal killing is hunting. What then distinguishes hunting from other forms of killing or interaction with animals? Marvin suggests two types of killing, domestic and wild. Domestic killing is not hunting, but is characterised by its purposefulness. There is no possibility for the animal to escape. Domestic killing can be of two types, cold and hot. Cold killing is when the animal is taken from its original space to be killed in a special place by professionals. The killing is highly mechanised and/or medicalised, and it is performed by professionals with no relation to the individual animal. This is what happens in the meat industry, or when pet animals are euthanised (Marvin 2006, 12–17).

What Marvin calls hot killing is instead when pest animals are killed. This process is not categorised as hunting but rather extermination or annihilation. It is done by professionals, but, contrary to the cold killing, this form requires the professional to have a relationship with the animal. The killer seeks out the animal in its habitat and is therefore required to establish some kind of relationship with it. The motivation for much 19th-century bear hunting was that of pest extermination. Large predators were relentlessly hunted, and authorities encouraged people to kill them. But when a large amount of people – sometimes several thousands – were gathered to drive the animals it was still considered to be a form of hunting, even though the purpose was to free the area of large predators (Marvin 2006, 17–18).

---

1 I have also discussed Falk and Lloyd in another paper about hunting narratives; see Dirke 2017.

Sport hunting fits into the second category: wild killing. This is when the hunter enters the space of the wild animal to kill it. The animal will not easily comply, and the hunter needs skill and knowledge to be successful. A close relationship with the prey animal is thus required. Wild killing is done by non-professionals for enjoyment (Marvin 2006, 18); the relationship with the animal can be emphasised. The hunter may express a kinship with the animal and describe it as an equal opponent.

Hunting, as it is defined by Marvin, is considered to be an entirely cultural process. Claims – mainly from a pro-hunting perspective – that hunting is a human's natural predation on other animals are refuted by Marvin with the argument that hunting requires planning and premeditation, which supposedly sets it apart from animal predation. Hunting by humans is therefore a highly organised and ritualistic endeavour. Marvin underlines the nature/culture-divide in order to argue for the specificity of hunting. He highlights how hunting by humans is a cultural endeavour, whereas hunting by animals supposedly is not. Hunting by humans mimics animal behavior – such as tracking and killing – but is really not the same, according to Marvin. He states, with reference to Tim Ingold, that animals only start the behaviour of predation when a prey is about (Marvin 2006, 12–14).

Yet, the question remains: How do we know an animal lacks premeditation when it sets out to hunt? Why should we make our categorisations dependent on a constructed division between culture and nature? Further, the 19th-century bear hunters do not fit perfectly into Marvin's categories. The categories need historisation in order to describe 19th-century hunting practices, when bear hunting was *both* sport hunting and pest extermination. The dual nature of the hunt seems to reflect the ambivalent perception of bears. Marvin's definition of hunting is useful to shed light on the ambiguities and complexities of human-animal relations, but needs to be nuanced when applied to historic source material.

In this chapter, my approach is to use Marvin's categories, but to investigate hunting without taking the constructed divide between nature and culture as point of departure. Hunting is undoubtedly a cultural practice; however, I think we need to accept that we do not know what the animal is thinking when it sets out to hunt. We can also state that what takes place between human and bear in the practice referred to as bear hunting is something quite different from the bears' perspective. Further, we cannot assume that bears were the same in the 19th century as they are today. How do we set about investigating hunting when all we have is source material left by humans? Hunting is, as Marvin shows, a very relational business. It is performed as an interaction between hunter and prey. Both influence the hunt and its outcome. How can hunting be understood if we also acknowledge the bear and its (involuntary) participation in it? Thus, I intend to discuss bear hunting during the 19th century in the light of Marvin's categories, but also by taking the perspective of the bear into account.

The bear killer's understanding of bears

Bears were prized hunting objects. One reason for hunting bears was that they damaged livestock. When a bear had killed cattle or sheep it was the owner's right to report this to the governor, who would arrange a chase. People from all around the county turned up to join the search party. Some of the hunts described by Falk and Lloyd are of this variety – "hot killing" in Marvin's terms. Bears were described as a great threat to farmers and their livestock. Falk complains in his book about bear hunting that commoners do not participate enough in extermination efforts. "Those who are affected by the damage done by bears complain, but rarely contemplate the measures required to free the area of these slaughterers and crop destroyers", he declares (Falk 1828, 21, author's translation from Swedish).

Written sources from earlier centuries speak of different types of bear, mainly a larger killer bear and a smaller grass bear. It was therefore debated whether the varieties belonged to the same species

(BERCH 1750, 13; ORRELIUS 1750, [no page number], 3; BRUMMER 1789, 35). Falk concluded that both he and Lloyd had never encountered more than one species of bear in Sweden (FALK 1834, 1138). However, the idea of a larger and more dangerous killer bear was ubiquitous. Bear killing was thus always motivated by the damage done by killer bears. Underlining the argument for killing bears was a moral obligation. Slain cattle (killed by someone other than the humans themselves) required retaliation in the form of community bear hunting (Fig. 3).

Bears were also hunted because of their size and assumed fierceness. In Sweden, as elsewhere, the bear was considered an impressive object to hunt. Sport hunters were interested because of the thrill of killing a dangerous animal. Falk bluntly asserted in his accounts of dramatic bear hunts in a hunting journal that only risky and eventful hunting operations were worth narrating (FALK 1834, 1145). One can thus conclude that hunting accounts generally provide an exaggerated view of the aggressiveness of the bears. Early on, bear hunting became linked to ideas of dramatic battle or heroic warfare. Together with the elk and the reindeer, the bear was the megafauna of Scandinavia in historic times and thus especially interesting to hunt. Humans seem universally inclined to kill large animals, making them overrepresented in extinctions. Everywhere humans have been present, large animals have, to a greater extent, become extinct (SVENNING 2017). Bears were no exception and, during the 19th century, the population of bears in Sweden was heavily decimated (DANELL et al. 2016, 217). Both Falk and Lloyd enhanced the size and fierceness of the animals in their stories.

Hunters during the 19th century did not differentiate when hunting bears of different age or gender. Males, females, or cubs could be hunted at any time and in any place, and hunting stories were commonly aimed at exaggerating the boldness of the hunter. Therefore, size, fierceness, and dangerousness of bears were often emphasised. Large bears were thus more attractive to hunters, and bears were often described as large. The killing of a large bear attracted attention. The size of the bear was always mentioned, and especially highlighted if it was a big one (Fig. 4).

Hunters tried to visualise the size of the animals in writing, even if it was not possible to weigh or measure in any other way. Lloyd tells us that a significant bear was shot by one of Falk's men; its fat would apparently have weighed 100 *marker* (1 *mark* = 212.5 g), and the circumference of its wrists was larger than the hunter could reach with both hands (LLOYD 1831, 33). On another occasion, Lloyd measured a killed elk with a string. Unfortunately, however, he later lost the piece of string from his pocket (LLOYD 1831, 279).

The gender of the bear was often of secondary importance. It was frequently mentioned, but the gender could be transformed within a story. Lloyd describes a female bear, who had been bereaved of her cubs earlier the same day. Despite this, the bear was referred to as "he" as soon as it attacked (LLOYD 1831, 101, 164–165). The gender of the bear was obviously linked to its behaviour rather than to its biology.

FALK'S HUNTING INSTRUCTIONS

However, when discussing how a bear should best be caught and killed, it was not its fierceness but rather its fear that was brought to attention. Falk demonstrated a great knowledge of bears, what scares them, and how to handle them. A frightened bear is dangerous. Therefore, much of Falk's instructions were aimed at keeping the bear calm. Bears could be shot in their winter dens, by silently seeking out their resting places. Falk teaches that, to catch a bear, it is necessary to walk quietly through the forest, noticing every scratched tree or other sign of bear. Scratch marks in the moss or broken spruce twigs will reveal the presence of the bear's winter quarters. If one finds the den, Falk instructs to leave it untouched, or the bear will move far away (FALK 1828, 22). Later, the hunter would return and attack the bear in the den.

Bears could also be hunted by large hunting parties during the summer, when great numbers of people were summoned to drive the bear. The goal was to, slowly and carefully, move the bear towards the hunters to enable them to shoot it. A circle, of sometimes hundreds of people, gradually drove the bear towards its execution. The bear was driven to the *skallplats*, a prepared open space, where it was killed. When the battue is beginning to close, Falk explains, the bear "shows itself" (Falk 1828, 7). The wording implies agency of the bear, but the reason for the bear to become visible is probably rather the stress the chase is causing; the bear may have been driven for hours or days. Uneven or bent lines of people are dangerous in the operation. If the bear enters such a bend it is almost impossible to keep within the line of people, according to Falk, because it comes too close to the humans. The lines of people have to be kept at such a distance that nobody risks being shot by the hunters or attacked by the bear, and at the same time close enough to keep the bear within the circle. During the hunt, the bear should always be kept at a distance of at least forty feet, otherwise it will easily break the line. In any disagreement with a bear, it is important to always be the aggressor, Falk declares. The grimmest bear will withdraw when directly attacked, but, if it is given time to strike first, it will break through any obstacles (Falk 1828, 24). The bear is, according to Falk, very reluctant to attack humans. It will only do so if it is forced to defend itself (Falk 1834, 1141).

The battue will certainly encircle all animals in the forest, but bullets may, according to Falk's instructions, only be fired at bear or wolf during the chase. Any other animals should be killed with clubs or spears (Falk 1828, 11). The difficult task was to hold the bear within the circle. Therefore, silence was necessary. If the bear was frightened, it would break the line and get away. It is of the essence to remain completely quiet unless the bear tries to escape, Falk explains, in which case intense noise must be made. At the moment the bear turns back, the people must be silent again. This way the bear is kept within the circle of people (Falk 1828, 12).

Communication with the bear was thus an important aspect of the hunt. The hunters were required to remain very attentive to the bear and to quickly respond to its behaviour. The hunt, Falk stated, should take place in the forest where the bear can find places to hide. If it is held in the open fields, the bear will easily panic and break through the line (Falk 1828, 13). Thus, the hunter primarily had to be aware of the timidity of the bear, and much of the chase was targeted on avoiding frightening the bear. The bear's behaviour was closely observed. The hunters emphasised the importance of knowledge about the bear in order to kill it, underlining the reciprocal characteristic of wild killing in Marvin's vocabulary.

For instance, both Falk and Lloyd noted (though confusing hunting or fighting with defending oneself) the bear's different combat methods. They wondered at the observation that the bear seemed to attack cattle with its front paws, while biting humans. Lloyd concluded that this is lucky since the power of the bear's strike could easily kill a human (Lloyd 1831, 33). More likely, the perceived difference in how bears attack is due to the fact that bears attack cattle as prey, but defend themselves against humans. Falk and Lloyd, however, considered all encounters the bear engaged in as a battle or warfare. For the bear, on the other hand, there was most likely a great difference between hunting for prey itself and attempting to defend itself against an aggressor. According to Falk, some speculate that the human is special, singled out by divine providence to be spared by the bear, and therefore the bear uses different strategies when attacking. On the contrary, Falk thought human dominance to be the reason for the bear refraining from using its paws when charging (Falk 1834, 1138).

The idea of the human as the dominant force was self-evident to both Falk and Lloyd. The bear hunters described the hunt as a form of legitimate warfare. Hunters chased the bear in a justifiable act of combat, in defense of livestock. The thought of being constantly at war with the bear was necessary for the legitimacy of this idea. The bear was described as a worthy opponent, powerful but deserving. Falk's way of hunting (in the form of large battues) in particular was closely linked to warfare and had its roots in 18th-century military strategy (Nyrén 2012).

Fig. 1.

Fig. 2.

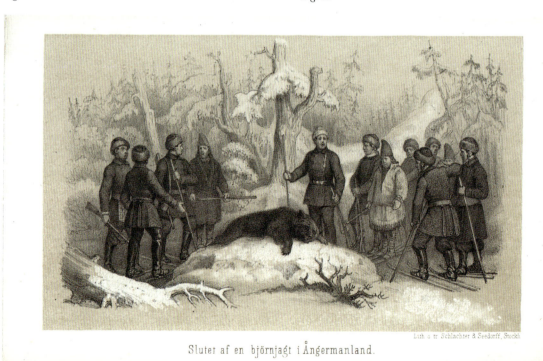

Slutet af en björnjagt i Ångermanland.

Fig. 3.

Fig. 4.

Fig. 1. Herman Falk (portrait from Svenska Jägareförbunderts Nya Tidskrift 1868, 190).

Fig. 2. Llewellyn Lloyd (portrait from Svenska Jägareförbundets Nya Tidskrift 1875, 119).

Fig. 3. This bear was killed, not by Falk or Lloyd, but during another hunt in Ångermanland in the north of Sweden, after having been followed for six days. It seriously injured two of the hunters before it was finally shot by a hunting party led by the local forester Wallroth (after Svenska Jägareförbundets Nya Tidskrift 1865, 129–134, picture on page 128).

Fig. 4. Lloyd being attacked by a bear, but saved by his companion Jan Finne (after Lloyd 1831, unpaginated).

Fig. 5. The bear illustrating Falk's instructions on how to kill bears looks weary and suspicious rather than aggressive and bold (after Falk 1828, title page).

Fig. 5.

At the same time, however, Falk and Lloyd (as well as the zoologist Sven Nilsson) described the bears as not being particularly prone to attack livestock. It is only when the domesticated animals themselves provoke the bears that the bears attack, they argued (FALK 1828, 27; 1834, 1138; LLOYD 1831, 28; NILSSON 1847, 199). The idea of the bear as a threat to both livestock and humans fits into the framework of domestic killing, of eradicating the bear. The emphasis on knowledge and a relationship with the bear, however, reveals other sides of the animal. It also appears as frightened, shy, and not particularly aggressive to the meticulous observer. Thus, both bear hunters displayed ambiguous ideas about the animals they hunted. Bears were, in the hunters' view, aggressive, threatening to livestock, large, and fierce, but at the same time fearful, placid, and possible to communicate with.

THE BEARS

Who were the bears being hunted by Falk, Lloyd, and other hunters? We can collect zoological and ethological information about bears, but we cannot be entirely sure that bears in the 1820s were the same as bears today. Therefore, we need to understand the particular bears being hunted in the accounts and take note of who they were and how they appear in the texts. Being heavily hunted – bear hunting did not decrease in Sweden during the 19th century – bears probably went out of their way to avoid humans. Hunting rights had changed in the generation before Falk and Lloyd walked the woods. From 1789, commoners were given the right to hunt for game on their own land (previously this had been the privilege of nobility and royalty). How this actually changed the hunting pressure on game animals is unclear. Since long before hunting rights were granted to commoners, there was consensus about the reasons for the decimation of wild game: poaching, hunting, and predation by large carnivores (NYRÉN 2012, 114). Poaching, though difficult to estimate, seems to have been ubiquitous (NYRÉN 2012, 168), and all people were also obliged to exterminate large carnivores (BERGSTRÖM et al. 2015, 58). Bears, thus, were continuously exposed to violent attacks from humans.

The bears in Falk's and Lloyd's stories, as we have seen, seem very violent and aggressive. Falk, however, states that he only tells us about bears that put up a fight. Bears that simply die are not worth mentioning (FALK 1834, 1145; KHEEL 2008, 84). The first bear to be awarded some sort of agency in Lloyd's account is one whose skull was kept by Lloyd. It was, according to Lloyd, shot by Falk and was found to be so large that ten people could only carry it a short distance. The bear put up a worthy fight and did not die without bravery, according to both hunters. Lloyd referred to Falk's account of the event: "He did not die tamely, for, after receiving several balls, he dashed at the cordon of people who encompassed him on all sides, and, according to the same author, severely wounded no less than seven of them in succession. [...] One of the men he bit in thirty-seven different places, and so seriously in the head, that his brains were visible" (LLOYD 1842, 97).

The animal was later shot by Falk (LLOYD 1831, 32–33). The making of a good story required the bears to be portrayed as vicious and dangerous. Bear hunters were most often remembered by the number of bears they had killed during their lifetime. In a similar manner, bears were remembered by the amount and precise descriptions of wounds they had inflicted upon their persecutors.

In his writings, Lloyd talks about a bear battue he joined in Dalarna, inland Sweden. Two bears had been seen to kill three horses; thus, a chase was ordered by the governor. One of the bears was described as particularly murderous. He could be recognised by a missing claw on his front paw and was in this way portrayed as an individual. According to Lloyd, approximately 1,500 people joined in the search party (LLOYD 1831, 39–40). The event was accounted as dramatic and purposeful; however, when the bear was spotted, Lloyd described him as "the friend" (LLOYD 1831, 43), and displayed affection for the bear. The position of the hunter is ambivalent, oscillating between the one-sided extermination of the bear, and the knowledgeable wild killing by the sports hunter, who is relating to the bear. Lloyd was suddenly surprised by a bear cantering by him, the story went, but he failed to shoot, not expecting the bear (Lloyd 1831, 47). When another bear was spotted by Lloyd it was not described with contempt, as an animal worthy of extermination, but with admiration: "A bear appeared with his head lifted high and with a proud appearance, fiery and brave like a warhorse" (Lloyd 1831, 47, author's translation from Swedish).

The animal was shot at, and, by the murmuring noise he made, Lloyd assumed he was wounded. According to Lloyd three bears were shot at the battue, one male and two females, the one Lloyd shot presumably being the male (Lloyd 1831, 48).

Again, the bears killed were described by Lloyd as being very large. They were assumed to be male, and they were killed after a long hunt. Another bear hunt described by Lloyd was, according to the hunter, a large organised battue. In his writings, he speculates about the cost of having such large amounts of people in comparison with the profit won from killing the bears (Lloyd 1831, 51), implying a vague, and otherwise quite rare, critique of bear hunting.

The operations described in the stories by Falk and Lloyd indicate highly stressed bears. Their cantering through the countryside and their aggressiveness are indicators of how hunting influenced the bears (apart from the ones actually being killed). As noted, bears had, and still have, a reputation for being aggressive and fierce. This idea generally stems from people observing bears while hunting them (Bradshaw 2017, 54). Lloyd seems to have understood this. From his other writings, a different view of large predators emerges. He describes how young girls and boys guard livestock with nothing but a stick and a birch-bark horn to scare bears off (Lloyd 1871, 133–134). According to Lloyd it was unheard of that bears attack children herding animals. Bears can be dangerous, he concludes, but only rarely if left undisturbed (Lloyd 1831, 98). When the bear is left in peace and is not followed by cubs, it will not attack. But if it is hurt or bothered, it will not back down (Lloyd 1831, 101).

Again, a dual perception of bears can be observed in the writings of the bear hunters, Falk and Lloyd. They described bears as dangerous and violent, but simultaneously as friends, placid souls, and admirable combatants (Fig. 5). The view of the bear was ambiguous, wavering between the violent contester of the noble huntsman and the placid inhabitant of the wilderness.

The extent of bear hunting, the large search parties involving sometimes thousands of people, the shots being fired at bears, sometimes hitting and wounding the animals, must have had an effect on the bears' minds. Lloyd claimed that, in wintertime, bears can be shot with several bullets without dying since shots in empty intestines do less harm (Lloyd 1831, 191). Bears were thus continuously chased, stressed, wounded (and killed) in Swedish forests during the 18th and 19th centuries. In a dissertation from 1750, it was noted that it was rare to come across bears that had not previously been injured by hunters (Berch 1750, 13). The number of inflicted injuries is hard to estimate, but was probably high. It is likely that this had an effect on bear welfare and behaviour. Recent observations on rescue animals indicate the damage that trauma does to animals, as well as humans, across generations. Stressed, wounded and traumatised bears are less capable of taking care of their offspring and upholding normal social relations within and across species (Bradshaw 2017, 79–82).

Further, recent ecological research points towards fear as a universal factor producing vigilance among animals. Fear truly has an impact on wildlife ecology (Laundré et al. 2010). Certainly, the bears' fear must have had an impact on the ecology of 19th-century Sweden, but to what extent is difficult to know. Research has shown how large carnivores change their behaviour significantly in response to fear of humans, whether they are in sight or in hearing (Ordiz et al. 2012; Suraci et al. 2019). It is not far-fetched to conclude that the relentless persecution of bears performed by Falk, Lloyd, and other bear hunters profoundly affected the bears, resulting in a traumatised population of large predators for generations to come.

Conclusion

Marvin's categorisation of different modes of killing is very useful in order to refine the ways we understand hunting. The categories are, however, somewhat inflexible when used in a historical context. In what category would one place the activities of the 19th-century bear hunters? Falk was certainly a professional bear killer with the obligation to exterminate bears as pest animals. He was called out when livestock had been killed by bears, and he argued that his bear killing was warranted by this (Falk 1828, 21). His companion in bear hunting, Llewellyn Lloyd, was a non-professional bear hunter looking for hunting opportunities and adventure. He explicitly justified the killing by referring to the damage the bears did to farmers' livestock, but at the same time expressly enjoyed the task. Hunting stories as a genre also had a logic of their own. Good stories, in the form of violent and adventurous accounts, were preferred, giving the perception of the bears a certain bias.

The hunting stories were filled with aggressive, attacking, and fierce bears. At the same time, another view of the bear was provided. It was also presented as a vigilant, fearful, and stressed creature. This twofold representation of the bear points towards a changing relationship between humans and bears in 19th-century Sweden. The hunting literature was aimed at inspiring hunting and legitimising hunting practices. At the same time, the very need for motivating narratives points towards another view of the bear.

Bears were hunted and relentlessly killed in 19th-century Sweden. At the same time, the writers of hunting narratives expressed a certain ambivalence towards the bear. In the eyes of the bear, they could also see a respected creature, a weary and frightened individual. Falk himself expressed this ambivalence when he stated that the bear is a majestic animal, which evokes both fear and awe, and which, despite its predatory rampage and crimes, should not be hunted to extinction. If one gets to know the bear, Falk concluded, one finds it worthy of protection (Falk 1834, 1137). When gaining knowledge about the bear, the hunters also built relationships with individual animals. The hunt sometimes went on for hours or days, and the particular animal may have been hunted before and was known to the hunters. This allowed them to gain a nuanced picture of the animals and to understand their way of life. In this manner, the perception of the bear grew from a shared experience of hunters and bears. Eventually, this alternative view of the bear entailed the relative protection of the animal in Sweden.

Bibliography

Berch 1750: A. Berch, Westmanlands Björn- och Wargfänge (Uppsala 1750).

Bergström et al. 2015: R. Bergström/K. Dirke/K. Danell, The wolf war in Sweden during the 18th century. Strategies, measures and leaders. In: P. Masius/J. Sprenger (eds.), A Fairytale in Question: Historical Interactions between Humans and Wolves (Cambridge 2015) 57–78.

Bradshaw 2017: G. Bradshaw, Carnivore Minds. Who These Fearsome Animals Really Are (New Haven, London 2017).

Brummer 1789: M. H. Brummer, Försök til et Swenskt Skogs- och Jagt-Lexicon (Göteborg 1789).

Danell et al. 2016: K. Danell/R. Bergström/L. Mattson/S. Sörlin, Jaktens historia i Sverige. Människa – samhälle – kultur (Stockholm 2016).

Dirke 2017: K. Dirke, Changing narratives of human-large carnivore encounters in nineteenth century Sweden. In: T. Räsänen/T. Syrjämaa (eds.), Shared Lives of Humans and Animals in the Global North (London, New York 2017) 163–178.

Falk 1828: H. Falk, Underrättelser om björnskall (Stockholm 1828).

Falk 1834: H. Falk, Något om Björn och Björnjag. Tidskrift för Jägare och Naturforskare 11/12, 1834, 1136–1145.

Kheel 2008: M. Kheel, Nature Ethics: An Ecofeminist Perspective (Lanham 2008).

Laundré et al. 2010: J. W. Laundré/L. Hernández/W. J. Ripple, The Landscape of Fear: Ecological Implications of Being Afraid. The Open Ecology Journal 2010(3), 1–7.

Lloyd 1831: L. Lloyd, Jagt-nöjen i Sverige och Norrige (Stockholm 1831).

Lloyd 1842: L. Lloyd, Field Sports of the North of Europe: Comprised in a Personal Narrative of a Residence in Sweden and Norway in the Years 1827–28, vol. 1 (London 1842).

Lloyd 1871: L. Lloyd, Svenska allmogens plägseder. Transl. by G. Swederus (Stockholm 1871).

Marvin 2006: G. Marvin, Wild Killing: Contesting the Animal in Hunting. In: The Animal Studies Group (eds.), Killing Animals (Champaign, Chicago 2006) 10–29.

Nilsson 1847: S. Nilsson, Skandinavisk fauna: en handbok för jägare och zoologer (Lund 1847).

Nyrén 2012: U. Nyrén, Rätt till jakt. En studie av den svenska jakträtten ca. 1600–1789. PhD thesis, Göteborg University (Göteborg 2012).

Ordiz et al. 2012: A. Ordiz/O.-G. Støen/S. Sæbø/J. Kindberg/M. Delibes/J. E. Swenson, Do bears know they are being hunted? Biological Conservation 152, 2012, 21–28.

Orrelius 1750: M. Orrelius, Historia animalum (Stockholm 1750).

Suraci et al. 2019: J. P. Suraci/M. Clinchy/L. Y. Zanette/C. C. Wilmers, Fear of humans as apex predators has landscape-scale impacts from mountain lions to mice. Ecology Letters 22(10), 2019, 1–9.

Svenning 2017: J.-C. Svenning, Future Megafaunas. A Historical Perspective on the Potential for a Wilder Anthropocene. In: A. Tsing/H. Swanson/E. Gan/N. Bubant (eds.), Arts of Living on a Damaged Planet: Ghosts and Monsters of the Anthropocene (Minneapolis, London 2017) 67–86.

Associate Prof. Karin Dirke
Department of Culture and Aesthetics
Stockholm University
Stockholm
Sweden
karin.dirke@idehist.su.se

From monster to endangered animal: Three bear stories by Selma Lagerlöf

By Claudia Lindén

Keywords: Bears, animals in literature, Selma Lagerlöf, Giorgio Agamben

Abstract: By the end of the 19th century, the bear was on the verge of extinction in Sweden. A growing opposition to the state's bounty for bears emerged, and several authors wrote stories about bears. Selma Lagerlöf's three bear stories cover the whole range of ways they were perceived by humans – from the bear as a mythical, dangerous animal, to the bear as man's neighbour in the Christian sense, to the bear as an endangered species. Up until the Middle Ages, the bear played a role in various cults throughout the whole of Europe. In the Nordic region, these cults continued for much longer. When Lagerlöf opens up the mythical as well as the religious and moral domain of the bear, it resonates with the long, ancient relationship humans have had with the bear.

Introduction

In the Nordic countries, the bear can, like no other wild animal, be traced in ancient folklore, folk tales, mythic imaginings of shapeshifting, place names, and linguistic expressions. Representations of bears are found in the heraldry of the nobility as well as on the coats of arms of cities and regions. The bear's unique place in Nordic culture is also evident in its recurring presence in literature. In *What Animals Mean in the Fiction of Modernity,* Philip Armstrong has pointed out that the way animals are understood and treated by humans should be considered in relation to the ways we feel about them, and in doing so the study of fiction has a special role to play: "Literary texts testify to the shared emotions, moods and thoughts of people in specific historical moments and places, as they are influenced by – and as they influence – the surrounding sociocultural forces and systems" (Armstrong 2008, 4).

In Sweden, bears had been hunted mercilessly since the late 18th century, based on a national extermination policy. When the new Swedish national hunting statute was passed in 1864, it was decided that the extermination of all the large predators – such as bear, wolf, wolverine, and lynx – should be intensified, and, accordingly, the bounty was increased tenfold. It was a successful policy: by the end of the 19th century, the bear was almost extinct.

What seemed a desired achievement in the 19th century would take on a different meaning at the dawn of the new century. When the bear was finally threatened by extinction, a greater opposition to the state's bounty on bears and its devasting effect emerged. In 1905, the Royal Academy of Science argued that the bear had to be protected, as so few bears were left that reproduction could no longer take place. According to the law on national parks from 1909 it was then forbidden to hunt bears, and in 1916 King Gustav V issued a ban on the cruel tradition of bear-dancing (Tünaydin 2013, 53).

Lagerlöf's three bear stories

Once it was on the verge of extinction, the bear seems to have spurred the cultural imagination in new ways, and several Swedish authors, among them Nobel laureate Selma Lagerlöf, wrote stories about bears at the *fin de siècle*.[1] Lagerlöf wrote three stories about bears. One is a chapter in her debut novel *The Story of Gösta Berling* (*Gösta Berlings Saga*) from 1891 (transl.: 1898), one a short story "The Truce of God" ("Gudsfreden") from 1899, and the last one a chapter in *The Wonderful Adventures of Nils* (*Nils Holgerssons underbara resa genom Sverige*) from 1906/1907 (transl. used here: 1913).

Lagerlöf's stories take place in the intersection of three, quite different, discourses about the relationship between bear and human. The first is the fear of the bear as predator, a monster with werewolf-like features. The second is the tradition from the bear cults with the importance of showing respect towards the bear. And the third is the post-Darwinist insight that man and animals are related, combined with an animal-rights perspective of the bear as an endangered species. These short stories represent a wide range of approach in relation to the bear, ranging from the bear as a dangerous animal of almost mythical proportions, over the bear as man's neighbour in the Christian sense, to the bear as a civilisation-critical being who understands that human modernity involves its own extinction.

Lagerlöf's bear stories differ from those of her contemporaries in that they make the bear the protagonist and a moral agent on the same level as humans. Even though Lagerlöf is ahead of her time with her awareness of the threat to the bear, her texts also reflect a general change of attitude in Sweden. They therefore bear witness to a change in the national attitude towards the bear at the turn from the 19th to the 20th century. In this article I will argue, through a reading of Lagerlöf's three bear stories, that she not only criticises the extermination politics, but actually problematises the anthropocentric paradigm and depicts the bear as a moral agent equal to humans.

Bear ceremonials and the bear in Christian tradition

Over a broad area of northern Scandinavia, stretching from northern Norway to Sweden and Finland and into Russia, indigenous peoples, such as the Sami, practised similar sorts of bear worship, where the hunting of the bear was regulated through a series of ritual events that were distinguished by the importance of showing the bear respect (Rydving 2010, 34; cf. reprint: Rydving, this volume). It seems that these cults did not spread outside the Sami (and Finnish-Karelian) culture. According to Olle Sundqvist, the bear was not a holy animal in the north, as far as can be ascertained from the written accounts (cf. Sundqvist, this volume). According to Juha Pentikäinen, the "specific knowledge about bear rituals comes from the south Scandinavian Sámi. According to Randulf, writing in 1723 on the basis of information relating to the southern Sámi, 'the Sámi consider all animals holy', and he continues: 'but the bear they consider the most holy of all'" (Pentikäinen 2015, 3).

In his book *Golden King of the Forest* Juha Pentikäinen writes about the Sami that there was an "unspoken agreement about what responsibilities pertained to each side" (Pentikäinen 2007, 44). The bear had to refrain from killing people, and they, in turn, had to honour the bear; they could kill it, but "only within a framework derived from rules established in myth, and with leave granted by the bear itself" (Pentikäinen 2007, 44). In Finland, the traces of these bear ceremonies are also very tangible among the non-Sami population, and several songs in the *Kalevala* are today considered

---

[1] Pelle Molin, Alfhild Agrell, and Helena Nyblom also wrote short stories at the *fin de siècle* ridiculing or criticising bear hunting, and stories where the bear plays a central role.

to be songs that were sung or read over the dead bear. The ceremonies were often called "the bear's funeral feast", or "the bear's wedding" (Pentikäinen/Poom 1999; Björklöf 2010; cf. Piludu, and Keinänen, this volume).

In his book *The Bear: History of a Fallen King* the French historian Michel Pastoureau argues that the Church struggled for over a millennium against bear cults: "Almost everywhere, from the Alps to the Baltic, the bear stood as a rival to Christ. The Church thought it appropriate to declare war on the bear, to fight him by all means possible, and to bring him down from his throne and his altars" (Pastoureau 2011, 3). This struggle against the bear took several forms, such as the demonisation of the animal, or the replacement of sacred spring rituals with Saints' days and church festivities. By the mid-13th century, the de-throning of the bear was complete, according to Pastoureau (2011, 167). The Church's struggle against the bear was connected to both older bear cults and the close resemblance between bear and human. It is noteworthy, Pastoureau continues, that in the bestiary of the seven major sins, the two animals considered closest to humans, bear and pig, are the most devalued. Too close kinships with animals seem to have been unbearable and were compensated for by devaluation and, in the case of the bear, extermination (Pastoureau 2011, 184).

In medieval bestiaries, the bear became connected to vice. Animals did not commit sins; they were imperfect creatures and fundamentally vicious. When, during the 13th century, the vices merged into the form of the seven deadly sins, as opposed to the seven virtues, each sin and virtue was associated with a certain number of animals. Lion, eagle, and horse were associated with virtue, whereas bear, fox, monkey, pig, and dog were always connected negatively with sin. But worst of them all was the bear, who was associated with five out of the seven deadly sins; lust, anger, gluttony, envy, and sloth: "From the thirteenth century, he was the star of this hateful bestiary, a sad fate for a wild animal who was once the king of the beasts" (Pastoreau 2011, 183–184).

The humiliating bear-dancing also commenced during the Middle Ages. Bear-dancing, with bears in chains and muzzles, was common throughout the whole of Europe up to the early 20th century and can be seen as a continuous tradition of humiliating the bear. Being the only four-legged animal that is also a plantigrade (putting down the foot heel first) – like humans –, the dancing bear becomes uncannily like a human in chains when moving on two legs (Tünaydin 2013, 52). Pastoureau points out that the Church, even though they condemned spectacles, tolerated dancing bears but looked down on the bear handlers, something that in turn made the bear an even more despised animal: "Associating the bear with them therefore effectively helped to devalue the animal and, therefore through a kind of osmosis, to project onto him all the vices imputed to his masters and companions in misfortunes" (Pastoureau 2011, 172–175).

The relationship between humans and animals in the latter part of the 19th century cannot be grasped without also taking into account Darwin's tremendous influence on thinking, with his conclusion in *The Descent of Man*: "The difference in mind between man and the higher animals, great as it is, certainly is one of degree and not of kind" (Darwin 1871, 105). Although Darwinism, on one level, was the most radical philosophical challenge to anthropocentrism, it did not have that effect. Instead, the notion of evolutionary development, from the primitive to the more evolved, could be used to create a difference between animals and humans, and also between people. Although man was an animal, he was at the top of the hierarchy, and the evolutionary doctrine could thus again secure man's supremacy.

Werewolves, werebears and outlawry

In her book on werewolf belief in Sweden, ethnologist Ella Odsted has pointed out that there are strong beliefs about transformation to both wolf and bear in Swedish and Nordic folklore. In the north, shapeshifting occurs with bears rather than wolves. It is reflected in expressions like "run bear", "go bear", "go into bear shape", "turn to bear" (ODSTEDT 2012, 95; cf. SUNDQVIST, this volume). In northern Sweden, the werewolf is thus what we could call a werebear. The bear one meets in the forest could be a human turned into a bear, or a bear that has been enchanted in some other way.

There is also a link in Scandinavian history between werewolves and outlawry (*Friedlosigkeit*[2]), which is attested in the old Icelandic expression *vargr i véum*.[3] As Olof Sundqvist has pointed out, a person who killed someone on sanctified ground was sometimes metaphorically described as a "wolf" in ancient Scandinavia: "In *Egils saga* 49, Queen Gunnhildr's brother, Eyvindr, was considered a 'wolf' after killing at a *vé* sanctuary: Because Eyvindr had committed murder at a sacred place he was declared a defiler [actually a wolf (*vargr*)] and had to go into outlawry at once" (SUNDQVIST 2016, 294). *Vargr i véum*, the human who becomes *friedlos*, an outlaw, is a hybrid of human and animal, a kind of shapeshifter, a werewolf – or a werebear. This transgression of the human-animal divide opens up the possibility of regarding the hunted animal as someone who has been made *friedlos*, an outlaw.

The philosopher Giorgio Agamben connects his concept of "homo sacer" with this Scandinavian notion of *Friedlosigkeit*, and the old Icelandic expression *vargr i veum*: "In the bandit and the outlaw (wargus, vargr, the wolf and, in the religious sense, the sacred wolf, vargr y veum[4]), Germanic and Scandinavian antiquity give us a brother of homo sacer beyond the shadow of any doubt" (AGAMBEN 1998, 63). This "monstrous hybrid of human and animal" is a man divided between the forest and the city, "the werewolf – is, therefore, in its origin the figure of the man who has been banned from the city" (ibid.). In this connection between homo sacer, outlawry, and werewolves, as pointed out by Agamben, the bear, with its complex tradition of werebears, holy bears, and bears as *friedlos* with a bounty on their heads can be understood as *ursus sacer* – and thereby also as a political category.

THE WEREBEAR IN *THE STORY OF GÖSTA BERLING*

The Story of Gösta Berling, Lagerlöf's debut novel, published in Sweden in 1891, is filled with adventures and romance and set in the Swedish province of Värmland in the 1820s, with the young and deposed priest Gösta Berling as its hero. On the surface, the chapter "The great bear in Gurlita cliff" is a story of a bear hunt, but, in the background, there is a love story. An old man kills a bear and pretends the deed was carried out by a young, poor man – in order to make the young man eligible as a husband so he can marry the woman he loves. In this story, the bear is at first depicted as an almost supernatural monster. It starts out with a description of the dark, dangerous forest where unholy creatures dwell "which long to sink themselves in a blood-filled throat, and whose eyes shine with murderous desires" (LAGERLÖF 1898, 122):

> And there lives the most terrible of them all, the bear, who has the strength of twelve men, and who, when he becomes a devil, can be killed only with a silver bullet. […] Terrible to say, dreadful to believe, this is no ordinary bear. No one can hope to kill him if he does not have a silver bullet in his gun. A bullet of silver and bell-metal cast on a Thursday evening at new moon in the church-tower […] (LAGERLÖF 1898, 123, 128).

---

2 In Swedish there are the same words: *fredlös* and *fredlöshet*, as in German *friedlos* and *Friedlosigkeit*. Since AGAMBEN 1998 refers to the German words, I will do the same here in this text.
3 I am grateful to Olof Sundqvist for pointing out that this expression is attested in *Óláfs saga Tryggvasonar* (Oddr): *gerir Hakon j. utlagþan oc scylldi hann heita vargr i veum. er hann hafði brotit hit ǿzta hof i Gautlandi*.
4 AGAMBEN (1998) uses "y" instead of "i"; the Old Icelandic spelling is *vargr i veum*.

The need for a bullet made out of silver and metal from the church bell suggests that this is some kind of werebear, a devilish bear. But, after the reader has been drawn into this gothic, scary mood, the story shifts the focus on to the bear himself. He is now called "the forest king" instead, and the reader gets to partake in his previous encounters with the hunters:

> He remembered how they had come on him another time, when he and his queen consort had just laid themselves down for their winter sleep in the old lair here on Gurlitta Cliff and had young ones in the hole. He remembered well how they came upon them unawares. He got away all right, throwing to either side everything that stood in his path; but he must limp for life from a bullet in his thigh, and when he came back at night to the royal lair, the snow was red with his queen consort's blood, and the royal children had been carried away to the plain, to grow up there and be man's servants and friends (LAGERLÖF 1898, 126).

When the bear becomes the narrator, he is shown as a *vargr i veum*, a werebear who is also an outlaw. When Gösta comes upon the bear, he cannot shoot him because he "sees him as he is – a poor, persecuted beast, whom he will not deprive of life, which is all he has left, since people have taken everything else from him" (LAGERLÖF 1898, 127). Within the same story, Lagerlöf goes from depicting the bear as devilish and murderous, to the opposite, a hunted and persecuted creature. Maria Karlsson has, in her assessment of the function of melodrama in *The Story of Gösta Berling,* pointed out that the man's change takes place in a meeting with a woman, where he takes her seriously and helps increase her influence upon him (KARLSSON 2002, 200). The melodrama functions in the same way with hunted animals in the story. It is in the existential meeting with the other in the form of the bear that both Gösta, and the reader, can see that the mythical werebear is only a "poor, persecuted beast".

THE SHORT STORY "THE TRUCE OF GOD"

In Selma Lagerlöf's short story "The Truce of God" ("Gudsfreden", 1899) not only the ethical relationship towards the bear, but the bear's moral agency is accentuated through the biblical parable of the good Samaritan (Luke 10:25–37). Lagerlöf explicitly refers to *Pax et treuga Dei* and the commandment that "You shall love thy neighbour as yourself", as it was used in the Gospels of Luke (10:27) and Mark (12:31).[5]

When Lagerlöf draws attention to the religious domain of the bear, a connection is also made to the ancient belief in the bear's holiness. As PASTOREAU (2011) has shown, the bear played a role in various cults throughout the whole of Europe, up until the Middle Ages. In the Nordic region, the belief in the importance of showing the bear respect carried on for much longer, and it seems to have co-existed with the Christian faith. There is an eyewitness account from such a ceremony as late

---

5 Lagerlöf is here using the Swedish word *Gudsfreden*, for the Latin *Pax et treuga Dei*, which translates as "The Peace and Truce of God". This tradition from the Middle Ages was created in reaction to the constant fighting, not only among noblemen, but between noblemen and everyone else. Peace was permanently proclaimed in certain buildings like churches, and certain persons, such as monks, clerics, and women, as well as cattle and horses should always be protected by this peace. The Truce of God or *Treuga Dei* concerned only special periods and had its origin in Normandy in the city of Caen in the 11th century. The Peace and Truce of God was required throughout Advent, the season of Lent, and from the beginning of the Rogation days until eight days after Pentecost, and during certain days of the week (cf. https://www.britannica.com/event/Truce-of-God; "Gudsfrid" in Nordisk familjebok 1909: http://runeberg.org/nfbj/0298.html). The Truce of God included domestic animals, but Lagerlöf chooses to also include the wild animals in the category protected by The Truce of God, which was probably an accepted interpretation of the wider concept to abstain from all kinds of violence during these days.

As often with Lagerlöf, it is possible to read her stories both religiously and secularly at the same time. By making bear and human equal moral beings before God, Lagerlöf rejects anthropocentrism, which places man, as an image of God, at the centre of the universe. Read from a cultural-historical perspective in a post-Darwinian era, Lagerlöf undermines the evolutionary logic that says that man is the crown of evolution and thus can do what he wants with more primitive beings. Also, from an animal rights perspective, it is possible to read the story as an argument in the contemporary debate, in which the Swedish newspapers of the 1890s made the appeal of "Do not let the bear be exterminated" (LÖNNBERG 1929, 11).

When Ingmarsson and his sons tried to kill the bear, they treated the bear as *friedlos*, as someone excluded from both profane and divine law. But old Ingmarsson was also made *friedlos* and expelled from both the divine and profane order, since he, according to Mrs Ingmarsson, was punished by God when the bear killed him, and he was deprived of a high-status funeral. In the end, the bear and old Ingmarsson share the same place in the universe, and the same condition of having become *friedlos*, that is "sacer" in Agamben's sense of being a hybrid between man and animal: "a realm of indistinction and of passage between animal and man, physis and nomos, exclusion and inclusion: the life of the bandit is the life of the *loup garou*, the werewolf, who is precisely neither man nor beast, and who dwells paradoxically within both while belonging to neither" (AGAMBEN 1998, 63).

Father Bear and Nils Holgersson

In *The Wonderful Adventures of Nils* (1913), the bear is also *friedlos*, because he is chased, and, because he is given a voice of his own, he transcends the border between man and beast. Written as a geography book for schoolchildren and published 1906–1907, it shows Nils travelling all over Sweden in its course. The story opens with the protagonist, Nils, being turned into a small elf as a punishment for disrespecting the house elf. He becomes a lilliput, but along with the transformation comes the ability to talk with animals and to have night vision. Anna-Karin Palm writes: "Exposed and vulnerable, he must now seek protection from the animals he previously despised" (PALM 2019, 374). Nils's penance is carried out through good deeds to different animals and, in the end, it is the cow Majros who gives him approval (EDSTRÖM 2020, 335). The animals become anthropomorphised and assume more or less human traits.

In the chapter called "Thumbietot and the Bears. The Ironworks" Nils is travelling across the mining district in Dalarna, when a gust of wind blows him off the back of the goose he is riding, and he falls to the ground. It turns out he has fallen into a shaft where a bear family lives. Mother Bear does not understand that Nils is human, and she gives him to her cubs to play with and eat. They have a lot of fun chasing poor Nils, so they do not eat him. They fall asleep with Nils between them in order to continue the fun in the morning.

When Father Bear, "this old monarch of the forest", comes home he detects the smell of a human. Mother Bear thinks this absurd and refers to the old regulations between men and bears:

> "It has been settled for good and all that we are not to harm mankind anymore; but if one of them were to put in an appearance here, where the cubs and I have our quarters, there wouldn't be enough left of him for you to catch even a scent of him!" (LAGERLÖF 1913, ch. "Thumbietot …").[7]

---

[7] I'm referring here to the edition at www.gutenberg.org, which is not paginated. I will therefore in the following just refer to this chapter (LAGERLÖF 1913, ch. "Thumbietot …").

It turns out that Father Bear has been away looking for a new residence for his family, because he is alarmed by the new ironworks: "I can't be content here now since the big noise-shop has been built right in our neighbourhood." He has been searching for other bears to see how they fared further away, "but I had my trouble for nothing. There wasn't a bear's den left in the whole forest." And Mother Bear answers resignedly: "I believe the humans want the whole earth to themselves. Even if we leave people and cattle in peace and live solely upon lingon and insects and green things, we cannot remain unmolested in the forest! I wonder where we could move to in order to live in peace?" (Lagerlöf 1913, ch. "Thumbietot …"). Father and Mother Bear both bear witness to how the bears are chased by humans to the extent that there are no bears left in the forest.

With its appeal of peace, Mother Bear's reflection both echoes Mrs Ingmarsson's understanding of peace between humans and animals during Christmas and shows her to be an animal with a moral agency in the same way as the bear in "The Truce of God". But Mother Bear's comment also links her to Lagerlöf's view of women's responsibility in the national project. Lagerlöf was very engaged in both women's suffrage and later, during World War I, the peace movement. In her famous speech *"Hem och stat"* ("Home and State") at the 6th international suffrage congress in Stockholm in 1911, she situated women in history as the makers of the "good home", but then went on to show how central the home has been for society as a whole, and she ended with the closing remark: "The little masterpiece, the home, was our creation with the help of man. The great masterpiece, the state, will be made by man when he seriously takes on woman as his helper."[8] Anna-Karin Palm points out that "Selma Lagerlöf's nationalism differs from the dominant male voices in that she sees so clearly that women must also be given a place in the national project if any real elevation is to take place, and women's hard work is given a large space in the school text book" (Palm 2019, 371). Mother Bear's comment that the bears keep their side of the bargain and maintain the peace with humans shows her to be a responsible and hard-working person, just as Lagerlöf envisions women in the national project.

When Father Bear discovers Nils he intends to kill him, but he is stopped when Nils lights some matches that he throws at the bear. When Father Bear understands that Nils can handle fire, he exclaims: "You shall render me a service. Now I'm very glad that I did not eat you!" He wants Nils to burn down the ironworks. Father Bear tells Nils that "my forefathers lived in this region from the time that the forests first sprang up", and that he has lived here in peace all his life:

> "In the beginning I wasn't troubled much by the human kind. They dug in the mountains and picked up a little ore down here, by the rapids; they had a forge and a furnace, but the hammers sounded only a few hours during the day, and the furnace was not fired more than two moons at a stretch. It wasn't so bad that I was unable to stand it; but these last years, since they have built this noise-shop, which keeps up the same racket both day and night, life here has become intolerable. Formerly only a manager and a couple of blacksmiths lived here, but now there are so many people that I can never feel safe from them" (Lagerlöf 1913, ch. "Thumbietot …").

Father Bear describes how technology and the process of modernity have invaded nature and disturbed the wildlife to the extent that the bears feel that they cannot live there anymore. They go to the ironworks, and when Father Bear asks Nils to look into these, he is not appalled, as the bear is, but enchanted: "The boy was completely charmed by the marvellous display and almost forgot that he was imprisoned between a bear's two paws" (Lagerlöf 1913, ch. "Thumbietot …").

---

8   Quoted from Alice Burman's English translation of Louise Vinge's introduction to Selma Lagerlöf and her work at https://litteraturbanken.se/författare/LagerlöfS/omtexterna/SelmaLagerlofEnglish.html.

Father Bear gives Nils an ultimatum: "If you will set fire to the noise-works, I'll promise to spare your life" – if not, he will be killed (Lagerlöf 1913 ch. "Thumbietot …"). Despite this ultimatum, and that Father Bear has said that "iron is the thing that has given men the advantage over us bears", Nils hesitates. Finally, he says: "You'll not get me to destroy the ironworks! Iron is so great a blessing that it will never do to harm it" (Lagerlöf 1913 ch. "Thumbietot …"). Just when the bear is about to eat him, Nils hears the clicking of a rifle from approaching hunters. He warns Father Bear, who lets go of Nils and runs off to safety. It is a very ambivalent ending. Even though Nils saves Father Bear from the hunters, he lets the industrial modernity that has destroyed the bears' habitat triumph.

When Father Bear is portrayed as a father trying to protect his family, he is anthropomorphised, and the reader is encouraged to think of him in the same way as a human father. The bears are portrayed as a family threatened by external forces they cannot control. They do not attack people and expect to be treated the same way, according to what they perceive as a mutual and equal agreement. On the other hand, if the humans violate this agreement, they will show no mercy, something which both Mother and Father Bear testify to. The potential moral conflict between saving the ironworks and saving the bears is solved partly because Father Bear is such an ambiguous figure himself. After all, he was just about to eat Nils when the hunters showed up.

Lagerlöf is ambivalent about the bears' situation. She sees and understands their needs and desires, but her strong belief in the processes of modernity makes her choose modern technology over the bears' habitat. Nils's desire to let the ironworks remain prevails, while Father Bear's desire to let nature return to a space uninhabited by humans does not. Lagerlöf's optimism for all kinds of development made her affirm the emerging industrialism to a higher extent than her contemporaries, according to Edström: "Industrialization is depicted without reservation. Although it is clear that the large ironworks in Bergslagen threatens the bear's wild natural habitat, Nils affirms an activity 'that gave bread to so many people'" (Edström 2020, 339).

Conclusion

Even though Father Bear, thanks to Nils's warning, escapes the hunters, the bear again becomes *friedlos*, an outlaw. Thus, the bear at Gurlitta, the bear in "The Truce of God", and Father Bear each end up as an *ursus sacer*, an animal that may be killed but not sacrificed: someone outside both human and divine law. In this sense, all three bears in Lagerlöf's bear stories continue to be, to quote Agamben again, a "monstrous hybrid of human and animal", i. e. a creature divided between the forest and the city (Agamben 1998, 63).

Lagerlöf's three bear stories show that she was familiar with the old folklore about the bear and at the same time was aware of how the bear was, in her own time, an endangered animal. Nevertheless, her view of the bear displays an inherent tension between the ethics of the individual and the ethics of culture as a whole. It is a tension that we also recognise in our own contemporary discussion on climate change and ecopolitics. At the individual level, when man and bear meet, they stand in a mutual, moral relationship. Therefore, it is immoral for humans to hunt the bear. At a societal level, on the other hand, Lagerlöf seems to say that human technological development must take precedence over the bears' interests, no matter how reprehensible they may be. In this sense, the bear becomes a political category, when the bear is the one who must adapt to the processes of modernity. Still, her stories show how devastating the state's bounty for bears was, and she opens up the Christian-ethical dimension of the bear and, in this, she problematises the anthropocentric paradigm.

Michel Pastoureau wrote that: "In killing the bear, his kinsman, his fellow creature, his first God, man long ago killed his own memory" (Pastoureau 2011, 239). In her bear stories, Lagerlöf reminds us that the bear is a fellow creature, and she restores some of this lost memory. Through literature, the bear's significance can thus continue to resonate into our own time.

Bibliography

Agamben 1998: G. Agamben, Homo sacer: Sovereign Power and Bare Life (Stanford 1998).

Armstrong 2008: P. Armstrong, What Animals Mean in the Fiction of Modernity (London 2008).

Björklöf 2010: S. Björklöf, Björnen: i markerna & kulturen (Möklinta 2010).

Darwin 1871: C. Darwin, The Descent of Man, and Selection in Relation to Sex. Vol. 1 (London 1871).

Edström 2020: V. Edström, Selma Lagerlöf: livets vågspel (Stockholm 2020).

Karlsson 2002: M. Karlsson, Känslans röst: det melodramatiska i Selma Lagerlöfs romankonst. Brutus Östlings Bokförlag Symposion. Unpubl. Dissertation, Uppsala University (Eslöv 2002).

Lagerlöf 1898: S. Lagerlöv, The Story of Gösta Berling. Transl. Pauline Bancroft Flach (Boston 1898): http://www.gutenberg.org/files/56158/56158-h/56158-h.htm.

Lagerlöf 1899: S. Lagerlöf, Drottningar i Kungahälla jämte andra berättelser (Stockholm 1899): https://litteraturbanken.se/författare/LagerlöfS/titlar/DrottningarIKungahälla/sida/III/etext.

Lagerlöf 1913: S. Lagerlöf, The Wonderful Adventures of Nils. Transl. Velma Swanston Howard (New York 1913): http://www.gutenberg.org/cache/epub/10935/pg10935.html.

Lönnberg 1929: E. Lönnberg, Björnen i Sverige 1856–1928 (Uppsala, Stockholm 1929).

Odstedt 2012: E. Odstedt, Varulven i svensk folktradition. Enlarged new edition (Täby 2012).

Palm 2019: A.-K. Palm, Jag vill sätta världen i rörelse: en biografi över Selma Lagerlöf (Stockholm 2019).

Pastoureau 2011: M. Pastoureau, The Bear: History of a Fallen King. Transl. G. Holoch (Cambridge 2011).

Pentikäinen 2007: J. Pentikäinen, Golden King of the Forest: The Lore of the Northern Bear (Helsinki 2007).

Pentikäinen 2015: J. Pentikäinen, The bear rituals among the Sami. In: E. Comba/D. Ormezzano (eds.), Uomini e orsi : morfologia del selvaggio (Turin 2015) 123–145.

Pentikäinen/Poom 1999: J. Pentikäinen/R. M. Poom, Kalevala Mythology. Enlarged edition (Bloomington 1999).

Rydving 2010: H. Rydving, The 'Bear Ceremonial' and Bear Rituals among the Khanty and the Sami. Temenos 46(1), 2010, 31–52.

Sundqvist 2016: O. Sundqvist, An Arena for Higher Powers: Ceremonial Buildings and Religious Strategies for Rulership in Late Iron Age Scandinavia (Brill 2016).

Tünaydin 2013: P. Tünaydin, Pawing through the history of bear dancing in Europe. Frühneuzeit-Info 24, 2013, 51–60.

Claudia Lindén
Professor of Comparative Literature
Södertörn University
Stockholm
Sweden
claudia.linden@sh.se

Bears as *pares*:
Some notes on bear stories in Zapinejie (Arkhangelskaya oblast, northern part of the Russian Federation) and the tendency to equality in human-bear relations

By Andrey V. Tutorski

Keywords: Bear, human, equality, intensity, narratives, Russian North

Abstract: The article assembles the narratives of Russian villagers in Zapinejie (Arkhangelskaya oblast, northern part of the Russian Federation) who encounter bears during their forest trips. The stories showcase that the bear, as opposed to other animals, is perceived not just as a beast but as a kind of forest owner who acts as humans do. In gatherers' stories, bears act as animals who are unwilling to harm people. In hunters' stories, bears are presented almost as human beings living in the woods, doing very similar things. Using the idea of "equality without equivalence", I will broaden the notion of "tendency to equality" proposed by H. Walker for interhuman relations and expand it to include human-animal relations. Such aspects as "singularity", "concreteness" or "love-pity-compassion" can be found in narratives about bears and in human-bear relations.

Introduction

The bear is a very important character in Russian literature and folklore, and it is also a widespread and well recognisable symbol of Russia. In this paper, I will demonstrate how the image is represented in the usual talk of people in a remote village in the northern part of European Russia. This region, which I call "Zapinejie", encompasses the three most vast and sparsely populated districts of Arkhangelskaya oblast, one of the most northern parts of the Russian Federation. Although they are part of Russian peasant agricultural civilisation, the people living there spend a long time in the woods, e.g. gathering berries, hunting, and fishing. Their knowledge of bears comes not only from tales and legends but primarily from their everyday experience. In talks, stories, and jokes, the bear is often depicted as a creature that has human traits rather than just as an animal.

I will use the concept of "equality without equivalence" to speak about humans and bears as opposed to other animals. This concept helps the further elaboration of the approach to bears as "other than human persons" (McClellan 1970; Brightman 1993; Smith 1998; Clark/Slocombe 2009; Dudeck 2018). This framework is chosen because this paper forms part of broader research on the definition of equality in the Russian North and how equality is produced and spoken about by the local people of that region. I realise quite clearly that the very notion of equality is a "western" concept. However, the inhabitants of Zapinejie (the region I study and which will be described in detail in the following pages) have had connections to the Russian State and its policy since at least the 16th century, and since the 18th century they have lived within the framework of the Imperial state policy of "imposing equality" (equal land sharing in *obshchina* [Russian for village community; author's note], and equal responsibility before the law [Russ.: *krugovaya poruka*]; Kolesnikov 1976; Kamkin 1995). A lot of observers (travellers, state revisors, and the like) have written that the Russian North is the place where equality can be seen in its pure form (Kachalov 2012, 434–435). My aim here is to showcase how equality works in relations between humans and bears as part of a larger study on how equality works in Zapinejie.

My theoretical framework will be the discussion about equality in contemporary anthropological literature (Beteille 1994; Artemova 2020). Before the 1980s, the term "equality" was widely used to designate several societies (especially those of the Melanesians and Bushmen), but, at the end of the decade, articles appeared that criticised the use of this kind of designation when the meaning of equality has not been especially addressed (Jolly 1987). In 1994, the American anthropologist Joel Robbins proposed the Dumontian approach to the notion of equality. He argued that equality was a value and could not be fully realised in everyday life (Robbins 1994). Unfortunately, this approach was not elaborated on by other anthropologists. On the threshold from the 20th to the 21st century, anthropological articles about equality mostly dealt with gender. In this case, equality is regarded more as a governmental project or more of a goal than as a status quo (Squires 2007; Rivkin-Fish 2010).

In my text, I will follow the main idea of Joel Robbins "to take equality seriously" and use the approach proposed by the British anthropologist Harry Walker. He writes that the very term "equality" is Eurocentric, and that it is almost impossible to find "true" equality anywhere else. However, he outlines ten specific types of interhuman relations that create "equality without equivalence" (Walker 2020), and he traces ten particular ways of dealing with interhuman relations among the Amazonian Urarina group. In this paper, I will demonstrate that at least some of the ways we deal with relationships can be applied to human-bear relations in the northern part of European Russia. Though these human-bear relations cannot be treated as "equal", they demonstrate a clear tendency to "equality without equivalence". And, although bears and humans cannot be treated as the same beings in respect of their relations, the tendency to see bears as *pares* is also evident (Clark/Slocombe 2009, 4).

What is Zapinejie?

As I said before, my field research took place in a village situated in the eastern part of Arkhangelskaya oblast, which I call Zapinejie (Fig. 1).[1] The settlement of Vozchiki that I studied is rather large in comparison to the average settlement in this area; it has more than 500 inhabitants, whereas an average settlement has about 50. Arkhangelskaya oblast is situated north of the 60th degree northern latitude, the same as Greenland or the northern territories of Norway. The lowest temperature in January is -53 °C. According to local tradition, the name *Vozchiki* is used both as a name of the settlement and, at the same time, as the name of its inhabitants. I will use this name in both senses, writing the former with the first letter capitalised, and the latter with it in lower-case.

Before proceeding directly to the subject, I would like to stress two features of vozchiki society that are important for this discussion – the multi-vector character of its household activities with a special focus on the forest, and the fractal character of the remoteness of the village in relation to Arkhangelskaya oblast and Russia in general.

---

1   All the geographic and personal names in this texts are pseudonyms.

As with the majority of the Russian village inhabitants, the vozchiki think of themselves as peasants and farmers. This point of view is ubiquitous throughout the country. However, Vozchiki is not an "average Russian village", which is usually imagined as a dozen small houses surrounded by steppe. While northern villages usually consist of the same dozen big timber buildings, each of which houses a large, extended family of six or more people, Vozchiki has more than 50 houses and a population of 500 people. The village is surrounded by endless taiga, and the distance between two inhabited places is usually about 40–50 kilometres. The villagers are Russians, but in lots of senses they are very close to indigenous peoples, as they themselves say "We – Mezenets – are like Nenets" (Russ.: "мы – мезенцы – те же ненцы" [the Mezenets: the inhabitants of Mezen town; the Nenets people are reindeer herders from northern parts of Russia; author's note]). This marginal or liminal position between agriculturalists and reindeer herders, and between Russians and indigenous peoples, can be seen in the following example, which illustrates the importance of fishing and hunting in vozchiki economics and concerns the place of agriculture among different types of household activities in the present-day village.

The most important crop for the vozchiki is the potato. This was typical for many Russian villages and *dacha* settlements all over the country in the 20th century (Ries 2009). Despite this, the potato is not treated very carefully. Many people plant it as soon as it is possible to dig the earth with a shovel. Usually, after that time the cold weather comes back for a week or two, so the potato, although it does not rot, stays dormant for a month or more. Those people who plant potatoes one or two weeks later see them sprout a week earlier than the impatient ones. The more patient and slower gardeners ultimately receive a better harvest.

So why do people hurry? What benefits do they receive? A friend of mine, named Andrey, told me that "the early potato planting frees people to go fishing". With the arrival of the first warm days, the middle part of the river usually melts, while the upper parts of the river (flowing from the south to the north; author's note) stay frozen. When the cold returns in the next one or two weeks, the temperature is still not low enough to stop ice floating, and the upper parts of the river also become free of ice. The next two or three weeks are the most convenient time for fishing: The water level is high enough to flood the rapids, and the fish come there for spawning. So, everyone has the choice – to get more potatoes or more fish – and many of the men choose to fish.

The second peculiarity of Vozchiki is its position in the area of the Russian North. The remoteness of this area plays a great role in the social life of the local inhabitants. In the middle of the 20th century, it took about two or three days to reach the *centr rayona* (district centre), about 70 km from the village. And from the centre of the district, it took another two or three days to reach the *oblastnoy centr* (land centre or county centre) – the city of Arkhangelsk. What is characteristic of Vozchiki is its remoteness from the centre. And this feature makes this particular village and the particular region of Zapinejie a fractal mirror of Russia. The Russian Federation is often seen as a very remote part of Europe, though within Russia there is Siberia, which is remote from the European part of the state. Within the European part of Russia, Siberia is situated in its north, which in Russian is simply called the "Russian North" (although in English these words denote all of the circumpolar part of the Russian Federation; author's note). Within the northern part of European Russia, Arkhangelskaya oblast is perceived as one of the most vast and remote regions. And finally, the most remote part of the Arkhangelsk region is Zapinejie.

This idea of remoteness is important for the following discussion because the inhabitants of Vozchiki are quite aware of who the bear is, what it means to be in the woods, and what it means to encounter this animal. Their knowledge comes from their real and personal experience rather than from tales, belief narratives, or other types of folklore texts. I will use the terms "talks", following Nancy Ries (Ries 1997) and "stories" to define different texts that are narrated from everyday life and only partly stem from folklore representations.

Widespread situations in which bears are involved

Coexistence
Gatherers' stories
The most common story about bears is "how one met a bear" while picking berries or mushrooms. Almost every person in Vozchiki can tell a story of this kind. The most simple one is: "We went to the forest to collect raspberries, and we made our way deep into the wood. I settled down on one side of a raspberry bush while my friend went to the other side. We were gathering berries silently for a while but then I could hear my friend moving and grousing on the other side. I started slowly moving towards her and suddenly saw that it was not her but a bear. I screamed loudly and ran away and my friend pretended to be a dog barking. The bear went away."

Typical parts of this particular story are the following: A person goes deep into the wood, s/he stays silent for quite a long time, and then s/he suddenly meets a bear. The person makes loud sounds and the bear goes away. Loud sounds can scare the bear, but they also let it know where the human is. Once, in 2010, I had to make a 40-km trip through taiga, and an old man who hosted our group told us: "If you don't want to meet a bear, take an empty tin and some other metal thing and knock them together. The bear will hear you from even more than a kilometre away." It is important to stress here that the bear is not perceived as a dangerous, aggressive animal who wants to attack every human, but rather as a creature that would not attack and even would keep at a distance if it knows that a human is nearby. Speaking about meeting with bears, a vozchiki woman said once: "We were so involved in picking berries that we stopped talking (singing/shouting/whistling). That's why a bear approached us." It is possible to recognise the same point here: The bear never attacks intentionally but only when the encounter is sudden.

There is a difference between bears and wolves in this regard. Bears most often attack in summer (because in winter they hibernate) and wolves almost always attack in winter. But the crucial difference is in the way these animals attack. If one wants to escape a bear, as local people convey, s/he should shout and run away or still shout and move away slowly. But, to escape a pack of wolves, it is only possible to climb a tree and wait for someone with a rifle. As one man put it: "The bear can turn away and *he* would leave a person; the wolf would never do so."

It is also important to stress two aspects: First, the bear is always mentioned as "he", not "she" in talks by women and men who are not hunters. I have never heard stories about a bear encounter in which a narrator noticed that an animal was female (though usually the use of the masculine form in Russian indicates in fact that the sex of the animal is indefinite). Second, a bear in the stories does the same things as a human; for example, he picks raspberries or just walks through the forest. This element is important because it shows the similarity between bears and humans.

Hunters' stories
In hunters' stories and talks, the bear appears in many and various situations. First of all, hunters, as opposed to mushroom and berry pickers, usually stay silent while walking through the wood. It is impossible for them to make a noise because this will scare the game. That is why, although hunters encounter bears more often, they are still not afraid of their aggressiveness. One of my best vozchiki friends, Dmitriy, told me a story of what he does if a bear is nearby. "Once I was walking along my *putik* (a local word denoting a path in the woods along which a hunter sets traps; author's note) and saw a piece of bear shit that was still steaming. I said silently to myself: 'Go your own way. I do my job; you do your job. I don't touch you; you don't touch me.' And that autumn I saw his footprints several times but never met him."

It is interesting to stress that the last phrase literally repeats the concept of indigenous peoples in Canada, who say to the bear: "I don't bother you; you don't bother me" (CLARK/SLOCOMBE 2009, 6). This hunter's story involves several ideas that are important for this text. First, the bear and the human each have their own, different, business [Russ.: свои дела] in the wood. Both of them live side by side and should not disturb and obstruct each other. In the gatherers' stories, a bear carries out the same activities as people; in the hunters' stories, one can find the same idea but the range of activities is broader: A bear can walk through the wood, collect berries, hunt, sleep and perform a lot of other activities in the same way as a hunter.

Second, it is important that a bear and a hunter are also similar in that they are able to hurt each other. On the one hand, a bear can kill a hunter, and local people use the Russian word *zadrat*, literally "tear to pieces". On the other hand, a hunter has a rifle and he (or rarely "she") can shoot a bear. However, an encounter between a hunter and a bear is not inevitable: They do not have to share anything except a territory. So, in many cases, they can coexist peacefully. I will turn to such stories in the next section of this paper.

Third, and maybe the most interesting aspect, is the possibility of speaking to a bear and, what is even more surprising, the ability of bear to understand people. In the story quoted above, the hunter speaks to the bear, but does it very silently. So, it is presumed that the recipient (the bear) does not hear the voice itself but rather captures the message of the hunter's inner voice. Once, I went with my friend to his hunting hut. It was locked by a padlock. Dmitriy knocked at the door and said so softly that I could hardly hear the words: "Hello my friend hut, we haven't seen each other for a long time". After that, he unlocked the door and entered. What he said was more important than whether he was heard by someone. In this case, it was not a bear but a hut that was greeted. I think that the most important thing about these two situations is that it is the very act of communication that acknowledges a relationship.

Intrusion
Villager's story
A lesser element in the killing stories concerns the bear who starts to disturb human possessions. I heard two types of this kind of account. The first, told by Tatiana, was the following: "In 2012, in March, a bear started to enter the village and regularly tried to find something to eat at the scrapyard. Our men were forced to shoot him. He was white from grey hair; he was very old. I suppose he hadn't got enough food in autumn, so he woke up very early (in Vozchiki region, the month of March is still the end of winter, though the meteorological spring starts, including the end of bear hibernation period, in mid-April; author's note). But if a bear comes to the village it is dangerous, it is necessary to kill him." This story, told by a woman, also depicts a bear as a being that can be understood, that can have its problems, very much like local people depict poor villagers who become drunkards or cannot find a job. The very act of killing is also depicted as unwilling (unintentional), and the narrator even shows compassion to the old animal. Another important thing is that the act of shooting is the consequence of breaking the rules: People had compassion for the bear, but were urged to shoot him because he had entered the village, i.e. had violated the border. A bear should not come to a human's place – to the village. If it does, then it should be punished.

Hunter's story
A hunter, Gennadiy, depicted the situation of returning to a hunting hut a bit differently. I would say that this situation was more folklorised (see below). His evidence was: "When one first enters a building one should greet its owner. If one comes to a village house when the owner is a *domovoy* (the Russian word for "brownie" or "boggart"; author's note), one should say: 'Hello, own[er]y-browny, let me in' (Russ.: привет, хозяюшко-домовеюшко, пусти пожить). But if one is coming to a hunting hut where the owner is a bear who lives nearby, one should also knock at the door and ask the bear for permission to dwell there. Near my hunting house lived a she-bear named Mashka. I asked her: 'Mashka, let me in' (*Mashka, pusti pozhit*)."

What we see here is "secondary folklorisation". This term was proposed by the Russian folklore researcher Kirill Chistov (CHISTOV 1986, 53–54) and means the returning of folklore texts, written down by professional ethnographers, to the village through the use of these texts in school education. I call Gennadiy's evidence more folklorised because he was not too shy to talk about boggarts, including a she-bear as an owner of his hunting place. Gennadiy explained in detail that a boggart in a village house was its last owner (in his case, Gennadiy's grandmother), who had died but still dwelt there as a spirit. He was not ashamed to tell me the name of the she-bear, the owner of his hunting place, and told me the name of his mother, who lived in his village house and was considered at that time to be a boggart in that house. It is important to stress that Gennadiy, born in Vozchiki, was taken at the age of seven to Arkhangelsk where he went to school, and he worked there for about ten years after he had finished school. He returned to the village at the age of 40, about five years ago, after he had lost his job in the city. Conversely, Dmitriy (see below, next story) has spent all his life in the village, going to the city only two or three times a year and spending just three or four days there. Dmitriy was always unwilling to talk about lower demonology characters. Also, he has never told me anything about hunting omens. He explained to me once: "If I tell you an omen, it won't work anymore". So, Dmitriy's attitude is more authentic and more vital from my point of view, whereas Gennadiy's attitude is more based on literature and folklore learnt at school. He is more alienated from vivid rural culture, that is why I label his stories as folklorised.

However, it is interesting to compare his account with that of Dmitriy and those of berry pickers (not hunters). First of all, his account is very clear and he gives comprehensible explanations about the details. For Dmitriy, all these details are taboo, especially if he has to discuss it with an outsider, like me. I can say that the same things that Dmitriy was doing silently, Gennadiy was explaining to outsiders. Second, a bear in hunters' stories becomes more individualistic. It is well known whether it is "he" or "she", it can have a special name, distinctive temper, and so on. Third, it is also possible to talk to the bear; hunters are not afraid of them as wild beasts (as simple villagers are) but they fear them as conscious actors who can remember bad things and seek revenge. Fourth, there is a special system of rules that a bear and a hunter know and obey. As Gennadiy put it, "if you greet the owner you will not have problems. If not, s/he will interfere with you".

To briefly sum up, the hunters' stories are much more detailed than those of the gatherers. Bears in hunters' stories may have names, temper, and a sex. The image (idea, picture, representation) of the bear as a being that dwells in the woods but is very similar to humans is also present. Moreover, in hunters' stories, it is possible to talk to a bear and it is assumed that there are several rules for interactions between humans and bears.

Sometimes, violation of human space and infrastructure was the reason for killing a bear. It is important to note that the bear was killed only when he had repeated his assault on the storehouse. So, as in the case with humans, the punishment was performed only when the offender did not stop. Thus the idea of the rules that were broken, and the shooting as a punishment is also obvious here.

Another typical case when a bear is killed at the end is the accounts of a bear who was used to destroying hunting huts and stealing food from them. Dmitriy once told me an interesting story of that kind. "It was about 20 years ago; I was working at the Arkhangelsk communications company. That winter, we were inspecting a telephone line from Leshukonskoye to Pinega (two big villages/small towns, centres of the neighbouring districts; author's note). When we were halfway to Pinega, we had to spend several nights at a small hunting hut in the taiga. Near the hut, there was a small storehouse in which tinned meat and condensed milk had been put just a week before our arrival. We found the storehouse destroyed (the door was knocked down; author's note) and all the tins were opened. We thought that people had done this, maybe escaped prisoners. But Makar, who was a good hunter, said: 'That is a bear'. The tins were not taken away, all of them were in the hut or right near it. They had been opened by being squeezed from two sides to push out the contents, especially those which contained condensed milk with sugar. He also likes sweet stuff! So we had to drink tea without milk because the bear had already used it. Next year, the same storehouse was destroyed again and the head of the communications company asked hunters to hunt him down and shoot him."

In this story, the parity of bears and humans is very sharp. First, the act of knocking down the door and destroying the food-stuff can be performed either by humans or bears, no other animal can do it. The first thought of the signalmen inspecting the line was that it was a theft. It was only after detailed analyses that they concluded that it was a bear who had done it. Second, the expression "he also likes sweet stuff" points to a human peculiarity: People like sweeties. This parallel is especially stressed by the word "also", which means "like us, humans". So, in this account we can see not only similarities between bears and humans, as in previous stories, but it is obvious that humans and bears are alike and other animals are a bit different.

Rivalry

Nevertheless, not all the talks and stories about bears report a more or less peaceful coexistence (in the case of the berry pickers) or one-time encounters. There are a lot of accounts about long-term rivalries and several temptations for a bear to kill a hunter. First, I will turn to stories about how vozchiki people narrate cases of killing a bear, and after that I will proceed to the narratives about bears who tried to kill humans.

Before I turn to the real stories about how a bear killed or tried to kill a hunter, I would like to mention two sayings about the intentional hunting of a bear. The first is, "You can kill 39 bears, but the 40th bear will kill you". The number 40 is a mystic one in Russian (Orthodox) tradition: A person who has died spends 40 days as a soul on earth before they go to another world, it is only 40 days after its birth that it is possible to baptize a child, the length of pregnancy is considered to be 40 weeks, etc. (ZELENIN 1992, 350–356). All these cases are linked with humans and human life and death. The linkage of the bear with the number 40 can be interpreted as the fact that the bear has a special place (that cannot be compared to the position of other animals but is more similar to a human position) in vernacular culture and that the bear is perceived as someone close to supernatural spiritual actors, such as "spirits", "souls" and "angels".

The other saying goes: "If you go hunting a bear you can prepare a place for its skin, if you go hunting an elk you should prepare yourself a coffin". The message here is that a bear is not as dangerous as an elk. It is important to have this in mind when we start with the accounts of bear-killers (people who were successful in killing bears).

The collection of stories about killing a bear can be divided into two parts. In the majority of cases, the bear is killed unintentionally. For example, one hunter said: "I went along my *putik* and suddenly saw a bear right in front of me. So, I couldn't stay apart and had to shoot him". It should be clarified that a bear is conceived to be very persistent: If a human goes straight up to him face-to-face it will never step aside and will fight. In this situation, the animal is understood not as cruel or bad but more as brave or tenacious. As one person said: "He is the same as us. If he meets you face-to-face he would never retreat". So, meeting a bear face-to-face means an inevitable fight and one of the participants will be killed. This idea urges hunters to kill a bear if they encounter him face-to-face. However, in the stories, hunters never boast of killing a bear. They usually say "I had to shoot him" or even "I was forced to do it". Or, if the story is told in the third person, the narrator says: "He couldn't do anything, he was compelled to shoot him". Dmitriy once told me: "The majority of bears are killed unintentionally, occasionally. If you meet him face-to-face you should shoot".

Another important idea here is "the sameness" of hunters and bears and their symmetric relations. These two ideas are crucially important for our discussion. The idea of sameness reflects the fact that the bear is intrinsically the same inside, though he is obviously not the same in appearance. It is not "animistic ontology" in Descola's sense (DESCOLA 2013), because this similarity of internal sameness and external differences binds only bears and humans and not all animals and natural objects. However, this very idea of sameness is very interesting and I will return to it at the end. The symmetric relation ("you have to shoot or s/he will kill you") is also implied in all bear-human relations.

The stories about bears who want to kill, try to kill, or have killed a person usually start with an unfinished fight, which is also a variant of symmetrical relations. Here are two very similar stories with different endings told by a hunter named Eugeniy. The first is: "Petr, one of my friends, encountered a bear and shot him down. He left the corpus (the body) in that place. But when he returned in the evening it wasn't there. He left that hunting place and didn't go hunting for two years. But the bears never forget and forgive. Two years later, at the time of haymaking, the man found that there was bear shit near his field. The next day, he found the shit right near the rick. He told his elder relative, and the latter said: 'He is chasing you, so we have to kill him'. They gathered five men from the village and shot him". The main idea of this account is that a bear always wants revenge on a hunter who has tried to kill him. There are some stories in which a bear found a hunter that missed him ten years later in another village. Here, the bear has its own temper and special traits. However, in different stories all the bears have almost the same temper: They are stubborn. And revenge becomes an idée-fixe for them. It is in this account, full of everyday life details, that it is possible to find the most typical description of a bear's temper.

The second story is an antipode to the first. Eugeniy related the following: "Another hunter, Mikhail, also encountered a bear. He tried to shoot but the rifle misfired. The bear attacked him and Mikhail lost consciousness. But bears don't like fresh meat, so he [the bear] dragged the man into the swamp not far from the road and covered him with moss and pieces of bark. They [bears] usually do this, and when the meat begins to rot they return and eat it. Mikhail's dog found him in the swamp, dug him up, and barked loudly until some other person came and took Mikhail home. The elder people said that the bear would return after Mikhail. The men from the village went to that place to kill that bear. But he [the bear] never returned".

The last account is not in line with the typical folklore story. We find the idea that a bear wants revenge, and even some men from the village start chasing him, but the bear was not found and did not appear in search of revenge. Interestingly, Mikhail does not want revenge on the bear. In this particular trait, bears are not like humans. It is possible to suppose that bears are more similar to bad humans, humans who cannot control their nature. This story is also very characteristic of the way these tales allow us to understand what happens when the hunter fails to kill. There are a lot of very short stories about someone who went to his hunting house and "has never returned" or "was not found" (Russ.: его больше не нашли). The account above is a happy-ending (more or less) variation of such stories. It is also interesting that, very much like a man, a bear leaves a killed rival and returns to the corpus sometime later. Though the intentions of bear and human in leaving the game behind are different – a bear wants it to be rotten, the human wants to return with some kind of vehicle to transport the meat – the descriptions of situations in the stories are quite similar.

All these stories about how someone has killed a bear and how a bear tried to kill someone show a lot of similar features with the narratives from the first two sections. The bear does things similar to humans, the bear has his temper, intentions, and individuality. The bear in the stories of the last section has a sex and special features in his appearance. New aspects of bear representation that we can see here are the following: First, the bear is very much like a human, but these two species are removed from other animals. In the story about the destruction of the storehouse, only bears and humans could potentially do that. Second, the bear (what it does, what it looks like) is very familiar to vozchiki people. They can draw conclusions from specific features of the bear's appearance and explain why it did this or that thing. And this "familiar attitude" is demonstrated not only in the accounts of hunters but in the accounts of the average inhabitants of the village.

What else do the inhabitants of Zapinejie know about the bear?

When collecting the stories and sayings, it happened several times that I acquired some information that did not fit into the notion of all the previous texts and talks. This includes times when bears were classified into "grass-eaters" and "predators", a very characteristic anecdote about a bear coming to God, and other peculiarities. All this lore was told to me by hazard in a very special situation that occurs only rarely.

In 2007, during my first visit to Vozchiki, I asked Dmitriy about a story I had read in an article dating from the end of the 19th century. The story was about a sorcerer who was so potent that he could bewitch cows in such a way that a bear did not kill them but just walked and grazed among them. This shocked me and I wanted to find an explanation. When I related this to Dmitriy he stayed very calm and said: "Perhaps it was a grass-eater". I tried to ask different people about grass-eaters and predators, but all of them could only agree that such a division exists; they could not add anything to it. I once got a very similar interpretation from another friend. When I was on a three-day trip with Nikolai to the remote lakes, he once saw two bears on the grassy river bank. They were eating something calmly from the ground, one about ten metres from the other. Nikolai pointed to them and said: "Bears there. They are grass-eaters". It is rather difficult to understand the meaning of this dual classification. The only two things I can say is that this grouping is well known among vozchiki people, but it does not play a great role in any of the talks.

The other interesting point is what a hunter does when he sees an animal in his way or near him. In the course of our trip to the lakes, I saw hares, elks, and bears on the banks and in the woods near the river. If a hunter meets a hare it means misfortune. Nikolai, who was driving a boat, rotated his cap on his head 180 degrees: "It is a way to beat misfortune", he explained. There are other ways to do it; for example, by making the fig-sign with the fingers of the hand in the pocket. The encounter with the hare required a quick reaction, but it was not something really important. Nikolai never told anyone about seeing the hare.

Once Nikolai saw an elk who was intending to cross the river. He slowed down the motor and waited a bit. The elk turned back to the forest. So Nikolai accelerated and we passed that place. It was clear that he did not want to encounter or even be in the vicinity of that kind of animal. In the evening of that day, we slept in a hunting hut of Nikolai's friend and they were talking about bears and ducks, but not about elks and hares.

Twice we saw bears. The first time was the two grass-eaters, and the second time Nikolai mentioned a bear cub who had climbed the precipitous pine tree that stood close to the bank. I took a camera and took a shot. This shot – not good and not professional at all – became the main topic of several subsequent conversations. Our boat passed by that place again and Nikolai asked jokingly: "Do you want to land and take another shot, from the vicinity…", and I said: "Of course not. I don't want to face his mummy". Nikolai was very satisfied with his joke and with my answer. However, in the evening talk he said to his friend as he pointed to me: "We saw a bear cub. And he" – pointing to me – "said, let's land and take other photos. I told him: Are you a fool? His mom is there. Let's keep navigating". That provoked a burst of laughter. After our return, a lot of people in the village learned about the photo with the bear cub. It is important to point out that the situation with the bears was less dangerous from the point of view of Nikolai; however, it was the most interesting to talk about and to joke about.

The last piece of information is composed of several narrative jokes (Russ.: *anecdote*) in which the bear is the main character. The word *anecdote* has two meanings in the Russian language. The first is a more historical meaning: In the 16th–19th centuries, it was a bizarre or outstanding situation that happened to a real person. The second meaning emerged in Soviet times and, in this sense, an anecdote means a narrative joke that ends with an unexpected semantic resolution, which makes people laugh (CHAMBERLIN 1957). Hereafter, I will use the term in the second sense.

In Vozchiki, I heard two anecdotes several times that I knew before my visit because city dwellers often narrate them. One of the anecdotes is very close to the stories presented in the previous part of the text. "A tourist is eating his breakfast in the forest, and suddenly a bear appears. The bear asks: 'Who are you? What are you doing?' The tourist answers (showing the bear a tin with the title 'Tourist's breakfast'): 'I am a tourist. I am eating my breakfast'. 'No', says the bear – 'the tourist is me, and you are the tourist's breakfast!'" I heard this anecdote several times. Once it was a quotation. A person addressed my students and me before we went into the forest: "So, you go to the forest, tourist's breakfasts?" This anecdote has a lot of common features with other stories about bears; the bear stands as a synonym of man, it is also a tourist, it also likes breakfasts. It is possible to talk to him. Moreover, he has a sense of humour, joking about the tourist and making a pun.

However, in other anecdotes, hares, wolves, foxes, and all other animals can do that. In Vozchiki, my friend Mikhail told me an anecdote that is very specific to the local culture. "A bear comes to God and asks: 'Why are humans so chatty? Every time I meet a human I don't run through the forest shouting: I have met a human!'". In fact, what we see in this story is not a question but a complaint. The bear says that humans are too chatty, and he is better than them (us). In this account, we can see the maximum degree of humanisation of the bear: It can come to God and paradoxically explain to him that humans are not as human as the bear. The bear is more fearless, sober, and equable. All these traits are perceived as human and are opposite to those of an animal. So, in this last story, the bear and the human change roles. The bear is depicted here as more human and more dignified than a human.

Is equality in the bear-human relationship possible?

In my speculations, I will follow the mainstream idea of a recent article by Harry Walker, entitled "Equality without equivalence", in which he writes that scholars should approach the idea of equality from a different direction. They/we should not start with the idea of sameness as a basis of equal relations, but should try to trace these relations in a situation such as we have with humans and bears: The counterparts are absolutely "not the same", but they are treated as the same. The Amazonian example of the Urarina people shows "not egalitarianism but a tendency towards what I shall call 'equality without equivalence'" (WALKER 2020, 149). This starting point is very important in Walker's speculation because it allows him to show that equal relations can be of a different kind; they cannot be reduced to sameness: same appearance, sameness in relations, mathematical identity in measurable characteristics (height, life length, or the area of habitation space). In my case everything is quite the opposite: Humans and bears cannot be the same, cannot have same appearance, cannot have identical characteristics. However, in the people's talks, the similarity and symmetry is usually stressed.

Elaborating on Walker's ten traits of "equality without equivalence" I will demonstrate that some of those tendencies can be found in "human-bear" relations as opposed to "human- other animal" relations. These ten particular traits of social interactions are, as he says, "interrelated ways" in which the Urarina people "refuse equivalence" and at the same time share "a particular conception of the common" (WALKER 2020, 150). Walker does not use the term "individual" because it is too Eurocentric, but writes about two "singularities" that can enter a relationship that creates "commonality". And thus, being a part of a whole and keeping their differences, the singularities become equal.

Out of Walker's ten points or traits, five are not relevant in my case concerning the bears (concreteness, partiality, immanence, predation, and homosubstitution). However, the other five points can be found in talks about bears, and thus these can confirm the idea that human-bear relations have a tendency to equality. These are 1) intensity, 2) singularity, 3) respect, 4) volatility, and 5) the love-pity-compassion complex. Here are Walker's explanations of these traits based on the Urarina:

1) Intensity, or the possibility "that people can often trace their genealogical relationship to each other in many different ways" (Walker 2020, 152). This means that one person may be to another not only, for example, brother/brother but also uncle/nephew and at the same time father-in-law/son-in-law;
2) Singularity, or the Urarina's "attention to individual rather than collective or group identities" (Walker 2020, 153);
3) Respect for the idea of the illegibility of others' singularities, the impossibility of understanding them clearly and, as a consequence, the necessity of respecting otherness and difference (Walker 2020, 153–154);
4) Volatility, or the idea that people often change their minds or, as Walker put it: "My friend one day is my enemy next day" (Walker 2020, 154); and
5) The "love-pity-compassion" complex is the set of interconnected "forms of emotional attunement" that lead people to share resources (Walker 2020, 156).

I will unite the five traits above into three groups. Intensity stands alone. Singularity and volatility are two aspects of the "rich inner world" of a person who can change his intentions, his ideas or expected way of behavior. The last two traits, love-pity-compassion and respect, can also be linked when we talk about bears. I have never heard a story from any Russian village inhabitants in which a human fell in love with a bear, though such a plot can be found in the folklore of indigenous peoples (McClellan 1970).

The intensity in human-bear relations is seen in almost every talk. For example, in gatherers' talks, a bear is the most frightening animal, and it is important to avoid an encounter by knocking on metal objects or singing songs. However, if an encounter takes place, it is possible to startle him by imitating a dog's barking or loud shouting. We see the same multi-layered relations in hunters' talks where the bear acts as a creature that can kill a hunter but where, however, in many cases the hunter kills the bear. It is also possible to meet him (if not face-to-face) and pass by. The bear can come from the forest and destroy a storehouse, though more often hunters come to the forest from villages to set traps and take the game, and in a sense they destroy the bear's world or "possessions". So, these relations are very different, and bears and humans often intrude on the other's territory. In human-bear relations, intensity can be demonstrated by a metaphor of a dense network of paths, each of which is two-way traffic. The relations with the hare are one-way: A hare can never kill a human. Relations with elks are more complex but not so diverse. What is also important is that hares and elks do not intend to harm people or spoil the hunting luck of humans; they are not actors but rather instruments in those situations. The bear, however, is a "person" who has his own intentions and character traits.

Singularity and concreteness are always stressed when people talk about bears. In gatherers' stories, the animal is always mentioned as "he" not "she", but also not "they", although in Russian it is usual to talk about a unknown class of subjects as "they". For example, hunters usually say about hares: "Hares, they bring bad luck". And these hares do not have personality and singularity. In hunters' stories about killing bears, people always focused on particular features of one or another bear. For example, in the story about the hunter Mikhail, the bear had to be killed because he had attacked a human and would attack him again. So the "elder people" in the account did not want revenge on any bear or on "all bears"; they understood that the bear who tried to kill Mikhail would return. People usually want to kill a specific singular bear, not every bear near the village. When talking about bears, vozchiki villagers talk about singularities, not species: The very old bear who came to the village was white, the other bear could destroy huts and open tins, another particular bear stopped his revenge and "never came back". So, we see here not just one specimen but a gallery of characters, persons, or singularities.

Respect and "love-pity-compassion" complexes are very similar when applied to bears. In some sense, and especially in my case, pity or compassion are addressed to a killed bear that was not treated with respect. So, pity-compassion is the tails side of the coin while respect is on the heads. It is possible to see pity in general in utterances about the unwillingness to kill an animal. "What could I do? I had to shoot!" This type of sentence presupposes that a person feels a bit guilty and has pity for a killed animal. But, most clearly, respect for the bear is demonstrated in the story about the old, white bear that came to the scrapyard. Tatiana knew that the animal was old. From that, she concludes that he did not have enough strength to collect food in autumn. Marking someone as hungry in Russian culture means marking him or her as worthy of compassion. The next sentence after denoting the bear as being hungry begins with "but". And the very act of killing is perceived as a kind of forced action. That bear had its own life history and destiny, and the narrator was interested in learning more about it and understanding it. The tendency to understand (not only the actions and goals but also their intentions and their state of mind) is exactly what Walker marks as respect.

In the paragraphs above, I have demonstrated that at least three specific aspects of human-bear relations in the Russian North can show their tendency to equality:

1) These relations are very diverse and cannot be reduced to a one-way road;
2) The bears act as singularities with volatile behaviour, which cannot always be foreseen and predicted;
3) The relations with bears are strongly linked with emotionally-coloured feelings (especially of respect and compassion).

I think that these kinds of relations are not just relations between two persons – human and non-human – these are relations that have "tendency to equality".

It is also important to stress that the idea of sameness that is rejected by Robbins (1994, 34) and Walker (2020, 148) is used by the local vozchiki people. I can say that, for me, "sameness" is not a researcher's epistemological notion, but a local term that creates or helps to create relations of equality.

The "sameness" demonstrated in the stories denotes those tendencies to equality in a performative way. I think that the very possibility of seeing "tendencies to equality" that emerge in human-bear relations can expand our understanding of equality.

Bibliography

Artemova 2020: O. Yu. Artemova, Equality as a human categorical imperative. Pervobytnaya archeologiya. Journal mezhdisciplinarnykh issledovaniy [Prehistorical archaeology. The journal of interdisciplinary studies] 1(1), 2020, 64–91.

Beteille 1994: A. Beteille, Equality and Inequality. In: Companion Encyclopedia for Anthropology. Humanity, Culture, and Social Life (London, New York 1994) 1010–1039.

Brightman 1993: R. A. Brightman, Grateful prey: Rock Cree human-animal relationships (Berkley 1993).

Chamberlin 1957: W. H. Chamberlin, The 'Anecdote': Unrationed Soviet Humor. Russian Review 1957, 27–34.

Chistov 1986: K. V. Chistov, Narodnye tradicii i fol'klor: Ocherki teorii (Leningrad 1986).

Clark/Slocombe 2009: D. A. Clark/D. S. Slocombe, Respect for Grizzly Bears: An Aboriginal Approach for Coexistence and Resilience. Ecology and Society 14(1), 2009. http://www.jstor.org/stable/26268040.

Descola 2013: Ph. Descola, Beyond Nature and Culture (Chicago 2013).

Dudeck 2018: St. Dudeck, Dialogical Relationships and the Bear in Indigenous Poetry. Sibirica 17(2), 2018, 114–120.

Jolly 1987: M. Jolly, The Chimera of Equality in Melanesia. Mankind 17(2), 1987, 168–183.

Kachalov 2012: N. A. Kachalov, Zapiski tajnogo sovetnika [Privy Counselor's Notes] (Moscow 2012).

Kamkin 1995: A. V. Kamkin, Krest'janskij mir na Russkom Severe. Materialy po istorii severorusskih krest'janskih soobshhestv XVIII veka [Peasant community {‚mir'} in the North of European Russia. Materials to the history northern peasant societies in the XVIII century] (Vologda 1995).

Kolesnikov 1976: P. A. Kolesnikov, Osnovnye jetapy razvitija severnoj obshhiny [Main stages in the history of development of Northern Community]. Ezhegodnik po agrarnoj istorii [Yearbook of agrarian history] 6, 1976, 3–35.

McClellan 1970: C. McClellan, The Girl who Married the Bear, a Masterpiece of Indian Oral Tradition (Ottawa 1970).

Ries 1997: N. Ries, Russian Talk: Culture and Conversation during Perestroika (Ithaka 1997).

Ries 2009: N. Ries, Potato Ontology: surviving postsocialism in Russia. Cultural Anthropology 24(2), 2009, 181–212.

Rivkin-Fish 2010: M. Rivkin-Fish, Pronatalism, gender politics, and the renewal of family support in Russia: Toward a feminist anthropology of' maternity capital. Slavic Review 2010, 701–724.

Robbins 1994: J. Robbins, Equality as a value: Ideology in Dumont, Melanesia, and the West. Social Analysis: The International Journal of Social and Cultural Practice 36, 1994, 21–70.

Smith 1998: D. M. Smith, An Athapaskan way of knowing: Chipewyan ontology. American Ethnologist 25(3), 1998, 412–432.

Squires 2007: J. Squires, The New Politics of Gender Equality (London 2007).

Walker 2020: H. Walker, Equality without equivalence: an anthropology of the common. Journal of the Royal Anthropological Institute 26(1), 2020, 146–166.

Zelenin 1991: D. K. Zelenin, Vostochnoslavianslaya etnographia [East-Slavic Ethnography] (Moskwa 1991).

Dr. Andrey V. Tutorski
Lomonosov State University Moscow
Department of Ethnology
Moscow
Russia
tutorski@gmail.com

Fig. 1. Area of research interest (map GIS department, ZBSA).

Chapter 9

Bears in philology (northern, central and eastern Europe)

One, if not the best, indication for the bear's special status is the taboo name-giving in different languages, such as "the brown one" in Germanic and "honey eater" in Slavonic. If the bear was called by its real name, so it was feared, it would overhear this and evoke disaster for humans (see PILUDU, The songs and rituals of the Finno-Karelian bear hunt, this volume; image P. Hanunen).

Bjørnestad, *Bjørnbåsen*, and *Godfardalen:* Bear/human relations as referred to in place names from southwestern Norway

By Inge Særheim

Keywords: Bears, place names, southwestern Norway, bear hunting, taboo, noa

Abstract: This chapter deals with Norwegian toponyms referring to bear/human relations, mainly based on examples from the southwestern part of the country. Bjørn m. "bear" is the most common word for wild animals found in Norwegian place names, denoting different types of bear locations. The original semantics of the appellative bjørn is "the brown" (animal), first used in Germanic languages as a noa word (i.e. a non-taboo substitute) for "bear", due to taboos about this animal. There are different words for "bear" in Norwegian place names, some of them found in names dating back to the 1st millennium AD, represented in ancient settlement names as well as in topographical names. Microtoponyms from southwest Norway reflect different kinds of bear/human relations in the past; a relationship that was often troublesome, as bears were regarded as a threat to cattle and people on Norwegian farms and summer mountain farms. Some names refer to old hunting methods (bear traps).

Introduction

Place names are an important source of information regarding language, nature, and cultural history that reflect past conceptions of the landscape. Microtoponyms map animal lives onto the landscape. They reflect special – sometimes ancient – traditions and concepts; for example, those relating to bear/human relations.

Bjørn m. "bear" is the most common word for wild animals found in Norwegian place names. These names denote different types of bear locations; for example, dens and places where people have seen traces of bears, places where people – according to local tradition – have encountered and killed bears, and places where cattle have been killed by bears. One example is the microtoponym *Der bjørnen drap* (in the local dialect: *Der bjødnen dreb*), "where the bear killed", from Ven farm in Bjerkreim (southwestern Norway), denoting a cleft in the mountain where a bear, according to local tradition, once killed a living being – probably referring to cattle (SÆRHEIM 1985, 24). From most farms in this area there are place names referring to bear/human relations, and sometimes stories told about encounters with bears.

This chapter will present some place names that refer to bears, and will discuss some aspects of bear/human relations that are reflected in toponyms, mostly microtoponyms. How numerous are place names referring to bears? Where are these names found? How old are the names? What do place names tell us about the relations between bear and human? Most of the examples mentioned come from southwestern Norway – Rogaland county and the bordering municipality of Sirdal (Agder county; cf. Fig. 1). In this area, a number of microtoponyms have been coined from oral sources since the 1980s.

Bear/human relations in the past

Encounters with, and the hunting of, bears in Norway is mentioned in medieval Icelandic sagas, e.g. in *Grettis saga*, which tells us that this animal is very strong and dangerous; it tears apart people's cattle, and no fence or gate is strong enough to provide protection (*Grettis saga* 1936, 74–75). In *Egils saga*, we are told that the bear is the most dangerous animal, killing both people and cattle. The cattle are therefore kept in enclosures, guarded by men with big dogs (*Egils saga* 1933, 167–168). Similar characteristics of the strength and threat of this animal are given in medieval legal documents, and in postmedieval literature. There was a medieval law order concerning the hunting of bears and special rules about hunting, e.g. about hunting bears in a den (*Gulatingslovi* 1981, 120–121).

In the past, bears were an important economical resource, providing meat, fat, and fur. However, bears have been regarded as a threat to people and cattle on Norwegian farms for a long time, especially on summer mountain farms (*seter*, *støl*), where members of the farming families used to stay with their cattle in the summer months to benefit from the mountain pasture (Reinton 1955). Bears were numerous in these areas in the past and, being strong, were regarded as very dangerous animals. Killer bears would, according to Reinton (1955), smash the shed doors or climb on top of the shed, tear off the roof and kill domestic animals such as cows, sheep, and goats, and they would often jump on domestic animals from behind. In 1745 Bendix Christian de Fine wrote that the threat of bears and wolves in some places in Rogaland was such a big problem at times that the farmers could not use their summer farms (de Fine 1952, 10).

There are many stories from all parts of Norway about people encountering bears, about bears attacking and killing cattle, and about bear hunting (Reinton 1955). According to Peder Claussøn Friis (1599), one winter, around 1560, a bear hunter from eastern Agder killed 15 bears with a steel bow. However, when he tried to kill the sixteenth one, he was attacked by the bear and thrown down a high mountain: *der falt hand i smaa Støcher*, i.e. "there he fell in small pieces" (Friis 1881a, 18; Reinton 1955, 364; Hodne 2008, 113). According to Hodne (2008, 113), the best bear hunter in Norway, Ola Olsen Messelt (1776–1869) from Stor-Elvdal (Hedmark), killed around 130 bears. There was a bounty for killing such animals. In the municipality of Hjelmeland (Rogaland), bounty was paid for 24 bears in the period 1760–1766; Per Vrålson from Vormeland shot seven big bears in 1766 (cf. Stavanger Aftenblad, 02.02.1969). According to Reinton (1955, 356), between 219 and 325 bears were shot in Norway every year in the period 1846–1850, i.e. on average more than 265 every year.

The bear is well represented in Norwegian folk tradition and folklore, often as a magical and mystical creature (see Böldl, this volume). In the past, it was believed that people skilled in magic were able to make bears kill other people's cattle and that some people, appearing in a bear hide, could carry out such evil-minded actions themselves (Hodne 2008, 113–116). In some fairy tales, the bear appears with human traits. Farmers and fishermen were in some situations unwilling to mention the bear, due to superstition; a non-taboo substitute – so-called noa – was used instead.

The appellative *bjørn* – first used as a noa word for "bear" due to the taboo

Several words for "bear" are represented in Norwegian place names. The most common one is, as mentioned, Norw. *bjørn* m. "bear". This word appears in microtoponyms like *Bjørndalen* "bear valley" and *Bjørnshi* "bear's den"; both names are common in the municipality of Sirdal and many other places in Norway.

The appellative Norw. *bjørn* developed from the Old Norse (ON) *bjǫrn*, Proto Nordic *bernuR*, which appears as an old *n*-stem in Germanic: **beran-*. The Scandinavian *bjørn* (Swedish *björn*) is related to the English *bear*, German *Bär* and Dutch *beer*. This appellative is most likely formed from the same root as the adjective Norw. *brun* "brown" (Indo-European **bhero-* "brown"), containing the Indo-European root **bher-* "shining, light brown", also represented in the Norw. *bever* (ON *bjórr* m.) "beaver" (Hellquist 1948, 76; Nielsen 1991, 54; Bjorvand/Lindeman 2000, 78).

The original semantics of the appellative *bjørn* is thus "the brown" (animal), most likely first used in Germanic languages as a noa word for "bear", due to the taboo connected with this animal. The word "noa" (from Polynesian) refers to lifting a taboo from a person, animal, or object by using a harmless non-taboo substitute. This and other noa words for "bear" are used to keep the animal away, which is an indication of superstitious beliefs (Hellquist 1948, 76; Nielsen 1991, 54; Bjorvand/Lindeman 2000, 78). One should not mention the bear with the normal – correct – word, which would be to summon the animal; if doing so, it was believed that the bear would come. In other Indo-European languages, an earlier word(stem) used for the bear was, like the Latin *ursus* and the Greek *árktos*, formed from a root with the meaning "the destroyer". There are noa words for "bear" in many Indo-European languages (Nielsen 1991, 54).

Some other words for "bear" in toponyms

A word for "she-bear" found in toponyms is ON *bera* f.; this word is related to *bjørn*. An example from southern Rogaland is the farm name *Berjod* (Sokndal), spelled *Berurjóðri* (dative) in the 1200s, with the final element ON *rjóðr* n. "open place, clearing in forest". Examples from northern Rogaland are *Berdalen* (ON **Berudalr*), a valley in Sauda, and *Berakvam* (ON **Berahvammr*), a farm in Suldal; *kvam* m. refers to a "grassy hollow, short valley". The first two names mentioned might, however, contain an old river name, ON **Bera* f., related to ON *bera* f. "female bear" (Særheim 2007, 28–29).

The word ON *bersi* m. "he-bear", which is also related to *bjørn*, most likely appears in the toponym *Bersetjørn*, denoting a small lake (*tjørn* f.) in Hå (southern Rogaland). An apparently similar toponym, the uncompounded *Berse* (ON **Bersi* m.), denotes a lake in Bjerkreim (southern Rogaland; cf. Fig. 2). However, this name most likely reflects an independent derivation with an *s*-suffix to the stem **ber-* in ON *bjǫrn* m. "bear"; in other words, it is probably not formed from the appellative ON *bersi* m. Parallels are the river names *Bessa* in Jotunheimen (Vågå) and *Besso* in Hardanger (Eidfjord), reflecting ON **Bersa* (**Bessa*) f. (Rygh 1904, 10–12). Such derivations with an *s*-suffix are believed to be very old.

ON *balti* m. "bear" probably appears in the coastal name *Balten* (Hovda 1941, 13), denoting a rock in the sea in the entrance to Hesbyvågen bay (Finnøy, northern Rogaland). The appellative *balti* m. is used in Old Norse poetry, and ON *Balti* is also a male nick-name. According to Hovda (1941, 13), the name is misspelled *Galten* on some maps, probably because *Galten* is a common name denoting sunken rocks in this area (cf. Norw. *galt* m. "hog").

Binna, ON **Birna*, is found as a river name in Norway (Rygh 1904, 13). The name is an old derivation from the stem in ON *bjǫrn* m. "bear". A similar derivation is the appellative Norw. *binne* f. "she-bear", ON *birna* f.

The first element of the farm name *Bangsberg* from Ringsaker (Hedmark), spelled *Bangsberg* in 1723, has been interpreted as ON *bangsi* m. "bear"; this word is still used in Icelandic, cf. also the related Norw. *bingse* f. "she-bear" (Rygh 1900, 19–20). The first part of the farm name *Bamsrud* from Eidsberg (Østfold), spelled *Bangsrud* approx. 1570, however, has been interpreted as the identical male name ON *Bangsi*, first used as a nick-name (RYGH 1897, 154). The appellative ON *bangsi* m., which originally was probably a pet name, is most likely derived from the verb *bangsa* "walk heavily and clumsily". The word *bamse* m., used as a pet name for "bear", is related to ON *bangsi* "bear", referring to something bulky and clumsy.

The Norw. *godfar* m. with the semantics "grandfather" (< "good father") is found in many Norwegian place names. This word is as an appellative also used as a non-taboo substitute – a noa word – for "bear". The use of *godfar* for "bear" is probably also represented in some microtoponyms from mountainous areas, such as *Godfardalen* (valley), found in Sirdal locations (SÆRHEIM 2019).

SETTLEMENT NAMES WITH *BJØRN-* AS THE FIRST ELEMENT – "BEAR LOCATION" OR "BJØRN'S FARM"?

The word *bjørn* m. (ON *bjǫrn* m.) "bear" and related word forms appear in quite a few old farm and settlement names in Scandinavia. In Norway, it seems to be represented in different types (classes) of farm names; for example, in names ending in ON *-vin*, *-heimr*, *-land*, *-staðir*, *-þveit*, *-bǿr/-býr*, *-holt*, and *-ruð*. It is sometimes difficult, however, to decide whether some of the names contain the appellative ON *bjǫrn* "bear" (Norw. *bjørn*) or the related – formally identical – male name ON *Bjǫrn* (Norw. *Bjørn*).

ON **-vin* "meadow" is the final element of the farm name *Bidne* from Voss (Hordaland), ON **Birnin* (RYGH 1910, 561, 568; cf. Fig. 3). The first element represents the stem **bern-* "bear", found in ON *bjǫrn* (< **bernuR*). Most of the 1,000 Norwegian settlement names of this type (**vin*-names) are believed to have been formed before approx. AD 800. Names belonging to this name class do not contain personal names as the first element.

ON *-heimr* m. "home, place of abode" is the final element of the farm name ON *Bjarn(h)eim(a)r*, found in Norway, in Bærum (Akershus; ON *Bjarn[h]eimr*; RYGH 1898, 147) and Sandefjord (Vestfold; ON *Bjarn[h]eimar*, plural; RYGH 1907, 263–264). The first element is ON *bjǫrn* m. "bear". Most of the 1,000 old Norwegian settlement names ending in ON *-heimr* are believed to have been formed before approx. AD 600. Names of this class do not contain personal names. There are also two parallel names in southern Scandinavia, i.e. *Bjørum* in northern Jutland (Jylland, Denmark), and the similar *Bjärnum* in Scania (Skåne, Sweden).

ON *-land* "(piece of) land" is the final element of the farm name *Bjørnland* from Tune (Østfold), spelled *Bjœrnaland* approx. 1430 (RYGH 1897, 294–295). A similar name is found in Bohuslän (Sweden), spelled *j Biarnlandum* approx. 1400. The first element has been interpreted as either the appellative ON *bjǫrn*, referring to bears, or the identical male name *Bjǫrn*. Most of the 2,000 old Norwegian farm names ending in *-land* are believed to have been formed before AD 1000.

Bjørn- is also represented as a first element in other types of old farm names, e.g. in names ending in *-by/-bø* (ON *Bjarna[r]býr/-bǿr*), *-set* (ON *Bjarna[r]setr*), *-tveit* (ON *Bjarnaþveit*), *-holt* (ON *Bjarna[r]holt*), and *-rud* (ON *Bjarna[r]ruð*). In many cases, however, it is difficult to decide whether the first element in such names refers to bears or to a man called *Bjørn*. Most of the 1,100 *by-/bø-* names, 900 *set*-names and 600 *tveit*-names date to the Viking Period (800–1050). However, some are older and some younger, whereas most of the 400 *holt*-names and more than 3,000 *rud*-names were formed in the following period (1050–1349).

ON *-staðir*, Norw. *-stad*, is the most common final element in old Norwegian farm names with the first element *Bjørn(e)-*. In Norway, there are approx. 47 *Bjørn(e)stad*, i.e. 27 *Bjørnstad* and 20 *Bjørnestad* (cf. Fig. 4). The first element of the *stad*-names has traditionally been interpreted as a personal name, most often a male name. There are, however, reasons to question this opinion. Researchers today tend to see fewer personal names and, more often, other types of first elements in the *stad*-names. Approx. 30–40 % of the *stad*-names from Rogaland probably contain a personal name

(Særheim 2006), not 80–90 % as earlier believed (Rygh 1915). It does not seem likely that as many as 47 Norwegian farm names ending in *-stad* have the male name *Bjørn* as the first element. The other frequent male names as a first element in *stad*-names have fewer than 20 representatives each, e.g. *Arne/Ørn*, *Eirik*, *Finn*, *Grim*, *Harald*, *Orm*, *Reidar*, and *Roald*. Some of the *Bjørn(e)stad*-names appear in typical bear areas, and it seems reasonable to interpret the first element in quite a few of them, probably more than half of the 47 *Bjørn(e)stad*-names, as the word *bjørn*, referring to bears, in other words "bear locations", i.e. ON *Bjarn(a)staðir* (not *Bjarnarstaðir*). Most of the 2,500 Norwegian farm names ending in *-stad* are believed to have been formed before AD 1000.

Such old settlement names containing a word for "bear" are indications that bear/human relations referred to in place names are very old. The same can be said about some old topographical names, e.g. a lake name like *Berse* (ON *Bersi*) and river names like *Binna* (ON *Birna*), *Bessa*, and *Besso* (ON *Bersa*); such names might have been formed in the first millennium AD.

Microtoponyms referring to bears from Morka farm (Gjesdal)

The municipality of Gjesdal (Rogaland) was, in the past, a typical bear area. A number of place names from Gjesdal refer to bear/human relations. Morka farm may serve as an example. From this farm, where approx. 300 microtoponyms have been coined from oral sources, as many as eight names refer to bears (Særheim 1985, 108). No other wild animal has that many representatives in toponyms from this farm. Most of the "bear"-names are found in the mountainous area, where people stayed in the summer months with their cattle.

The toponym *Bjørnshi* (pronounced *Bjønns-* in the local dialect) refers to a den (*hi* n.), whereas *Bjørnahola* (*Bjødna-*; found in two places) refers to a "hollow" and *Bjørndalen* (*Bjønn-*) to a "valley". *Bjørnberget* (*Bjønn-*) denotes a mountain (*berg*), *Bjørnasteinen* (*Bjødna-*) a rock, and *Bjørnaklampen* (*Bjødna-*) a small mountain (big rock). There are also microtoponyms from Morka referring to other wild animals, e.g. wolf (*Ulvaberget*, *Ulvakjelda*, *Skrubbatørne*), fox (*Revasteinen*, in two places), and lynx (*Gaubestørne*).

Stories are told about the killing of bears in the early 1800s, linked to a couple of the "bear"-names from Morka. One story refers to the rocks of *Bjørnasteinen* and *Bjørnaklampen* on the mountain. It tells us that a bear was shot, and the hunter was about to stab it. However, the bear livened up and set off for the man. He managed to climb *Bjørnasteinen* rock where he loaded his gun (muzzle-loader) and shot the bear dead.

An interesting name from Morka is *Godfarbekken*, denoting a stream coming from *Bjørndalen* ("bear valley"). The first element, *godfar* "good father", normally means "grandfather" when used in Norwegian place names. However, this word is also used as a noa word (name) for "bear" in Norwegian dialects, i.e. a word used instead of the normal word in order to keep bears away from people and cattle. This use of the word seems to be represented in some microtoponyms from southwestern Norway, and the interpretation of the name seems reasonable, due to the many occurrences of "bear"-names from this farm, and the fact that this stream comes from *Bjørndalen*. A somewhat parallel example from Fintland farm (Sirdal) is the *Bjørnbekk* stream (*Bjønn-*; "bear stream"), which runs through the *Bjørndalen* valley (*Bjønn-*; "bear valley").

The many "bear"-names coined from oral sources for this single farm (Morka) reflect how important and close bear/human relations were on many Norwegian farms in the past. The local tradition linked to some of the names indicates that this relationship was regarded as problematic.

BJØRN-, BJØRNS- AND BJØRNE- IN MICROTOPONYMS FROM SIRDAL (AGDER)

The valley – and municipality – of Sirdal is a mountainous area in northwestern Agder, southern Norway. Sirdal borders Bjerkreim, Gjesdal, Forsand, and Hjelmeland in Rogaland. There are hundreds of summer mountain farms in this area. In the different volumes of the history books of Sirdal, there is some information about people who have hunted bears. A bear hunter from Tveiten farm, Atlak Svensen (1842–1901), called *Atlak bjødneskyttar* ("bear-shooter") shot 19 bears, whereas Ole Olsen (1822–1912) from Valevatn shot 15 (SELAND 1980, 382, 438). The last bear was killed in Sirdal (Ausdal) in 1910, whereas the wolves had already disappeared *c*. 1860 (SELAND 2011, 200). No person is reported to have been killed by bears or wolves in Sirdal (SELAND 2011, 56). Some bears were given a name; for example, *Skakkefot* ("crooked foot, tilted foot"), who had hurt his leg (SELAND 1987, 216). Approx. 12,000 microtoponyms are coined from oral sources in Sirdal. As many as 80 of them have a first element *Bjørn-* (pronounced *Bjønn-* or *Bjødn-*), *Bjørns-* (*Bjønns-*), or *Bjørne-* (*Bjødne-*), referring to bears (SÆRHEIM 1992, 209–213). No other word for wild animals is used this much in Sirdal place names.

Twenty-eight different words for topographical features are used as final elements. Most common are names like *Bjørnetona* and *Bjørnstørnan(e)* (plural), containing *-to* ("mountain shelf"), found in 18 places, often referring to a den. *Bjørndalen* and *Bjørnedalen*, with the final element *-dal* ("valley"), is found in ten places, whereas *Bjørnshi* (*-hi* "den") and *Bjørnehola* (*-hola* "hole, cave", often referring to a den) are each registered in seven places. There are five names ending in *-kvæv* ("short valley, grassy hollow"), *Bjørnekvæven*, and also five ending in *-hom* ("short valley, grassy hollow"), *Bjørnhom, Bjørnehommen, Bjørnehomma*.

The final element *-bakke* ("hill") is found in four toponyms referring to bears: *Bjørnsbakken, Bjørnebakken* and *Bjørnebakkan(e)* (plural). Three names contain *-li* ("mountain side"): *Bjørnsli, Bjørneli(a)*, whereas there are two names ending in *-ås* ("mountain ridge"), *Bjørnås(en)*, and *-sti* ("path"), *Bjørnestien*, the latter referring to a path used by bears, according to the local tradition.

Some names contain a topographical word for a wet element: *-bekk* ("stream"; *Bjørnbekk*), *-høl* ("hole in a river or stream"; *Bjørnhøl*), *-myr* ("bog"; *Bjørnmyra, Bjørnemyran*), and *-øy* ("island"; *Bjørnøyna*), each of these is found in two places. There is also one example of the final elements *-å* ("river"; *Bjørnåna*), *-tjørn* ("small lake"; *Bjørnstjørn*) and *strond* ("lake side"; *Bjørnestronda*).

Other final elements represented (each used in one name) are: *-stein* ("rock"; *Bjørnesteinen*), *-urd* ("pile of rocks, rocky mountain side"; *Bjørnurda*), *-fjell* ("mountain"; *Bjørnefjellet*), *-haug* ("mound"; *Bjørnehaugen*), *-hei* ("mountain plateau"; *Bjørneheia*), *-klamp* ("big rock"; *Bjørneklampen*), *-red* ("ridge"; *Bjørneredet*), *-skard* ("mountain pass"; *Bjørnskardet*) and *-sprang* ("leap"; *Bjørnespranget*).

In one toponym ending in *-støl* ("summer mountain farm"; *Bjørnestøl*), and one ending in *-slåtte* ("hayfield"; *Bjørneslåtta*), it is difficult to decide whether the first element is the word *bjørn* m. "bear" or the similar male name *Bjørn*; most likely it is "bear" due to the form of the first element (*Bjørne-*).

Three to four names with the first element *Bjørn(s)-* do not refer to bears, but to a male person with the name *Bjørn*, e.g. *Hagen av Bjørn* ("Bjørn's enclosure"). Other examples are *Bjørnåger* ("Bjørn's field") and *Bjørnsflekk(en)* ("the small field of Bjørn").

Thirteen microtoponyms from Sirdal have *Godfar-* as the first element. Most of these seem to refer to bears, containing a noa word for "bear", *godfar* ("grandfather", < "good father"), due to the taboo. This might also be the case with the name of the mountain named *Gamlefarknuden* in Sirdal.

The many microtoponyms from Sirdal containing a word for "bear", i.e. approx. 80 with the appellative *bjørn* "bear" and 13 with the word *godfar*, give an impression of how important – and challenging – bear/human relations were in earlier days in Sirdal and many other places in Norway. As mentioned, *bjørn* is by far the most numerous word for wild animals appearing in Norwegian place names (INDREBØ 1924, 22; SLYNGSTAD 1951, 74). There are also words for other wild animals (of prey) represented in the microtoponyms from Sirdal, e.g. wolverine (one name, *Erveknuden*), wolf (24 names, *Ulvshidalen, Skrubbsodden, Gråbeinbakken*), fox (37 names, *Melrakkhaugane, Revurda, Mikkjelskardet*), and lynx (two names, *Gaubåsen*). Names like *Gråbeinbakken* and *Mikkjelskardet* probably contain noa words for wolf (*gråbein* "grey legs") and fox (*mikkel*). Many other types of wild animals are also represented in Sirdal place names; for example, *c.* 40 names referring to reindeer, including *c.* 20 names with the first element *Dyr(a)-*, containing *dyr* n. "animal", in this context referring to "reindeer".

Bjørnbåsen – a bear trap

An old hunting method (the bear trap), called *bjørn(e)bås* ("bear stall", "bear box") is mentioned in Old Norse literature as ON *bjarn(ar)báss* "enclosure, pit for catching a bear, bear trap", e.g. in *Jomsvikingar saga*, *Magnus Lagabøte's law* and the Swedish law, *Helsingelagen*. A so-called *bås* ("stall", "box") was used to catch bears, wolves, wolverines, lynxes, and foxes. FRIIS (1881b, 311; original text from 1613) mentions *bås* to catch wolves.

This hunting technique, as used by Swedish farmers, is described by KEYLAND (1906, 12–13) as a small house (wooden enclosure) with an opening at one end where the animal could enter. The roof was loose, not fastened to the two long walls and rested on the back wall; the roof was piled (stacked). When the animal took the bait, the front part of the roof would fall down and the animal was shut in and trapped. However, wooden or stone-built bear traps are not known from southwestern Norway, only naturally formed rock enclosures.

The word *bjørnbås* appears in place names – *Bjørnbåsen*, pronounced *Bjønnbåsen* in the local dialect – in several places in southern Rogaland, e.g. in the municipalities of Sokndal, Eigersund, Sandnes (from Høyland, Høle and Osaland farms), Forsand (from Hatleskog and Underberge farms), and Gjesdal (from Østabø, Mjåland and Brådland farms). Eldar Molaug comments on the four last-mentioned places in his cand. philol. thesis (MOLAUG 1934) on place names from Forsand (and Gjesdal). He writes that there is an old, oral tradition about hunting bears linked to such locations and names. Some places with this name are found close to summer mountain farms; for example, close to Brådlandsstølen (Gjesdal) and Bergestølen (Forsand), probably reflecting the great challenge that bears and other wild animals represented in such places. However, in Rogaland, *Bjørnbåsen* is a more common place name in lowland landscapes than in mountain and summer farm areas. Some of the places are situated close to the sea (fjord). The known hunting sites in this area are mountain clefts or rock enclosures into which bears were chased, trapped and killed.

One of the sites in Rogaland is mentioned by Svein Molaug (MOLAUG 1956, 669). This is probably *Bjørnbåsen* in Mjåland (Gjesdal), which is situated at the foot of a big mountain cleft (ravine), called *Bjørnbåsjuvet* (Figs. 5–6). There is a huge pile of rocks below the ravine. Molaug points out that the ravine probably served as walls of the *bjørnbås*. In this enclosure, there was no place from which the bear could escape.

Bjørnbåsen in Hatleskog (Forsand, now Sandnes Municipality) is a cleft or pass on *Bjørnbåsknuden* Mountain, close to the River Dalaåna in the valley of Daladalen (Songesandsdalen). This *bjørnbås* is located quite close to the farm houses, less than 1 km from Hatleskog farm, and close to Lysefjorden Fjord.

Bjørnbåsen in Trodal (Høle, Sandnes) is a narrow and steep valley and mountain cleft with big rocks. Bears were probably chased into this cleft, where they were trapped and killed. This *bjørnbås* is located right on the fjord (in the bay of Trodalsvågen in the fjord Hølefjorden). Another *Bjørnbåsen* can be found not far from the one in Trodal, on Osaland farm (Høle, Sandnes). This is a narrow mountain pass, between the mountains of Tverrfjellet and Litlelifjellet (Fig. 7). As was the case at

Trodal, the bears were most likely chased into this pass, where they were trapped and then killed. According to the local oral tradition in Høle, the two *Bjørnbåsen*-names refer to locations where bears were hunted and killed.

The uncompounded place name *Båsen* ("the stall", "the box") might, in some cases, refer to a bear trap (*bjørnbås*). This name is coined from several places; for example, from Ven farm (Bjerkreim), denoting a cleft in the mountain close to the mentioned *Der bjørnen drap*. This cleft might have been a bear trap. Several toponyms from Ven have *Bjørna-* (pronounced *Bjødna-*) as a first element, referring to bears; for example, *Bjørnatona* (a mountain shelf, maybe referring to a den), *Bjørnakråna* ("crook, corner"), and *Bjørnadølda* ("hollow"), which refers to a location where a she-bear is said to have given birth to cubs. Wild animals such as bears and wolves represented a big problem in this area. Ven's old summer mountain farm, Gamlestølen (also called Heimre stølen), was moved from the River Raudåna to a safer location, due to the threat by wild animals.

In Rogaland (and Sirdal), there are also place names referring to a *bås* to catch other types of animals, e.g. *Revabåsane* in Hompland (Sirdal; referring to "fox"). Other types of old hunting techniques are mentioned in toponyms too; for example, *lem* "shutter" in *Skrubbelemshovudet* ("wolf") and *Ørnelemmen* ("eagle"), both from Sirdal. Another example is *stokk* "log", e.g. in *Jasastokken* (from Suldal; "hare"). Some microtoponyms from different parts of Norway containing the words *bjørn* "bear" and *stokk* "log", e.g. *Bjørn(e)stokken*, most likely refer to places where bears have been hunted and killed by a log falling down and hitting the animal when it was taking the bait. Another method is referred to in the place name *Sjavskotet* ("self" + "shot"; a spring gun) from Setesdal (northern Agder); the bear was shot by a mounted gun when he took the bait (SKJEVRAK 1953, 13). These and other trapping and killing methods for bears and other wild animals are mentioned by FRIIS (1881a, 18–24), KEYLAND (1906), SKJEVRAK (1953, 12–15), REINTON (1955, 364–365), and MOLAUG (1956, 664–670).

GODFAR – A NOA WORD FOR "BEAR" IN MICROTOPONYMS?

Several Norwegian toponyms contain a first element *Godfar-*, often pronounced *Goffa(r)-* or *Gofa(r)-*. In most cases, this is interpreted as the appellative *godfar* m. "grandfather", with the adjective *god* "good" and the appellative *far* m. "father". However, as mentioned above, the appellative *godfar* is, in Norwegian dialects, also used as a noa word for "bear" (SOLHEIM 1940, 78; REINTON 1955, 356; NO 4, 555). Is there reason to believe that this use of the word is also represented in some microtoponyms; for example, denoting topographical features in areas where it is well known that bears would appear and that they were a big problem to people and cattle?

SOLHEIM (1940, 79) writes that there is much evidence that people, especially cattle farmers, from the Nordic countries were unwilling to mention the bear. According to the tradition from Sande (Sunnmøre), one should not mention the bear when working in the forest, but say *svarten* ("the black") instead (cf. the original meaning and use of the appellative *bjørn* ["the brown"]). Fishermen from Sunnmøre did not mention the bear, but said *svartekaren* ("the black guy") instead. In Hordaland, SOLHEIM (1940) tells us that people in Masfjorden said *den svarte* ("the black") when referring to the bear, and people in Bruvik said *gofar* (i.e. *godfar* "good father", "grandfather"); these terms were used as noa words.

In Setesdal (Agder), bordering Sirdal, it is said (SKAR 1961, 156) that shepherd boys watching the cattle in the mountains used the word *goffa* (i.e. *godfar*) when a bear was approaching the cattle; they spoke gently to the bear asking him to go away (cf. also SKJEVRAK 1939, 52). One saying was: "*Lo'ne goffa gakk ifrå! Mi e små*", i.e. "Woolly goffa walk away, we are small", or "*Å goe goffa, gakk at bakkjen å set seg!*", i.e. "Oh, good goffa, walk to the hill and sit down" (SKAR 1961, 156). In his book on the life on summer mountain farms in Norway, REINTON (1955, 356) writes that people tried to keep the bear away by using other words; for example, *bamse* and *goffa* ("good father", "grandfather"). Other examples are *bamsefar* ("-father", "he-bear") and *bamsemor* ("-mother", "she-bear").

As mentioned, the word *godfar* m. appears in several Norwegian place names; for example, referring to fields on a farm. Normally, this word is interpreted as "grandfather", which seems reasonable. But, in some cases, there is reason to believe that the word refers to "bear", used as a noa word (SÆRHEIM 2019).

In an area in Ognaheia, the southern, upper and somewhat higher parts of the low-lying coastal area of Jæren (Rogaland), there are four locations with a name containing the first element *Godfar-*, i.e. *Godfarfjellet* (*fjell* n. "mountain"; Fig. 8) and *Godfartona* (*to* f. "mountain shelf", often referring to a den) – the last mentioned of which is found in three places (SÆRHEIM 2019, 63–66). Ognaheia is a rocky landscape used as pasture for sheep, goats and cows. In the past, wild animals such as bears, wolves, lynx, and fox were numerous in this area, and several microtoponyms refer to hunting; for example, *Fella* ("the trap") and *Skrubbaglepsa* ("wolf snap"). There are many microtoponyms referring to bears and other wild animals in this area; for example, *Bjørnkula* ("bear mountain top") and *Bjørndalen* ("bear valley"), both situated close to Godfarfjellet (Fig. 8). Another name from this area is *Bjørnshivatnet* ("bear's den" + "lake"), referring to a bear den. The location where the den is situated is now called *Bjørnahola* ("bear hole"). A small lake close to Bjørnshivatnet is called *Bersetjørn*, containing another word for "bear": *berse* m., ON *bersi* m. "male bear". The local tradition and the microtoponyms from the area – i.e. the toponymic milieu – support the interpretation of the first element in the toponyms *Godfarfjellet* and *Godfartona* as a noa word for "bear".

In the mountains of Sirdal Municipality there are 13 place names with the first element *Godfar-*, coined from five different places (SÆRHEIM 2019, 66–69). From the mountain plateau of Ådneram farm, there are four names: *Godfardalen* ("valley"), *Godfarskardet* ("mountain pass"), *Godfartjørnene* ("small lakes") and *Godfarlonene* ("narrow pools, river extensions"; Fig. 9). Three names are coined from the mountain plateau of Liland: *Godfartjørna* ("small lake"), *Godfarknuden* ("mountain top") and *Godfarlega* ("resting place for cattle"). Two names are known from the mountains of Ausdal: *Godfarsdalen* ("valley") and *Godfarstjørna* ("small lake"). The names *Godfarlia* ("mountain side") and *Godfarmyra* ("bog") are found on the mountain farm of Finsnesheia, and *Godfartjørn* ("small lake") and *Godfarslåtta* ("hayfield") on Espetveit farm.

All of the *Godfar*-names from Sirdal are found in the mountains; the names from Ådneram and Liland approx. 14 km (as the crow flies) from the farm houses and the ones from Ausdal 8.5 km from the farm. In Sirdal there are, as mentioned, more than 80 microtoponyms with the first element *Bjørn-* referring to "bear", and several names refer to other wild animals, such as wolverine, wolf, mountain fox, and lynx, in addition to eagle, falcon, hawk, etc. The *Godfar*-names seem to be linked to areas with mountain pasture and mountain farms, where people stayed during the summer months with their cattle. In these locations they were especially exposed to the threat of the wild animals.

It seems most reasonable to interpret the first elements of the 13 *Godfar*-names from Sirdal, or at least most of them, as a noa word for "bear". They all denote locations in the mountains, in typical bear-areas, far from the respective farms. It does not seem likely that the first elements of these names could be interpreted as "grandfather". The place names containing words for family relations, for example, *Godmorsåger* ("grandmother's field") and *Farbrorsåger* ("uncle's field") from Sirdal, most often denote locations close to the farm houses. This is an indication that *godfar* in the mentioned names from the mountains does not refer to "grandfather" but rather to "bear".

Another microtoponym from Sirdal that should be discussed in this context is the mountain name *Gamlefarknuden* (Fig. 10), located in the area with summer mountain farms at Suleskard, only 5 km from the area with five *Godfar*-names on the neighbouring Ådneram farm (SÆRHEIM 2019, 69). Close to *Gamlefarknuden*, there are eight toponyms with the first element *Melrakk(e)-*, containing the appellative *melrakke* m. "mountain fox". The appellative *gamlefar* (*gamalfar*) "the old father (man)" is also used with the semantics "grandfather". In the toponym *Gamlefarknuden,* this word should probably be interpreted as a noa word for "bear", due to the location of the mountain, the toponymic milieu, and the local tradition and superstition concerning this animal.

The mentioned *Godfar*-names from Sirdal, Ognaheia (Hå), and Morka (Gjesdal), as well as the *Gamlefar*-name from Suleskard most likely reflect the troublesome bear/human relations on Norwegian farms and summer farms in the past. The bear was a permanent threat to people and cattle. For this reason, people often used noa words instead of the real word, to keep the animal away.

Words for "bear" that are taboo in toponyms denoting rocks in the sea

Words for "bear" are also found in coastal names, often denoting dangerous sunken rocks in sailing routes. Examples from Sørlandet, the southern coast of Norway, are *Bjørnen* "the bear" and *Bjørneskjer* "the bear rock". These names are found in several places; for example, in Mandal, denoting rocks in the sailing route entering a harbour. *Bjørn* m. "bear" is the first element of names like *Bjørnbåen* (Sunnmøre), a very dangerous sunken rock (*båe* m.) with the biggest breakers in the area, and *Bjørnafluna* (Jæren), a rocky bank (*flu* f.) with big breakers in the sailing route. A rock in the sailing route entering the bay and harbour of Hesbyvågen in Finnøy (northern Rogaland) is, as mentioned, called *Balten*. The name probably contains the Old Norse word *balti* m. "bear", a word used in Old Norse poetry.

Some words for animals – domestic as well as wild – were taboo among seamen; using them was believed to lead to bad fishing or shipwreck. Mentioning certain landmarks by their normal name was also believed to lead to disaster. Such names were taboo as well. Another word or name, i.e. a noa word (noa name) was used instead. This tradition is recorded in literature from the Norwegian coast from the early 17th century (Friis 1881b, 298; original text from 1613) to the present day, and from many other places in northern Europe (Solheim 1940). From Shetland, Jakobsen (1928–1932, 63, 66, 69, 74) mentions different Norn noa words used by seamen, e.g. 18 words for "horse", seven for "cow", 11 for "sheep", 13 for "pig", five for "dog", and 22 for "cat".

For some parts of the Norwegian coast, Solheim (1940, 141–145, 162–163) mentions traditions and behaviour linked to certain locations, for example, to (pretend to) stab, greet, or make sounds like a pig, and to stab or roar like a bull. To prevent disaster, one should e.g. pretend to stab the bear when sailing past *Bjørnøy* ("bear island") in Talvik (Finnmark). *Bjørneflua* rock ("bear" + "rocky bank") in Hustad (Romsdal) is regarded as a very dangerous location. In the local tradition this used to be a taboo name. Mentioning the real name was believed to lead to shipwreck.

Concluding remarks

The word *bjørn* m. "bear" is by far the most used word for a wild animal found in Norwegian place names. Several words for "bear" and derivations of word stems referring to "bear" appear in Norwegian place names. The number of local microtoponyms in oral tradition containing a word for "bear" has been illustrated by examples from Morka farm (Gjesdal, Rogaland) and the municipality of Sirdal (Agder). The high number is due to the fact that bears in the past were a continuous threat to people and cattle on Norwegian farms. Such names are also numerous in mountain areas, where farmers stayed in the summer months with their cattle on farms. The name *Bjørnbåsen* appears in several places, often close to farm houses, reflecting an ancient method of trapping and killing bears. Some microtoponyms, e.g. some names with the first element *Godfar*-, seem to contain a noa word for "bear", which was probably used as a non-taboo substitute in order to keep the bear away. This way of naming reflects the troublesome bear/human relations. Some toponyms containing words for "bear" are very old, e.g. some settlement names (*vin-names, heim-names, etc.), and lake and river names formed as derivations with old suffixes (*Berse*, *Bessa*, etc.). Some of these names date back to the 1st millennium AD.

Bibliography

Primary sources:
Egils saga: Egils saga Skalla-Grímssonar. Ed. Sigurður Nordal. Íslenzk fornrit 2 (Reykjavík 1933).

Grettis saga: Grettis saga Ásmundarsonar. Ed. Guðni Jónsson. Íslenzk fornrit 7 (Reykjavík 1936).
Gulatingslovi: Gulatingslovi. Norrøne bokverk 33 (Oslo ⁴1981).

Secondary sources:
Bjorvand/Lindeman 2000: H. Bjorvand/F. O. Lindeman, Våre arveord. Etymologisk ordbok (Oslo 2000).

De Fine 1952: B. Chr. de Fine, Stavanger Amptes udførlige Beskrivelse. Ed. P. Thorson (Stavanger 1952).
Friis 1881a: P. C. Friis, Om Djur, Fiske, Fugle oc Trær udi Norrig. In: G. Storm (ed.), Samlede Skrifter af Peder Claussøn Friis (Kristiania 1881) 1–138.
Friis 1881b: P. C. Friis, Norrigis Bescrifuelse. In: G. Storm (ed.), Samlede Skrifter af Peder Claussøn Friis (Kristiania 1881) 243–409.

Hellquist 1948: E. Hellquist, Svensk etymologisk ordbok (Lund ³1948).
Hodne 2008: Ø. Hodne, Trolldom i Norge (Oslo 2008).
Hovda 1941: P. Hovda, *Okse, galt, hund* og andre dyrenemne i skjernamn. In: Bergens Museums årbok, Hist.-antikv. rekke 7 (Bergen 1941) 3–16.

Indrebø 1924: G. Indrebø, Norske innsjønamn 1. Upplands fylke (Kristiania 1924).

Jakobsen 1928–1932: J. Jakobsen, An etymological dictionary of the Norn language in Shetland 1–2 (London, Copenhagen 1928–1932).

Keyland 1906: N. Keyland, Om den svenska allmogens jakt. Fataburen 1986, 1–16, 115–125.

Molaug 1934: E. Molaug, Stadnamni frå 28 gardar i Fossan (Forsand) herad i Rogaland. Unpubl. cand. phil. thesis, University of Oslo (Oslo 1934).
Molaug 1956: S. Molaug, Björnejakt. In: Kulturhistorisk leksikon for nordisk middelalder 1 (København, Helsingfors, Reykjavík, Oslo, Stockholm 1956) 664–670.

Nielsen 1991: N. Å. Nielsen, Ordenes Historie. Dansk Etymologisk Ordbog (København ⁴1991).
NO 4: Norsk Ordbok 4. Ed. L. S. Vikør (Oslo 2002).

Reinton 1955: L. Reinton, Sæterbruket i Noreg 1. Sætertypar og driftsformer (Oslo 1955).
Rygh 1897: O. Rygh (ed.), Norske Gaardnavne 1 (Kristiania, Oslo 1897).
Rygh 1898: O. Rygh (ed.), Norske Gaardnavne 2 (Kristiania, Oslo 1898).
Rygh 1900: O. Rygh (ed.), Norske Gaardnavne 3 (Kristiania, Oslo 1900).
Rygh 1904: O. Rygh (ed.), Norske Elvenavne (Kristiania 1904).
Rygh 1907: O. Rygh (ed.), Norske Gaardnavne 6 (Kristiania, Oslo 1907).
Rygh 1910: O. Rygh (ed.), Norske Gaardnavne 11 (Kristiania, Oslo 1910).
Rygh 1915: O. Rygh (ed.), Norske Gaardnavne 10 (Kristiania, Oslo 1915).

Seland 1980: P. Seland, Sirdal. Gard og ætt 1 (Tonstad 1980).
Seland 1987: P. Seland, Sirdal. Gard og ætt 2 (Tonstad 1987).
Seland 2011: E. Seland (ed.), Sirdal 7. Kultursoge 2 (Tonstad 2011).
Skar 1961: J. Skar, Gamalt or Sætesdal. Ed. O. Bø/R. Djupedal (Oslo ²1961).
Skjevrak 1939: M. Skjevrak, Bygland i gamal tid. In: Byglands soge (Kristiansand 1939) 11–56.
Skjevrak 1953: M. Skjevrak, Bjørn og varg på Agderbygdene og Vest-Telemark i eldre tid (Kristiansand 1953).
Slyngstad 1951: A. Slyngstad, Skjergardsnamn frå Sunnmøre (Oslo 1951).
Solheim 1940: S. Solheim, Nemningsfordomar ved fiske (Oslo 1940).
Særheim 1985: I. Særheim, Stadnamn fortel historie (Stavanger, Oslo, Bergen, Tromsø 1985).
Særheim 1992: I. Særheim, Sirdal. Namn og stader 1. Namnetolkingar (Tonstad 1992).
Særheim 2006: I. Særheim, Busetnadsnamn på -*staðir* i Rogaland. Busetnadsnamn på -*staðir*. NORNA-rapporter 81, 2006, 161–174.

Særheim 2007: I. Særheim, Stadnamn i Rogaland (Bergen 2007).

Særheim 2019: I. Særheim, *Godfar* – noanemne for 'bjørn' i nokre sørvestnorske stadnamn? Namn och Bygd 107, 2019, 61–74.

Prof. emer. Dr. philos. Inge Særheim
Institutt for kultur- og språkvitskap
Universitetet i Stavanger
Stavanger
Norway
inge.saerheim@uis.no

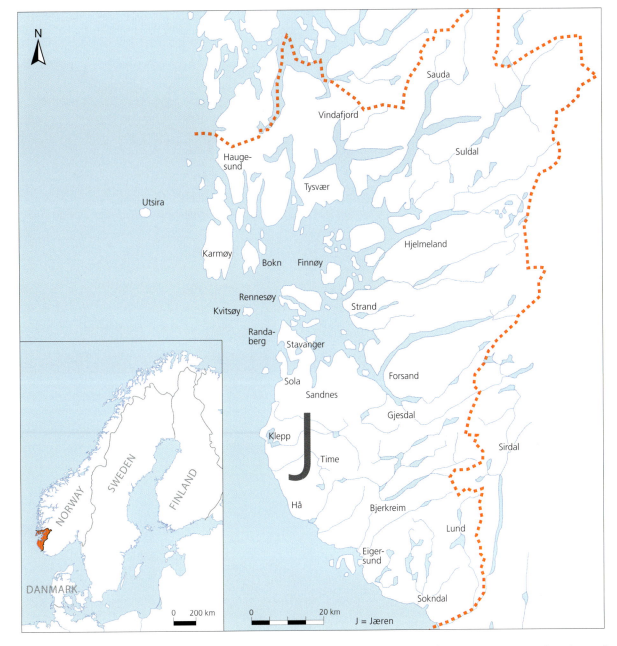

Fig. 1. Rogaland county in southwestern Norway, with its municipality names (up until 31.12.2019). J = coastal territory of Jæren (map J. Schüller, Landesmuseen Schleswig-Holstein, after a draft by I. Særheim).

Fig. 2. Lake Berse *(Bjerkreim). The northeastern side of the lake (on the right hand) is called* Bjørnsstronda *("bear lake side"). The name Berse is probably an old s-derivation of the stem in the word* bjørn *"bear", Proto-Nordic* *bernuR *(photo I. Særheim).*

Fig. 3. Bidne farm *(Voss, Hordaland). The farm name* Bidne *is an old* vin-*name (Old Norse* *vin *"meadow"), Old Norse* *Birnin. *The first part represents the stem* *bern- *"bear", found in the appellative* bjørn *"bear", Proto-Nordic* *bernuR *(photo A. Rødstøl).*

Fig. 4. Bjørnestad *farm in Sirdal. The first element of this farm name has been interpreted as the male name* Bjørn, *i.e.* "Bjørn's farm". *It seems more likely, however, that this name contains the appellative* bjørn *"bear", referring to an area where bears were numerous. Among microtoponyms coined from this farm are* Bjørndalen *("bear valley") and* Bjørnetørnan *("bear mountain shelves"; pronounced* Bjødnetødnan*). There are 47 Bjørn(e)stad-names in Norway (photo I. Særheim).*

Fig. 5. Bjørnbåsjuvet *is a ravine in Mjåland, Gjesdal. This name refers to a so-called* bjørnbås, *an old hunting method (trap) for bears. The bears were chased into an enclosure and killed (photo P. S. Særheim).*

Fig. 6. Bjørnbåsjuvet *(Mjåland, Gjesdal) is a big mountain cleft (ravine) with a pile of rocks below. The name refers to a* bjørnbås, *a trap for bears (photo I. Særheim).*

Fig. 7. Bjørnbåsen *mountain pass (Osaland, Sandnes). The name refers to a* bjørnbås, *an old hunting method for bears. The bears were chased into an enclosure and killed (photo I. Særheim).*

Fig. 8. The Godfarfjellet *Mountains (the middle of the picture, to the right) and* Bjørnkula *(to the left) on Lake Homsevatnet (Hå and Bjerkreim). Both names most likely refer to bears,* Bjørnkula *with the semantics "bear mountain top", and* Godfarfjellet *probably containing a noa word for "bear",* godfar *("grandfather" < "good father"; photo I. Særheim).*

Fig. 9. Lakes Godfarlonene *(lon f. "narrow pool, river extension") and* Godfartjørnene *(tjørn f. "small lake") from Ådneram (Sirdal). In this area, one also finds* Godfardalen *valley and* Godfarskardet *mountain pass. These names probably contain a noa word for "bear",* godfar *("grandfather" < "good father"; photo I. Særheim).*

Fig. 10. Gamlefarknuden *Mountain (Suleskard, Sirdal). The first element of this toponym,* gamlefar *("old father (man), grandfather"), is probably a noa word for "bear", similar to* godfar *("grandfather" < "good father"; photo E. Smedvig).*

Germanic "bear" and Germanic personal names before *c.* AD 1000 with elements referring to "bear"

By Robert Nedoma

Keywords: Bear, Germanic, word taboo, personal names, theriophoric anthroponyms

Abstract: This paper presents a brief survey of "bear" words in old Germanic languages. The inherited word, for example continued in Gk. ἄρκτος árktos, was replaced by a noa-name "the brown one". The main reason for this word taboo is, according to animistic conceptions, to prevent the dangerous animal being summoned when its real name is uttered by humans. Many old Germanic anthroponyms refer, if motivated in morphosemantic respects, to the sphere of reign, power, strength and warfare. "Bear" is a frequent name element that is attested as early as the 4th* century AD. Of particular interest are bitheriophoric formations, e.g. WFranc. Ber-ulf "bear" + "wolf", that are supposed to have an operative-additive sense.*

"Bear" in Germanic

The Proto-Indo-European term for the (species of the) animal genus *Ursus* is to be reconstructed *$*h_2$ŕtḱo-* m.(/f.).[1] Reflexes are preserved in most daughter languages: Hitt. *ḫartakka-*, OI *ŕkṣa-* (< **ŕtśa-*), Avest. *arša-*, Gk. ἄρκτος *árktos* (involving regular inversion *-kt-* < **-tḱ-*), furthermore – exhibiting phonological peculiarities – Lat. *ursus* (with unexpected *u-* and *-s-* from **-ḱþ-* ~ **-ḱt-* < **-tḱ-*), OIr. *art*, Arm. *arǰ* and Alb. *arí*. All these cognates denote "bear" rather than generic "large predator" or "wild animal".[2]

In Germanic, however, the inherited term has been tabooed and replaced by a noa-word, namely PGmc. **beran-* m. "the brown one", which is reflected in OHG *bero*, etc. (see below).[3] This is an *n*-stem nominalisation of the adjective PGmc. **bera-* < PIE **bʰer-ó-* "brown"[4] which is continued

---

1 Cf. Wodtko et al. 2008, 343–345 s.v. **h₂ŕtḱo-* (with discussion and literature). However, there are no further *comparanda* (in the best of cases a verbal root as derivation base) so that we cannot trace the PIE "bear" word back to its creation and establish the original meaning of the formation. – PIE **h₂* was probably realised as a voiceless uvular (or velar) fricative, thus phonetically [χ] (or [x]).
2 This applies most probably to Hittite as well (Puhvel 1999, 201 s.v. *hartak(k)a-*; Kloekhorst 2008, s.v. *ḫartakka-*). The assumption that *ḫartakka-* (phonemic /hart(a)ka-/) denotes "wolf" or any "large predator" is no longer held.
3 Accordingly, the bear in the 12th century MHG beast epic *Reinhart Fuchs* by Heinrich (called *der glîchezære*) is named *Brûn* "Brown" corresponding to *Bruun* in Willem's MDu. version *Van de vos Reynaerde.* – ME *brune*, E *bruin* "bear" is a loan from MDu. *bruun, bruin* "brown".
4 This is the majority view. An alternative approach was put forward with some reserve by Seebold (1967, 115) and more confidently by Bammesberger (1990, 176 s.v. **ber-an-*) and Ringe (2017, 128), who connect PGmc. **beran-* with Gk. θήρ *thḗr* "(wild) animal", Lith. *žvėrìs* "wild animal", OCS *zvěrь* "(wild) animal" etc., < PIE **ǵʰueh₁r-*, cf. Lat. *ferus* "wild,

as Lith. *bė̃ras*, Latv. *bẽrs* "(reddish) brown".⁵ A related formation is OHG OE *brūn*, OIcel. *brúnn* "brown, dark-coloured", etc., which can be traced back to PGmc. **brūna-* (< PIE **bʰr-uII-no-*, cf. Gk. φρύνη, φρῦνος "toad").⁶ Two comparable cases of deadjectival formations with individualising *n*-suffix (these are, however, not noa-words) are OHG OS *haso*, OE *hare*, OIcel. *heri* m. "hare" (PGmc. **hasan-* ~ **hazan-*), literally "the grey one" (cf. OE *h(e)asu*, OIcel. *hǫss* "grey" < PGmc. **has-wa-*)⁷, and OHG *rōto* m. "red fish (char?)", literally "the red one" (cf. OHG *rōt* "red", etc.); both are derivatives from an adjective denoting colour. There is another animal name derived from PIE **bʰer-ó-* "brown", i.e. OHG *bibar, bibur*, OS *bibar, beber*, OE *beb(e)r, be(o)fer, be(o)for*, etc. m. "beaver" < PGmc. **bebru-*, which reflects reduplicated PIE **bʰe-bʰr-ú-* (> OI *babhrú-* "[reddish] brown; chestnut, sorrel", Lith. *bebrùs, bẽbrus* "beaver", etc.).⁸

A different taboo prevailed in the Slavic languages. The word for "bear" is OCS *medvědь*, Russ. медведь *medved'*, Cz. *medvěd* (and *nedvěd*), Sloven. *medved*, etc., which literally means "honey-eater".⁹ The name of the legendary hero in the eponymous OE epos *Bēowulf* is a comparable metaphor, if it is interpreted as "bee-wolf", which is the standard view.¹⁰ In the Baltic languages, which are related to the Slavic, the inherited word has also been supplanted, namely by OPruss. *clokis*, Lith. *lokỹs* and Latv. *lâcis*. However, these formations are etymologically unclear; presumably we are dealing with dysphemistic "lacerater, mauler".¹¹

Word taboo seeks to avoid anxiety or adversity that might arise when addressing somebody or something by its real name.¹² Thus speakers censor their language due to social necessities by replacing lexical items or phrases which they consider
- too sacred (E *[may] God be with you* → *goodbye*, F *par dieu* "by God" → *parbleu* "but of course!");
- too offensive, particularly relating to sexuality and excretion, nowadays also due to political correctness (OIcel. *lag* "laying, companionship" → "sexual intercourse", E *to go to the toilet* → *to powder one's nose*,¹³ E *red Indian* → AmE *Amerindian*, CanE *First Nations*);
- too fearful (G *sterben* "to die" → *den Löffel abgeben*, literally "to give up the spoon");
- too dangerous, either tabooing terms for mighty persons, creatures and supernatural beings (Icel. *djöfull* "devil" → *kölski*, literally "mocker, vituperator", PIE **h₂ŕ̥tḱo-* "bear" → PGmc. **beran-*

---

savage". This provides a decent etymology, but only if one accepts Seebold's sound change PIE **g/ǵʷʰ-* (and **g/ǵʰu̯-*) > PGmc. **b-* before vowels except *u*, a proposition which remains quite controversial. KROONEN (2013, 60 s.v. **beran-*) leaves etymological matters unresolved.

5 As for the Baltic words, there is no need to postulate PIE **bʰerH-o-* with a root-final laryngeal (as per DERKSEN 2015, s.v. *bė̃ras*) which would yield PGmc. **berr-a-* after loss of the laryngeal with compensatory lengthening of the preceding resonant (LÜHR 1976, 73–75; MÜLLER 2007, 88). PBalt. **bḗro-* can be explained easily as a so-called vr̥ddhi-derivative (characterised by lengthened vowel and, mostly, also by shifted accent); see LÜHR 2000, 194.
6 LLOYD et al. 1998, 374–377 s.v. *brūn, prūn*.
7 The PGmc. *n*-stem "hare" (**hásō* nom. sg., **haznéz* → **hazenez* gen. sg., etc.) compares to OI *śaśá-* "hare", Av. *saŋha-* "id.", etc. (PIE **ḱas-ó-*); cf. SCHAFFNER 2001, 545–546; WODTKO et al. 2008, 410–411 s.v. **ḱas-*.
8 MAYRHOFER 1996, 210 s.v. *babhrú-*; KÜMMEL 2004, 106–116. – PIE **bʰe-* is the reduplicated element, *-bʰr-* the zero grade of **bʰer-*.
9 VASMER 1955, 110 s.v. медвѣ́дь; KLOTZ 2017, 152 s.v. **medwē̂di*. For more details on Slavic "bear" words see UDOLPH, this volume.
10 See recently e.g. EITELMANN 2010, 180–181; HAUBRICHS 2017, 246. *Bēowulf* seems to be a byname (nickname) since such a metaphoric name is unique in Germanic heroic poetry and it does not alliterate with his father's name, *Ecgþēow*.
11 SMOCZYŃSKI 1999 (compares Lat. *lacer* "mangled, lacerated", etc.). For different views see FRAENKEL 1962, 384–385 s.v. *lokỹs* and DERKSEN 2015, s.v. *lokys* (with references).
12 Standard works on word taboo are MEILLET 1906 ("bear": 7–12), SAHLGREN 1918 ("bear": 4–5), HAVERS 1946 ("bear": 34–37) and EMENEAU 1948 ("bear": 56–63); see furthermore PANAGL 1984 ("bear": 148–149, 154), SCHRÖDER 2001 (research report: 231–238), ALLAN/BURRIDGE 2006 (political correctness: 90–111) and HOBERG 2019 (sexuality and political correctness), to name just a few. – On the concept of taboo in general, see SUNDQVIST 2005, 4–5.
13 Another example: Having read a German-language manuscript of mine, a helpful colleague objected to the hyphenation of *des Urin- | dogermanischen* ("of Proto-Indo-European"), for *Urin-* can be mistaken as "urine" in speed reading. Thus we decided to refrain from word division entirely (see NEDOMA 1995, 65 line 2 from below).

"the brown one", as explained above) or even distorting one's own name (Pre-OS/Pre-OHG *Būriso → **buirso**)[14].

Tabooed words can undergo deformation (E *hell* → mispronounced *heck*) or replacement (by means of euphemistic or dysphemistic noa-names as for "bear", see examples above). Terms for wild and dangerous animals like bears seem predestined for tabooistic substitution. The main reason was obviously fear that the beast which in an animistic view is able to understand human speech might be summoned and attack should its real name be uttered. *Wenn man den Wolf nennt, kommt er g'rennt* ("speak of the wolf, and it will come") is a southern German and Austrian proverb that reflects this idea. Another motive for using a noa-name is that hunters did not want to warn their prey which might then escape. Finally, religious motives may also have been an issue when the bear was regarded as sacred by prehistoric worshippers, and its real name was tabooed out of reverence.[15]

To return to linguistics: the early Proto-Germanic *n*-stem noun "bear" inflected as follows:

sg.	nom.	*berõ	pl.	nom.	*beranez
	gen.	*bernez → *berenez		gen.	*bernõ^n → *beranõ^n
	dat.	*bereni (loc.)		dat.	*berunm^a/$_i$z → *berum^a/$_i$z
	acc.	*beranu^n		acc.	*bernunz

OHG *bero* masc. (G *Bär*), MLG *bēr(e)*, *bāre*, MDu. *bēre* (Du. *beer*), OE *bera* (E *bear*), WFris. *bear*, NFris. *båår*, OIcel. *ber-* (first element in poetic compounds)[16] reflect the regular *n*-stem paradigm. In addition, there are the *Motionsfeminina* OHG *birin*, OE *byren* (*ber-injō-*), OIcel. *bera* (*ber-ōn-*) and OIcel. OSwed. *birna* (*ber-ijōn-*) "she-bear", respectively, as well as hypocoristic OIcel. *bersi* masc. (*ber-s-an-*) "(male) bear" with its by-form *bessi*.[17] Two case forms of the old paradigm, namely the dative pl. in *-um^a/$_i$z and the accusative pl. in *-unz, were formally identical with the corresponding *u*-stem endings. These pivots triggered a declension shift in North Germanic where the "bear" word was then inflected as an *u*-stem *bernu-* m., continued in OIcel. *bjǫrn* (gen. *bjarnar*), OSwed. *biorn*, *biørn*, etc.[18] There is debate as to whether OE *beorn* m. "warrior, hero, noble, prince", which is attested only in poetry, corresponds to OIcel. *bjǫrn* or to Lith. *bérnas* "boy, lad, farm servant", Latv. *bȩ̀rns* "child" (< PIE *bʰer-no-*).[19] In Gothic (and in the other East Germanic languages), "bear" is not attested.

There is a particular expression for "bear cub" in Old West Norse, *húnn* m. (figuratively also "boy"), that has no counterpart in any Old Germanic language. Moreover, Old Icelandic provides

---

14 By the use of anagrammatic **buirso**, the carver of the runic inscription on the Beuchte fibula (*c.* AD 560–585; SG-12) shelters from the buried female who is thought to be a revenant; see NEDOMA 1998, 44–45. The idea of gaining control over somebody through knowing his name – "I called you by name, you are mine", as said in the Old Testament (*Isaiah* c. 43,1) – is old and widespread.

15 On the tabooing of "bear" for reasons relating to anxiety, hunting or religion, see the literature cited above (note 12). – On bear cults among North Eurasian pagan cultures, see e.g. WAMERS 2009, 14–15 (with further references).

16 *Vǫlundarkviða* 10,1 (*Sat á berfialli* "[He] sat on a bear-skin") and *Atlakviða* 38,7 (*brœðr sína berharða* "[she never wept for] her brothers, bold like a bear"): *Edda*, 118, 246. Both lays have their origin in old German and/or English traditions; maybe these two *ber*-compounds are West Germanic poetic relics. – OIcel. *berserkr* m., denoting a frenzied warrior, is ambiguous: "bear-shirt (someone who wears a bear skin [in fight])", which is the majority view, or "bare-shirt (someone who wears a mere shirt, has no armour [in fight])"; cf. recently SAMSON 2020, 48–71 and also SUNDQVIST, this volume.

17 In modern Icelandic, *bangsi* (derived from *banga* "to beat, forge") is a common term of endearment for "bear", especially for "teddy bear".

18 This was already recognised by VAN HELTEN 1905, 225; see also LÜHR 1988, 200, 211.

19 Cf. LLOYD/SPRINGER 1988, 564 (with references). The metaphorical shift "bear" → "warrior, prince" on the one hand would have a parallel in OE *eofor* "boar" → OIcel. *jǫfurr* "warrior, prince", and "boy" → "warrior" on the other hand in OE *hyse* m. "young man" → "warrior". – Lith. *bérnas* "boy, lad, farm servant" and Latv. *bȩ̀rns* "child" (**"the carried one, burden") are derived from the root PIE *bʰer-* "to carry, bear"; cf. DERKSEN 2015, s.v. *bernas*; cf. WODTKO et al. 2008, 18.

lists of poetic "bear" synonyms which are to be found in *Snorra Edda*, namely in the *Skáldskaparmál* section and in the versified *Þulur*.[20] Leaving aside *bjǫrn*, *bersi*, *bera* and *húnn* mentioned above, the circumlocutions attested in *Snorra Edda* are either *hapax legomena* or restricted to skaldic poetry. It is not possible to determine whether they were used in everyday language, as were *Gullfot*, *Storgubben*, *Fulingen*, *Myrtaßen*, *Storfar*[21] in mid-18th-century Sweden.

"Bear" in Germanic personal names before c. AD 1000

Many of the stems that appear in Old Germanic[22] dithematic anthroponyms and (originally) derivated short forms belong to the spheres of reign, power, strength and warfare. Frequently used name elements refer to PGmc. **rīk-* m. "ruler", **aþala-* n./m.? "kin, noble descent", **þrūþijō-* f. "power, strength", **mǣra-* adj. "famous, glorious", **balþa-* ~ **balda-* adj. "bold, brave", **gunþijō-* f. "fight, battle", **heldijō-* f. "id.", **harja-* n. "army", **gaiza-* m. "spear" and **segaz/-ez-* n. "victory".[23] In addition, there is a group of name elements relating to strong and violent animals (see Müller 1970) that are also associated with heroic-martial concepts. An anonymous commentary on the Gospel of Matthew, *Opus imperfectum in Matthaeum*, which dates to the 5th or 6th century, mentions Germanic theriophoric anthroponyms and their martial background:[24]

> *Sicut solent et barbarae gentes nomina filiis imponere ad devastationem respicientia bestiarum ferarum vel rapacium volucrum, gloriosum putantes filios tales habere, ad bellum idoneos et insanientes in sanguinem.*

> And so the barbarian tribes [of the Danube area, thus presumably Germanic peoples] also use to give names to their sons according to the devastations of wild beasts or of rapacious birds, thinking it glorious that their sons have such names, suitable for war and raving in blood.

The most common name elements of this kind refer to PGmc. **wulfa-* m. "wolf", **ebura-* m. "boar", **aran-* ~ **arna-* ~ **arnu-* m. "eagle",[25] and **beran-* ~ **bernu-* m. "bear" as well. Of these, **wulfa-* is the most frequent. It seems that particular fighting strength and violence was ascribed to the wolf, as

---

20 *Snorra Edda*, *Skáldskaparmál*, cap. 64 [51], 72 [58] (ed. Finnur Jónsson [diplomatic], 157, 169 = ed. A. Faulkes [normalised], 75, 88) and *Þulur* IV cc (st. 510–511: *bjarnar heiti*; ed. Finnur Jónsson, 211 = ed. A. Faulkes, 132). – Curiously, *hlébarðr* "leopard" is also included in the third (poetical) list! More than a few circumlocutions in these poets' mnemonics are etymologically ambiguous or unclear, for example, *bolmr* (var. *blómr*) could be "sleeper, snorer", "fierce one" or "thick one"; cf. Magnússon 1989, 70 s.v. *bolm(u)r*. See also Lombardi; Ney, this volume.

21 Wallner 1746, 3 note [b]. *Myrtassen* (ß ß taken as *ss*) is difficult to explain ("the ant-paw", "the marsh-paw"?); the remaining items mean "gold-foot", "the big old man", "the ugly guy, rascal" and "grandfather". Well-motivated noa-names of this kind are also attested in many non-Indo-European languages; cf. for example, Peuckert 1927, 881–182; Knüppel 2017, 606–608.

22 According to LaN (I, ix–x), the term *Old Germanic* refers only to the period up to *c.* AD 600. Note that I deal in this article with names attested until AD 1000 or somewhat later.

23 Cf. LaN II, 595–601, 469, 633, 571–576, 476, 527–530, 542–545, 535–539, 513–514, 609–610. Early examples include: (1) Boian (ethnically Celtic!) *Aino-rix* 1st c. BC (Bratislava): LaN I, 17; (2) Ostrogoth. *Athala-ricus* 6th c.: LaN I, 81–83; (3) Pre-OHG (Franc.) *Þ"ruþ-hild* f., 6th c. (Friedberg): SG-35; (4) WGmc. (Cherusc.) *Segi-merus* 1st c. BC/AD: LaN I, 595; (5) and (6) WFranc. *Gunde-baldo* dat., 5th c.: LaN I, 395; (7) *Þ"ruþ-hild*: see example (3); (8) PGmc. *Hari-gasti* 3rd/2nd c. BC (Ženjak-Negau): Nedoma 1995, 51–56, 70–72; (9) Goth. *Rada-gaisus* 3rd/4th c.: LaN I, 546–547; (10) *Segi-merus*: see example (4).

24 *Opus imperfectum*, 626 (homily 1, ad Manasseh); cf. Haubrichs 2017, 244–245.

25 Early examples: 1. Visigoth. Ούλφίλας, *Ulfila(s)* 4th c.: LaN I, 795; 2. Visigoth. *Evervulfus* 5th c.: LaN I, 264; 3. WGmc. (Quad.) *Araharius* 4th c.: LaN I, 56. – On Old Germanic "eagle" names, see recently Nedoma 2018, 1590–1591.

confirmed by circumlocutions like OE *hilde-wulf* "fight-wolf: warrior"[26]. Interestingly, theophoric formations do not belong to the oldest layer of Germanic anthroponyms; they emerge only since the 4th century, following a general onomastic trend of late antiquity (cf. Lat. *Ursus* "bear", *Lupus* "wolf", *Leo* "lion", *Aper* "boar", *Aquila* "eagle" and derivatives).[27]

In dithematic personal names, "bear" occurs far more frequently (and in the Old Germanic period exclusively) in first components. According to the appellative *relata*, there is a well-stocked onomastic nest:[28]

- **Bera-* (the compound version of **beran-*): Ostrogoth. *Bere-mud, -mund* 5th c., Vandal. *Beremuda* f., 6th c., WFranc. *Bere-trudis* f., 6th c., OHG *Bere-heri* 9th c., Langob. *Per-prand* 8th c., OS (isolated) *Ber-mer* 9th c., Hisp.-Goth. *Ber-ildi* f., 9th c.; furthermore PN **bera** = *Ber-ǣ* 5th c. (= OHG *Bero*, etc., see below);[29]
- **Bernu-*: OHG *Bern-hild* f., 8th c., Langob. *Bern-ardus* 10th c., WFranc. *Berne-hildis* 9th c., OS *Bern-heri* 8th c., OE *Beorn-redus* 8th c., OWN *Bjǫrn-ulfr* 8th c. and *Bjarn-grímr* 11th c., OSwed. **biarnulfr** = *Bjarn-ulfr* 11th c.;[30]
- *Berin-* (based on gen. sg. **berenez > *beriniz* ≥ OHG *berin*[31] or exhibiting anaptyxis of *i*) and, less frequent, *Beran-* (with anaptyxis of *a*):[32] OHG *Berin-ger* 9th c., *Peran-ger* 9th c., Langob. *Berengarius* 9th c., WFranc. *Berin-gildis*, *Beran-suuind* f., 9th c.;[33]

26 Cf. *Genesis*, v. 2050–2051 (p. 62): *Hildewulfas herewīcum nēh / gefaren hæfdon* "The fight-wolves (warriors) had travelled near the army camps". Does the metaphor *hildewulfas* – either the Elamites or Abraham's troops are meant here – allude to the battle-hardened Vikings that invaded England from the late 8th century?

27 KAJANTO 1965, 329–330 (*Ursus, -a*, and derivatives), 327–328 (*Lupus, -a* etc.), 327 (*Leo, Lea* etc.), 325 (*Aper, Apra* etc.), 330 (*Aquila* etc.); OPEL IV, 186–188 (*Ursus, -a* etc.), III, 38–39 (*Lupus* etc.), III, 22–23, 21 (*Leo, Lea* etc.), ²I, 63, 68–69 (*Aper, Apra*, etc.), ²I, 70 (*Aquila*, etc.); cf. JOCHUM-GODGLÜCK 2011, 455–467; HAUBRICHS 2017, 231, 252–253. – *Ursus, -a*, and derivatives seem to be the most popular Latin theriophoric cognomina (KAJANTO 1965, 88).

28 Apart from a few remarkable formations, I exclude here monothematic formations such as OHG *Bero*, OWN *Bjǫrn*, OHG *Pirin* f. (created by zero-derivation, i.e. conversion; see below), OS *Bern-o* (created by derivation) and WFranc. *Ber-il-a* f., OWN *Bjar-k-i* (created by derivation and diminution).

29 Ostrogoth. *Beremud, -mund*: LaN I, 134; Vandal. *Beremuda* (Carthage): ENNABLI 1975, 215 no. 82; WFranc. *Beretrudis*: LaN I, 134; OHG *Bereheri*: FÖRSTEMANN 1900, 262; Langob. *Perprand*: BRUCKNER 1895, 233 (relating Langob. *Ber(e)-, Bern(e)-, Beren-* incorrectly to PGmc. **bera-* "to carry"); OS *Bermer*: SCHLAUG 1962, 58 (listed as "Bernmâr" by mistake); Hisp.-Goth. *Berildi*: PIEL/KREMER 1976, 100; PN *Berǣ* (Kragehul): RäF 28. – Goth. *Berig* (var. *Berich*, Jordanes; LaN I, 135), the name of a legendary king who led the Goths from Scandinavia to the southern Baltic Sea coast, is morphologically opaque (if indeed Germanic: **Ber-ig-a-?, *Ber-ik-an-?*). The person may be a fictitious one, for whom Jordanes uses the Germanic-looking name of the Hunnish noble Βέριχος *Bérichos* 5th c. (Priskos; LaN I, 135). – PN *Berǣ* has only a few and mostly late analogues: ON (?) *Biari* 8th c. (Saxo, *Gesta Danorum* VIII,2,6), ONorw. *Biere* 15th c. (LIND 1905–1915, 135) and OSwed. *Biæri* 15th c. (LUNDGREN et al. 1892–1934, 27); **biari** on the Rök stone (*c*. AD 800; SR-Ög 136) is dubious because of the numerous readings possible; cf. REICHERT 1998, 90–92; HOLMBERG et al. 2020, 29.

30 OHG *Pernhart*: FÖRSTEMANN 1900, 270; Langob. *Bernardus*: BRUCKNER 1895, 233; WFranc. *Bernehildis*: MORLET 1968, 53; OS *Bernheri*: SCHLAUG 1962, 58; OE *Beornredus*: SEARLE 1897, 102; OWN *Bjǫrnulfr*: LIND 1905–1915, 147–148; OIcel. *Bjarngrímr*: ibid., 135; OSwed. *Bjarnulfr* (Svista): SR-U 1103. – There is also Latinised *Bern-* in early England, e.g. *Bern-uuini* 7th c. (Bede, Hist. eccl.): STRÖM 1939, 9, 161. – OWN *Bjarn-* did not undergo *u*-umlaut either owing to early loss of the compound vowel *u* or was influenced by gen. *bjarnar* etc. (cf. WIDMARK 1991, 44, 51, 175; PETERSON 2007, 42); *Bjarn-* was the derivation base for the popular short-form *Bjarni* (LIND 1905–1915, 136–137) = OSwed. *Biærne* (LUNDGREN et al. 1892–1934, 27). – The runic sequence **b̦irnaʀ** gen. sg. on the Rök stone (SR-Ög 136; formerly read **airnaʀ**) may be explained either as a WGmc.-NGmc. hybrid anthroponym *Bern-aʀ* or as misspelled ON (Swed.) *Bjarnaʀ* with regular *a*-breaking *e > ja*.

31 The root vowel *e* should have been raised due to *i*-umlaut (PGmc. **berenez* gen. sg. > **beriniz* > late PGmc. †*biriniz*), though *e* was restored by intraparadigmatic levelling.

32 Cf. MÜLLER 1970, 13–14. Nonetheless SCHRAMM (1957, 151 sub 7.; 2013, 138 sub 7.), KAUFMANN (1965, 90; 1968, 57) and GEUENICH (1976, 141, 167) presume secondary vowel insertion ("Vokaleinschub") in general. However, anaptyctic vowels between *r* and *n* are very rare in OHG appellatives (BRAUNE/HEIDERMANNS 2018, 103 § 70 note 1). – Kaufmann's (supplementary?) assumption that *Berin-, Beran-* would exhibit a "Nebensilben-Ablaut" (KAUFMANN 1968, 58) is misleading.

33 OHG *Beringer, Peranger* (e.g. St. Gallen a. 819 *bis*): cf. FÖRSTEMANN 1900, 267–268; Langob. *Berengarius*: BRUCKNER 1895, 233; WFranc. *Beringildis, Beransuuind*: MORLET 1968, 53, 54.

- *-bera- (rare): OHG *Isin-per* 8th c., WFranc. *Lan-berus* 9th c.;[34]
- *-beran- (in exception to the old principle of avoiding *n*-stems as second elements): OHG *Sigi-pero* 8th c., Langob. *Siue-bero* 10th c., OS *Athel-bero* 10th c.;[35]
- *-bernu-: OHG *Egil-bern* 9th c., Langob. *Geri-bernus* 9th c., WFranc. *Hilde-bernus* 9th c., OE *Wig-beorn* 9th c., OWN *Arin-bjǫrn* 9th c.;[36]
- *-berinjō- f.: OHG *Hrod-pirin* 8th c., OS *Athal-birin* 9th c.;[37]
- *-berōn- f. (only West Norse): OWN *Hall-bera* 9th c.[38]

There is a clear preference for the first component *Bera-* in East Germanic dithematic names, and for *n*-formations (*Bernu-*) in Scandinavia, England and Saxony. In southern West Germanic (Old High German, West Franconian and Langobardic) anthroponyms, however, both variants are well represented, as well as *Berin-*, *Beran-* (for details, see MÜLLER 1970, 10–18). It seems that these onomastic variants were to some extent interchangeable; for example, *Athelbero* (transposed PGmc. *Aþala-berō*), palsgrave of Saxony († 982), was also known by *Bern* (*Bernuz*) and *Berno* (*Bernō*).[39]

As to monothematic formations, there are a number of personal names which are formally identical with the corresponding appellatives, e.g. OHG *Bero* 8th c., Langob. *Pero* 5th c. (?), WFranc. *Bero* 9th c., Visigoth. *Bera* 7th c., OE *Beorn* 8th c., and ON *Bjǫrn* 9th c., as well as OHG *Pirin* f., 8/9th c., OWN *Birna* 10th c. and OWN *Bersi* 10th c.[40] In most cases this was the result of conversion (also called zero-derivation). This is a process of word formation in which the morphology of a pre-existing item remains unaltered whereas word-class and referential semantics are changed, for example, appellative BEAR "bear" → proper noun BEAR denoting "person" and just connoting "being (like) a bear". Yet some of the aforementioned anthroponyms could equally emerge as original short-forms by truncating one component of a dithematic name without suffixation;[41] cf. OE *Leof-heah* "dear, beloved" + "tall, high-class, illustrious" → *Leof{}* (Latinised *Leofus*; curly brackets {} indicate deletion) 9th century.[42]

In many dithematic anthroponyms, "bear"-components are connected with name elements relating to strength and warfare. Examples already mentioned include WFranc. *Bere-trudis* f. (PGmc. *þrūþijō-* "strength, power"), OE *Beorn-heard* (*hardu-* "hard, severe, bold"), WFranc. *Hilde-bernus* (*heldijō-* "fight, battle"), OE *Wig-beorn* (*wīga-* "id."), OHG *Bere-heri* (*harja-* "army"), OHG *Ber-in-ger* (*gaiza-* "spear") and OHG *Sigi-pero* (*segaz/-ez-* "victory"). Bears are strong and powerful animals, and it seems that many compound names of this kind are morphologically and semantically

---

34 OHG *Isinper* (*Breves notitiae*): FÖRSTEMANN 1900, 974; WFranc. *Lanberus* (Reims): MORLET 1968, 156 ("*Landberus*" by mistake).
35 OHG *Athalbero*: FÖRSTEMANN 1900, 1320; Langob. *Siuebero*: BRUCKNER 1895, 305; OS *Athelbero*: SCHLAUG 1962, 48.
36 OHG *Egilbern*: FÖRSTEMANN 1900, 29; Langob. *Geribernus*: BRUCKNER 1895, 256; WFranc. *Hildebernus*: MORLET 1968, 129; OE *Wigbeorn*: SEARLE 1897, 487; OWN *Arinbjǫrn*: LIND 1905–1915, 31–32.
37 OHG *Hrodpirin*: FÖRSTEMANN 1900, 892; OS *Athalbirin*: SCHLAUG 1962, 48.
38 OWN *Hallbera*: LIND 1905–1915, 455–456.
39 SCHLAUG 1962, 48. – There are more instances of variability, e.g. OHG *Bernhart* = *Perenhart* 9th c. (Fulda): GEUENICH 1976, 167. Dithematic *Bernu-* + *harda-* "hard, severe, bold" became very popular in medieval Europe even before the lifetime of the famous Cistercian monk and abbot of Clairvaux; cf. OHG *Ber(e/a)nhart* 9th c. (FÖRSTEMANN 1900, 269: "unendlich häufig" ["endlessly common"]), Langob. *Ber(e)nardu(s)* 9th c. (BRUCKNER 1895, 233), WFranc. *Bern(eh)ardus* 9th c. (MORLET 1968, 53), OS *Bernhardus* 8th c. (SCHLAUG 1962, 57–58), OE *Beornheard* 8th c. (SEARLE 1897, 100). OWN *Bjarnharðr*, as far as I have been able to determine, is not attested until later times (LIND 1905–1915, 135).
40 OHG *Bero* (e.g. St. Gallen a. 752): cf. FÖRSTEMANN 1900, 260 ("sehr häufig" ["very common"]), Langob. *Pero* (only attested in Hist. Langob. cod. Goth., written 9th c.): BRUCKNER 1895, 233; WFranc. *Bero*: MORLET 1968, 53; Visigoth. *Bera*: LaN I, 134; OE *Beorn*: SEARLE 1897, 98–99; ON *Bjǫrn*: LIND 1905–1915, 143–147; PETERSON 2007, 44; OHG *Pirin* f. (e.g. Salzburg): FÖRSTEMANN 1900, 266; OWN *Birna*: LIND 1905–1915, 142; OWN *Bersi*: ibid., 132–133.
41 See NEDOMA 2015, 299, 303; 2018, 1587–1588.
42 SEARLE 1897, 328, 333. – The vast majority of unisegmental short-forms was however coined by reduction and derivation, cf. Alem.-Langob. *Droct-ulf* → *Droct{}-o* 6th c. (BRUCKNER 1895, 243).

motivated (G *Primärkombinationen* as per HÖFLER 1954, 33, *passim*), even if we cannot precisely determine the semantic relationship between the two components. For instance, WFranc. *Hilde-bernus* "fight" + "bear" may refer to ancient animal-warriors, i.e. fighters who identify with (and change mentally into) mighty beasts;[43] but does this express the wish that the bearer of the name be a "bear in fight" or a "fight-hardened bear (i.e. warrior)"? Of course, there is no great difference between these two possibilities. However, it is difficult to explain martial names such as OHG *Berin-ger* "bear" + "spear". In any case, there are secondary dithematic formations (G *Sekundärkombinationen* as per HÖFLER 1954, 34, *passim*) that are clearly demotivated in morphosemantic respects, e.g. OHG *Berefrid* "bear" + "peace, security" and WFranc. *Ber-lindis* "bear" + "gentle, mild".[44] In some of these anthroponyms, so-called variation names, one component of a parent's (mostly the father's) name is repeated.[45]

Finally, there are bitheriophoric anthroponyms, which are almost entirely restricted to males. Formations such as WFranc. *Ber-ulf* 6th c. or, conversely, OHG *Wolf-bero* 8th c.[46] may emerge as *Primärkombinationen* (apparently copulative compounds "bear and wolf" or "wolf and bear")[47] or *Sekundärkombinationen* (usually variation names).

As to details of theriophoric personal names, there are different views in onomastic research. According to SCHRAMM (1957), the core of Germanic anthroponyms reflect heroic diction, or more precisely, old – indeed even Indo-European – poetic metaphors for "hero, warrior". Theriophoric name elements would be based on the identification of fighting men with mighty beasts.[48] It seems that both suppositions are valid albeit not exclusively.[49] However, WERNER (1963) argued that animal symbols on weapons and jewellery (apotropaic "Heilsbilder") and theriophoric name elements refer to animals associated with gods, e.g. eagle – *Wōdanaz (OIcel. *Óðinn*). Thus (bi)theriophoric personal names could be regarded as quasi-theophoric names.[50] Werner's *ad hoc* suppositions were rightly criticised (MÜLLER 1968, 202–211; REICHERT 1992, 561–563; HAUBRICHS 2017, 245). In a more balanced manner, MÜLLER (1968; 1970) stated that ideas of animal-warriors as well as animal masking and religious concepts were relevant to naming. A theriophoric anthroponym would, if motivated, intend to address the bearer of the name as the animal in order that they become virtually

---

43 On animal-warriors, see e.g. SPEIDEL 2004 (bear-warriors: 34–40); 2005 (bear-warriors: 580–581); cf. MÜLLER 1970, 178–179, 194–195; HAUBRICHS 2017, 247; SUNDQVIST, this volume.

44 OHG *Berefrid*: FÖRSTEMANN 1900, 261; WFranc. *Berlindis*: MORLET 1968, 52.

45 Two examples: The first element varies in Alem. *Mede-richus* ("reward [?]" + "ruler"), father of *Agena-richus* ("?" + "ruler"), 4th c. (LaN I, 499 and 13); the second element varies in OS *Ail-bertus* ("?" + "bright"), father of *Ail-bern* ("?" + "bear") 10th c. (SCHLAUG 1962, 75). – For "meaningless" *Agina- and *Agila- (> OS *Ail-*) see NEDOMA 2004, 149–150. On *Primärkombinationen* (in every case characterised by intended meaning) and *Sekundärkombinationen* (in many cases concerned with marking genealogical relationship), see recently NEDOMA 2015, 295–299 (with references).

46 WFranc. *Berulf*: LaN I, 139; OHG *Wolfbero*: FÖRSTEMANN 1900, 1646. – Note that only "bear" + "wolf" is used frequently and widely: Ostrogoth. *Berevulfus* 6th c. (Voghera; LaN I, 134, cf. LaN II, 21), Hisp.-Goth. *Ber-, Uerulfus* 9th c. (PIEL/KREMER 1976, 101), OHG *Perolf* 9th c., *Bernulf* 8th c. (FÖRSTEMANN 1900, 266, 273), OS *Bernulf* 9th c. (SCHLAUG 1962, 59), OE *Beornwulf* 8th c. (SEARLE 1897, 103–104), OWN *Bjǫrnólfr* 8th c. (LIND 1905–1915, 147–148). In contrast, "wolf" + "bear" is far less common: OHG *Wolfbero, -bern* 9th c. (FÖRSTEMANN 1900, 1646), OE *Wulfbeorn* 11th c. (SEARLE 1897, 506).

47 Cf. SCHERER 1953, 12 ("Identitätskomposita"); MÜLLER 1970, 167–168; BECK 1986, 312; HAUBRICHS 2017, 251. Several name formations, e.g. OHG *Suan-olf* 10th c. "swan" + "wolf" (FÖRSTEMANN 1900, 1378), cannot be explained in this way, nevertheless this is a clear *Sekundärkombination*.

48 SCHRAMM 1957, 77–83 ("Der Mann als Tier"), 106–107; cf. also SCHRAMM 2013, 67–73, 121–123. This view is supported by ANDERSSON 2003, 592–593.

49 See BECK 1965, 96–98; MÜLLER 1970, 192–195. Yet numerous Germanic theriophoric names echo theriophoric formulae for "man, warrior" known from Old English poetry (e.g. *hilde-wulf* "fight-wolf: warrior"; see above, note 26). Furthermore there are no extant correspondences within Old Icelandic skaldic poetry which is full of *kenningar*. Hence it is doubtful that theriophoric warrior metaphors are common Germanic.

50 WERNER 1963, 379–383. Earlier, MEYER (1913, 145) assumed that name elements relating to "wolf" and "raven" would refer to *Wōdanaz, which is likewise arbitrary.

Sprachwissenschaft und germanische Sprachen. Innsbrucker Beiträge zur Sprachwissenschaft 113 (Innsbruck 2004) 105–117.

LaN I–II: H. Reichert, Lexikon der altgermanischen Namen. Thesaurus Palaeogermanicus 1. I: Text. II: R. Nedoma/H. Reichert, Register (Wien 1987–1990).

Lind 1905–1915: E. H. Lind, Norsk-isländska dopnamn ock fingerade namn från medeltiden (Uppsala, Leipzig 1905–1915).

Lloyd/Springer 1988: A. L. Lloyd/O. Springer, Etymologisches Wörterbuch des Althochdeutschen. I: -a–bezzisto (Göttingen, Zürich 1988).

Lloyd et al. 1998: A. L. Lloyd/R. Lühr/O. Springer, Etymologisches Wörterbuch des Althochdeutschen. II: bî–ezzo (Göttingen, Zürich 1998).

Lühr 1976: R. Lühr, Germanische Resonantengemination durch Laryngal. Münchener Studien zur Sprachwissenschaft 35, 1976, 73–92.

Lühr 1988: R. Lühr, Expressivität und Lautgesetz im Germanischen. Monographien zur Sprachwissenschaft 15 (Heidelberg 1988).

Lühr 2000: R. Lühr, Die Gedichte des Skalden Egill. Jenaer indogermanistische Textbearbeitung 1 (Dettelbach 2000).

Lundgren et al. 1892–1934: M. Lundgren/E. Brate/E. H. Lind, Svenska personnamn från medeltiden. Nyare bidrag till kännedom om de svenska landsmålen och svenskt folkliv 10,6–7 (Uppsala 1892–1934).

Magnússon 1989: Ásgeir Blöndal Magnússon, Íslensk orðsifjabók (Reykjavík 1989).

Mayrhofer 1996: M. Mayrhofer, Etymologisches Wörterbuch des Altindoarischen II (Heidelberg 1996).

Meillet 1906: A. Meillet, Quelques hypothèses sur des interdictions de vocabulaire dans les langues indo-européennes (Chartres without year [1906]).

Meyer 1913: R. K. Meyer, Theophore und theriophore Namen in der germanischen Mythologie. In: Actes du IVᵉ Congrès International d'Histoire des Religions, Leiden 1912 (Leiden 1913) 145.

Morlet 1968: M.-Th. Morlet, Les noms de personne sur le territoire de l'ancienne Gaule du VIᵉ au XIIᵉ siècle. I: Les noms issus du germanique continental et les créations gallo-germaniques (Paris 1968, repr. 1971).

Müller 1968: G. Müller, Germanische Tiersymbolik und Namengebung. Frühmittelalterliche Studien 2, 1968, 202–217.

Müller 1970: G. Müller, Studien zu den theriophoren Personennamen der Germanen. Niederdeutsche Studien 17 (Wien, Köln 1980).

Müller 2007: St. Müller, Zum Germanischen aus laryngaltheoretischer Sicht. Mit einer Einführung in die Grundlagen. Studia Linguistica Germanica 88 (Berlin, New York 2007).

Nedoma 1995: R. Nedoma, Die Inschrift auf dem Helm B von Negau. Möglichkeiten und Grenzen der Deutung norditalischer epigraphischer Denkmäler. Philologica Germanica 17 (Wien 1995).

Nedoma 1998: R. Nedoma, Zur Problematik der Deutung älterer Runeninschriften – kultisch, magisch oder profan? In: K. Düwel/S. Nowak (eds.), Runeninschriften als Quellen interdisziplinärer Forschung. Reallexikon der Germanischen Altertumskunde, Ergänzungsband 15 (Berlin, New York 1998) 24–54.

Nedoma 2004: R. Nedoma, Personennamen in südgermanischen Runeninschriften. Studien zur altgermanischen Namenkunde I,1.1 (Heidelberg 2004).

Nedoma 2015: R. Nedoma, Wege und Probleme der areal- und sozioonomastischen Auswertung von Personennamen in älteren Runeninschriften auf Fibeln. In: O. Grimm/A. Pesch (eds.), Archäologie und Runen. Fallstudien zu Inschriften im älteren Futhark. Schriften des Archäologischen Landesmuseums, Ergänzungsband 11 (Kiel, Hamburg 2015) 291–332.

Nedoma 2018: R. Nedoma, Germanic personal names before AD 1000 and their elements referring to birds of prey. With an emphasis upon the runic inscription in the eastern Swedish Vallentuna-Rickeby burial. In: K.-H. Gersmann/O. Grimm (eds.), Raptor and human – falconry and bird symbolism throughout the millennia on a global scale. Advanced studies on the archaeology and history of hunting 1 (Kiel, Hamburg 2018) 1583–1602.

OPEL I–IV: B. Lőrincz[/F. Redő], Onomasticon provinciarum Europae Latinarum ²I, II–IV (²2005, 1999–2002).

Panagl 1984: O. Panagl, Was man ungern ausspricht. Altes und Neues vom Sprachtabu. Jahrbuch Universität Salzburg 1981–1983, 1984, 147–158.

Peterson 2007: L. Peterson, Nordiskt runnamnslexikon (Uppsala ⁴2007).

Peuckert 1927: W.-E. Peuckert, "Bär (Tier)". In: Handwörterbuch des deutschen Aberglaubens I (Berlin, Leipzig 1927) 881–905.

Piel/Kremer 1976: J. M. Piel/D. Kremer, Hispanogotisches Namenbuch. Der Niederschlag des Westgotischen in den alten und heutigen Personen- und Ortsnamen der Iberischen Halbinsel (Heidelberg 1976).

Puhvel 1991: J. Puhvel, Hittite Etymological Dictionary. III: Words beginning with H. Trends in Linguistics, Documentation 5 (Berlin, New York 1991).

RäF [+ no.]: W. Krause/H. Jankuhn, Die Runeninschriften im älteren Futhark. I: Text. II: Tafeln. Abhandlungen der Akademie der Wissenschaften Göttingen, Philologisch-historische Klasse, 3. F., 65 (Göttingen 1966).

Reichert 1992: H. Reichert, Altgermanische Personennamen als Quellen der Religionsgeschichte. In: H. Beck/D. Ellmers/K. Schier (eds.), Germanische Religionsgeschichte. Quellen und Quellenprobleme. Reallexikon der Germanischen Altertumskunde, Ergänzungsband 5 (Berlin, New York 1992) 552–574.

Reichert 1998: H. Reichert, Runeninschriften als Quellen der Heldensagenforschung. In: K. Düwel/S. Nowak (eds.), Runeninschriften als Quellen interdisziplinärer Forschung. Reallexikon der Germanischen Altertumskunde, Ergänzungsband 15 (Berlin, New York 1998) 66–102.

RGA²: H. Beck et al. (eds.), Reallexikon der Germanischen Altertumskunde. ²I–XXXV and Register I–II (Berlin, New York [1968/] 1973–2008).

Ringe 2017: D. Ringe, From Proto-Indo-European to Proto-Germanic. A Linguistic History of English 1 (Oxford ²2017).

SAHLGREN 1918: J. SAHLGREN, Förbjudna namn. Namn och bygd 6, 1918, 1–40.

SAMSON 2020: V. SAMSON, Die Berserker. Die Tierkrieger des Nordens von der Vendel- bis zur Wikingerzeit. Reallexikon der Germanischen Altertumskunde, Ergänzungsband 121 (Berlin, Boston 2020).

SCHAFFNER 2001: St. SCHAFFNER, Das Vernersche Gesetz und der innerparadigmatische grammatische Wechsel des Urgermanischen im Nominalbereich. Innsbrucker Beiträge zur Sprachwissenschaft 103 (Innsbruck 2001).

SCHERER 1953: A. SCHERER, Zum Sinngehalt der germanischen Personennamen. Beiträge zur Namenforschung 4, 1953, 1–37.

SCHLAUG 1962: W. SCHLAUG, Die altsächsischen Personennamen vor dem Jahre 1000. Lunder germanistische Forschungen 34 (Lund, Kopenhagen 1962).

SCHRAMM 1957: G. SCHRAMM, Namenschatz und Dichtersprache. Studien zu den zweigliedrigen Personennamen der Germanen. Zeitschrift für vergleichende Sprachforschung, Ergänzungsheft 15 (Göttingen 1957).

SCHRAMM 2013: G. SCHRAMM, Zweigliedrige Personennamen der Germanen. Ein Bildetyp als gebrochener Wiederschein früher Heldenlieder. Reallexikon der Germanischen Altertumskunde, Ergänzungsband 82 (Berlin, Boston 2013).

SCHRÖDER 2001: H. SCHRÖDER, Sprachtabu und Euphemismen – Sprachwissenschaftliche Anmerkungen zu Stefan Schorchs „Euphemismen in der hebräischen Bibel". In: A. Häcki Buhofer/H. Burger/L. Gautier (eds.), Phraseologiae Amor. Aspekte europäischer Phraseologie. Festschrift für Gertrud Gréciano. Phraseologie und Parömiologie 8 (Hohengehren 2001) 229–246.

SEARLE 1897: W. G. SEARLE, Onomasticon Anglo-Saxonicum. A list of Anglo-Saxon proper names from the time of Beda to that of King John (Cambridge 1897).

SEEBOLD 1967: E. SEEBOLD, Die Vertretung idg. $g^{u}h$ im Germanischen. Zeitschrift für vergleichende Sprachforschung 81, 1967, 104–133.

SG- [+ no.]: K. DÜWEL/R. NEDOMA/S. OEHRL, Die südgermanischen Runeninschriften I–II. Reallexikon der Germanischen Altertumskunde, Ergänzungsband 119 (Berlin, Boston 2020).

SMOCZYŃSKI 1999: W. SMOCZYŃSKI, Lit. *lokỹs*, lat. *lacer* und gr. ἀπέληκα. In: E. Eggers/J. Becker/J. Udolph/D. Weber (eds.), Florilegium Linguisticum. Festschrift für W. P. Schmid (Frankfurt/Main etc. 1999) 425–435.

SPEIDEL 2004: M. P. SPEIDEL, Ancient Germanic Warriors. Warrior Styles from Trajan's Column to Icelandic Sagas (London, New York 2004).

SPEIDEL 2005: M. P. SPEIDEL, "Tierkrieger". In: Reallexikon der Germanischen Altertumskunde ²30 (Berlin, New York 2005) 579–581.

SR-Ög [+ no.]: E. BRATE, Östergötlands runinskrifter. Sveriges runinskrifter 2 (Stockholm 1911–1918).

SR-U [+ no.]: E. WESSÉN/S. B. F. JANSSON, Upplands runinskrifter I–IV. Sveriges runinskrifter 6–9 (Stockholm 1940–1957).

STRÖM 1939: H. STRÖM, Old English Personal Names in Bede's History. An etymological-phonological investigation. Lund Studies in Old English 8 (Lund 1939).

SUNDQVIST 2005: O. SUNDQVIST, "Tabu". In: Reallexikon der Germanischen Altertumskunde ²30 (Berlin, New York 2005) 249–260.

VASMER 1955: M. VASMER, Russisches etymologisches Wörterbuch II: L – Ssuda (Heidelberg 1955, repr. 1979).

WAGNER 2008: N. WAGNER, Zum Tier in den zweigliedrigen germanischen Personennamen. Beiträge zur Namenforschung N.F. 43, 2008, 397–406.

WALLNER 1746: M. E. WALLNER, Kolare Konsten Uti Swerige (Stockholm 1746).

WAMERS 2009: E. WAMERS, Von Bären und Männern. Berserker, Bärenkämpfer und Bärenführer im frühen Mittelalter. Zeitschrift für Archäologie des Mittelalters 37, 2009, 1–46.

WERNER 1963: J. WERNER, Tiergestaltige Heilsbilder und germanische Personennamen. Bemerkungen zu einer archäologisch-namenkundlichen Forschungsaufgabe. Deutsche Vierteljahrsschrift für Literaturwissenschaft und Geistesgeschichte 37, 1963, 377–383.

WIDMARK 1991: G. WIDMARK, Fornvästnordiska förleder i omljudsperspektiv. Acta Universitatis Upsaliensis, Studia philologiae Scandinavicae Upsaliensia 19 (Uppsala 1991).

WODTKO et al. 2008: D. S. WODTKO/B. IRSLINGER/C. SCHNEIDER, Nomina im Indogermanischen Lexikon (Heidelberg 2008).

Prof. Dr. Robert Nedoma
Universität Wien
Institut für Europäische und Vergleichende Sprach- und Literaturwissenschaft,
Abteilung Skandinavistik
Wien
Österreich
robert.nedoma@univie.ac.at

The Slavic word for "bear"

By Jürgen Udolph

Keywords: Bear, taboo word, Slavic names, place names, field names, personal names, family names

Abstract: The Slavic word for "bear" (Russian medvéd´, Polish niedźwiedź, etc.) means "honey eater" and is a taboo word. In Slavic names, it is well-documented. In the case of hydronyms, place and field names, the frequent appearance of the animal close to rivers and other places motivated the naming. Concerning Slavic family names, it is often about a man supposed to be a "bear" in life and battle. Here, the wish of a human being to be similar to the animal or the belief to be intrinsically linked to the animal is of importance. The bear is also common in personal and family names in other languages and cultures. In the German language alone, the basic word Bär "bear" is attested c. 15,000 times as a family name.

General remarks

The Indo-European languages had a mutual word for "bear" early on. It is still preserved in Latin *ursus*, Greek *árktos* and Old Indic *ŕkṣa-*. Even so, its original meaning cannot be determined.
In the Germanic, Baltic, and Slavic languages, this original word was apparently replaced by a taboo word through several periphrases (EWAhd I, Sp. 565). A taboo word springs from the speaker's dread to use the term for a certain topic, a certain sector, or – as it is in the case of the bear – the actually quite well-known word for an animal, which is then replaced by a different expression. Likewise, names of gods or demons were only to be uttered under certain circumstances. According to W. Havers (1946, 34), A. Meillet (1906) constituted all subsequent works on language taboo in the Indo-European language, since Meillet's classic paper starts with animal names in general and the name for the bear specifically. The Austrian linguist O. Panagl (1984) calls this name taboo the "Rumpelstiltskin-effect". H. Schröder (2001) gives a good overview of studies on language taboo, also of the time after the publication of W. Havers' paper. The belief in name magic even covers wild animals. One tried to avoid the perceived incantation of wild animals by calling them by their name, since people and their herds should not be endangered by them (see also Nedoma, this volume).

The terrifying and strong bear falls into this category. In the Germanic languages, the original word inherited from prehistoric Indo-European times was replaced by a term with the meaning "brown", namely an original adjective **beran-*, to a basis **bher-* "shiny, (light)brown" (Old High German *bero, pero*, Middle High German *ber*, Modern High German *Bär*, an old *-n*-stem); besides, in the Nordic languages a *-u*-stem *björn* "bear" exists, derived from **bernu-*. Here, the bear is actually "the brown one", and, correspondingly, in the Baltic languages, Lithuanian *béras* and Latvian *bęrs* "brown". Additionally, the word for "beaver" belongs here, too. The Old Icelandic language knows another word for bear: *bolmr* "the big, fat one", even though this term is only documented twice and poetically. According to a Swedish account from the year 1746 by one M. E. Wallner, the bear should not be called by its real name. Instead, he should for example be called "gold foot" (Swedish *Gullfot*) or "the great old man" (Swedish *Storgubben*; cf. SUNDQVIST 2005, 256).

MEILLET (1906) had already pointed out Estonian, Finnish, and Lappish (Sami) substitute names for the bear such as "forest fame", "the old one", "the mighty paw of honey", "the ragged one", "the wide foot", or "termite devourer". Other taboo names for the bear exist among the Astrakhan-Russians, on Kolim, among the Hutsuls and the Caucasians, as HAVERS (1946, 35, comment 3) states. He also indicates that the bear is worshipped as a God among several peoples and Finno-Ugric peoples in particular, and one should for instance not say: "We killed the bear", but rather: "The holy animal died".

In the Baltic languages, instead of the original term a word group exists as a substitute. It consists of Lithuanian *lokỹs*, Latvian *lâcis* and Old Prussian *clokis*. Its etymology and original form are controversial, though. A connection to the Slavic word for "animal hair, fur" is most likely. As a consequence, this would mean that the original "bear"-word was also replaced in this language family. This is supported by another word in Lithuanian, because Lithuanian *meškà* "bear" was loaned from the Slavic language, which I will deal with hereafter.

In the Slavic languages, the original word was replaced by a connection between *med* "honey" (related to the Germanic word *Met* "mead, honey wine") + **ed-* "eat" (this word, too, has a German equivalent in *essen*); here, the bear is the "honey-eater". Thus, the Old Slavs paraphrased the dangerous animal to not send for it by calling it by its real name.

In the Slavic languages, the word appears as follows: Russian *medvéd'*, Ukranian *médvid'*, *védmid'*, Church Slavonic *medvěd'*, Bulgarian *medvě'd*, Serbian, Croatian *mèdvjed*, Slovenian *medved*, Czech *medvěd*, *nedvěd*, Polish *niedźwiedź*, dialect *miedźwiedź* (the *n-* in the prefix was created through remote assimilation to *d* and dependence on *ne-* "no, not"), Upper Sorbian *mjedwjédź*, Lower Sorbian *mjadwjeź* (VASMER 1958, 110–111), with reference to further literature).

Related creations exist in Old Indic (*madhuvád-* "eating sweets") and in the Lithuanian language (*mės-ėdis* "meat eater"). In Sorbian, the colloquially loaned *bar*, *barica* "bear" from German is now used (SCHUSTER-ŠEWC 1981–1984, 919–920). The bear's gusto for honey was also the basis for the disguising paraphrase *Beowulf* which is to be understood as "bee wolf" (MÜLLER 1970, 211).

Apart from the widespread "honey eater" word, there is one more term for the bear which is also fostered by taboo – *meška*, *mečka* ("bear") which today is mainly distributed among the southern Slavs. In the past, this term was also known among the east and west Slavs, which is supported by a loan to the Baltic (Lithuanian *meškà* "bear").

THE BEAR IN SLAVIC NAMES

Introduction

It has been known for a long time that there are several geographic and personal names in which the name of the bear lives on. Thus, H. REICHSTEIN (1976, 46) justifiably states: "Die Häufigkeit des Braunbären in älteren Zeiten, seine Beliebtheit als Jagdobjekt und seine Stellung als König der Wälder spiegeln sich in zahlreichen Orts- und Personennamen wider" ("The frequency of the brown bear in older times, its popularity as an object of hunting and its position as the king of the forests is reflected in several place and personal names", author's translation).

This is also true for the large geographical area which the Slavic languages cover today (and, to some extent, areas covered earlier, e.g. in today's eastern Germany, and parts of the southern Balkans). To begin with, the basis of the detailed list of respective names in the following is a profound collection of geographical names. Since 1970, I have been noting down these geographical names – in those days without a PC, electronic data processing, or internet. Today, they have been electronically recorded. All data – which cover *c.* 450,000 co-entries – are accessible through the website of the Academy of Sciences in Göttingen (https://adw-verwaltung.uni-goettingen.de/ortsnamen/images_lightbox.php), labelled *Nomina Geographica Europaea*. The sources of most of the family and place names, hydronyms, and field names stated below can be found there. Apart from that, I have analysed large collections of place names and hydronyms, such as given in M. Vasmers *Russisches geographisches Namenbuch* (Vasmer 1968–1989), the Polish *Słownik Geograficzny* (Słownik Geograficzny 1880–1902), and others, again. These collections contain hundreds of geographical names which I will not cite in detail here.

The striking occurrence of bears in the respective areas must be the reason for the development of most of the following names. I will not comment further on the formation of names by adding suffixes which is the preferred method among Slavic languages. The collection of names speaks for itself, also in this respect.

Hydronyms
Medveđa in former Jugoslavia; *Medvjednica* in the region around the Sava River (Croatia); *Medved'* in Ukraine; *Medvedevo* in the region around the lower Desna (Ukraine); *Medvedka* in the region around the Seym (western Russia, northeastern Ukraine); *Medwedżiv* in the region around the San (southeastern Poland); *Medveja*, place name and hydronym in the Bukovina (northern Ukraine, southern Romania); *Medvežyj*, *Medvidka*, hydronyms in Ukraine; *Medvedica* in the region of the former Tver Governorate, *Medvedka*, several times in middle and northern Russia.

Place and field names
When examining the following names, one cannot be certain if one is dealing with a place name (in terms of a settlement name) or a field name (mostly to be understood as names for unsettled areas). Thus, I have not always made a separation.

Medvedce, Medwedicze, Medwedowce, Medwedówka, Medvedza, Medwedzek, Medvedzi, Medweża, place and field names in southeast Poland (partly in former Galicia); *Medvedev*, about 15 place names in the eastern Slavic settlement area; *Medvedija, Medvedivka, Medvedok, Medvedova, Medvedyč, Medvedža*, in Ukraine; *Medved, Medvěd'*, field names in the whole of the Slavic world, such as e.g. in Czechia and in Slovakia (here also as *Medveďová, Medvedzie, Meďvetka, Medvetka*), or in Bulgaria, also as *Medvědí, Medvědice*; the Czech annalist Cosmas (*c.* 1045–1125) already mentions a name for a mountain *Medvěz*; furthermore, *Medvež, Medvjednica*, field names in Croatia; *Medveja, Medviđa, Medviđak, Medvidina, Medvija, Medvijak, Medvije, Medviš*, field and place names at the Adriatic Sea and in Croatia; as *Medvedac, Medvedište* in Serbia, there also as a field name *Medvednjak*; *Medvedjek, Medvednjek, Medvedova*, several field names in the Julian Alps (Slovenia); *Medvednoe*, field name in Belarus; *Medveajde, Medvež´e, Medvjažd* in Bulgaria; *Medvedja, Medvěžda*, Slavic place names in Greece (Vasmer 1958). Names can also be detected in formerly Slavic areas: *Groß* and *Klein Medewege* close to Schwerin, Germany, 1186 *Medewede,* 1285 *maior Medeuuede*; as Slavic relicts also in Austria (*Medvedina, Medvejak*), and in Hungary, for example as *Medwednyk* near Veröce, and as a field name *Medvigye* in the comitatus Scabolcs.

The allocation of *Medewitz* and *Medewitzsch* near Bischofswerda, near Bautzen, Germany, and in Pomerania (Poland) is not certain, because here, a derivation from the Slavic word for honey, *med*, might also be possible.

As noted above, there has been a change from the initial sound *m-* (*medved*) to *n-* (Czech *nedvěd*, Polish *niedźwiedź*) in the Slavic word for bear. Consequently, especially geographical names with *N-* as initial sound are to be expected in Czechia, Poland, and Slovakia. In the above noted collection of names from central Europe, I found the following including others: *Nedvěd, Nedveka*, hydronyms in Czechia; *Nesdvědička*, a tributary to the Svratka in the southern Moravian region. Close to it, the town *Nedvědice* is situated; *Nedweis*, German variety of place names near Olomouc, Czechia, Czech *Nedvězí*; *Nedvěk*, field name near Pilsen; *Nedvězí*, several field names in Czechia, and in Slovakia.

Due to Polish settlements even in Ukraine, traces can also be found in that country (*Nedwedyś*, group of houses close to Lviv). The farm name (a specific name given to a farm) *Nedved* in Carinthia, 1348 *Medbeg*, is probably also to be classified here. It is not always determinable for sure if a name is derived directly from the word bear or from the name of a person which in turn contained the word. Thus, it is more likely that the Czech place names *Nedvědič*, 1209 (fabrication, 13th century) *de Medwediz*, 1251 *Medeweditsch* were derived from the respective name of a person *Medvědík*. The transition from *M-* to *N-* in the initial sound is clearly visible here.

Several geographical names are to be found in Poland as well: *Niedźwiad*, several place names near Posen, Gnesen and in southeast Poland; *Niedźwiada*, hydronym in the area around the River Oder, also a place name near Lublin; *Niedźwiadek*, field name near Gdansk and near Posen; *Niedzwiadka*, *Niedźwiadka*, hydronyms and field names in Galicia; *Niedźwiadna*, place name near Suwałki and in Silesia; *Niedźwiady*, *Niedźwiada*, *Niedźwiedź*, several place and field names in Poland; *Niedźwieck*, place name in southern Poland; *Niedźwiednia*, place names in Galicia; *Niedźwiedziak*, place names near Gniezno and near Gdansk; *Niedźwiedza*, several place names in southern Poland; *Niedźwiedzianka*, hydronym near Szczecin; *Niedzwiedzica*, several place and field names in different regions in Poland.

Personal and family names
"Der Mann als Tier" ("Man as an animal", author's translation) – this expression can be found in the works of G. SCHRAMM (1957) as well as those of K. KUNZE (2003) and other authors. G. MÜLLER (1970, 178–195) deals with this topic in a subchapter, "The animal warrior".

Animal denotations in personal and family names are common in Indo-European languages; such names are called theriophoric. The most common animal is the *wolf*, besides that also the *lion*. In the Germanic languages, the *bear* is attested in large numbers. According to MÜLLER (1970, 10) "können nach den Wolf-Namen [...] die Bär-Namen [...] als die bedeutendste Gruppe theriophorer Anthroponymica im Germanischen gelten" ("Next to names relating to the wolf, the most important group of theriophoric anthroponyms in the Germanic languages contains names attributable to the bear", author's translation).

The Old English language proves how narrow the conception of "man as an animal" was in the case of the bear in the Germanic languages, since Old English *beorn* means "warrior". Without any doubt, this word is etymologically identical with Old Nordic *bjǫrn* "bear".

Among the first names, one comes across several animals – eagle, bear, buck, boar, falcon (hawk), deer, dog, crow, marten, raven, horse, swan, bull, ram, wolf, snake, dragon. Many of these animals are strong and aggressive, etc.; thus, such names are probably to be seen in connection with the idea of man as a warrior (KUNZE 2003, 25).

The most comprehensive paper about animal denotations was presented by MÜLLER (1970, 179). Here, he produces an extensive list of names that show "wie produktiv jene Komposita waren, die den Mann als 'Wolf' oder 'Bär' des Kampfes und der 'Waffen' bezeichneten" ("[...] how productive these compounds were when they denoted man as 'wolf' or 'bear' of battle and of 'weapons'", author's translation). In detail, he attends to the question why animal denotations were used in personal names. It is mainly "der Wunsch des Menschen, diesem Tier ähnlich zu werden, oder der Glaube, mit ihm wirklich wesenhaft verbunden zu sein" ("the wish of the human being to resemble the animal, or the genuine belief to be intrinsically linked to it"; MÜLLER 1970, 201, author's translation). In addition, there is "die Bewunderung der Kraft des Bären und der Glaube an sein übernatürliches Wesen" ("the admiration of the power of the bear and the belief in its supernatural character"; MÜLLER 1970,

201, author's translation). At the same time, the bear was seen as a fighting animal, just as the wolf, boar, and others, which man wanted to resemble to be able to lead a better life – with the help of his name. MÜLLER (1970, 211) summarises: "Die theriophoren Anthroponymica sollten vor allem ihre Träger selbst als Tiere benennen" ("The bearers of the theriophoric anthroponyms should be identified with the animals themselves", author's translation).

In accordance with these explanations, it is not surprising that the bear plays an important role in Slavic family names, whereas there is no safe evidence for first names. Unfortunately, it is only possible to present more precise details about the family names in some of the western Slavic countries, since the eastern and southern Slavic territories do not provide sufficient material. Even telephone books on CD-ROM or in similar form are rarely found. To begin with, I present the well-investigated countries of Czechia and Poland (since the data basis is also missing in Slovakia).

In Czechia, almost 2,000 surnames can be attested, more specifically: *Nedvěd* – 1,662, *Nedvědický* – 77, and *Nedvídek* – 281. Still, names with *M-* as the initial do also exist. Examples are *Medvec* – 52, *Medvecký* – 38, *Medveď* – 84, *Medvěd* – 48, mainly in Moravia and partly in northern Bohemia. To some extent, dialect words are of importance. In Moravia, the Slovakian language is very close, and here, the transition from *m-* to *n-* in the initial sound is missing, since the bear is called *medveď* in Moravia.

In Poland, even more surnames could be recorded, but this is due to the excellent data basis. The following remarks are based on a profound collection of Polish surnames, which is unique on a global scale. It contains the names of approximately 38.5 million inhabitants and is based on the documents of a state insurance company in Poland, in which almost every citizen of Poland was registered until the time of the German reunification in 1990 (published by K. RYMUT, first in book form [1992–1994], after that also on CD-ROM [2003]).

The collection contains thousands of names; not all of them can be listed here. The number of names beginning with *Nedv-* and *Nedwid-*, respectively, is small (*Nedved* – 9, *Nedvidek* – 1, *Nedwed* – 18, *Nedwidek* – 45). Here, German administrators at the time influenced the originally Polish spelling.

With regard to the names with *Niedzw-* and *Niedźw-* as initial sounds, I only mention: *Niedzwiadek* – 14, *Niedzwiecka* – 478, *Niedzwiecki* – 495, *Niedzwiedzka* – 166, *Niedźwiadek* – 842, *Niedźwiecka* – 3,602, *Niedźwiecki* – 3,364, *Niedźwiedzińska* – 159, *Niedźwiedziński* – 159, *Niedźwiedziuk* – 109, *Niedźwiedzka* – 1,350, *Niedźwiedzki* – 1,245, *Niedźwiedź* – 4,046.

Besides, there are also names with *M-* as initial sound, such as – among others – *Medwecka, Medwed, Medwedew*. These are not very common. One will probably come across them in the east of the country, and for the most part, these contain the Eastern Slavic form of the bear-word in which the initial *m-* has not turned into *n-*. These include *Medvid, Medwedczuk, Medwedecki*, and yet others.

Due to the events at the end of World War II, several of the listed Polish and Czech surnames exist in the border region between Germany, Poland and former Czechoslovakia. As a result of flight, forced migration, and relocation, there are Poles, Czechs, and Slovaks of German descent bearing names of Slavic origin.

Some of the names below also come from the Soviet Union and the southern Slavic Balkan states. Due to the fact that the basis for an examination is very good – because of the computerised versions of telephone book-CDs – the following names can be attributed to the Slavic bear-word (in parentheses, the number of names that come from a CD with family names from the year 2002). To begin with, the names with *M-* as initial sound: *Medvecky* (5), *Medved* (98), *Medvedenko* (4), *Medvedev* (46), *Medvedeva* (19), *Medvedik* (4), *Medvedovskyy* (3), *Medvidovic* (17).

With initial *N-*, but probably derived from Czech: *Nedved* (25), *Nedvidek* (13).

The largest group with initial *Niedz-* which is derived from Polish includes several hundred surnames: *Niedzwecki* (16), *Niedzwedzki* (12), *Niedzwetzki* (152), *Niedzwetzky* (3), *Niedzwiadek* (5),

Niedzwicki (10), *Niedzwiecka* (9), *Niedzwiecki* (26), *Niedzwiedz* (78), *Niedzwiedzinski* (6), *Niedzwiedzki* (12), *Niedzwietz* (10), *Niedzwietzki* (3), *Niedzwitz* (8).

This collection makes it clear that the bear has played an important role in the Slavic languages. As a side note: The basic surname *Bär* can be attested 15,000 times in the German language. In this respect, the bear proportions in the Slavic countries are comparable to those in the German language.

Bibliography

EWAhd I: Etymologisches Wörterbuch des Althochdeutschen. Band I: *-a - bezzisto* (Göttingen, Zürich 1988).

Havers 1946: W. Havers, Neuere Literatur zum Sprachtabu (Wien 1946).

Kunze 2003: K. Kunze, dtv-Atlas Namenkunde – Vor- und Familiennamen im deutschen Sprachgebiet (München ⁴2003).

Meillet 1906: A. Meillet, Quelques hypothèses sur des interdictions de vocabulaire dans les langues indoeuropéennes (Chartres 1906).

Müller 1970: G. Müller, Studien zu den theriophoren Personennamen der Germanen (Köln, Wien 1970).

Panagl 1984: O. Panagl, Was man ungern ausspricht. Altes und Neues vom Sprachtabu. Jahrbuch der Universität Salzburg 1981–1983 (1984), 147–158.

Reichstein 1976: H. Reichstein, „Bär. Zoologisches. Kulturhistorisches". In: Reallexikon der Germanischen Altertumskunde 2 (Berlin, New York ²1976) 45–46.

Rymut 1992–1994: K. Rymut, Słownik nazwisk współcześnie w Polsce używanych 1–10 (Kraków 1992–1994).

Rymut 2003: K. Rymut, Słownik nazwisk używanych w Polsce na początku XXI wieku (Kraków 2003).

Schramm 1957: G. Schramm, Namenschatz und Dichtersprache. Studien zu den zweigliedrigen Personennamen der Germanen (Göttingen 1957).

Schröder 2001: K. Schröder, Sprachtabu und Euphemismen – Sprachwissenschaftliche Anmerkungen zu Stefan Schorchs „Euphemismen in der hebräischen Bibel". In: A. Häcki Buhofer/H. Burger/L. Gautier (eds.), Phraseologiae Amor – Festschrift für Gertrud Gréciano (Hohengehren 2001) 229–246.

Schuster-Šewc 1981–1984: H. Schuster-Šewc, Historisch-etymologisches Wörterbuch der ober- und niedersorbischen Sprache 2 (Bautzen 1981–1984).

Słownik Geograficzny 1880–1902: Słownik geograficzny Królestwa Polskiego jako źródło do badań rozmieszczenia sił wytwórczych kapitalizmu w Polsce 1–14 (Warszawa 1880–1902).

Sundqvist 2005: O. Sundqvist, Tabu. In: Reallexikon der Germanischen Altertumskunde 30 (Berlin, New York ²2005) 249–260.

Vasmer 1958: M. Vasmer, Russisches etymologisches Wörterbuch 2 (Heidelberg 1958).

Vasmer 1968–1989: M. Vasmer, Russisches geographisches Namenbuch 1–11 (Wiesbaden 1968–1989).

Prof. Dr. Jürgen Udolph
Zentrum für Namenforschung
Rosdorf
Germany
Udolph@t-online.de

Chapter 10

Bears in image science (northern Europe)

The earliest iconic bear imagery comes from the rock art in Alta, northern Norway, acknowledged as a World Heritage site by UNESCO. There are many bear depictions known from Alta, some of which are woven into narrative cycles, with a dating to the late Mesolithic and the early younger Stone Age. One particular case is the hunter with a spear who waits for the bear to come out of the den. Next to the hunter there is a person without a weapon; who is this person? (see Grimm, Summary, this volume; graphics K. Tansem, World Heritage Rock Art Centre, Alta, Norway).

Stone Age amber bear figurines from the Baltic Sea area

By Daniel Groß and Peter Vang Petersen

Keywords: Amber, figurative art, stray finds, dating problems, animism, Palaeolithic, Neolithic

Abstract: Amber bear figurines are a small group of objects found all around the Baltic Sea. They are usually naturalistically shaped and come in a variety of forms and wealth of detail. A major issue when dealing with this topic is the fact that the figurines are mostly stray finds, so they lack archaeological context. Furthermore, they are not directly datable due to their material. This makes it difficult to contrast them against their archaeo-cultural background. This paper will therefore discuss the dates that have been assigned to these figurines and the reasons why bears were depicted. As it turns out, there are less indications for a Mesolithic date than for an earlier or later phase, based on comparable art and styles. This also aligns well with other naturalistic amber finds from the Palaeolithic and Neolithic, and hence questions the traditional attribution of the amber bear figurines to the Mesolithic. The function of these artefacts, however, remains unclear due to the lack of contextual information.

Introduction

Ever since humans resettled the areas around the modern Baltic Sea in the Late Glacial, bears have been an element of their natural environment. Having already been represented in cave paintings during the Upper Palaeolithic in French caves, such as Grotte Chauvet, and by the headless clay sculpture in the Grotte de Montespan (France), in later times representations of bears, made of amber (and other materials), apparently "became mobile". The phenomenon of amber bear figurines is so far little understood, as they are not regular finds and are very limited in numbers. However, they are not a single and unprecedented phenomenon. Ceramic bear figurines were already manufactured at the Upper Palaeolithic site of Dolní Věstonice (*c.* 29,000–25,000 cal BP; Vandiver et al. 1989).

The bear figurines in the focus of this contribution are exclusively manufactured of amber and thus made of collected raw material. Amber in Europe is mainly distributed in the coastal regions of the southern Baltic where it is often found on the shore, since it has been redeposited from Eocene glauconitic sediments by fluvial and glacial processes.

As amber is a material that, similar to other organic substances, deteriorates rather quickly when exposed, only low numbers of such finds are known from the Stone Age around the Baltic Sea – even when compared with finds made of other organic materials (e.g. bone, antler, and wood). Figuratively bear-shaped amber objects are even scarcer, so only 15 specimens are known – including one modern fake and five ambiguous finds –, six from the territories of modern Denmark, four from Latvia, two from Poland, and one from Norway, Estonia, and Finland, respectively (Fig. 1). The discussed find group is extremely small and thus not very well suited for comparative analysis or general conclusions, but traditionally the figurines are attributed to the Mesolithic (*c.* 11,600–6500

its bridged legs might represent a genuine Palaeolithic ornament, which was washed up during the Neolithic after lying in an inundated Palaeolithic site for thousands of years and then found by amber collectors and exported to Norway together with the four perforated pendants.

Słupsk/Stolp, voivodeship Pomerania, Poland (N 54.454°; E 17.029°)
A bear figurine[2] (Fig. 8) was found in 1886 or 1887 in peat, close to the city of Słupsk (cf. VIRCHOW 1887, 401). In keeping with the other finds, this one also lacks more information on the find situation and a reliable dating. It has repeatedly been discussed whether the figurine was remodelled, but TERBERGER/ANSORGE (2000) concluded that it is in its original condition, with only a polishing of the surface having been applied. The figurine has a natural hole in the distal part of the body and detailed facial attributes. The reinvestigation by TERBERGER/ANSORGE (2000) showed that the facial details are original, thus rejecting earlier assumptions that they might have been modern additions.

Brześć Kujawski, voivodeship Kujawia-Pomerania, Poland (N 52.603°; E 18.899°)
In 1976, a small amber figurine was excavated at the Neolithic site of Brześć Kujawski, site 3 (*c.* 5000 cal BP; CYREK et al. 1986, 121–122). Its shape cannot clearly be identified as a bear but generally it appears to be zoomorphic (Fig. 9.1).

Zvidze, Madona, Latvia (N 56.852°; E 26.910°)
A small amber bear (?) figurine was found at the Middle Neolithic settlement of Zvidze in Latvia (*c.* 5500–4300 cal BP; LOZE 2000, 75; 2003, fig. 3.2). It is carved out of a flat piece of amber and has a perforation in its back (Fig. 10.3). This hole was probably used for suspending it as a pendant in a similar way to other amber pendants from the site, which depict birds. According to LOZE (2003, 82), another amber find from the site depicts the frontal part of a bear (Fig. 10.2).

Sārnate, Ventspils, Latvia (N 57.110°; E 21.429°)
At the Neolithic settlement of Sārnate (*c.* 5900–5000 cal BP) in Latvia, an amber figurine was found in one of the dwellings (BĒRZIŅŠ 2008, 365). Only the rear part is preserved (Fig. 11) but, due to the shape of the back, and after comparison with the Słupsk/Stolp specimen, the find was interpreted as a bear (VANKINA 1970, 111). Its association with several amber beads and pottery dates the assemblage to the Neolithic Comb Ware group (BĒRZIŅŠ 2008, 407–408).

Suļka, Rēzekne, Latvia (N 56.713°; E 26.785°)
A figurine of a furry animal has been reported from the Comb Ware and Pitted Ware settlement of Suļka (LE-752: 4060 ± 60 bp; 5580–5150 cal BP) in Latvia (GIMBUTAS 1985, 234; LOZE 2003, fig. 2.2; OTS 2006, 50). However, it is unclear whether the figurine actually depicts a bear (Fig. 10.1).

Tamula, Võru, Estonia (N 57.844°; E 26.981°)
At the Estonian cemetery of Tamula (*c.* 5900–2600 cal BP; TÕRV 2018, 138–140), a bear-shaped pendant (Fig. 12) was found in a grave (burial XII; cf. GIMBUTAS 1985, endnote 5; OTS 2003, fig. 3.1; 2006, 32, 42). Similar to the specimen from Zvidse, it is manufactured from a flat piece of amber and perforated in its back. On its bottom part, small incisions have been identified (OTS 2003, 102).

---

2   GIMBUTAS (1956, plate 32) confused the find location of this figurine with an amber boar figurine from Gdansk (VIRCHOW 1884; cf. TERBERGER/ANSORGE 2000, 345–346; TERBERGER 2003). Due to the anatomical details, TERBERGER/ANSORGE (2000, endnote 15) agree with the interpretation as a wild boar or pig (Fig. 9.2). The find is therefore not further discussed here.

Astuvansalmi, Ristiina, in Mikkeli, Finland (N 61.420°; E 27.523°)
A piece of amber, interpreted as part of a bear (?) figurine (Fig. 13), is reported from Astuvansalmi (LAHELMA 2008a, fig. 10). It was found in the water in front of rock paintings at the site, depicting humans and other animals, including a possible bear (LAHELMA 2008b, 27; HELSKOG 2012, 218). These are most likely associated with the Neolithic Comb Ware culture (*c.* 6200–4000 cal BP), which is also underlined by some other amber pendants found at the site (LAHELMA 2008b, 34–40). The amber piece, however, does not show any clear signs of deliberate shaping, so the possibility of it being a part of a figurine must be rejected.

NON-AMBER BEAR FIGURINES

Even though they are not within the scope of this study, some non-amber bear figurines will be briefly mentioned in the following. This compilation is not meant to be exhaustive, but is intended to present a selected assemblage of finds from various contexts and timeframes.

Geißenklösterle, Baden-Württemberg, Germany (N 48.398°; E 9.771°)
From the upper Aurignacien layers of the Geißenklösterle cave in the Swabian Alps (*c.* 42,000–38,000 cal BP), a small figurine made of mammoth ivory was found that probably depicts a cave bear (*Ursus spelaeus*). Even though the *c.* 5 cm tall figurine is not a depiction of a brown bear, this find is worth mentioning here as it represents one of the oldest bear figurines in the countries under consideration (WAMERS 2015, 26–27).

Dolní Věstonice, Moravia, Czech Republic (N 48.892°; E 16.626°)
At the Gravettian site of Dolní Věstonice, the world's oldest ceramics have been found with an age of 24,000 years, consisting of more than 6,500 burnt fragments, which include almost 4,000 fragments of ceramic figurines. Among these is the famous "venus" of Dolní Věstonice, but there is also a bear figurine (Fig. 14) with a total length of 7.6 cm (VANDIVER et al. 1989). In contrast to the amber figurines, which appear to have been worn as pendants, there are no perforations or other evidence that the bear and other clay figures from Dolní Věstonice were used as such.

Bonn-Oberkassel, North Rhine-Westphalia, Germany (N 50.712°; E 7.158°)
Several grave goods that are of interest for this contribution have been found in the burial from Bonn-Oberkassel (Late Pleistocene/Federmesser groups, *c.* 14,200 cal BP): One ornamented object made of elk antler most likely represents an animal, but its head and legs are broken off. Its body shape shows striking similarities with that of the figurine found at Weitsche, Germany (cf. GIEMSCH et al. 2015, fig. 1). Another object that was found in the grave is a *c.* 20 cm long bone artefact, commonly identified as a hairpin (Fig. 15). On its proximal end it shows a modelled animal head that can be interpreted as that of a bear, based on its rounded ears, but other interpretations cannot be fully ruled out as the shape of the head itself is not so reminiscent of a bear (GIEMSCH et al. 2015, 242–246). Furthermore, a baculum of a bear (*Os penis*) with several cutmarks and haematite colouring was found in the grave (GIEMSCH 2017; STREET 2002, 280).

Äleby, Östergotland, Sweden (N 57.886°; E 15.861°)
At the Swedish Iron Age site of Äleby, several small clay figurines were found in connection to a possible stone-setting. Among these, two 1–2.6 cm large figurines resembling bears "could be a good deal earlier than the Iron Age material" (JANZON 1983, 3–4). However, it remains unclear how Janzon arrived at this conclusion.

Tråsättra, Uppland, Sweden (N 59.472°; E 18.354°)
At the Neolithic Pitted Ware culture site of Tråsättra (*c.* 4800–4200 cal BP), Sweden, more than 250 clay figurines, and fragments thereof, of different animals as well as humans were found. They mainly represent humans (n = 204), but at least six pieces have also been identified as bears (BJÖRCK et al. 2019, 121–123; cf. LINDSTRÖM, this volume).

Hietaniemi, Kuhmo, in Kainuu, Finland (N 64.280°; E 29.457°)
At the Finish site of Hietaniemi, an animal clay figurine was found that can be interpreted as a bear representation (cf. JANZON 1983, fig. 6.4). It measures 3.6 cm, but was originally longer as it is damaged on its rear part.

Lake Lubāns lowland, Rēzekne, Latvia (N 56.766°; E 26.861°)
From the lowland of Lake Lubāns, a small bear figurine made of clay has been reported (LOZE 1979, pl. 52.5). The sites that are compared by LOZE (1979, 138–139) all date to the Late Neolithic and Early Bronze Age Corded Ware culture (*c.* 5100–4400 cal BP).

Zvejnieki, Vecate parish, Latvia (N 57.776°; E 25.229°)
A small clay figurine from the Late Mesolithic/Early Neolithic Latvian burial ground of Zvejnieki (*c.* 9500–4600 cal BP) was originally interpreted as a bear. This resemblance is, however, limited, so ZAGORSKA et al. (2018, 113) consider it to be a representation of a human embryonic figure.

Çatalhöyük, Konya province, Turkey (N 37.666°; E 32.825°)
Two small bear figurines made of clay, dating to *c.* 8400–8000 cal BP, were found at the Neolithic/Chalcolithic site of Çatalhöyük in Turkey (MARTIN/MESKELL 2012, 407, tab. 1.4). Many of the figurines were made identifiable by the explicit formation of their tails (MESKELL 2015, 7–8). In the depictions at the settlement bears also play a role, but they are not frequently depicted. Remarkably, these are most frequent in an excavation area where other quadrupeds and wild animals, like boars or foxes, are more often depicted (MESKELL 2015, 2).

INTERPRETATION

As has been shown above, bear figurines are not only connected to hunter-gatherer groups, nor are they exclusively made of amber. As this contribution is focussed on the figurines made of the ancient tree resin, the other examples mentioned here are mainly used for opening a wider perspective. Since humans have long created representations of the world around them, from cave paintings to founder myths, this provokes the question of the role the figurines played and what possible implementations for understanding prehistoric human behaviour, environmental perception, and social interactions they provide.

First of all, it is important to critically evaluate the dating of the finds before further analyses regarding their use or meaning are conducted, as the amber bear figurines existed against a specific socio-environmental and cultural background. Through understanding when and where they were produced and utilised, further insights may be gained for contextualising them. Consequently, if the finds are not chronologically placed, hence lacking all contextual information, they become nothing but bibelots from the past.

How old are they – really?
Due to the difficulties associated with dating the bear figurines from the Baltic Sea region, their decoration has been a classic means for placing them into an archaeo-cultural context. According to the typological approach, the decoration patterns of well-dated artefacts are used for indirectly dating the contextless amber bear figurines. Simplistic patterns, however, like boreholes, zig-zag lines, and chevrons, have repeatedly been used throughout time, so using them as a sole chronological identifier inevitably comes with pitfalls. As direct dating of other artefacts has repeatedly shown, typo-chronology is a rather insecure tool. For instance, with respect to the chronological precision of specific Mesolithic bone points, it has recently been suggested that the duration of certain types is far more unreliable (and less archaeo-culturally determined) than previously implied; an example is the so-called Duvensee-type bone point, which had once been recognised as a typical element of the Early Mesolithic toolkit in the southwestern Baltic area. Direct dating of such implements, however, has shown that they were already present in the Late Palaeolithic and are regionally divergent with respect to chronology (e.g. CZIESLA/PETTIT 2003; cf. HARTZ et al. 2019, fig. 10). LARSSON (2000, 33) further states, discussing decorative elements on Mesolithic objects from Scandinavia, that "[a]ttempts to find chronologically related motifs or combinations of motifs have been made without achieving convincing results".

VANG PETERSEN (2013, 229) discusses similarities between the patterns of a bone rod from Fogense Enge, Denmark, which was directly dated to the Late Palaeolithic, and the bear figurine from Resen. He concludes that, contrary to earlier assumptions, the pattern on the bear figurine cannot "be taken as an indication of a (Late) Mesolithic date". Hence, several of the figurines with zigzag and chevron patterns should possibly be dated to the Late Palaeolithic. As the decoration patterns, especially the densely packed zigzag-lines, appear more often on reliably dated Late Palaeolithic art objects (VANG PETERSEN 2018, fig. 11; cf. PŁONKA et al. 2011), one should rather place them in earlier contexts. The chronological range of the use of such patterns remains nonetheless unclear; at Wustermark 22, Germany, such a pattern has been found on an elk antler tool, which has been dated to the Pleistocene/Holocene transition (around 11,600 cal BP; Ua-20962: 10,005 ± 70 BP; 11,800–11,260 cal BP; GRAMSCH et al. 2010). Yet it is unclear whether this tool, which is formally not securely identifiable (GRAMSCH et al. 2010, 114), can be understood as being Mesolithic after all (cf. GRIMM et al. 2020, 15).

Consequently, this shows that relying on the decoration for dating objects using such universal patterns is too insecure to apply. Additionally, VANG PETERSEN (2013, 233–234) underlines that amber bear figurines "differ completely from other (well-dated) Mesolithic works of art, which primarily consist of non-figurative decorations or heavily stylised anthropomorphic or zoomorphic drawings incised into the surface of tools, implements and personal objects". PŁONKA (2003, footnote 191) even excludes the Linnes figurine from his considerations on Mesolithic portable art "because the style of this piece and the pendants discovered in the same assemblage […] are more likely to be associated with the Neolithic". However, even for the Scandinavian Neolithic, figurative art is extremely rare. Mesolithic art of this kind is mainly known from eastern Europe, with different (naturalistic) representations of elk, but the main patterns used are abstract decoration (PŁONKA 2003, *passim*).

As a conclusion and based on the current state of research, too few indications speak for the association of the amber bear figurines from around the Baltic Sea with the Mesolithic. Therefore, their association with the Late Palaeolithic becomes slightly more likely. It can be argued (VANG PETERSEN in press) that the amber bear figurines might represent the last phase of Palaeolithic figurative art. Along with figures of other important big game species such as horse and elk, these amber bears represent a continuation of the Magdalenian tradition of amulets and personal objects decorated with naturalistic depictions of animals. The findings also indicate that, along with the transition from the Magdalenian to the Azilian/Federmesser culture in the Allerød period, a shift from reindeer and horse of the open tundra towards forest species like elk and bear took place in the hunters' ideology.

However, this picture may change if reliably dateable finds are unearthed, or methods for direct age determination for the production of the pieces are developed. To this date, the only amber bear figurines from the circum-Baltic region that can be somewhat reliably dated are the finds from Brześć Kujawski, Poland, the burial in Tamula, Estonia, and the settlement site of Sārnate, Latvia – all these are from Neolithic contexts.

Why amber?
Amber has always been regarded as a special material. As can be seen from the detailed Late Palaeolithic figurines or its use for pendants and ornaments during later times, the material has been so valuable that whole industries have been built around it. This is exemplified by the long-distance trade in the Bronze Age which already then ensured that Baltic amber was transported to the Mediterranean (e.g. KRISTIANSEN/SUCHOWSKA-DUCKE 2015).

The use of amber has a long tradition around the Baltic Sea, as its outcrops have been eroding since the Ice Age, so fragments of the prehistoric tree resin are regularly washed ashore. Its use for prehistoric mobile art and ornaments can be traced back to the Late Palaeolithic (see above), where the origin of creating naturalistic amber figurines can most probably be found. A re-use of ancient artefacts is also indicated by finds that are discovered out-of-context. For instance, the Linnes bear was found in an area without amber outcrops. So, either its raw material, or the figurine itself, must have been imported from further south. If a Late Glacial dating of the figurines is accepted (see above), this indicates the use of traded ancient artefacts in a Pitted Ware culture context. An example of Neolithic reuse of a clearly Mesolithic decorated pendant has recently been described (VANG PETERSEN in press).

The use of, and fascination for, amber may have had many different reasons: The material has several remarkable characteristics, it is lightweight and floats in salty water, even though it appears to be a stone. It feels nice and warm against the skin; when polished, it becomes translucent and shiny, and if burnt, the scent is unmistakable and fragrant. The most fascinating feature is probably that when rubbed against hair or fur it develops a static electrical charge and can emit green sparks visible in the dark. Furthermore, we can expect that even during the Stone Age the fossil material was understood as such and therefore utilised on purpose; people probably even then understood that amber is ancient tree resin. So, it is possible to imagine that amber may have provided a magic connection to the metaphysical world.

Why bears?
While the bear figurines' chronological association with any timespan remains foggy, the question of why bears were depicted might be addressed more easily. "What these figurines might mean or how they were socially active depends a great deal on how animals were seen, how humans saw themselves and how the relations between them were perceived" (VALERA et al. 2014, 16). To fully understand the function and purpose of the figurines, the relevance of the socio-cultural environment into which they have been placed must be recognised. However, as long as this remains a challenge, the question of why bears have been depicted in a special material like amber can nonetheless be discussed. Currently, only bears and elks appear most regularly in the amber art around the Baltic from the Stone Age (IRŠĖNAS 2000). The compilation by IRŠĖNAS (2001) also shows some other quite clearly identifiable species (e.g. beavers), but those that are most common are elks, bears, and humans. Disregarding the latter, the question arises of why elks and bears are such a predominant subject for prehistoric amber figurines.

The depiction of the tangible and intangible worlds is a very old tradition in human societies: Ever since the Upper Palaeolithic, cave paintings and rock carvings were used for conserving perceived situations in real life. But they most likely also invoked powers and spirits with magic abilities to help (feed) and protect the people who made them and carried them around. Big animal species like elk and bear may have been regarded as especially magically gifted, as they were able to move across the borders between different worlds – swimming in the sea, shedding antlers, or sleeping all winter underground, etc. (cf. HELSKOG 2012). The depiction of daily life situations, as for instance in the Alta rock carvings, Norway (GJERDE 2019), does not contradict any magical elements, as a strict separation between profane and sacred is a modern construct. Hence, they possibly appear more as representative scenes from everyday life than metaphysical representations.

In contrast, the amber figurines appear as depictions of single individuals. Hence, it is significant that they are more directly bound to the animal as such, rather than to the scenery, making them individual tokens. Consequently, their meaning is created through them themselves, instead of through them interacting with the environment. They might therefore be considered as providing a function or acquiring a social role (cf. VALERA et al. 2014, 16–19), like an amulet or manifestation. However, as bear bones are a rather rare element at Mesolithic and Neolithic sites (cf. KLASSEN/GREGERSEN, this volume), it seems rather unlikely that they can be seen in connection with hunting magic on a regular basis. It is worth mentioning, however, that the Neolithic Pitted Ware culture might have had a special relationship to bears, as is represented by comparably high numbers of such bones in the assemblages (cf. KLASSEN/GREGERSEN, this volume), and indications for the transport of commodities made of them to areas out of their natural range.

Generally, the bear's effective relevance for human subsistence, excluding its value for identity or cultural practices, must be considered rather low, which might partly relate to the expectable lower density and visibility of the animals in the landscape, compared with, for instance, that of deer or wild boar. Nonetheless, the nutritional value of bear meat is very good (SCHMÖLCKE et al. 2017).

Can amber bear figurines therefore be seen in a more metaphysical context? As has been shown, bears are a very elusive species even when not hibernating (cf. contributions on Bears in biology, this volume). Therefore, the chances for an encounter with a bear are extremely low today and were most likely low in the dense forests of the past as well. As dogs were already part of hunting communities at that time, they will have supported the chase and discovery of prey and thus probably played a relevant role in occasional (?) bear hunts. The fact that humans hunted and exploited bears for meat as well as material for tools and ornaments is undoubted, but it should be considered that bears were usually not their main prey but an exception, as can be seen from the zooarchaeological studies. It can be assumed that bears were a rather uncommon and exceptional sight also for prehistoric hunter-gatherers, so the depiction of them might have served as a memorial or ornament. This might even be reflected by their value as totems for specific groups (cf. HELSKOG 2012, 214).

Another perspective proposed by HILL (2013) overcomes the rather modern dichotomy of humans and animals and thus recognises that some societies, especially foragers, do not clearly differentiate between the two. Larger predators, especially, tend to be seen as relational to people,[3] according to ethnographic studies, and might have received similar treatment (e.g. HILL 2013, 119–120; LOSEY et al. 2013). Consequently, the amber bear figurines (and those made of other materials) can rather be understood as representations of other-than-human persons than as mere depictions of a more distant environment. Hence, the figurines incorporated a means for directly interacting and exchanging with these other parts of the societies (i.e. the bears). Therefore, it is not relevant that bear bones are represented in relatively low numbers in the archaeological assemblages, as "interaction is sensual, ongoing, and may involve the living and the dead, in addition to prey and hunter" (HILL 2013, 126).

---

3   This might also be true for smaller animals and could explain the depiction of these as amber figurines as well (e.g. a beaver figurine from Valma [Estonia]: OTS 2003, 104). SCHMÖLCKE et al. (2017, 903) propose „the beaver was potentially regarded as a kind of water spirit – comparable to the bear as a spirit of the forest".

IRŠĖNAS (2007, 7–8) highlights that the mobile art[4] from the Baltic region has received far less attention than, for instance, Upper Palaeolithic cave paintings, and consequently it has also been less subject to diverging interpretations. Many of the interpretations of such objects have been focussed on their sacrificial meaning, so understanding them as having a more profane use, e.g. as toys, is more uncommon (cf. IRŠĖNAS 2007, 8–10). Iršėnas agrees with LARSSON's (2000, 33) statement that "the present-day division between sacred and profane is a delusion", which thus accentuates that a dualistic view on many prehistoric artefacts is a modern post-Renaissance perspective. Different ethnographic examples further underline how inseparable (perceived) profane and metaphysical functions are (IRŠĖNAS 2007, 11–14). There is no reason to believe that categorising artefacts based on richness of detail or elaborateness, be it tool vs. toy, sacred vs. profane, mature vs. child (cf. IRŠĖNAS 2007, 15–17), brings us any closer to reality.

Conclusion

For understanding the role and purpose of amber bear figurines, as well as those made of clay, it is important to be aware of general preconceptions underlying such endeavours: while "parallel functions do *not* necessarily imply parallelism of form" (DOWNS/STEA 1973, 5), parallel forms do not necessarily imply parallelisms of function either. Keeping this in mind, the role of amber bear figurines from (potential) hunter-gatherer contexts might have served a very different purpose than the later figurines from Neolithic sites. As the use-wear traces on some of the amber figurines show, they had been attached to strings and thus might have been used in an ornamental form, perhaps as an amulet or an attachment to clothing. Yet this neither excludes nor implies a single function as an ornament or ritual token.

Even if the amber bear figurines are low in number, they represent a sparse but re-occurring find type. Therefore, deciphering underlying patterns and possible meaning is necessary to integrate them into the physical and metaphysical landscapes of the past. As archaeology is a time- and space-bound research field, understanding the cultural context of objects is mandatory for interpreting their purpose. Alas, the chronological information is completely lacking for most of the amber bear figurines under consideration, and it has been shown that indirect dating techniques only work to a limited extent.

As demonstrated, the amber bear figurines from around the Baltic Sea are an archaeological phenomenon that finds some parallels in other materials as well as in other depicted species. They can therefore be understood as a rare but widespread find type that most likely had some inherent metaphysical purpose and social context, even if these cannot be reconstructed due to their lack of archaeological context. All things considered, the current evidence highlights that the amber bear figurines might very well be associated with Late Palaeolithic groups or the early Neolithic, as comparable mobile art is known from these periods. The previous practice of assigning them to the Mesolithic must be critically questioned. Interpreting more generously, amber bear figurines are unique evidence of ritual activities and are potentially magic tokens from prehistory. Displaying the most powerful and human-like creature of the forest (cf. SCHMÖLCKE et al. 2017, 902), by recreating it from an ancient material with unique characteristics, must have resulted in the most powerful amulet-types known from the Stone Age of northern Europe.

---

4   The term „mobile art" has to be understood in this context as a *terminus technicus* for all shaped and decorated objects that are not stationary. This includes, for instance, ornamented tools, figurines, and decorated pendants. This explicitly does not incorporate any functional or meaning-related interpretation (cf. LARSSON 2000, 33).

Acknowledgements

Thanks to Patricia Tellhelm and Grzegorz Osipowicz for help with the acquisition of literature, and Valdis Bērziņš for helping with information on the Sārnate specimen. Mari Tõrv helped us tremendously with acquiring a photograph of the Tamula specimen.

D. Groß: This research is part of the Centre for Baltic and Scandinavian Archaeology's (ZBSA) strategic research topic "Human and Animal Studies". Research on this topic was carried out under the umbrella of the CRC 1266 (funding by the Deutsche Forschungsgemeinschaft [DFG, German Research Foundation] – project no. 2901391021 – SFB 1266), which is thankfully acknowledged.

Bibliography

Andersen 1980: S. H. Andersen, Jægerrav. Skalk 1980(5), 28–29.

Bērziņš 2008: V. Bērziņš, Sārnate: Living by a coastal lake during the East Baltic Neolithic (Oulo 2008).

Björck et al. 2019: N. Björck/M. Artursson/K.-F. Lindberg, Tråsättra – aspekter på säljägarnas vardag och symbolik. Rapport 2019:40 (Stockholm 2019).

Brumm et al. 2021: A. Brumm/A. A. Oktaviana/B. Burhan/B. Hakim/R. Lebe/J.-X. Zhao/P. H. Sulistyarto/M. Ririmasse/S. Adhityatama/I. Sumantri/M. Aubert, Oldest cave art found in Sulawesi. Science Advances 7(3), 2021: doi.org 10.1126/sciadv.abd4648.

Cyrek et al. 1986: K. Cyrek/R. Grygiel/K. Nowak, The basis for distinguishing the ceramic Mesolithic in the Polish lowland. In: T. Malinowski (ed.), Problems of the Stone Age in Pomerania (Warsaw 1986) 95–126.

Cziesla/Pettitt 2003: E. Cziesla/P. Pettitt, AMS-14C-Datierungen von spätpaläolithischen und mesolithischen Funden aus dem Bützsee (Brandenburg). Archäologisches Korrespondenzblatt 33, 2003, 21–38.

Downs/Stea 1973: R. M. Downs/D. Stea, Theory. In: R. M. Downs/D. Stea (eds.), Image and Environment. Cognitive mapping and spatial behavior (Chicago 1973) 1–7.

Giemsch et al. 2015: L. Giemsch/J. Tinnes/R. W. Schmitz, Comparative studies of the art objects and other grave goods from Bonn-Oberkassel. In: L. Giemsch/R. W. Schmitz (eds.), The Late Glacial Burial from Oberkassel Revisited (Darmstadt 2015) 231–251.

Giemsch 2017: L. Giemsch, Bear necessities? On potential uses of the ursine baculum (Os penis) in archaeological and ethnological contexts. In: P. Fasold/L. Giemsch/K. Ottendorf/D. Winger, Forschungen in Franconofurd. Festschrift für Egon Wamers zum 65. Geburtstag (Regensburg 2017) 41–53.

Gimbutas 1956: M. Gimbutas, The Prehistory of Europe, Part 1. Mesolithic, neolithic and copper age cultures in Russia and the Baltic Area (Cambridge [Mass.] 1956).

Gimbutas 1985: M. Gimbutas, East Baltic amber in the fourth and third millennia BC. Journal of Baltic Studies 16(3), 1985, 231–256.

Gjerde 2019: J. M. Gjerde, Alta (Norway), Rock Art of. In: C. Smith (ed.), Encyclopedia of Global Archaeology (Cham 2019) 1–10.

Gjessing 1926: H. Gjessing, Linnes. In: Reallexikon der Vorgeschichte 7 (Kleinasien-Malta) (Berlin 1926) 299.

Gramsch et al. 2010: B. Gramsch/J. Beran/K.-U. Heussner/F. Brose/H. U. Thieke/S. Hanik/S. Jahns, Spätaltsteinzeitliche Funde von Wustermark, Fundplatz 22, Lkr. Havelland. Veröffentlichungen zur brandenburgischen Landesarchäologie 41/42, 2010, 95–141.

Grimm et al. 2020: S. B. Grimm/D. Gross/K. Gerken/M.-J. Weber, On the onset of the Early Mesolithic on the North German Plain. In: A. Zander/B. Gehlen (eds.), From the Early Preboreal to the Subboreal period – Current Mesolithic research in Europe. Studies in honour of Bernhard Gramsch on his 85th birthday (Kerpen-Loogh 2020) 15–37.

Hartz et al. 2019: S. Hartz/H. Lübke/D. Gross, Early Mesolithic bone points from Schleswig-Holstein. In: D. Groß/H. Lübke/J. Meadows/D. Jantzen (eds.), Working at the Sharp End: From Bone and Antler to Early Mesolithic Life in Northern Europe (Kiel, Hamburg 2019) 203–238.

Helskog 2012: K. Helskog, Bears and Meanings among Hunter-fisher-gatherers in Northern Fennoscandia 9000–2500 BC. Cambridge Archaeological Journal 22(2), 2012, 209–236.

Hill 2013: E. Hill, Archaeology and Animal Persons: Toward a Prehistory of Human-Animal Relations. Environment and Society: Advances in Research 4, 2013, 117–136.

Iršėnas 2000: M. Iršėnas, Elk Figurines in the Stone Age Art of the Baltic Sea. In: A. Butrimas (ed.), Prehistoric Art in the Baltic Region (Vilnius 2000) 93–105.

Iršėnas 2001: M. Iršėnas, Stone Age Amber Figurines from the Baltic Area. Acta Academiae Artium Vilnensis 122, 2001, 77–85.

Iršėnas 2007: M. Iršėnas, Stone Age Figurines from Baltic Region: Toys or Sacred Objects. In: G. Mickūnaitė (ed.), Art and the Sacred (Vilnius 2007) 7–17.

Janzon 1983: G. H. Janzon, Zoomorphic clay figurines and beads from Ire, Hangvar parish, Gotland. Fornvännen 78, 1983, 1–20.

Kabaciński et al. 2011: J. Kabaciński/S. Hartz/T. Terberger, Elks in the early Stone Age art of the northern Lowlands. Praehistorische Zeitschrift 86, 2011, 151–164.

Kristiansen/Suchowska-Ducke 2015: K. Kristiansen/P. Suchowska-Ducke, Connected Histories: the Dynamics of Bronze Age Interaction and Trade 1500–1100 BC. Proceedings of the Prehistoric Society 81, 2015, 361–392.

Lahelma 2008a: A. Lahelma, Communicating with 'stone persons': anthropomorphism, Saami religion and Finnish rock art. ISKOS 15, 2008, 121–142.

Lahelma 2008b: A. Lahelma, A Touch of Red: Archaeological and Ethnographic Approaches to Interpreting Finnish Rock Paintings. ISKOS 15, 2008, 6–76.

Larsson 2000: L. Larsson, Expressions of Art in the Mesolithic Society of Scandinavia. In: A. Butrimas (ed.), Prehistoric Art in the Baltic Region (Vilnius 2000) 31–61.

Losey et al. 2013: R. J. Losey/V. I. Bazaliiskii/A. R. Lieverse/A. Waters-Rist/K. Faccia/A. W. Weber, The bear-able likeness of being: ursine remains at the Shamanka II cemetery, Lake Baikal, Siberia. In: C. Watts (ed.), Relational Archaeologies: Humans, Animals, Things (London 2013) 65–96.

Loze 1979: I. A. Loze, Pozdnij neolit i rannjaja bronza Lubanskoj Ravniny (Riga 1979).

Loze 2000: I. Loze, Late Neolithic amber from the Lubana wetlands. In: A. Butrimas (ed.), Prehistoric Art in the Baltic Region (Vilnius 2000) 63–78.

Loze 2003: I. Loze, Middle Neolithic Amber Workshops in the Lake Lubans Depression. In: C. W. Beck/I. Loze/J. M. Todd (eds.), Amber in Archaeology. Proceedings of the fourth international conference on amber in archaeology, Talsi, 2001 (Riga 2003) 72–89.

Martin/Meskell 2012: L. Martin/L. Meskell, Animal Figurines from Neolithic Çatalhöyük: Figural and Faunal Perspectives. Cambridge Archaeological Journal 22(3), 2012, 401–419.

Meskell 2015: L. Meskell, A society of things: animal figurines and material scales at Neolithic Çatalhöyük. World Archaeology 47(1), 2015, 6–19.

Ots 2003: M. Ots, Stone Age Amber Finds in Estonia. In: C. W. Beck/I. B. Loze/J. M. Todd (eds.), Amber in Archaeology. Proceedings of the fourth international conference on amber in archaeology, Talsi, 2001 (Riga 2003) 96–107.

Ots 2006: M. Ots, Merevaiguleiud Baltimaade kivi- ja pronksiaegsetes muististes. Unpublished MA-thesis (Tallinn 2003).

Płonka 2003: T. Płonka, The portable art of Mesolithic Europe (Wrocław 2003).

Płonka et al. 2011: T. Płonka/K. Kowalski/M. Malkiewicz/J. Kuryszko/P. Socha/K. Stefaniak, A new ornamented artefact from Poland: final palaeolithic symbolism from an environmental perspective. Journal of Archaeological Science 38(3), 2011, 723–733.

Schmölcke et al. 2017: U. Schmölcke/D. Gross/E. Nikulina, Bears and Beavers – The Browns in daily life and spiritual world. In: B. V. Eriksen/A. Abegg-Wigg/R. Bleile/U. Ickerodt (eds.), Interaktion ohne Grenzen. Beispiele archäologischer Forschungen am Beginn des 21. Jahrhunderts. Interaction without borders. Exemplary archaeological research at the beginning of the 21st century (Kiel, Hamburg 2017) 901–917.

Street 2002: M. Street, Ein Wiedersehen mit dem Hund von Bonn-Oberkassel. Bonner Zoologische Beiträge 50(3), 2002, 269–290.

Terberger 2003: T. Terberger, Decorated objects of the older Mesolithic from the northern lowlands. In: L. Larsson/H. Kindgren/K. Knutsson/D. Loeffler/A. Åkerlund (eds.), Mesolithic on the Move. Papers presented at the Sixth International Conference on the Mesolithic in Europe, Stockholm 2000 (Oxford 2003) 547–557.

Terberger/Ansorge 2000: T. Terberger/J. Ansorge, Der Bernsteinbär von Stolp (Slupsk, Polen): ein mesolithisches Amulett? Archäologisches Korrespondenzblatt 30, 2000, 335–352.

Toft/Brinch Petersen 2013: P. A. Toft/E. Brinch Petersen, Five Thousand Years of Decorated Amber Pendants from the Danish Mesolithic. Die Kunde 64, 2013, 197–217.

Tõrv 2018: M. Tõrv, Persistent Practices. A Multi-Disciplinary Study of Hunter-Gatherer Mortuary Remains from c. 6500–2600 cal. BC, Estonia. Untersuchungen und Materialien zur Steinzeit in Schleswig-Holstein und im Ostseeraum 9 (Kiel, Hamburg 2018).

Valera et al. 2014: A. C. Valera/L. S. Evangelista/P. Castanheira, Zoomorphic figurines and the problem of human-animal relationship in the Neolithic and Chalcolithic of Southwest Iberia. Menga: Revista de prehistoria de Andalucía 5, 2014, 15–41.

Vandiver et al. 1989: P. B. Vandiver/O. Soffer/B. Klima/J. Svoboda, The Origins of Ceramic Technology at Dolni Vecaronstonice, Czechoslovakia. Science 246(4933), 1989, 1002–1008.

Vang Petersen 2013: P. Vang Petersen, Amber pendants, bears and elks. Die Kunde 64, 2013, 219–237.

Vang Petersen 2018: P. Vang Petersen, Ravfigurer og zigzagmønstre – jægerkunst fra istiden. Nationalmuseets arbejdsmark 2018, 2018, 136–156.

Vang Petersen 2019: P. Vang Petersen, Zigzag lines and other protective patterns in Palaeolithic and Mesolithic art. Quaternary International 573, 2019, 66–74.

Vang Petersen in press: P. Vang Petersen, Amber, animal figures and protective art in Scandinavia. Die Kunde (Festschrift Stephan Veil), in press.

Vankina 1970: L. Vankina, Torfânikovaâ stoânka Sarnate (Riga 1970).

Veil et al. 2015: S. Veil/K. Breest/P. Grootes/M.-J. Nadeau/M. Hüls, A 14 000-year-old amber elk and the origins of northern European art. Antiquity 86, 2015, 660–673.

Virchow 1884: R. Virchow, Alte Thierfigur aus Bernstein. Verhandlungen der Berliner Gesellschaft für Anthropologie, Ethnologie und Urgeschichte 1884, 566–569.

Virchow 1887: R. Virchow, Thierstück aus Bernstein von Stolp. Verhandlungen der Berliner Gesellschaft für Anthropologie, Ethnologie und Urgeschichte 1887, 401–402.

Wamers 2015: E. Wamers (ed.), Bärenkult und Schamanenzauber. Rituale früher Jäger (Regensburg 2015).

Zagorska et al. 2018: I. Zagorska/J. Meadows/M. Iršėnas, New dates from Zvejnieki burial ground graves with anthropomorphic and zoomorphic figurines. Archaeologia Baltica 25, 2018, 100–124.

Dr. Daniel Groß
Museum Lolland-Falster
Nykøbing F
Denmark
dag@museumlollandfalster.dk

Peter Vang Petersen M.A.
Nationalmuseet
København
Denmark
peter.vang.petersen@natmus.dk

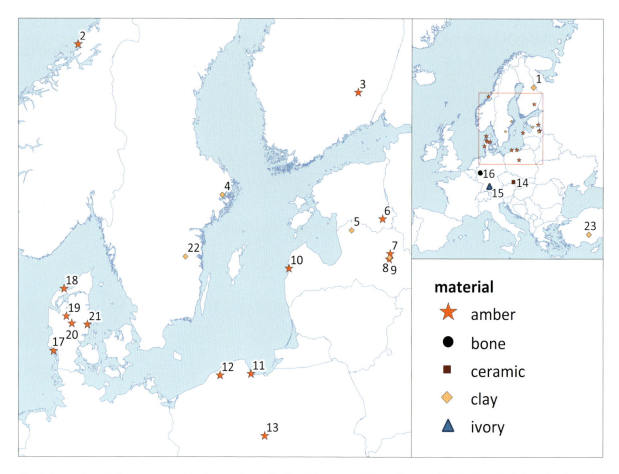

Fig. 1. Overview of sites mentioned in the text (map: D. Groß, based on: © EuroGeographics for the administrative boundaries). 1: Hietaniemi (Finland); 2: Linnes (Norway); 3: Astuvansalmi (Finland); 4: Tråsättra (Sweden); 5: Zvejnieki (Latvia); 6: Tamula (Estonia); 7: Zvidze (Latvia); 8: Lake Lubāns (Latvia); 9: Suļka (Latvia); 10: Sārnate (Latvia); 11: Gdansk (Poland); 12: Słupsk/Stolp (Poland); 13: Brześć Kujawski (Poland); 14: Dolní Věstonice (Czech Republic); 15: Geißenklösterle (Germany); 16: Bonn-Oberkassel (Germany); 17: Fanø (Denmark); 18: Lild Strand (Denmark); 19: Resen (Denmark); 20: Bølling Sø (Denmark); 21: Tangkrogen (Denmark); 22: Äleby (Sweden); 23: Çatalhöyük (Turkey).

Fig. 2. The elk figurine from Weitsche, Lüchow, Lower Saxony, Germany (after Veil 2012, fig. 3; photos U. Bohnhorst, compilation: S. Veil, © Landesmuseum Hannover).

Fig. 3. The Fanø bear, Jutland, Denmark (NM A 52089; after Vang Petersen 2018, figs. 2; 13a,b).

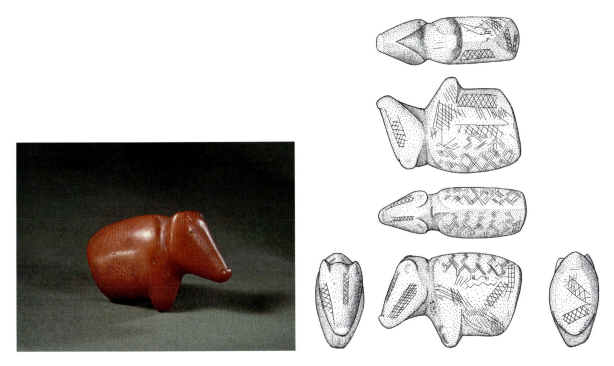

Fig. 4. The Resen bear, Jutland, Denmark (NM A 8411; after Vang Petersen 2019, fig. 5).

Fig. 5. The Tangkrogen figurine, Jutland, Denmark (NM A 49747; after Vang Petersen 2018, fig. 13c,d).

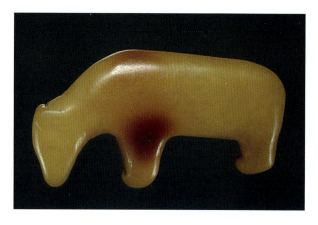

Fig. 6. The Lild Strand figurine, Jutland, Denmark. Modern fake (NM A 52274, photo © Danish National Museum).

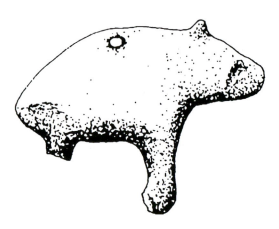

Fig. 7. The Linnes bear, Trøndelag, Norway (after Terberger/Ansorge 2000, fig. 4.9).

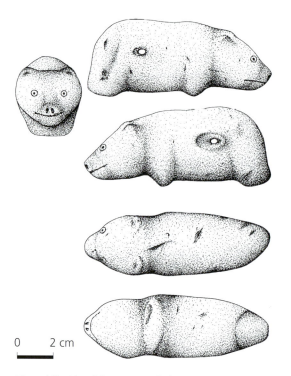

Fig. 8. The Słupsk bear, voivodeship Pomerania, Poland (after Terberger/Ansorge 2000, fig. 1).

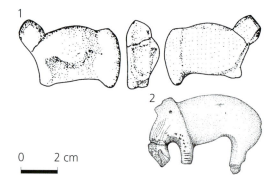

Fig. 9. Amber figurines from Poland. 1: Brześć Kujawski, site 3 (after Cyrek et al. 1986, fig. 16.2); 2: Boar/pig (?) figurine from Gdansk (after Terberger/Ansorge 2000, fig. 5.1).

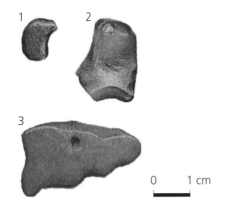

Fig. 10. Amber figurines from Latvia. 1: Suļka; 2–3: Zvidze (after LOZE 2000, figs. 2.2; 3.2,8).

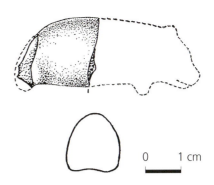

Fig. 11. The possible bear figurine fragment from Sārnate, Ventspils, Latvia (after VANKINA 1970, plate LV.1).

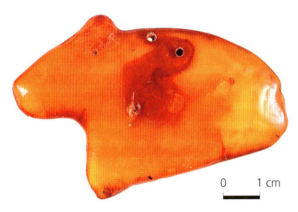

Fig. 12. The Tamula specimen, Võru, Estonia (photo M. Usler, Archaeological Research Collection, Tallinn University, AI 4118 1751).

Fig. 13. The natural amber lump resembling a bear-head from Astuvansalmi, Ristiina in Mikkeli, Finland (Finnish Heritage Agency, Inventory ID KM27146:1, CC BY 4.0).

Fig. 14. The ceramic bear figurine from Dolní Věstonice, Moravia, Czech Republic (Moravian Museum, ID: DV 30014, used with permission by the museum).

in the form of an in-depth study, but as a brief overview, in chronological order, as a basis for future discussion.

When searching for visual representations of bears in Late Iron Age and Viking iconography, we have to be aware of the fact that, usually, this kind of art is non-naturalistic. Animal depictions during these periods can be primitive, fanciful or highly stylised, and they can be hybrid, mixing up different species (PESCH 2012). Naturalistic, or even semi-naturalistic animal depictions, featuring crucial characteristics that clearly indicate a certain species, are exceptions. Our zoological identification – if we believe that the depiction of a certain animal species is intended at all – is dependent on the presence of typical physical features and crucial animal characteristics. However, in many cases, there will always be some subjectivity in the zoological interpretation. What physical features do brown bears have that can be used by an artist to indicate this specific predator? Brown bears are generally bulky, they have large bodies with relatively short and stocky limbs and a strong neck, massive heads (the forehead is high and rises steeply), a longish but relatively short snout with predator teeth and small rounded ears, they have fur, shaggy hair, large paws with five long non-retractile claws, and short tails. As a special feature, bears are able to stand and walk on their hind feet.

The earliest relevant visual representation featuring bear-like characteristics known to date is a small figurine made of soapstone from Modvo in Sogn og Fjordane, Norway, which is about 7 cm wide and 6 cm high (Fig. 1; cf. HVOSLEF KRÜGER 1988, 362). It was found in a house foundation dating to the Roman Period, which means roughly between the birth of Christ and AD 400. Its actual context and function are unknown. From this period, the first centuries AD, indigenous pictorial art in Scandinavia is relatively scarce; only a small number of animal depictions are known, and these are very much dependent on Roman influences and models (BLANKENFELDT 2015).

The "age of the image" starts in the Migration Period, about AD 400, when Germanic art became increasingly independent and started to establish itself in large parts of central and northern Europe, with a sheer explosion and flood of pictures (PESCH 2012, 651). The main motifs of this newly-established art, usually very stylised but sometimes more or less semi-naturalistic, are animals. However, there is no Migration Period depiction that can, with any certainty, be regarded as a bear. Some depictions seem to represent predators with more or less bear-like body shapes, but they lack any clear features. Mention can be made of the beasts on the famous gilded silver relief brooch from Hol in Sør-Trøndelag, Norway (Fig. 2; cf. MAGNUS 1975, 66; 2014, 75–77, figs. 4–5; BEERMANN 2016, 64), and a carved figure on a wooden piece of unknown function from Evebø in Sogn og Fjordane (Fig. 3; cf. STRAUME 1962, pl. 16; MAGNUS 1975, 64–66). Interestingly, this enigmatic item belongs to the grave of a high-ranking warrior who was buried lying on a bear skin (see GRIMM, this volume). Nevertheless, animal images like these remain completely unclear, though they possibly represent bears or another kind of beast or mythical monster (BEERMANN 2016, 64–65 with more such dubious cases).

From the following era, the Vendel or Merovingian Period, i.e. between the 6th and 8th centuries, more relevant depictions are preserved. A well-known and very famous one is the cast bronze die A from Torslunda on the isle of Öland in the Baltic (Fig. 4). The four Torslunda dies (HAGBERG 1976; HAGBERG et al. 1991, 257; recently SAMSON 2020, 252–264) were used for the production of stamped bronze sheets (*Pressbleche*; Fig. 5), decorating high-quality warrior helmets, which are known from several Swedish elite burials (BÖHNER 1994). Die A depicts a warrior flanked by two bears (HELMBRECHT 2011, 102 fig. 11a, cat. no. 591). The beasts are standing on their hind legs, grasping the man's arms and touching his head with their snouts, as if they were kissing or biting him. The warrior is holding a knife or dagger in his hand and, with the other hand, he is stabbing one of the bears with his sword.

There is a surprisingly similar depiction from the Continent, on a silver *phalera*[2] found in an Alamannic elite burial in Eschwege in Hessen, Germany (Fig. 6; cf. Böhner 1991). However, in this case, the man is apparently naked and completely unarmed, presenting his empty hands, holding them close to his body. There is good reason to believe that this image is influenced by the iconography of Daniel the prophet (ibid., 702–704; cf. Holzapfel 1973, 26–28; Holmqvist 1938, 92). On finds from the Merovingian Period, Daniel in the lions' den is frequently depicted, in particular on belt buckles from the Burgundian territory in present-day France and Switzerland (Jörg/Martin 1984). The praying prophet is always flanked by two lions. The fact that the beasts, defeated only by God's will, are bowing to the (of course unarmed) prophet, is represented by specific gestures, mainly by licking Daniel's garment or his feet[3] – or sometimes by touching his face with their snouts, licking, kissing, snuggling like a faithful dog, as on a Merovingian Period belt buckle from unknown find spot in Spain (Oxenstierna 1956, 76–77 fig. 156; Holzapfel 1973 12–13 fig. 2), and a Merovingian Period disc brooch from Sprendlingen in Rheinland-Pfalz, Germany (Kühn 1941/1942, 158; Klein-Pfeuffer 1993, no. 293; 2015, 266 fig. 2,4; see Figs. 7–8). Later, in Romanesque art, this gesture is depicted more frequently (e.g. Debidour 1961, fig. 209; Holzapfel 1973, fig. 1).

The same Christian motif appears to be depicted on the *phalera* from the Alamannic grave in Eschwege. Many Alamannic elite burials of this period contain Christian features. First and foremost, the group of gold sheet crosses must be mentioned (Böhme 1998; Terp-Schunter 2017), but also further Merovingian Period *phalerae* decorated with Christian imagery, such as equestrian saints or The Mother of God (Fingerlin 2010; 2012). On the Eschwege *phalera*, however, the biblical lion turned into the most impressive and most dangerous indigenous predator species – the bear. Nevertheless, the meaning of the motif is still the same: The defenceless human figure is rescued only by prayers and divine help. Probably, this Christian message is also intended in the case of the gold and garnet *cloisonné* mounts on the purse lid from the Anglo-Saxon princely ship burial of Sutton Hoo, East Anglia, England, dating to the 7th century AD (Bruce-Mitford 1975–1983). The Anglo-Saxon "Daniel" is depicted without weapons, performing a hand gesture that could be regarded as praying or as a sign of helplessness, while the lions have been transformed into wolves instead of bears (Fig. 9; cf. Holzapfel 1973, 26–28; Hauck 1982, 352–359; Böhner 1991, 702–703).

The artist from Torslunda, however, equipped his "pagan Daniel" with weapons, transforming the Christian motif into a heroic image. Some scholars even tried to name the depicted hero – e.g. Ragnar *loðbrók* (Schück 1926, 75–77) and Beowulf (Hauck 1982, 352–356) have been suggested, which is hardly convincing; others tried to argue that the bear killer represents a young man performing a kind of initiation ritual, supposedly connected with the cult of Óðinn (cf. Ström 1980, 266–267) and the *berserkir* (e.g. Nordberg 2003, 233–235; regarding the *berserkir*, see also Sundqvist, this volume). Worth mentioning at this point are the *Res gestae*, written by Ammianus Marcellinus at the end of the 4th century AD. According to Ammianus, among the probably eastern Germanic tribe of the Taifals, young men had to kill a boar or a bear in order to be admitted to the warrior community (liber XXXI, cap. XVI: Seyfarth et al. 1978, 31; Ranke/Reichstein 1976; see Hurka, this volume).

A similar "heroisation" or "paganisation" of the Daniel motif appears to have been implemented in the imagery of a sculptured Viking tombstone (a so-called hogback, see below) from Sockburn in Durham County, dating back to the 9th or 10th centuries;[4] an apparently naked man is surrounded by

---

2   A *phalera* is a decorated disc, originally representing a Roman military award but later belonging to luxury horse harnesses.
3   Concerning this specific element see e.g. Kühn 1941/1942, fig. 1–18; Werner 1977, 310–311 fig. 22 pl. 99; Klein-Pfeuffer 1993, 202–203; Oehrl 2019a, 141.
4   Cramp 1984, 143–144 fig. 146; most recently, regarding interpretation and research history: Oehrl 2011, 154–162; 2019a, 139–142.

a group of animals and flanked by a pair of huge and bulky four-legged beasts. One of the beasts is touching the man's hand with its mouth. Most probably, the image should be regarded as an imitation of the Daniel motif as it is depicted on many other early medieval insular stone monuments, such as the hogback from Lythe in North Yorkshire (KOPÁR 2012, fig. 28), the high cross from Dysert O'Dea in Ireland (HENRY 1933, fig. 49), and the stone slab no. 14 from St. Vigeans in Scotland (ALLEN 1903, fig. 285A; see also OEHRL 2019a, pl. 112). However, as a specific feature, the Anglo-Scandinavian Daniel from Sockburn is equipped with a battle axe that can be seen behind his right arm. Interestingly, the "lions" on the Sockburn hogback could be described as bear-like. As will be discussed below, bears actually play an important role in the iconography of the hogback group.

On the background of the bear killing scene from Torslunda A, it is plausible to regard the strange looking beast on Torslunda die B (and the *Pressblech* images from Vendel) as a bear as well, as has been supposed by numerous scholars (e.g. BECK 1964, 45–50; STRÖM 1980, 266–267; NYMAN/HAGBERG 2006, 77), but in a much more stylised version, of course (Figs. 10–13). The Torslunda die B depicts a warrior with a battle axe, fettering the stylised bear-like monster, holding it like a dog on a leash, while the creature's limbs seem to be knotted together.[5]

The Swedish archaeologists Wilhelm Holmqvist (HOLMQVIST 1939, 151–152; cf. ROMDAHL 1922, 221) and Birgit Arwidsson (ARWIDSSON 1977, 122–123), and recently German archaeologist Egon Wamers (WAMERS 2009, 25–42) argued that the bear-fettering and bear-killing motifs from Torslunda may be influenced by depictions of bear hunts and bear fights in Roman amphitheatres, where bears were actually killed with weapons, fettered and kept on leashes.[6] Impressive images of bear fettering and bear fighting can be seen in, for instance, the 6th-century Byzantine mosaics from Kissoufim in Israel and Umm-al-Rasas in Jordan, as well as on the Mediterranean bronze pan from a Merovingian burial in the *Reihengräberfeld* of Güttingen, Kt. Thurgau, Switzerland (Figs. 14–16), to name just a few examples (WAMERS 2009, fig. 33.2; SCHMAUDER 2000, fig. 7; GARSCHA 1933; for more parallels see OEHRL 2013, 308–309). Wamers also considers depictions of the *damnatio ad bestias*, the killing of Christians by wild animals in the arena, as possible iconographic models.

Whatever the foreign models of the Torslunda images might have been, hunting, fighting, killing and capturing a bear must have been regarded as an extraordinarily heroic deed (OEHRL 2013), and thus it makes sense to place it on a warrior chieftain's helmet. Heroes fighting bears are often mentioned in medieval written sources (ibid. 298–310; RANKE/REICHSTEIN 1976), in the Middle High German *Nibelungenlied* (stanzas 939, 962: GROSSE 1999, 286–287, 292–293; see also OBERMAIER, this volume), in several Icelandic sagas, and in the heroic Edda poem *Helgakviða Hundingsbana ǫnnor* (stanza 8: NECKEL/KUHN 1983, 152). In this poem, the hero Helgi is asked by his wife Sigrun where he fought. Helgi answers that he fed eagles with weapons and took, i.e. captured, bears in Bragalund (*er ec biorno tók í Bragalundi / oc ætt ara oddom saddac*). "Feeding eagles" is a poetic phrase (*kenning*) and means "to slay enemies in battle" (see e.g. JESCH 2002). The kenning "taking/capturing bears" is to be understood in the same way, as a poetic code meaning "to defeat enemies heroically" (BECK 1964, 45–50; 1970). Even if these written records are much younger than the Torslunda dies, they can give us an impression of the significance of the bear motif in Germanic heroic tradition. The images may actually refer to those poetic metaphors (ibid.).

On another bronze die from Torslunda (die D) and several other Scandinavian, Anglo-Saxon and Alamannic artefacts from the Merovingian Period, warriors with animal heads can be seen, probably

---

5   For the relevant photos and drawings of the material see HAUCK 1978, pl. VIII, 10a–c, 11–14; BÖHNER 1991, figs. 18, 19. See also HELMBRECHT 2011, 104 fig. 11b, d, e, cat. nos. 592, 1090, 1101. Regarding the numerous previous interpretations of the fettering and bear fighting motifs: ibid., 102–106; OEHRL 2011, 172–175; 2019a, 131–136.

6   On show hunts and animal fights in the Roman amphitheaters, in which bears were often included, see e.g. KELLER 1909, 178–179; TOYNBEE 1983, 84–90; HUGHES 2007, 66–68; PASTOUREAU 2008, 55–56. See also HURKA, this volume.

masked men, dressed up as beasts of prey, most probably as wolves (Figs. 17–19). These depictions are commonly – and quite convincingly – interpreted as *úlfheðnar* (most recently SAMSON 2020, 252–264), warriors mentioned in early Skaldic poetry who, according to Snorri Sturlusson's *Ynglinga saga* (ch. 6; BJARNI AÐALBJARNARSON 1941, 17), are closely linked to the chief god, Óðinn. The god himself, with horned headgear, seems to accompany the warrior with the wolf mask on the Torslunda die (in particular BECK 1968, 247–249; HAUCK 1994, 218, 220–222; most recently OEHRL 2017; 2019a, 230–242). These figures represent wolf warriors, rather than bear warriors.[7] Nevertheless, they should be introduced here, because I want to present a previously unknown recent find from England, which is comparable to the Torslunda motif, depicting the horned warrior (i.e. Óðinn), together with a bear – not a human warrior with an animal mask, but a real bear, standing on its hind legs (Fig. 20). This bronze figural application is 31 mm high and has two application lugs on its flat backside. It was offered at TimeLine Auctions with only the information that it is in "very fine condition" and "professionally cleaned and conserved" and, concerning the provenance, it was stated: "Property of a professional collector, acquired before 1990". The piece was sold on the 28th of November 2018 for 500 British pounds. Unfortunately, there is currently no way of checking whether this figure is authentic or not.

Further relevant Vendel Period depictions are the bronze bear figurines placed on the socket of a spearhead from grave 12 in the burial ground of Vendel (STOLPE/ARNE 1927, pl. 34.5) and a similar but fragmented piece from By in Nord-Trøndelag, Norway (PETERSEN 1912; GJESSING 1934, 54–55 fig. 54; see Figs. 21–22).[8] The latter represents a single find, while the Vendel spear was found in a very richly furnished warrior's grave. The Vendel spearhead's blade is relatively broad and could therefore be regarded as a boar spear or a bear spear. Based on this, the small beast heads on Migration Period spears, probably hunting spears,[9] from Rheden in Gelderland, Netherlands, and Vermand in Aisne, France, can be regarded as bears as well (BÖHME 1974, 101–102, pls. 65, 137; BEERMANN 2016, 65–66 fig. 50). In the light of these Migration and Vendel Period "spear bears", a group of animal-shaped sword and knife pommels from Uppland, Södermanland and Finland must be considered as well (LAMM/RUNDKVIST 2005; 2011; BEERMANN 2016, 66–67, fig. 51–53; see Figs. 23–26). Actually, most of them seem to depict beasts of prey; however, clear bear-like features cannot be observed. The sword pommel from Ed parish in Uppland, incidentally, originates from a 6th-century warrior's grave, which also contained bear claws, possibly originating from an entire bear skin.

The next group to be considered here are the so-called *guldgubbar* or gold foil figures,[10] which are very tiny (mostly between 1 and 2 cm²) paper-thin gold foils with stamped images of human figures or, forming a relatively small sub group, individually produced figures cut out of gold foil. There are more than 3,000 gold foil figures known today, most probably dating to the Vendel Period, all of them found in Scandinavia, usually at so-called central places, i.e. settlements of political, economic and cultic significance and seats of rulership, such as Helgö and Uppåkra in Sweden, Sorte Muld on Bornholm and Gudme in Funen, Denmark, and Borg in Lofoten, Norway. Many of the gold foil

---

7 On the 9th century Oseberg tapestry, several warriors with animal masks can be seen, such as boars, birds and other beasts (VEDELER 2019) – a bear warrior, however, cannot be identified with certainty. Regarding the Old Norse bear warriors (*berserkir*), see SUNDQVIST, this volume.

8 BÖHNER 1991, 697 fig. 13; LAMM/RUNDKVIST 2005, 110 fig. 8–9; OEHRL 2013, 312 fig. 32; BEERMANN 2016, 65 fig. 49.

9 The protruding animal heads on the weapons' socket may form a kind of guard (or "stopper"). This "guard" and the broad shape of the blade make these weapons comparable to boar spears and bear spears known from medieval and modern times.

10 In 2017, the gold foil figure phenomenon was discussed in the course of an international and interdisciplinary workshop at the ZBSA in Schleswig, the conference papers of which represent the most extensive and up-to-date publication on the topic so far: PESCH/HELMBRECHT 2019. See also the papers by K. Hauck and M. Watt in HAUCK 1992, which are still considered essential. The main monograph and edition of the material by M. Watt, which has long been announced, is still in preparation.

figures were found in hoards, in connection with great hall complexes. Most likely, these gold foils were used as sacrificial offerings, probably as a kind of cultic currency or temple money, similar to figuratively decorated votive plaques and figurines known from temples of the Roman Empire and its provinces (Watt 1992, 221–224; 1999, 138–140; Hauck 1993, 411–421; 1998, 318–320; Lamm 2004, 130).

A very small group of figures are individually cut-out animals (Watt 1992, 218–219 fig. 11a–e), which possibly represent animal sacrifices (Oehrl 2019b, 401–402). One of them, found in Sorte Muld, can be identified as a boar because of its curled tail, its bristle crest, and its tusks (Watt 1992, 218; Fig. 27). A further animal figure from Sorte Muld has very similar body proportions; however, the typical boar features are completely missing, so it should actually be regarded as the depiction of a bear, as already stated by Watt (ibid.; see Fig. 28). The same is true, in my and Watt's opinion, in the case of a newly discovered, previously unpublished gold foil figure from the settlement of Smørenge on Bornholm, a metal detector find from 2006 (Fig. 29).[11] The shape and proportions of the body, the massive head with small rounded ears resemble a bear, while typical boar-like features are missing.

There is another relevant recent find I would like to present here – a stamp for producing gold foil figures found in Bjerringbro in Jutland, Denmark, a metal detector find from April 2018 (Mortensen 2018; see Fig. 30).[12] The bronze stamp is 1.5 cm wide and 1.8 cm high, and it represents, strangely enough, the same sitting and fettered beast, probably a stylised bear, who is depicted on the Vendel Period helmet plaques (Figs. 10–13). The gold foil stamp is corroded and worn, but the crucial elements can be observed – the sitting body posture, the outstretched and apparently crossed arms and long claws, as well as the longish snout.

On Late Vendel and Viking Period Gotland, there is a special type of brooches of the female costume, called "animal head brooches" (Carlsson 1983), which have been brought into play and considered as representations of bears as well. As a matter of fact, many of these brooches are reminiscent of boar or rather bear heads (Beermann 2016, 63–64 fig. 45); with some imagination, a massive forehead with small rounded ears, a longish snout, sometimes with nostrils, and a pair of eyes can be observed (Fig. 31). However, whether all of these brooches were actually intended to represent bears, another beast or anything else remains unclear. In a considerable number of cases this interpretation seems at least conceivable. Even if this type of brooch was not intended to represent bear heads right from the beginning (as its form clearly derives from the shape of crossbow fibulae, which have nothing to do with animal heads), it is, in my opinion, hard to imagine that the Viking Age Gotlanders did not associate their characteristic form with certain animals at all. However, this must of course remain unproved.

From the Early Viking Period, an exceptional item from Klinta on Öland must be mentioned – an iron staff of at least 82 cm in length with a basket-like end, consisting of four rods and crowned by a bronze miniature of a house (Fig. 32).[13] There are 37 iron staffs of this kind known from Viking Scandinavia, and they have been interpreted in different ways – as sceptres, weapons, roasting spits, musical instruments, lamps, but commonly as magic staffs (Gardeła 2016). The staff from Klinta was found in a very complex female cremation grave, which contained a large amount of grave goods, including such remarkable items as copper sheets with runic inscriptions, Thor's hammer amulets, the remains of a horse harness, and bear claws, possibly originating from an entire bear skin (ibid., 342–343). The top of the basket-like element is decorated with four animal heads, one at each side, which very much resemble the heads of bears.

---

11 I would like to warmly thank Margrethe Watt for this information and the photos as well as for the permission to publish the find in the current volume.
12 My thanks go to Rasmus Birch Iversen, who provided me with information about the find and photos.
13 Further thanks go to Leszek Gardeła, who provided me with pictures of the item.

I would also like to mention the silver ear spoon from the 9th century grave no. 507 in the burial ground at Birka (ARBMAN 1940, pl. 173.lb; HOLMQVIST 1960, 113 fig. 19; HELMBRECHT 2011, cat. no. 525 fig. 23j), depicting a woman with a drinking horn on one side of the handle, and an upright-standing beast with outstretched paws on the other side (Fig. 33). The fact that this beast is standing on its hindlegs could indicate the depiction of a bear. However, the animal has a long tail, which is more suited to a dog or a wolf. It is difficult to decide whether a bear or a wolf is depicted – or something in between.

Furthermore, there are two Early Viking Period brooches kept at the Archaeological Museum in Stavanger, Norway, which deserve to be called "bear brooches" (KRISTOFFERSEN 2014, 32–37). The oval brooch from Friestad in Rogaland was found in 1898, its context is unknown (Fig. 34a). The brooch is made of gilded bronze with inlaid silver. On both long sides of the brooch, a massive bear-like head with small rounded ears protrudes, flanked by a pair of broad paws, which seemingly belong to the animal (Fig. 34b). The second bear brooch from Stavanger is an equal-armed brooch from Ragje in Rogaland, which was found in 1935 during work on cultivated land, probably near a large mound (Fig. 35a). The brooch is made of gilded bronze and features four small figurines clearly depicting a teddy bear-like creature, hugging itself (Fig. 35b) – which is a typical feature in the Early Viking gripping beast style, to which this same brooch belongs. Such brooches were part of the female costume.

There are at least two further Norwegian gripping beast figurines, which in my opinion are intended to represent bears, both single finds without a find context – a bear-like beast made of amber that is gripping its own neck and feet from a farmstead in Råde, Østfold (Fig. 36), and a pair of two bears gripping each other, which are made of jet (so-called "black amber") and were found somewhere on the shore of Tresfjord in Vestnes, Møre og Romsdal (Fig. 37).[14] There is no information about the original contexts and function of these figurines available; however, they are quite tiny, only a few centimetres, so they could have been used as pendants. The gripping beast from Råde is kept in the University Museum Oslo, and the Vestnes bears in Bergen.

Based on these finds from Ragje, Råde, and Vestnes, which represent individual three-dimensional gripping beasts (which could be subsumed under the term "naturalistic sculptural gripping beast style"), it may be concluded that the frequently-depicted (two-dimensional) animals in the Viking Period gripping beast styles must be regarded as bears in general, instead of cats or lions as has so far been suggested (STEUER 1994). In this case, the bear would represent one of the most frequently represented animals in Viking Period art. The connection between the Viking gripping beast motif and the bear might also be seen in the fact that the animal head brooches from Gotland (see above) are frequently decorated in the gripping beast style. Are they representing bear heads decorated with "gripping bears"? There are even more three-dimensional gripping beasts to be mentioned: In 2004, a pair of exceptional gilded copper-alloy oval brooches with silver inlay was found in Finglas, northwest of Dublin, in a richly furnished 9th-century female grave (SIKORA 2010; Fig. 38). These Scandinavian brooches are decorated with gripping beast-style elements and feature four protruding bear figurines, sitting and with outstretched gripping paws, as well as four beautiful big bear heads.[15]

There are more monuments from the Viking settlement area on the British Isles to be considered, in particular a special Scandinavian type of tombstone, the so-called hogbacks (BAILEY 1980, 85–100; LANG 1984) already mentioned above. The common term for these monuments, referring to their longish and convex shape, is misleading – actually, the stones imitate the shape of Scandinavian long houses or hall buildings; sometimes even the roof with wooden shingles and other constructional

---

14 These figures are often depicted in publications on Viking Period art but have not been discussed in more detail so far (e.g. GRAHAM-CAMPBELL 1980, 104, 137; 2013, 69 fig. 68; LINDOW 2001, 6).
15 I shall discuss what I call "naturalistic sculptural gripping beast style" and the connection between the Early Viking Period gripping beast styles and bear symbolism in more detail elsewhere.

details are depicted. In most cases, these stone houses are flanked by so-called end beasts, which clearly represent bears. The stones 17A from Brompton and 4A from Ingleby Arncliffe, both in North Yorkshire, are good examples (Figs. 39–40; cf. Lang 2001, pls. 82, 335). That a house-shaped tombstone refers to the idea of a house or hall of the dead is obvious; however, the meaning of the bear-shaped end-beasts is difficult to determine. Do they protect the building as a kind of apotropaic element? Interestingly, some of the bears are wearing a muzzle, which appears to characterise them as captured and more or less tamed and harmless. Ultimately, the Early Viking Period hogback end-beasts can be regarded as three-dimensional gripping beasts and should thus be ascribed to the "naturalistic sculptural gripping beast style" as well (see above).

Concerning the muzzle, there is a conclusive parallel depicted on the famous Bayeux tapestry dating to the late 11th century (Fig. 41); a man with sword and shield is attacking a bear who is fettered to a tree, wearing a muzzle (Wilson 1985, 12). Man-beast combat performances and bear-baiting blood sport events were quite popular during the Middle Ages and were practised in England from Anglo-Saxon times onwards, specifically connected with royal courts (see O'Regan, this volume; Strutt 1801, 204–207; Bartlett 2000, 669–670; Brunner 2005, 73–76; 2010, 139–141; Kiser 2007, 117–118; Pastoureau 2008, 211–212; Oehrl 2013, 309). The bear was chained to a pole, often handicapped by having been blinded or having had its claws cut off, and a pack of dogs was set on it (Fig. 42). Man-bear combats and bear baiting are also mentioned in Old Norse saga literature (Oehrl 2013, 309). This blood sport (as well as bear dancing performances) might be the background of the hogback end-beasts wearing muzzles, indicating a tamed bear used for an aristocratic pastime – performed in the ruler's hall that is represented by the tombstone itself.

There are two extraordinarily interesting but very strange belt buckles from England that should also be mentioned at this point. These are (or have been) available on the internet, one on TimeLine Auctions (Fig. 43) and the other one on the UK Detector Finds Database (Fig. 44a–b). The rectangular plate of these bronze buckles is formed by a bear seen from a bird's-eye view, with outstretched paws grasping the bar of the D-shaped frame. The plate and frame are made in one piece. In addition, there is a pair of wings on the bear's neck. According to the two mentioned websites, the buckles date to the Viking Age. However, there are no dateable parallels known so far. The hogback end-beasts could be mentioned as remotely similar comparative material. Remarkably, the English "buckle bear" is *gripping* the bar of the buckle's frame. As a matter of fact, there are several belt buckles decorated in the gripping beast style known from Scandinavian find spots, some of which depict small hands gripping the buckle's frame (e.g. Arbman 1940, pls. 86.2, 87.2, 87.5). However, these parallels are not yet fully convincing, and in any case the naturalism and the entire conception of the English buckles are without direct comparison. Possibly, however, they have to be assigned to the "naturalistic three-dimensional gripping beast style" characterised above.

Another depiction of a bear from the British Isles can be found on a Viking stone cross from the Isle of Man. The Manx Crosses were made by Scandinavian settlers, combining elements of indigenous insular and Viking traditions. The runic cross from Kirk Andreas (no. 103) depicts an elaborate hunting scene, consisting of a horseman and a dog chasing a red deer, as well as a procession of further wild animals, including a bear (Fig. 45a–b; cf. Kermode 1994, 194–195 fig. 38.9 pl. LIII; Wilson 2018, 75–77 fig. 30).[16] The bear also occurs among the frequent hunting scenes depicted on the 8th/9th-century Pictish carved stone slabs of Scotland, which are of a similar design to the Manx Crosses, featuring a central cross symbol and plenty of animals arranged around it (Fig. 46; cf. Allen 1903, 235–239 fig. 250B; Fraser 2008, cat. no. 67.1).

---

16 The runic inscription on the slab's narrow side reads as follows: *Sandulf the black erected this cross to the memory of his wife Arinbjörg.*

Surprisingly enough, as far as I can see, there are no striking depictions of bears from the Late Viking Age to be presented. The only relevant figures that I know of are some small and very simple, not to say primitive, quadrupeds on a handful of Upplandic rune stones (in particular U 241, 860, 969). As a matter of fact, the body proportions and short tail seem bear-like; however, the figures are not very detailed and are difficult to interpret (Figs. 47–49). However that may be, these beasts are depicted as elements of a Christian imagery, as the sign of the cross on the top of the stones appears to indicate. In the case of the Måsta stone (Balingsta sn), the "bear" belongs to a group of animals eating from the cross, i.e. the *arbor vitae* (cf. U 1140).

To sum up: This paper constitutes a survey of bear depictions from Sweden, Norway, and Denmark, as well as Scandinavian-influenced parts of England, including certain or almost certain bear depictions, probable bear depictions and also a group of possible depictions that are not verifiably bears. Chronologically, the relevant material ranges from the Roman Iron Age (i.e. the soap stone figurine from Modvo, Norway, sometime between the 1st and 4th centuries) to the Late Viking Age (11th century); however, most of it by far dates to the Merovingian/Vendel Period and the Early Viking Age (6th–9th/10th centuries). The images occur in quite different contexts, on different objects, bearing different possible meanings – military and heroic contexts, commemoration and sepulchral contexts, female jewellery, figurines or pendants (amulets?), and more. My main conclusion is: The bear is definitively not "non-existing" in Late Iron Age and Viking art, though it is not very frequent and is hard to identify with certainty. However, there is a respectable amount of probable and more or less certain bear depictions to be considered. The number of clearly identifiable depictions of wolves and boars in Late Iron Age art, for instance, is not very much higher. The major problem remains that the bear has almost no crucial physical features that are suitable for indicating pictorial representations clearly and unambiguously – unlike red deer, raptors, and boars, which are often clearly marked by antlers, a hooked beak, or tusks and a bristle crest, respectively. The limited possibilities of zoological identification are a main problem of Germanic pictorial art and animal style in general, which does not necessarily indicate the lack of a certain species but rather a lack of clarity and naturalism, which is typical for this kind of imagery – and probably a lack of understanding of this playful, and sometimes primitive and alien-like style of art.

Postscriptum

Through the kind help of Peter Pentz, National Museum Copenhagen, Denmark, for which I would like to express my sincere thanks, another find from Denmark came to my attention after the completion of the manuscript. A recent detector find from Herringe sogn, Sallinge Herred, Svendborg Amt (South Fyn; Fig. 50) represents the same bound beast that is also depicted on the plate from Torslunda (Fig. 10), the foils from Vendel (Fig. 11–13), and the small patrice from Borre Vestergård (Fig. 30). It is a rectangular pendant with two small eyelets for hanging at each of the top edges. The rectangular plate has a frame around a recessed area with an embossed, furry animal figure with four pairs of twisted legs ending in long curved claws. The inlay is made of silver.

Bibliography

Allen 1903: J. R. Allen, The Early Christian Monuments of Scotland. A classified, illustrated, descriptive list of the monuments, with an analysis of their symbolism and ornamentation (Edinburgh 1903).

Arbman 1940: H. Arbman, Birka. Untersuchungen und Studien I. Die Gräber. Tafeln. Kungl. Vitterhets Historie och Antikvitets Akademien (Stockholm 1940).

Arwidsson 1977: B. Arwidsson, Valsgärde 7. Die Gräberfelder von Valsgärde III. Acta Musei Antiquitatum

Septentrionalium Regiæ Universitatis Upsaliensis V (Uppsala 1977).

BAILEY 1980: R. N. BAILEY, Viking Age Sculpture in Northern England (London 1980).

BARTLETT 2000: R. BARTLETT, England Under the Norman and Angevin Kings 1075–1225 (Oxford 2000).

BECK 1964: H. BECK, Einige vendelzeitliche Bilddenkmäler und die literarische Überlieferung. Bayerische Akademie der Wissenschaften, Philosophisch-Historische Klasse, Sitzungsberichte 1964:6 (München 1964).

BECK 1968: H. BECK, Die Stanzen von Torslunda und die literarische Überlieferung. Frühmittelalterliche Studien 2, 1968, 237–250.

BECK 1970: H. BECK, Tiere der Jagd und der Walstatt in den eddischen Liedern. In: U. Schwab (ed.), Das Tier in der Dichtung (Heidelberg 1970) 55–73.

BEERMANN 2016: S. BEERMANN, Bärenkrallen und Bärenfelle in Brand- und Körpergräbern der vorrömischen Eisenzeit bis Völkerwanderungszeit in Mittel- und Nordeuropa. Universitätsforschungen zur Prähistorischen Archäologie 279 (Bonn 2016).

BJARNI AÐALBJARNARSON 1941: BJARNI AÐALBJARNARSON, Snorri Sturluson. Heimskringla 1. Íslenzk Fornrit 26 (Reykjavík 1941).

BLANKENFELDT 2015: R. BLANKENFELDT, Fünfzig Jahre nach Joachim Werner: Überlegungen zur kaiserzeitlichen Kunst. In: W. Heizmann/S. Oehrl (eds.), Bilddenkmäler zur germanischen Götter- und Heldensage. In: RGA-E 91 (Berlin, Boston 2015) 9–81.

BÖHME 1974: H. W. BÖHME, Germanische Grabfunde des 4.–5. Jahrhunderts zwischen Elbe und Loire (München 1974).

BÖHME 1998: H. W. BÖHME, Goldblattkreuze. In: RGA 12 (Berlin, New York 1998) 312–318.

BÖHNER 1991: K. BÖHNER, Die frühmittelalterlichen Silberphaleren aus Eschwege (Hessen) und die nordischen Pressblech-Bilder. Jahrbuch des RGZM 38(2), 1991 (1995), 681–743.

BÖHNER 1994: K. BÖHNER, Die frühmittelalterlichen Spangenhelme und die nordischen Helme der Vendelzeit. Jahrbuch des RGZM 41, 1994, 471–549.

BRUCE-MITFORD 1975–1983: R. BRUCE-MITFORD, The Sutton Hoo Ship Burial. Vol. I: Excavations, Background, the Ship, Dating and Inventory; Vol II: Arma, Armour and Regalia; Vol. III: Late Roman and Byzantine silver, hanging bowls, drinking vessels, cauldrons and other containers, textiles, the lyre, pottery bottle and other items (London 1975–1983).

BRUNNER 2005: B. BRUNNER, Eine kurze Geschichte der Bären (Berlin 2005).

BRUNNER 2010: B. BRUNNER, Bär und Mensch. Die Geschichte einer Beziehung (Darmstadt 2010).

CARLSSON 1983: A. CARLSSON, Djurhuvudformiga spännen och gotländsk vikingatid: text och katalog. Stockholm Studies in Archaeology 5 (Stockholm 1983).

CRAMP 1984: R. CRAMP, Corpus of Anglo-Saxon Stone Sculpture I: County Durham and Northumberland (Oxford 1984).

DEBIDOUR 1961: V.-H. DEBIDOUR, Le Bestiaire Sculpté du Moyen Age en France. Collection Grandes Études d'Art et d'Archéologie, N° 2 (Paris 1961).

FINGERLIN 2010: G. FINGERLIN, Die ältesten christlichen Bilder der Alamannia. Zu Herkunft und Ikonographie der drei silbernen Phalerae aus dem Kammergrab von der „Gierhalde" in Hüfingen, dem Hauptort der frühmittelalterlichen Baar. In: V. Huth/J. Regnath (eds.), Die Baar als Königslandschaft (Sigmaringen 2010) 25–46.

FINGERLIN 2012: G. FINGERLIN, Die ältesten christlichen Bilder der Alamannia – zur Herkunft und Ikonographie der drei silbernen Phalerae aus dem Grab in Hüfingen. Schriften des Vereins für Geschichte und Naturgeschichte der Baar 55, 2012, 7–26.

FRASER 2008: I. FRASER (ed.), The Pictish Symbol Stones of Scotland (Edinburgh 2008).

GARDEŁA 2016: L. GARDEŁA, (Magic) Staffs in the Viking Age. Studia Medievalia Septentrionalia 27 (Wien 2016).

GARSCHA 1933: F. GARSCHA, Die Bronzepfanne von Göttingen. Germania 17, 1933, 36–42.

GJESSING 1934: G. GJESSING, Studier i norsk merovingertid. Kronologi og oldsakformer. Skrifter utgitt av Det Norske Videnskaps-Akademi i Oslo II. Hist.-filos. klasse 1934:2 (Oslo 1934).

GRAHAM-CAMPBEL 1980: J. GRAHAM-CAMPBELL, Das Leben der Wikinger (München 1980; original edition: The Viking World [London 1980]).

GRAHAM-CAMPBEL 2013: J. GRAHAM-CAMPBELL, Viking Art (London 2013).

GROSSE 1999: S. GROSSE, Das Nibelungenlied. Mittelhochdeutsch/Neuhochdeutsch. Nach dem Text von Karl Bartsch und Helmut de Boor ins Neuhochdeutsche übersetzt und kommentiert von Siegfried Grosse. Reclam Universal-Bibliothek 644 (Stuttgart ²1999).

HAGBERG 1976: U. E. HAGBERG, Fundort und Fundgebiet der Modeln aus Torslunda. Frühmittelalterliche Studien 10, 1976, 323–349.

HAGBERG et al. 1991: U. E. HAGBERG/B. STJERNQUIST/M. RASCH, Ölands järnåldersgravfält II. Högsrum, Glömminge, Algutsrum, Torslunda, Långlöt, Runsten, Norra Möckleby och Gårdby (Stockholm 1991).

HAUCK 1957: K. HAUCK, Alemannische Denkmäler der vorchristlichen Adelskultur. Zeitschrift für Württembergische Landesgeschichte 16(1), 1957, 1–40.

HAUCK 1978: K. HAUCK, Bildforschung als historische Sachforschung. Zur vorchristlichen Ikonografie der figuralen Helmprogramme aus der Vendelzeit. In: K. Hauck/H. Mordek (eds.), Geschichtsschreibung und geistiges Leben im Mittelalter. Festschrift für Heinz Löwe zum 65. Geburtstag (Köln, Wien 1978) 27–70.

HAUCK 1982: K. HAUCK, Zum zweiten Band der Sutton Hoo-Edition. Frühmittelalterliche Studien 16, 1982, 319–362.

HAUCK 1992: K. HAUCK (ed.), Der historische Horizont der Götterbild-Amulette aus der Übergangsepoche von der Spätantike zum Frühmittelalter. Bericht über das Colloquium vom 28.11.–1.12.1988 in der Werner-Reimers-Stiftung, Bad Homburg. Abhandlungen der Akademie der Wissenschaften in Göttingen, Philologisch-Historische Klasse, Dritte Folge 200 (Göttingen 1992).

HAUCK 1993: K. HAUCK, Die bremische Überlieferung zur Götter-Dreiheit Altuppsalas und die bornholmischen Goldfolien aus Sorte Muld. Zur Ikonologie der Goldbrakteaten LII. Frühmittelalterliche Studien 27, 1993, 409–479.

Hauck 1994: K. Hauck, Altuppsalas Polytheismus exemplarisch erhellt mit Bildzeugnissen des 5.–7. Jahrhunderts. Zur Ikonologie der Goldbrakteaten LIII. In: H. Uecker (ed.), Studien zum Altgermanischen. Festschrift für Heinrich Beck. RGA-E 11 (Berlin, New York 1994) 197–230.

Hauck 1998: K. Hauck, Goldblechfigürchen. In: RGA 12 (Berlin, New York 1998) 318–322.

Hedeager 2004: L. Hedeager, Dyr och andre mennesker – mennesker og andre dyr. In: A. Andrén/K. Jennbert/C. Raudvere (red.), Ordning mot kaos. Studier av nordisk förkristen kosmologi. Vägar till Midgård 4 (Lund 2004) 219–252.

Hedeager 2011: L. Hedeager, Iron Age Myth and Materiality. An Archaeology of Scandinavia AD 400–1000 (Abingdon 2011).

Helmbrecht 2011: M. Helmbrecht, Wirkmächtige Kommunikationsmedien. Menschenbilder der Vendel- und Wikingerzeit und ihre Kontexte. Acta Archaeologica Lundensia Series Prima in 4°, no. 30 (Lund 2011).

Henry 1933: F. Henry, La Sculpture Irlandaise 1–2. Pendant les douze premiers siècles de l'ère chrétienne. Études d'art et d'archéologie (Paris 1933).

Holmqvist 1938: W. Holmqvist, Zur Herkunft einiger germanischer Figurendarstellungen der Völkerwanderungszeit. IPEK 12, 1938, 78–94.

Holmqvist 1939: W. Holmqvist, Kunstprobleme der Merowingerzeit (Stockholm 1939).

Holmqvist 1960: W. Holmqvist, The dancing gods. Acta Archaeologica 31, 1960, 101–127.

Holzapfel 1973: O. Holzapfel, Stabilität und Variabilität einer Formel. Zur Interpretation der Bildformel „Figur zwischen wilden Tieren" mit besonderer Berücksichtigung skandinavischer Beispiele. Mediaeval Scandinavia 6, 1973, 7–38.

Hughes 2007: J. D. Hughes, Hunting in the Ancient Mediterranean World. In: L. Kalof (ed.), A Cultural History of Animals 1: In Antiquity (Oxford, New York 2007) 47–70.

Hvoslef Krüger 1988: S. Hvoslef Krüger, Bjørneklør fra vestlandske graver. In: Festskrift til Anders Hagen. Arkeologiske Skrifter fra Historisk Museum, Universitetet I,4 (Bergen 1988) 357–366.

Jesch 2002: J. Jesch, Eagles, ravens and wolves: Beasts of battle, symbols of victory and death. In: J. Jesch (ed.), The Scandinavians from the Vendel Period to the Tenth Century. An Ethnographic Perspective. Studies in Historical Archaeoethnology (Woodbridge 2002) 251–280.

Jörg/Martin 1984: Ch. Jörg/M. Martin, Danielschnallen. In: RGA 5 (Berlin, New York 1984) 244–248.

Keller 1909: O. Keller, Die antike Tierwelt 1: Säugetiere (Leipzig 1909).

Kermode 1994: Ph. M. C. Kermode, Manx Crosses. With an introduction by David M. Wilson (London ²1994; first edition 1907).

Kiser 2007: L. J. Kiser, Animals in Medieval Sports, Entertainment, and Menageries. In: B. Resl (ed.), A Cultural History of Animals 2: In the Medieval Age (New York 2007) 103–126.

Klein-Pfeuffer 1993: M. Klein-Pfeuffer, Merowingerzeitliche Fibeln und Anhänger aus Pressblech. Marburger Studien zur Vor- und Frühgeschichte 14 (Marburg 1993).

Klein-Pfeuffer 2015: M. Klein-Pfeuffer, Zur Deutung der Pressblechscheiben von Eschwege-Niederhone Gr. 17. In: W. Heizmann/S. Oehrl (eds.), Bilddenkmäler zur germanischen Götter- und Heldensage. RGA-E 91 (Berlin, Boston 2015) 261–293.

Kopár 2012: L. Kopár, Gods and Settlers. The Iconography of Norse Mythology in Anglo-Scandinavian Sculpture. Studies in the Early Middle Ages 25 (Turnhout 2012).

Kristoffersen 2014: E. S. Kristoffersen, Touched by a Viking. Tales of Treasures (Stavanger 2014).

Kühn 1941/1942: H. Kühn, Die Danielschnallen der Völkerwanderungszeit. IPEK 15/16, 1941/1942, 140–169.

Lamm 1980: J. P. Lamm, En storman i Ed. Historiska Nyheter 13, 1980, 3.

Lamm 2004: J. P. Lamm, Figural Gold Foils found in Sweden: a study based on the discoveries from Helgö. In: H. Clarke/K. Lamm (eds.), Excavations at Helgö XVI. Exotic and Sacral Finds from Helgö. Kungl. Vitterhets Historie och Antikvitets Akademien (Stockholm 2004) 41–142.

Lamm/Rundkvist 2005: J. P. Lamm/M. Rundkvist, Björnen i Ägget. En vapengrav i Eds socken, Uppland och vendeltidens vapen med djurfiguriner. Fornvännen 100, 2005, 101–113.

Lamm/Rundkvist 2011: J. P. Lamm/M. Rundkvist, Makthavare i grannskapet – en vapengrav från 500-talet nära Runsa. In: M. Olausson (red.), Runnhusa: bosättningen på berget med de många husen. Skrifter från projektet Runsa borg, Eds socken, Uppland 1 (Stockholm 2011) 111–127.

Lang 1984: J. T. Lang, The Hogback. A Viking Colonial Monument. Anglo-Saxon Studies in Archaeology and History 3, 1984, 85–176.

Lang 2001: J. Lang, Corpus of Anglo-Saxon Stone Sculpture VI: Northern Yorkshire (Oxford 2001).

Lindenschmid 1870: L. Lindenschmid, Die Alterthümer unserer heidnischen Vorzeit. II. Band (Mainz 1870).

Lindow 2001: J. Lindow, Norse Mythology: A Guide to Gods, Heroes, Rituals, and Beliefs (Oxford 2001).

Magnus 1975: B. Magnus, Krosshaugfunnet. Et forsøk på kronologisk og stilhistorisk plassering i 5. årh. Stavanger Museums Skrifter 9 (Stavanger 1975).

Magnus 2014: B. Magnus, Kvinnene fra Krosshaug i Klepp og Hol på Inderøya. In: E. S. Kristoffersen/M. Nitter/E. S. Pedersen (red.), Et Akropolis på Jæren? Tinghaugplatået gjennom jernalderen. AmS-Varia 55 (Stavanger 2014) 71–87.

Mortensen 2018: S. T. Mortensen, Et lille blik på guldgubbepatricernes udbredelse. Fund & Fortid, Arkæologi for alle 2018(3), 17.

Neckel/Kuhn 1983: G. Neckel/H. Kuhn (eds.), Edda. Die Lieder des Codex Regius nebst verwandten Denkmälern I: Text. Germanische Bibliothek, 4. Reihe: Texte (Heidelberg ⁵1983).

Nordberg 2003: A. Nordberg, Krigarna i Odins sal. Dödsföreställningar och krigarkult i fornnordisk religion (Stockholm 2003).

Oehrl 2011: S. Oehrl, Vierbeinerdarstellungen auf schwedischen Runensteinen. Studien zur nordgermanischen

Tier- und Fesselungsikonografie. RGA-E 72 (Berlin, New York 2011).

Oehrl 2013: S. Oehrl, Bear hunting and its ideological context (as a background for the interpretation of bear claws and other remains of bears in germanic graves of the 1st millennium AD). In: O. Grimm/U. Schmölcke (eds.), Hunting in northern Europe until 1500 AD. Old traditions and regional developments, continental sources and continental influences. Papers presented at a workshop organized by the Centre for Baltic and Scandinavian Archaeology (ZBSA), Schleswig, June 16th and 17th, 2011. Schriften des Archäologischen Landesmuseums, Ergänzungsreihe 7 (Neumünster 2013) 297–332.

Oehrl 2017: S. Oehrl, Der göttliche Schiffsbegleiter mit dem "Hörnerhelm". Ein bislang unbekanntes wikingerzeitliches Bildsteinfragment aus St. Valle im Kirchspiel Rute auf Gotland. Zeitschrift für deutsches Altertum und deutsche Literatur 146, 2017, 1–40.

Oehrl 2019a: S. Oehrl, Die Bildsteine Gotlands – Probleme und neue Wege ihrer Dokumentation, Lesung und Deutung. Studia archaeologiae medii aevi 3 (Friedberg 2019).

Oehrl 2019b: S. Oehrl, Karl Haucks Studien zu den skandinavischen Goldblechfigürchen (guldgubbar): Zusammenfassung, Kritik und neue Überlegungen. In: A. Pesch/M. Helmbrecht (eds.), Gold foil figures in focus. A Scandinavian find group and related objects and images from ancient and medieval Europe. Schriften des Museums für Archäologie Schloss Gottorf, Ergänzungsreihe 14 (München 2019) 389–426.

Oxenstierna 1956: E. Oxenstierna, Die Goldhörner von Gallehus (Lidingö 1956).

Pastoureau 2008: M. Pastoureau, Der Bär. Geschichte eines gestürzten Königs. Aus dem Französischen übersetzt von S. Çorlu (Neu-Isenburg 2008).

Pesch 2012: A. Pesch, Fallstricke und Glatteis: Die germanische Tierornamentik. In: H. Beck/D. Geuenich/H. Steuer (eds.), Altertumskunde – Altertumswissenschaft – Kulturwissenschaft. Erträge und Perspektiven nach 40 Jahren Reallexikon der Germanischen Altertumskunde. RGA-E 77 (Berlin, Boston 2012) 633–687.

Pesch/Helmbrecht 2019: A. Pesch/M. Helmbrecht (eds.), Gold foil figures in focus. A Scandinavian find group and related objects and images from ancient and medieval Europe. Schriften des Museums für Archäologie Schloss Gottorf, Ergänzungsreihe 14 (München 2019).

Petersen 1912: Th. Petersen, Et magisk dyrebillede i Trondhjems museum. Kgl. Norske Videnskabers Selskabs Skrifter 1912:5 (Trondheim 1912).

Piccirillo 1993: M. Piccirillo, The Mosaics of Jordan. American Center of Oriental Research I (Amman 1993).

Pollington et al. 2010: St. Pollington/L. Kerr/B. Hammond (eds.), Wayland's Work. Anglo-Saxon art, myth and material culture from the 4th to the 7th century (Ely 2010).

Ranke/Reichstein 1976: K. Ranke/H. Reichstein, Bär. In: H. Beck et al. (eds.), RGA 2 (Berlin, New York 1976) 45–48.

Romdahl 1922: A. L. Romdahl, Vendel und Byzanz. Byzantinisch-orientalische Einflüsse in einem schwedischen Grabfund der Völkerwanderungszeit. In: H. Glück (ed.), Studien zur Kunst des Ostens. Josef Strzygowski zum sechzigsten Geburtstage von seinen Freunden und Schülern (Wien, Hellerau 1922) 217–226.

Samson 2020: V. Samson, Die Berserker. Die Tierkrieger des Nordens von der Vendel- bis zur Wikingerzeit. RGA-E 121 (Berlin, Boston 2020).

Schmauder 2000: M. Schmauder, Vielteilige Gürtelgarnituren des 6.–7. Jahrhunderts: Herkunft, Aufkommen und Trägerkreis. In: F. Daim (ed.), Die Awaren am Rand der byzantinischen Welt. Studien zu Diplomatie, Handel und Technologietransfer im Frühmittelalter. Monographien aus Frühgeschichte und Mittelalterarchäologie 7 (Innsbruck 2009) 15–44.

Schück 1926: H. Schück, Illustrerad svensk litteraturhistoria 1: Forntiden och medeltiden (Stockholm ³1926).

Seyfarth et al. 1978: W. Seyfarth/L. Jacob-Karau/I. Ulmann (eds.), Ammiani Marcellini Rerum gestarum libri qui supersunt, Vol. 2: Libri XXVI–XXXI (Leipzig 1978).

Sikora 2010: M. Sikora, The Finglas burial. In: J. Sheehan/D. Ó Corráin (eds.), The Viking Age. Ireland and the West. Papers from the Proceedings of the fifteenth Viking Congress, Cork, 18–27 August 2005 (Dublin 2010) 402–417.

Steuer 1994: H. Steuer, Zur Herleitung des nordischen Greiftierstils. In: H. Uecker (ed.), Studien zum Altgermanischen. Festschrift für Heinrich Beck. RGA-E 11 (Berlin, New York 1994) 648–676.

Stolpe/Arne 1927: H. Stolpe/T. J. Arne, La nécropole de Vendel. KVHAA Monografier 17 (Stockholm 1927).

Straume 1962: E. Straume, Nordfjord i eldre jernalder. Årbok for universitetet i Bergen, human. ser. no. 4, 1961 (Bergen 1962).

Ström 1980: Å. V. Ström, Björnfällar och Oden-religion. Fornvännen 75, 1980, 266–270.

Strutt 1801: J. Strutt, The Sports and Pastimes of the People of England. The sports and pastimes of the people of England from the earliest period, including the rural and domestic recreations, May games, mummeries, pageants, processions and pompous spectacles (London 1801).

Terp-Schunter 2017: M. Terp-Schunter, In signo crucis. Eine vergleichende Studie zu den alamannischen und langobardischen Goldblattkreuzen. Tübinger Forschungen zur historischen Archäologie 8 (Büchenbach 2017).

Toynbee 1983: J. M. C. Toynbee, Tierwelt der Antike. Bestiarium romanum. Übersetzt von M. R.-Alföldi und D. Misslbeck. Kulturgeschichte der Antiken Welt 17 (Mainz am Rhein 1983).

U + Nr. = S. B. F. Jansson/E. Wessén, Upplands Runinskrifter 1–4. Sveriges Runinskrifter VI–IX (Stockholm 1940–1958).

Vedeler 2019: M. Vedeler, The Oseberg Tapestries (Oslo 2019).

Wamers 2009: E. Wamers, Von Bären und Männern. Berserker, Bärenkämpfer und Bärenführer im frühen Mittelalter. Zeitschrift für Archäologie des Mittelalters 37, 2009, 1–46.

Watt 1992: M. Watt, Die Goldblechfiguren („guldgubber") aus Sorte Muld, Bornholm. In: K. Hauck (ed.), Der historische Horizont der Götterbild-Amulette aus der Übergangsepoche von der Spätantike zum Frühmittelalter. Bericht über das Colloquium vom 28.11.–1.12.1988 in der Werner-Reimers-Stiftung, Bad Homburg. Abhandlungen der Akademi der Wissenschaften in Göttingen, Philologisch-Historische Klasse, Dritte Folge 200 (Göttingen 1992) 195–227.

Watt 1999: M. Watt, Gubber. In: RGA 13 (Berlin, New York 1999) 132–142.

Werner 1977: J. Werner, Die Ausgrabungen in St. Ulrich und Afra in Augsburg 1961–1968. Textband und Tafelband. Münchner Beiträge zur Vor- und Frühgeschichte 23 (München 1977).

Wilson 1985: Sir D. M. Wilson, Der Teppich von Bayeux (Frankfurt am Main, Berlin 1985).

Wilson 2018: Sir D. M. Wilson, Manx Crosses. A handbook of stone sculpture 500–1040 in the British Isles (Oxford 2018).

Prof. Sigmund Oehrl
Universitetet i Stavanger
Arkeologisk museum
Stavanger
Norway
sigmund.oehrl@uis.no

Fig. 1. Figurine made of soapstone from Modvo in Sogn og Fjordane, Norway; Roman Iron Age (© Universitetsmuseet i Bergen, B11436, Norway).

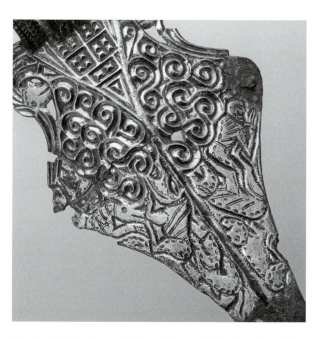

Fig. 2. Gilded silver relief brooch from Hol in Sør-Trøndelag, Norway; Migration Period (© NTNU Vitenskapsmuseet, T9822, Trondheim, Norway; photo K. Dahl, VM).

Fig. 3. Wooden piece of unknown function from Evebø in Sogn og Fjordane, Norway; Migration Period (© Universitetsmuseet i Bergen, B 4590, Norway; photo S. Skare).

Fig. 4. Bronze die A from Torslunda on the isle of Öland, Sweden; Vendel Period (© Statens historiska museer, Stockholm, Sweden).

Fig. 5. Bronze foil (Pressblech) fragment from Valsgärde in Uppland, Sweden; Vendel Period (after ARWIDSSON 1977, fig. 142).

Fig. 6. Silver phalera from Eschwege in Hessen, Germany; Merovingian Period (© Museumslandschaft Hessen Kassel, Sammlung für Vor- und Frühgeschichte).

Fig. 7. Belt buckle from unknown find spot in Spain; Merovingian Period (after OXENSTIERNA 1956, fig. 156).

Fig. 8. Disc brooch from Sprendlingen in Rheinland-Pfalz, Germany; Merovingian Period (after LINDENSCHMIT 1870, Heft VII, Tafel 6.1; cf. also KLEIN-PFEUFFER 1993, no. 293).

Fig. 9. Gold and garnet cloisonné mounts on the purse lid from the Sutton Hoo ship burial in East Anglia, England; Vendel Period (graphically reworked by M. Bolte, ZBSA).

Fig. 10. Bronze die B from Torslunda on the isle of Öland, Sweden; Vendel Period (© Statens historiska museer, Stockholm, Sweden).

Fig. 11. Bronze foil (Pressblech) from Vendel I, Uppland, Sweden; Vendel Period (after STOLPE/ ARNE 1927, pl. VI.2).

Fig. 12. Bronze foil (Pressblech) from Vendel I, Uppland, Sweden; Vendel Period (after HAUCK 1978, pl. 9.12).

Fig. 13. Bronze foil (Pressblech) from Vendel XI, Uppland, Sweden; Vendel Period (after HAUCK 1978, pl. 8.10a).

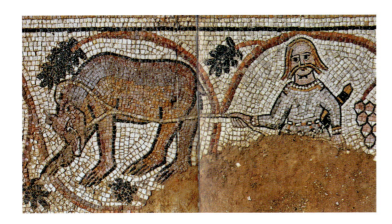

Fig. 14. Byzantine mosaic from Umm-al-Rasas, Jordan; 6th century (after PICCIRILLO 1993, fig. 389; © The American Center of Research [ACOR], Amman, Jordan).

Fig. 15. Byzantine mosaic from Kissoufim, Israel; 6th century (© The Israel Museum, Jerusalem / Bridgeman Images).

Fig. 16. Mediterranean bronze pan from a burial in the Reihengräberfeld *of Güttingen, Kt. Thurgau, Switzerland; Merovingian Period (after* GARSCHA *1933).*

Fig. 17. Bronze die D from Torslunda on the isle of Öland, Sweden; Vendel Period (© Statens historiska museer, Stockholm, Sweden).

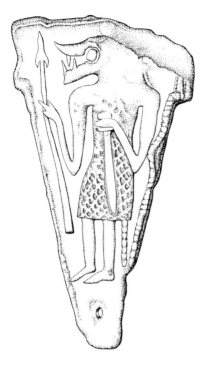

Fig. 18. Bronze foil (Pressblech) die (?) from Fen Drayton in Cambridgeshire, England; Vendel Period (after POLLINGTON et al. 2010, fig. 8.56).

Fig. 19. Scabbard mount from Gutenstein in Baden-Württemberg, Germany; Merovingian Period (after HAUCK 1957, pl. III.5).

Fig. 20. Bronze application from unknown find spot, England; Vendel Period (?). As the only available image from the internet was unsuitable for a reproduction, a drawing had to be made. The image is not online any longer, but can be made available by the author (drawing A. C. Lange, after a template).

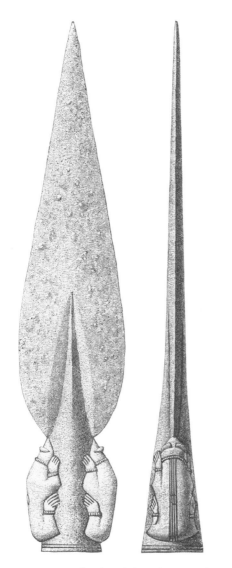

Fig. 21. Spear head with bear figurines from Vendel grave 12, Uppland, Sweden; Vendel Period (after BÖHNER 1991, fig. 13).

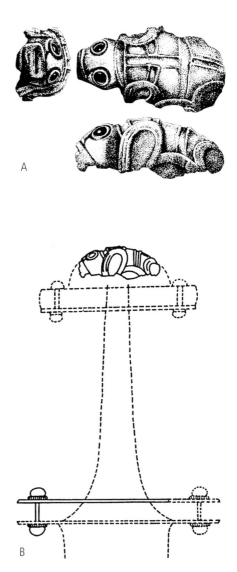

Fig. 23a–b. Sword pommel from Ed in Uppland (prästgården), Sweden; Vendel Period (after LAMM/RUNDKVIST 2005, figs. 2–3.

Fig. 22. Spear head with bear figurine from By in Nord-Trøndelag, Norway; Vendel Period (© NTNU Vitenskapsmuseet, T1269, Trondheim, Norway; photo O. B. Pedersen, VM).

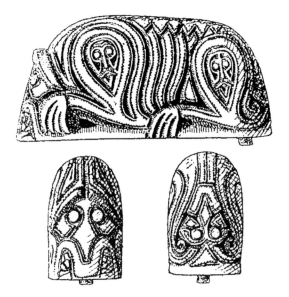

Fig. 24. Sword pommel from Birkaland in Vesilahti (Kirmukarmu), Finland; Vendel Period (after Lamm/Rundkvist *2005, fig. 4).*

Fig. 25. Sword pommel from Husby, Trosa-Vagnhärad, in Södermanland, Sweden; Vendel Period (after Lamm/Rundkvist *2005, fig. 7).*

Fig. 26. Sword pommel from Vörå in Österbotten (Gulldynt), Finland; Vendel Period (after Lamm/Rundkvist *2005, fig. 5).*

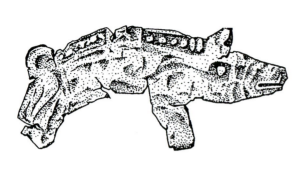
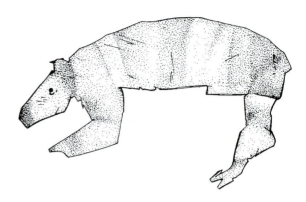

Fig. 27. Gold foil figure from Sorte Muld in Bornholm, Denmark; Vendel Period (?) (after WATT 1992, fig. 11c).

Fig. 28. Gold foil figure from Sorte Muld in Bornholm, Denmark; Vendel Period (?) (after WATT 1992, fig. 11d).

Fig. 29. Gold foil figure from Smørenge in Bornholm, Denmark; Vendel Period (?) (© Bornholms Museum, Denmark; photo R. Laursen).

Fig. 30. Patrice? for producing gold foil figures from Borre Vestergård, Bjerringbro, Denmark; Vendel Period (?) (© Moesgård Museum, Denmark).

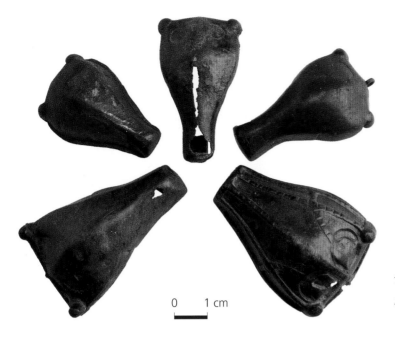

Fig. 31. Early Gotlandic animal-head brooches; Vendel Period (© Statens historiska museer, Stockholm, Sweden, inv. no. 7157).

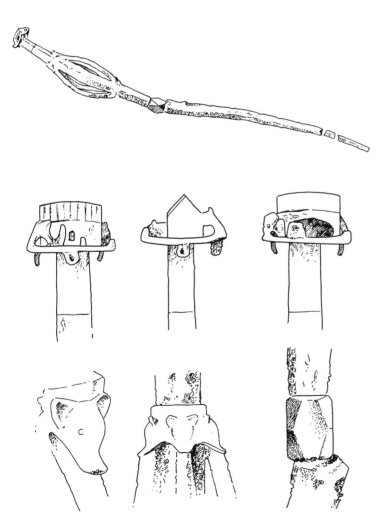

Fig. 32. Iron staff from Klinta, Öland, Sweden; Early Viking Age (© Statens historiska museer, Stockholm, Sweden, object 107776, inv. no. 25840).

Fig. 33. Ear spoon from Birka grave 507, Stockholm, Uppland, Sweden; Early Viking Age (after ARBMAN 1940, pl. 170.1).

Fig. 34a–b. Oval brooch from Friestad in Rogaland, Norway; Early Viking Age (© Arkeologisk Museum Stavanger, Norway, inv. no. S2095; photo T. Tveit).

Fig. 35a–b. Equal-armed brooch from Ragje in Rogaland, Norway; Early Viking Age (© Arkeologisk Museum Stavanger, Norway, inv. no. S11240; photo T. Tveit [a], no information [b]).

Fig. 36. Amber figurine from Råde in Østfold, Norway; Early Viking Age (© Kulturhistorisk Museum Oslo, Norway, inv. no. C4033; photo E. I. Johnsen).

Fig. 37. Jet (black amber) figurine from Vestnes in Møre og Romsdal, Norway; Early Viking Age (© University Museum Bergen, Norway, inv. no. B290).

Fig. 38. Oval brooch with silver inlay from Finglas near Dublin, Ireland; Early Viking Age (© National Museum of Ireland).

Fig. 39. Hogback from Brompton (17A) in North Yorkshire, England; Early Viking Age (© Corpus of Anglo-Saxon Stone Sculpture; photo T. Middlemass).

Fig. 40. Hogback from Ingleby Arncliffe in North Yorkshire, England (4A); Early Viking Age (© Copyright Corpus of Anglo-Saxon Stone Sculpture; photo T. Middlemass).

Fig. 41. Bear-baiting scene on the Bayeux tapestry; late 11th century (© Musée de la Tapisserie, Bayeux, France / Bridgeman Images).

Fig. 42. Illumination from the Luttrell Psalter, British Library Add. 42130, fol. 161r; AD 1325–1335 (© British Library Board. All Rights Reserved / Bridgeman Images).

Fig. 43. Bronze buckle from unknown find spot, England; Viking Age (?). As the only available image from the internet was unsuitable for reproduction, a drawing had to be made. The image is not online any longer, but can be made available by the author (drawing A. C. Lange, after a template).

Fig. 44a–b. Bronze buckle from unknown find spot, England; Viking Age (?). As the only available image from the internet was unsuitable for reproduction, a drawing had to be made. The image is not online any longer, but can be made available by the author (drawing A. C. Lange, after a template).

Fig. 45a–b. Stone cross from Kirk Andreas on the Isle of Man; Early Viking Age (© Manx National Heritage / Bridgeman Images [a]; drawing by the author [b]).

Fig. 46. Pictish stone slab (Drosten stone) no. 1 from St. Vigeans in Angus parish, Scotland; 8th/9th centuries (© HES: Historical Environment Scotland).

Fig. 47. Rune stone from Lingsberg in Vallentuna sn, Uppland, Sweden; Late Viking Age (Upplands Runinskrifter [U] 241).

Fig. 48. Rune stone from Måsta in Balingsta sn, Uppland, Sweden; Late Viking Age (Upplands Runinskrifter [U] 860).

Fig. 49. Rune stone from Bolsta in Vaksala sn, Uppland, Sweden (Late Viking Age) (Upplands Runinskrifter [U] 969).

Fig. 50. Pendant (recently discovered detector find, see postscriptum) from Herringe sogn, Sallinge Herred, Svendborg Amt, Denmark; Vendel Period (?) (© National Museum Copenhagen, Denmark [NM C58867], photo A. Jæger Manøe Schäfler, license CC-BY-SA).

Bears in Swedish imagery, AD 1000–2000

By Åsa Ahrland and Gert Magnusson

Keywords: National landscapes, Sweden, bears, hunting, education, folk tales, symbols

*Abstract: The brown bear (*Ursus arctos arctos*) is the single largest predator in Europe. In Sweden, it is mainly found in forest areas in the northern part of the country – from the county of Dalarna in the south to the Lapland region in the north. Occasional bears occur outside the deep forests and can also be found in southern Sweden. The many narratives and wide range of imagery over time testify to the fascination and mixed emotions generations have felt for this animal. In this article, we will provide examples of how the bear has been perceived and depicted in Swedish imagery and the various themes that have arisen over the centuries. Our aim is not to be comprehensive, but to highlight some reoccurring and sometimes conflicting perceptions projected onto this animal. Bears have been a very real part of Swedish life throughout history, while at the same time evoking people's imaginations.*

The bear and the national landscape

Periodically, bear hunting in Sweden has been brutal. From medieval times and onwards, entire villages were obliged to participate in large battues (Swedish: *skalljakt, drevjakt*), where the bears were driven into lakes to be killed. As late as the 19th century, ten such hunts were carried out in Dalarna alone. Such hunting was also used for wolves, lynxes and wolverines with the aim of exterminating animals that were perceived as damaging and/or as dangerous predators. At times there was even a bounty paid for killing bears (see for instance BJÖRKLÖF 2010, 121–125, 129–134, 146–157, 187–197; NYRÉN 2012). One technique used by smaller groups of hunters was to attack the bears in winter during their hibernation. As we can see in Wilhelm Wallander's painting from 1858, this was not without its risks (Fig. 1). The image was one in a series of 24 paintings created, according to the publisher Albert Bonnier, in order to "collect the more and more disappearing most characteristic idiosyncrasies and customs of Swedish folk life, and at the same time show the people's national costumes, which are also increasingly beginning to disappear" (WALLANDER/WETTERBERGH 1864/1865, back cover [authors' translation]). Bonnier had commissioned them for *Svenska Folket, sådant det ännu lefver vid elfvom, på berg och i dalom* (The Swedish people, as they still live, by rivers, on mountains and in valleys), published in 1864–1865 (WALLANDER/WETTERBERGH 1864/1865).[1] Encounters between humans and bears in a more humorous way are related in an album from c. 1859, in which the artist Fritz von Dardel (1817–1901), by using drawings, tells a story about two English travellers. The album reflects the interest of foreign well-to-do 19th-century tourists in the wider Scandinavian

---

1   https://digitaltmuseum.se/011023447969/tavla (accessed 20.07.2021).

landscape, particularly the mountains and forests where fishing and hunting could be pursued (Fig. 2; cf. Garnert/Rundquist 1991; Von Dardel 1991). Towards the end of the century, the goal of exterminating the bear had almost been achieved. The risk was imminent that the bear would become extinct in Sweden. However, the understanding of nature and the relationship between humans and the wilderness with its animals and plants was by then changing among a learned élite as well as among larger groups of the Swedish population. There was a growing interest in what was perceived as typical Swedish nature and a sense of urgency to protect its specific landscape features, habitats, animals and plants. The Romantic movement with its emphasis on the close bond between man and nature, in particular untouched nature, was an important influence. In the eyes of the Romantics, nature was seen as dynamic and animated with a strong correlation between the micro- and macro-cosmos. The movement was not only a major leverage in the arts, but affected many other areas such as education, as well as the social and natural sciences (Ödman et al. 1982, 92, 188, 198; Johannison 1984, 21–22, 32–33, 52–55, 70–72).

As in many Western countries, the protection of nature formed part of a larger nation-building process (Sörlin 1988, 107–110; Mels 2022, 138–139). Since the 1880s, Sweden had experienced major socio-economic changes following industrialisation and rural mechanisation. Though urbanisation was slower than in many other European countries, the urban population doubled in 1870–1900, and one million Swedes emigrated to the United States. In addition, the union with Norway ended in 1905 (Eskilsson 2008, 368). There was a political need to identify Swedish self-identity in an unstable social situation and to find symbols that would unify the people. The love of nature and the scientific knowledge of nature proved to be strong unifying symbols (Ödman et al. 1982, 154; compare to Löfgren 1979, 45–69; 1992, 150–157; 1993, 51–52; Sörlin 1988, 82–92, 100–109; 1992). Schools, books, journals and new institutions, such as museums, were important vehicles to convey the Swedish identity. In Stockholm, Skansen, the world's first open-air museum was founded in 1891. Interlinked was the nearby Nordiska museet, devoted to Nordic cultural history, which was inaugurated in 1907. The liberal politician and botanist Karl Starbäck's remark that nature conservation would result in "the creation of a large national outdoor museum" in 1915 shows that the museum idea was also relevant in relation to landscapes (Starbäck 1915, 33 [authors' translation]). Starbäck submitted a proposal to the Swedish Parliament in 1904 for a new legislation on the protection of areas of wilderness and was a member of the working group that was suggesting park areas (Mels 2002, 139–140). As the first country in Europe to do so, Sweden designated nine national parks in 1909. These were mainly monumental landscapes in the north and thus well suited for the formation of a national identity.

Sonfjället, a part of the mountain range in the county of Härjedalen, was included in this first group. Important considerations for the protection of the area were the somewhat unspoilt lichen heaths, a small and rare group of wild reindeer and the claim that bears had lived in the area since time immemorial (Bergström et al. 1990). In 1989, the park was expanded from its original 27 to 104 km^2 (10,400 hectares). Today, Sonfjället National park is coined "bear country" on its official web page, as "the mountain and surrounding forests are known as one of the most important bear refuges in Scandinavia".[2] The area also hosts a large elk population, lynxes, and occasionally wolves, wolverines and musk oxen. Common bird species are heather piper, mountain piper, willow ptarmigan and ptarmigan, as well as buzzards, ospreys and golden eagles. All these species are associated with, and in a sense define, the wilderness in northern Sweden.

---

2 Sveriges nationalparker, https://www.sverigesnationalparker.se/en/choose-park---list/sonfjallet-national-park/ (accessed 26.05.2022).

Fights between man and bear

Encounters between bears and humans are relatively uncommon. The bear often discovers the human being before he or she is at risk of coming in close proximity, or is even aware of the animal. A compelling example is that of a local woman in Ängersjö village in Härjedalen, who in the 1990s was out alone picking berries in a clearing in the woods, unaware of being surrounded by five bears among the trees. The incident was observed from a helicopter, which unsuccessfully tried to scare the bears off. The berry picker thought the helicopter behaved strangely, but never discovered the bears. Confrontations between bear and human are usually provoked by hunters and their dogs. The encounter is often described in dramatic terms as a duel between the hunter and the bear. This ties in with a long tradition of narratives and illustrations in which the bear is presented as a fighter with whom a grown man has to measure himself. This was also valid for royalty. Though a keen hunter of animals such as elks, wolves, foxes and grouse, King Charles XI (1655–1697) preferred the thrill of bear hunts. According to his own notes, his hunting team killed 27 bears during the period 1681–1697.

One of these hunts took place in January 1681 at the Sickelsjö estate outside Arboga in the county of Närke. The king's short note in his diary does not reflect the drama of the hunt: "The 20th to Sikelsjö for bear hunting" (Karl XI 1808, 72 [authors' translation]). Other accounts of the event relate how the king met a large and strong bear, whom he tried to shoot, but only wounded and angered. The animal got up on his hind legs and to everybody's horror was about to knock the king over, when two hunting dogs ran forward and attacked the bear. Whether the king himself subsequently managed to kill the bear or if somebody else did is somewhat unclear. In the tribute poems written about the event, the king was described as a hero. The 200-kg-bear was stuffed and depicted in a large painting by the court painter David Klöcker Ehrenstrahl (1628–1698; see Fig. 3). The future Charles XII (1682–1718) was, like his father, an enthusiastic hunter all his life and had already participated in many hunts as a child (Karl XII 1893, 30–32, 34, 214, 227). He killed his first bear at the age of eleven. His proud father wrote in his diary about the event in much more detail than the bear hunt in Sickelsjö: "His Royal Highness shot the bear on the right side by the neck, so that that the bullet went all the way through the heart and out on the left side again" (Karl XI 1808, 306 [authors' translation]; see Fig. 4).

In most images depicting a fight between man and bear, the latter is depicted standing on its hind legs opposite the opponent. The upright position humanises the large animal, but has at the same time been seen as a sign of the animal's hostility. However, standing up could simply be a means for the bear to get a better view in certain situations. Bears' eyesight is not poor, probably the equivalent of that of a human being, but their smell and hearing are very well developed and are probably the senses they trust the most (Björklöf 2010, 83). The association with strength and sudden rage has made the bear a frequent motif in Swedish heraldry. There are, however, many more connotations of the bear in a Swedish context, which reflect the lasting fascination for this large furry animal, such as cleverness, wisdom and playfulness, all traits that humans have been able to identify with and admire.

The bear in folk tradition

Many Swedish folktales relate to a situation where a bear is deceived by a cunning and exasperating fox. Despite being the strongest of all animals and rather wise, the bear's strength is not sufficient in relation to the much smaller and weaker animal's ability to cheat (Sahlgren/Sahlgren 1981, 131–134).

Other stories take their point of departure from the fact that some humans can transform into a bear or a wolf, which in effect is a werewolf tradition. One such narrative is linked to Dödmansudden on Lake Håckren in the county of Jämtland. In 1691, a so-called *hamnbjörn*, a werebear, i.e. a shapeshifter who can change between bear and human form, killed a farmer and his farm-worker. Interestingly, the evidence that secured its identity as a werebear were the belts and other parts of clothing that were found under the skin after the bear had been killed (Fig. 5). Men could also be transformed into bears by others. In one such story, a Sámi girl turned her unfaithful fiancé into a bear and tricked settlers into shooting the beast (ODSTEDT 1943, 96, 98–100).There are also folk tales around the bear's habit of entering a barn or a storehouse through a hole in the roof in order to provide himself with supplies, which he would then take away with him. In this case, it seems that the narrative sometimes refers to a *hamnbjörn,* sometimes to a real animal.

Another widely-spread tale appears in the works of the Danish historian Saxo Grammaticus (*c.* 1150 – *c.* 1220) and the Swedish archbishop and historian Olaus Magnus (1490–1557). This is a narrative of a bear who abducted a beautiful young girl and brought her to his hiding place but, instead of eating her, he was enraptured by her beauty and seduced her. In the end, the beast was hunted down and killed, but his blood lived on in their son, called Björn (Bear) after his father (Fig. 6). Björn's son was the Danish chieftain, Thorgil Sprakling, whose grandsons became kings of Denmark and England (OLAUS MAGNUS 1976, 18:30). The notion that the qualities of the bear were transferred to the Viking Age Danish royal family might have a parallel in the appearance of bears in older Swedish heraldry. The bear is even more frequent in the folk tradition in Finland, which constituted the eastern part of the Swedish realm until 1809. This might partly be explained by the many references to the deep forests in the national epic, the Kalevala, compiled from old ballads, lyrical songs, and incantations that were a part of the Finnish oral tradition (BJÖRKLÖF 2010, 254–267, 272–273, 379–380).

Towards the end of the 19th century, the image of the bear became increasingly "peaceful" in popular culture. One example is the children's song "Mors lilla Olle" (Mother's little Olle), published in 1895 in the song book *Sjung med oss, mamma!* (Sing with us, mother!), which to this day is sung by all Swedish children. It was written by Alice Tegnér (1864–1943), a music teacher, poet and the foremost composer of Swedish children's songs during the late 19th and the first half of the 20th century (TEGNÉR 1895; 1903). The song is about the little boy, Olle, who is out in the woods picking blueberries and feeling a little lonely. Suddenly he encounters a big brown bear, and he pats it and feeds it berries. When his mother comes screaming the bear runs off. Olle gets sad and wants mummy to call his new friend back. The story is inspired by a real event outside the village of Särsjön in Dalarna in 1850. The boy's name was Jon Ersson; he was one year and seven months old. Together with his older siblings, Jon was picking lingonberries a couple of hundred meters from home. He seems to have been on his own when he encountered a female bear with two cubs. Jon fed the little ones with lingonberry sprig and when tired he lay down to rest with the cubs at the bear crib. When his eldest sister discovered him, she fetched her mother, who scared the bears away. The song's immediate popularity is evident from the title of the song book *Mors lilla Olle och andra visor av A. T.* (Mother's little Olle and other songs by A. T.), published in 1903 (Fig. 7). The illustrations by the artist Elsa Beskow (1874–1953), one of the best-known Swedish authors of children's books, contributed to the book's status as an icon. The 2001 facsimile edition is described by the publishers as "one of the most famous and beloved in our children's literature". They also state that Alice Tegnér's lyrics and music with Elsa Beskow's images "have become part of our cultural heritage".[3] The motif also appeared in the textile *Barnkretong* (children's cretonne) created by the internationally renowned Swedish textile designer, Astrid Sampe (1909–2002; see Fig. 8). In this context, the work of the Swedish artist Jenny Nyström (1854–1946) must also be mentioned. Due to her husband's illness, she had to provide for the family, so, from the 1880s onwards, she illustrated books, journals, newspapers, calendars, and Christmas and Easter cards (WERKMÄSTER 1996). Nyström is particularly known for her imagery in

---

3   https://www.rabensjogren.se/bok/9789129640458/mors-lilla-olle (accessed 06.02.2022).

children's books and magazines, which often feature friendly bears playing with children and helping them out (Fig. 9).

The bear in education

In 1842, the Elementary School Act/Statute on Common Schools (*1842 års folkskolestadga*) was issued in Sweden. The main purpose was to establish that every parish was obliged to set up and finance at least one school. It was not really introducing a compulsory school system that stated at what age the children would have to start, what they should learn and for how long they would go to school. To a large extent the population was already literate through homeschooling and the education the Church provided (Johansson 1972). The Act has been described as a "late official confirmation and legitimation of a practice developed throughout the country" (Petterson 1991, 27).

With time, the state exerted a stronger hold over education itself, in order to make it more uniform. Professional teachers, common textbooks and new subjects such as history, geography and science were introduced (Hultén 2008, 58–60). An important step in the process of controlling the content of the subjects was a national curriculum. The first curriculum, which appeared in 1878, specified for instance the 102 animals Swedish children should learn about. The brown bear was one of the beasts and remained on the list in the 1900 curriculum, despite the fact that the number of bears in the country had been reduced to 74, under the pretext that "only the most typical and most important to man" were to be included (Hultén 2008, 65). The rapid expansion within the natural sciences during the 19th century was reflected in the appearance of special textbooks in the field (Hultén 2008, 58–60). The books mainly focused on descriptions of animals, plants, stones and other objects, inventions and natural phenomena (Hultén 2008, 64). Thanks to new techniques in the mass production of images, access to visual material in schools increased radically during the 19th century. The fact that images became an important part of the pedagogy was based on the conviction that reality could really be captured in an image. While school posters do not seem to have become common until the turn of the 19th century, textbooks were illustrated early on. This applied not the least to those on natural science, where text and image were combined "to describe individual natural objects in more detail and to convey the most important images" as the author of *Folkskolans naturlära* (The natural science of the Elementary School) put it (Bäckman 1871 [authors' translation]; cf. Johannesson 1997, 15–17, 22–29, 144, 147–148; Hultén 2008, 74–76).

As mentioned, the period from the 1870s and onwards was a time of great change on many levels in Swedish society. The production of *Läsebok för Folkskolan* (The elementary School reading book) coincided with this period. The first edition appeared in 1868 as a response to the lack of a comprehensive textbook that covered the important subjects in school. It was initiated and financed by the state. Education was seen as a way to shape children's perception of the world (Edgren 2017, 95–103). *Läsebok för Folkskolan* proved to be a useful tool to move towards a more secular and modern school, where the individual and its relation to the state and the nation were more pronounced and qualities such as morality, discipline and patriotism would be at the core in order to create good citizens.[4] The elementary school could help to bind together and shape "The new Sweden".

With this in mind, let us look at the extensive description of the brown bear in the revised 1878 edition of *Läsebok för Folkskolan*, as it conveys what was considered to be important knowledge for a child at that time. In the text the bear is a "he", which gives the animal a human quality. Being suspicious and shy, he prefers remote and desolate forest areas in the north. He is also described as wild

---

4 Ch. Florin, Från folkskola till grundskola: http://www.lararnashistoria.se/2010, 8–10 (accessed 22.07.2022).

and angry by nature. He eats roots, juicy plants and berries such as blueberries and lingonberries. Together with ants, this constitutes a young bear's food. When older he gets a taste for meat and attacks horses and cattle. This is always done in an upright position in order to have his paws free. His strength is underlined. When attacking a strong horse, he grabs its shoulders with one paw and gallops next to the horse until he can grab a tree with the other paw and pull the horse to the ground. He can also carry a horse or a cow between his front paws (LÄSEBOK FÖR FOLKSKOLAN 1878, 93–95). The impression that this is somebody we as humans can identify with in a way is shown in the following lines: "He swims fast and long distances and often takes a swim during hot summers. When he gets scared, he runs fast but clumsily. His external senses are very well developed: his eyesight is sharp, his hearing good and his sense of smell extremely fine, which is why he seldom meets people during his summer walks in the forests" (LÄSEBOK FÖR FOLKSKOLAN 1878, 94 [authors' translation]). In the last paragraph, we recognise a narrative of the Swedish folk tradition: "Although the bear usually withdraws from human habitation, it sometimes happens at night that he visits farms, goes up on the cowshed roof, tears up a piece of it, goes down through the opening, kills a cow and carries her away the same way he came down" (LÄSEBOK FÖR FOLKSKOLAN 1878, 94–95 [authors' translation]). The fact that children were informed about different techniques in bear hunting as well (LÄSEBOK FÖR FOLKSKOLAN 1878, 96) shows that these animals were still present in farming societies.

The description is a mixture of observations of the wild animal and the folk tradition's image of the bear – its strength, the fact that it can stand on its hind legs and that the paws are deadly weapons. Yet there are obvious similarities between man and bear. Also noticeable are the omissions. There is no mention of lifespan, mating or how many cubs a female bear can have. The illustration of an adult brown bear on the other hand aligns with the scientific tradition. It shows striking similarities with zoological images such as Johan Wilhelm Palmstruch's brown bear in *Svensk zoologi* (Swedish Zoology) published in 1804 (Figs. 10–11). Palmstruch's illustration is described as a "faithful copy of Maréchall's excellent drawing in Ménag. du Mus. nat. d'Hist. nat., though here proportionally adapted to the format" (QUENSEL/SWARTZ 1806, 39, 50 [authors' translation]). The remark refers to the artist Nicolas Maréchal's depiction of a brown bear in *Le ménagerie du Muséum d'histoire naturelle* (DE LACÉPÈDE/CUVIER 1801), the extensive and luxurious catalogue of the public zoo in Paris founded in 1794. Most of animals in the zoo came from former royal and aristocratic menageries and were skillfully depicted from life by Maréchal (BARATAY/HARDOUIN-FUGIER 1998, 99–105, 160–161; PAULSON 2009, 95–96). The ethnologist Orvar Löfgren points out that the discovery of the national landscapes took place in a collaboration between authors, travellers and political debaters and was a process that transformed "highly mundane pieces of nature into homelands, loaded with history and national symbolism" (LÖFGREN 1993, 90–91). *Läsebok för folkskolan* was one important step in the construction of a Swedish landscape. The former emphasis on nature as a creation of God, and as such something universally valid, beautiful and good, had faded away. Now, children were taught the qualities that characterised the landscape of their own local area and other regions, the characteristics and uniqueness of Swedish nature and geography as well as those of foreign countries (HULTÉN 2008, 116–117). Instrumental in this process was the novel *Nils Holgerssons underbara resa genom Sverige* (The Wonderful Adventures of Nils) by the author Selma Lagerlöf, who was commissioned to write a geography and history reader for elementary schools. The book is about a young mischievous boy, who is unkind to animals and is therefore transformed into an elf-sized boy with the ability talk to animals. On the back of a goose, he goes on a journey all around Sweden. On his return to the farm, he has learnt his lesson to respect others and is turned into a real boy again (LAGERLÖF 1931). The book was published in 1906/1907 and read by many generations of school children during the 20th century. Selma Lagerlöf received the Nobel prize in 1909 (Fig. 12).

The Swedish Museum of Natural History was inaugurated in 1916. The monumental building complex took almost a decade to build in what was then the outskirts of Stockholm and is to this day the country's largest museum. The architect, Axel Anderberg (1860–1937), also designed the new headquarters of the Swedish Academy of Science opposite the museum (BECKMAN 1999, 14–15, 134–135, 207–208). The high granite pillars on either side of the main gate are adorned with brown bears, symbolising Sweden and Swedish nature. In comparison to earlier imagery, the presentation reflects a change in the perception of the bear (Fig. 13a–b). On one side of the entrance, a female bear is sitting down with her cubs playing at her feet, on the other we find a large, watchful male in a seated position. What we see is a bear family depicted as part of nature without any interaction with humans. A few years earlier, in 1906, the Swedish sculptor Carl Milles (1875–1955)[5] got a commission for Berzelii Park, a small public park in the city center of Stockholm dedicated to the memory of the famous Swedish chemist Jöns Jacob Berzelius (1779–1848; see Fig. 14). Milles chose to create two naturalistic sculptures of playful bear cubs on each side of the entrance, as well as otters, another species of Swedish fauna, along the surrounding wall. The rather heavy, yet simplistic style without too many details represented a new direction among artists around Europe. Carl Milles' stay in Munich, Germany, a few years earlier had proved influential in this respect. His interest in animals had developed during his study period in Paris, when he used to go to the public zoo in the Jardin de Plantes to capture its inhabitants on paper as truthfully as possible. He would make several animal sculptures in the years to come. The choice of material in Berzelii park, the stern granite, was not only in accordance with the artistic expression, but constituted an apt connection to the Swedish landscape (CORNELL 1963, 22–24, 28, 41–44). The bear family theme reoccurred in many images in the 20th century; school posters were no exception there (Fig. 15). Aligning with the family theme is also the hugely popular tradition introduced by the Skansen open-air museum in Stockholm in the 1930s, namely, the yearly announcement of the names of the new bear cubs as they emerge from the den with their mother in spring. No other animals in the museum are subject to this kind of attention.

BEARS IN RELIGIOUS ART

Almost all surviving medieval art in Sweden is religious. Bears are not common among the motifs, but they occur in exceptional cases. The oldest known bear depiction is from Dädesjö church in the county of Småland, famous for its 13th-century paintings on the wooden ceiling (Fig. 16). It is situated on the base of the eave and shows a bear standing on its hind legs in front of a ram. The roof has been dated to *c.* 1180 by dendrochronology and the scene was probably carved shortly afterwards. It forms part of a series of 42 carved images on the eaves representing the power struggle between a number of fabled creatures, riders and other human figures. The bear was seen as the most dangerous animal in the forest and as a symbol of anger, while the ram represented a domestic and sacrificial animal (ULLÉN 2020, 11, 45, 54).

The other medieval bear image to be mentioned is in Härkeberga church in the county of Uppland. The building has been dated to *c.* 1300, and the paintings of the 1480s have been attributed to the artist Albertus Pictor (1440/45–1509). The picture shows the biblical David defending his domestic animals by clubbing a bear (Fig. 17; cf. KILSTRÖM 1968, 32).

The biblical motifs also include the vernacular wall painting from Backhans's cottage in Svärdsjö from 1781, in the county of Dalarna, which shows Paul on the road to Damascus. Here, the bear symbolises Paul's rage against the Christians before his conversion (Fig. 18).

---

5 W. Carl E. Milles: https://sok.riksarkivet.se/sbl/artikel/9354. Svenskt biografiskt lexikon, article by U. Abel (accessed 18.11.2022).

The perception of bears in Olaus Magnus' *Historia de gentibus septentrionalibus*

The most remarkable Swedish description from the 16th century is Olaus Magnus' work – the *Carta Marina* map published in the 1530s and the large book *Historia de gentibus septentrionalibus* (History of the Nordic peoples) from 1555, which included a substantial number of illustrations (see BÖLDL, this volume). The *Historia* was written in Rome by Olaus Magnus (1490–1557), a high-ranking clergymen who was forced to go into exile after the Protestant Reformation had been initiated by the Swedish king, Gustavus I, in 1527. In his portrayal of the Nordic countries, Olaus Magnus mixed ethnology, history and cartography with tales and folklore. It was partly based on his own observations during travels in northern Sweden, but also, as if to underline the relevance of older history, legends and historiographies such as that of the 12th-century Danish author, Saxo Grammaticus.[6] As they were written long before the concept of source criticism was established among historians, some of the "truths" in the *Historia de gentibus septentrionalibus* can of course be questioned. He assumed for instance that runic stones were created by giants in ancient times and described sea creatures in the Bothnian Sea, as well as the Sámi using skies and women with hunting skills that far exceeded those of men. He emphasised the courage of the Nordic peoples and their war achievements as well as the importance of access to natural resources, such as mining and ironworks. Olaus Magnus' narrative has played an important role in the understanding of daily life in the 16th century Nordic countries and was used within general education as late as in the 19th century. His observations and interpretations are still of interest to scholars within ethnology and other disciplines today, but are evidently read and interpreted in a different way. Not least, the *Historia de gentibus septentrionalibus* conveys the beliefs among the learned at the time about nature in general, as well as specific phenomena. It is an added value that it is richly illustrated.

The description of the Nordic bears covers 11 chapters in the 18th book (on wild animals) and each chapter includes an illustration. One chapter elaborates on how to hunt a bear with cunning as well as the anger with which a bear can attack the hunter if injured. A female bear when she "breastfeeds" her young is even more dangerous (OLAUS MAGNUS 1976, 18:25). As we have seen, through history this wrath is a reoccurring motif in narratives relating to human encounters with bears and the duel between man and bear.

Olaus Magnus describes the brown bears as omnivores, who eat, among other things, ants, crayfish and meat. When hunting deer, they jump up on its back and let the deer fight until it can no longer bear it. The female bears are supposedly more brutal than the males. His account of how a bear kills a hedgehog is even more imaginative. The poor animal was squeezed to death with a tree trunk, but remained dangerous to the bear when swallowed, as its spines could pierce holes in the bear's stomach (Fig. 19). Despite their cunning techniques, bears had a hard time catching a wild boar. The fight would be brutal and time and again would end with a victory for the boar (OLAUS MAGNUS 1976, 18:26–27). The bears' fondness of honey and the eagerness with which they plunder bees' nests is particularly commented on (Fig. 20). The fierce competition for bees and honey is put forward as one of the reasons for killing bears (OLAUS MAGNUS 1976, 18:28–29). The theme of the honey-loving bear has lived on until modern times, not least through Winne the Pooh, perhaps the world's most famous bear. Pooh, who really is a teddy bear but lives in the woods, is always thinking about food, particularly "hunny". The stories by A. A. Milne from the 1920s had already been translated and published, together with the iconic drawings by Ernest H. Shepard, in Sweden in the early 1930s. Possibly, Pooh served as an inspiration to the Swedish cartoonist and illustrator Rune Andréasson

---

6   Olaus Magnus: https://sok.riksarkivet.se/sbl/artikel/7681. Svenskt biografiskt lexikon, article by G. Broberg (accessed 23.07.2022).

(1925–1999), who created children's comics around various animals, often bears, from the 1940s onwards. In the 1960s, he developed the Bamse character in comics and animated cartoons, which are still today well-known to most Swedish children. Bamse is a small and friendly brown bear who needs to consume grandma's special honey (*dunderhonung*) in order to become the strongest bear in the world and solve all the problems that he and his friends encounter. As in the Pooh stories, his friends are other kinds of animal species (Fig. 21).

Among the aspects of bears' behaviour that are described by Olaus Magnus is the abduction of people. A bear could kidnap a shepherd when attracted to the music he played on his bagpipe for the cows to assemble, whereas the sound of a goat horn would scare the beast away (OLAUS MAGNUS 1976, 18:31). One chapter deals with the already-related story of the beautiful virgin who became the subject of a bear's unnatural lust. To alleviate the incongruousness of the "strange procreation" between the girl and the bear, Olaus Magnus argues that nature let it "bear fruit in an ordinary fetus and the blood of the beast be absorbed in a body with human features" (OLAUS MAGNUS 1976, 18:30 [authors' translation]). The thought among early historians, such as Saxo Grammaticus and Olaus Magnus, that the extraordinary bear-child Björn was the ancestor of the royal family in Denmark makes the bear a totem, i.e. a symbol that serves as an emblem for the family or clan. We will see that this identification with bears, their strength and cunning, is expressed in Swedish heraldry as well.

Olaus Magnus also reports how bear tamers from Russia and Lithuania would make bears dance and do tricks, such as collecting money from spectators. They could also be used for fishing and running in pedal wheels (OLAUS MAGNUS 1976, 18:32–34; Fig. 22). The techniques to "teach" the bears were often brutal (BJÖRKLÖF 2010, 230–231). By the mid-18th century, travelling menageries from abroad with dancing bears, among other animals, regularly appeared in markets and squares in Swedish towns (SVANBERG 2010, 115). From the early 16th century and onwards, Swedish royalty would keep bears and use them in fights with other animals for entertainment. Others would be tamed and played with, as Queen Ulrika Eleonora the elder's painting from 1682 suggests. This was placed above a door in her husband King Charles XI's bedchamber in the royal estate, Kungsör, which was often used as a hunting lodge (Fig. 23; BERG 1965, 98). Bear motifs also form part of the décor in Queen Christina the elder's 1590s bedchamber in Gripsholm castle south of Stockholm. Christina Holstein Gottorp (1573–1625) was the second spouse of Duke Charles, later Charles IX (1550–1611). The rich decorations of the room include a large painted frieze with a sequence of portraits, animals such as deer and bears and still life depictions of fruit (Fig. 24; VON MALMBORG 1971, 26–28; LINDGREN 1996, 235–236.

In the 1680s, a polar bear, a gift from Tzar Peter I of Russia, was kept at the court in Stockholm. A separate building was erected close to the royal stables, which could be observed from the windows in the royal chambers as well as seen by passers-by. According to a contemporary source, the polar bear was taken for swims in the stream next to the palace and fed with fish. The spectators were impressed by the actual time the exotic animal could stay under water. When the polar bear died, he was stuffed, and, like many other animals during Charles XI's reign, he became the subject of a painting by the court painter, David Klöcker Ehrenstrahl (Fig. 25). However, the bear was not depicted as a captured animal, but in a wild and mountainous landscape, obviously perceived as his "natural" habitat (BENGTSSON 2005, 192–205; STÅHLBERG/SVANBERG 2016, 113). Even Olaus Magnus has a passage on polar bears, in which he describes and illustrates their fishing techniques on the ice around Iceland. They are depicted on the *Carta Marina* map as well. He mentions that hunters would offer the white skins to churches in order to keep the priests' feet warm in front of the high altar (OLAUS MAGNUS 1976, 18:24; see also JAHNSEN and KORHONEN, this volume).

There are numerous accounts of private citizens keeping one or two tamed grown brown bears or cubs as pets or guardians up until early 20th century in Sweden (BERG 1965, 93–98). The temptation to force this strong and dangerous animal into submission and to make it interact with humans seems to have been fascinating and almost irresistible (see for instance SCHWARTZ 1798, 10). As we can see, bears were present in different ways in everyday life during the first half of the 16th century and would continue to be so during the following centuries.

The bear as a heraldic motif

Bears have been used as heraldic symbols in Sweden since medieval times. The oldest known coat of arms with a bear belonged to a family called Björnsson, who owned estates in the county of Södermanland not far from Stockholm in the late 13th century and onwards. Their coat of arms depicts an upright bear showing anger. Similarly, the 14th-century coat of arms of the Björnram family in the county of Uppland depicts a furious and threatening bear. Other coats of arms would depict a bear's head as well as its legs or paws with their claws (BJÖRKLÖF 2010, 321–322). To emphasise the bear's own "weapon" and – as was often done later – to combine them with cannon balls or maces, such as in the Björnram's 16th-century coat of arms, can be interpreted as an expression of anthropomorphisation. In coats of arms of the 17th and 18th centuries, bears are even depicted carrying weapons, which makes the parallel all the more obvious. One example is the 17th-century coat of arms of Mathias Björnclou (sometimes spelt Biörenklou), with a collared bear with a crown on his head, a lifted sword and his paw with his big claws hanging over the edge of a balustrade (Fig. 26). These bears are less threatening than their medieval counterparts, but they express the brown bear's well-known strength. The number of titled families increased quickly in Sweden during the 17th and 18th centuries. There were several reasons for this, one being the requirement that only members of the nobility could occupy posts in the growing state administration. The fact that quite a few would choose to associate their new name and the design of their coat of arms with a bear shows that its validity as a symbol of a noble family remained intact.

Bears also appeared in quite a few coats of arms of municipalities during the last century. In Finland, which formed a part of the Swedish realm until 1809, the bear has been and is still perhaps an even more common heraldic symbol (BJÖRKLÖF 2010, 336–339).

The four sons of King Gustavus I (1496–1560) received a duchy each. Duke John's (1537–1592) first coat of arms was based on established symbols of his new duchy of Finland – a crowned gilded helmet with Swedish flags and a standing bear with a raised sword (Fig. 27). Perhaps his decision to change his ducal coat of arms later was an action that was intended to declare that he was second in line to the Swedish crown. The duke settled for a design where the coat of arms included the Finnish symbols, the three crowns of the national Swedish coat of arms, the medieval Folkung (House of Bjelbo) lion and a central shield mark with the Vasa family's sheaf symbol. The Finnish provinces of Satakunda in Turku and Pori counties still had a crowned bear threatening to raise a sword in their coat of arms in 1997 (BJÖRKLÖF 2010, 336). Bears are also found on several municipal coats of arms in both Sweden and Finland from the 20th century.

It is notable that in heraldry the bear is portrayed as threatening and sometimes furious. It is largely the scary aspect of the bear that has been emphasised since the Middle Ages and which is repeated even in the municipal coat of arms of the 20th century. It is more exceptional that the bear is depicted as natural, calm or playful and less frightening.

The bear in Swedish imagery – concluding remarks

As we have shown, the brown bear has been a part of people's reality and imagination, at least in Sweden, during the last millennium and this has been reflected in images in different contexts. They appear in frescos and sculptures, in coats of arms of nobility and modern municipalities as well as in wall paintings and other art in homes, in commemorative stones, as works of art and architecture in institutions and public places, in books on history and science and stories for children. Many images focus on the encounter between humans and bears. Depending on the situation and the actors involved, the perception of the bear can be very different. When depicted in a meeting with a grown man, the bear is standing on his hind legs and the two appear as counterparts. The bear is strong, dangerous and easily angered, while the man needs to be brave, quick and cunning. Heraldry shows that Swedish royalty and nobility have identified themselves with these particular qualities associated with the wild animal. The connotations of bears interacting with children are completely different, regardless of whether the scene is in the wild or in a domestic environment. The bear is friendly, docile and good natured. Imagery showing bears in captivity also underlines these qualities together with their supposed willingness to learn, as well as their playfulness and ability to adjust to and perform for humans. The latter could, on the other hand, be interpreted as an expression of human superiority over animals, who have lost their power and have to live according to human rules. Common to all these motifs is an anthropomorphism, the readiness to identify with this animal, but we can also see that it is not one stereotypical view, but reflects many different human traits that people have seen and still see in a brown bear. In some sense, the bear can be perceived as symbolic and as crossing borders.

There is also imagery that depicts bears in the wild seemingly on their own terms without having to interact with humans. This type of depiction increases with the growing interest in nature conservation and the Swedish landscape. The protection of a bear habitat was a major reason for establishing one of the first nature reserves, Sonfjället in Härjedalen, in 1909. It is striking how depictions of bears changed around this time. While earlier works presented lonely males in the wild, now artists showed family units, often mummy bear with her cubs playing around her. This represented a new way of relating to bears, by which they were given a natural rather than mythological role. At the same time, it is still a case of anthropomorphism, as the family motif probably made it easier for people to relate to bears, to accept them and their right to exist.

It is noteworthy that so many ideas and perceptions of the brown bear have remained fairly unchanged over time. Despite the Enlightenment and the breakthroughs of modern science, traditions and folklore have lived on and have remained important. In educational material from the 1800s, scientific observations are mixed with folkloristic traditions regarding the character and habits of the brown bear. Even today, some of these folk beliefs linger in society and influence the view of this animal, not least in popular culture. The bond between bears, often cubs, and the emerging human generation seems particularly strong. Around children, these large and potentially dangerous animals, who avoid human beings if possible, are depicted as friends – unthreatening, willing to learn and help out.

Nature and culture are in many ways connected in the physical landscape and in our minds. The imagery in the study shows that this also applies to a "wild" animal such as the bear. We understand the bear through ourselves and nature through culture. Sometimes we perceive the bear – nature – as a peril to us and our resources. In other instances, we recognise ourselves – culture – as a threat to the survival of bears. The question is still on the agenda: Who has a right to live in the landscape and use it and, if there are clashes, who has priority?

Bibliography

Ambrosius 1906: J. M. Ambrosianus, Läsebok för småskolan. Omarb., rikt ill. uppl. (Lund 1906).

Bäckman 1871: J. Bäckman, Folkskolans naturlära. Första boken: Om menniskan och djuren, tredje upplagan (Stockholm 1871).

Baratay/Hardouin-Fugier 1998: E. Baratay/E. Hardouin-Fugier, Zoos: Histoire des jardins zoologiques en Occident (XVIe-XXe siècle) (Paris 1998).

Beckman 1999: J. Beckman, Naturens palats: Nybyggnad, vetenskap och utställning vid Naturhistoriska riksmuseet 1866–1925 (Stockholm 1999).

Bengtsson 2005: E.-L. Bengtsson, Kungliga djur. In: E.-L. Bengtsson/R. Millhagen (eds.), Strömsholms slott (Stockholm 2005) 192–205.

Berg 1965: G. Berg, Tama björnar, dansande björnar och björnförare. Fataburen, Nordiska museet och Skansens årsbok 1965, 93–112.

Bergström et al. 1990: E. J. Bergström/G. Magnusson/J. Raihle, Härjedalen: natur och kulturhistoria (Östersund 1990).

Björklöf 2010: S. Björklöf, Björnen: i markerna & kulturen (Möklinta 2010).

Cornell 1963: H. Cornell, Carl Milles: hans verk (Stockholm 1963).

Von Dardel 1991: F. von Dardel, Herrar Black & Smith på väg till Skandinavien (Stockholm 1991).

Edgren 1917: H. Edgren, Kungsådran av den lilla läroanstaltens bildning. In: J. Westberg/E. Larsson/A. Berg/M. Michaëlsson/A. Åkerlund (eds.), Utbildningens revolutioner: till studiet av utbildningshistorisk förändring (Uppsala 2017) 95–117.

Eskilsson 2008: L. Eskilsson, Friluftsliv. In: J. Christensson (ed.), Signums svenska kulturhistoria. Det moderna genombrottet (Stockholm 2008) 363–391.

Garnert/Rundquist 1991: J. Garnert/A. Rundquist, Efterskrift. In: F. von Dardel, Herrar Black & Smith på väg till Skandinavien (Stockholm 1991) 146–231.

Hultén 2008: M. Hultén, Naturens kanon: formering och förändring av innehållet i folkskolans och grundskolans naturvetenskap 1842–2007 (Stockholm 2008).

Johansson 1972: E. Johansson, En studie med kvantitativa metoder av folkundervisningen i Bygdeå socken 1845–1873 (Umeå 1972).

Johannesson 1997: L. Johannesson, The mass-produced image (Stockholm ²1997).

Johannisson 1984: K. Johannisson, Det sköna i det vilda: en aspekt på naturen som mänsklig resurs. In: T. Frängsmyr (ed.), Paradiset och vildmarken: studier kring synen på naturen och naturresurserna (Stockholm 1984) 15–81.

Karl XI 1808: Karl XI, Konung Carl den XI:tes dag-bok (Härnösand 1808).

Karl XII 1893: Karl XII, Konung Karl XII:s egenhändiga bref (Stockholm 1893).

Kilström 1968: B. I. Kilström, Härkeberga kyrka (Stockholm 1968).

De Lacépède/Cuvier 1801: E. De Lacépède/G. Cuvier, Le ménagerie du Muséum d'histoire naturelle: ou Les animaux vivants, peints d'après nature, sur vélin, par le citoyen Maréchal, peintre du Muséum et gravés au Jardin des plantes, avec l'agrément de l'Administration, par le citoyen Miger. Avec une note descriptive et historique pour chaque animal, par les citoyens Lacépède et Cuvier (Paris 1801).

Lagerlöf 1931: S. Lagerlöf, Nils Holgerssons underbara resa genom Sverige (Stockholm 1931).

Läsebok för folkskolan 1878: Läsebok för folkskolan 8, revid. ed. (Stockholm 1878).

Lilleputt 1916: Lilleput, barnens minsta julbok: en samling sagor och berättelser. Folkskolans barntidning (Stockholm 1916).

Lindgren 1996: M. Lindgren, Måleriet. In: G. Alm (ed.), Signums svenska konsthistoria 5. Renässansens konst (Lund 1996) 217–264.

Löfgren 1979: O. Löfgren, Människan i naturen. In: J. Frykman/O. Löfgren, Den kultiverade människan (Lund 1979) 45–73.

Löfgren 1992: O. Löfgren, Landskabet. In: O. Löfgren/B. Berggren/K. Hastrup (eds.), Den nordiske verden 1 (Copenhagen 1992) 109–192.

Löfgren 1993: O. Löfgren, Nationella arenor. In: B. Ehn/J. Frykman/O. Löfgren (eds.), Försvenskningen av Sverige: det nationellas förvandlingar (Stockholm 1993) 22–119.

Von Malmborg 1971: B. von Malmborg, Gripsholm. In: B. von Malmborg/A. S. Svensson/S. T. Kjellberg, Slott och herresäten i Sverige: ett konst- och kulturhistoriskt samlingsverk. De kungliga slotten 2. Gripsholm, Haga, Rosendal, Rosersberg, Strömsholm och Tullgarn samt Solliden, Stenhammar och Prins Eugens Waldemarsudde (Malmö 1971) 9–84.

Mels 2002: T. Mels, Nature, Home, and Scenery: The Official Spatialities of Swedish National Parks. Environment and Planning D: Society and Space 20(2), 2002, 135–154.

Nyrén 2012: U. Nyrén, Rätt till jakt: en studie av den svenska jakträtten ca 1600–1789 (Göteborg 2012).

Ödman et al. 1982: E. Ödman/E. Bucht/M. Nordström, Vildmarken och välfärden: om naturskyddslagstiftningens tillkomst (Stockholm 1982).

Odstedt 1943: E. Odstedt, Varulven i nordisk folktradition (Uppsala 1943).

Olaus Magnus 1976: Olaus Magnus, Historia om de nordiska folken D. 4 Sjuttonde-tjuguandra boken (Stockholm 1976).

Paulson 2009: N. Paulson, Natural Disorder: The Animal Image in French and British Art before Darwin, c. 1790–1859. Unpubl. PhD thesis, Washington University, St. Louis (St. Louis 2009). https://openscholarship.wustl.edu/etd/273.

PETTERSON 1991: L. PETTERSON, 1842, 1822 eller 1882? Vad är det som bör firas? Forskning om utbildning. Tidskrift för analys och debatt 18(4), 1991, 22–27.

QUENSEL/SWARTZ 1806: C. QUENSEL/O. SWARTZ: Svensk zoologi, utgifven af J. W. Palmstruch. Med text börjad af C. Quensel. Och fortsatt af O. Swartz (Stockholm 1806).

SAHLGREN/SAHLGREN 1981: G. SAHLGREN/J. SAHLGREN (red.), Svenska folksagor 3 (Stockholm 1981).

SCHWARTZ 1798: O. SCHWARTZ, Tal, om hushålls-nyttan af de däggande djuren; hållet för kongl. vetenskaps academien vid præsidii nedläggande den 19 julii 1797 (Stockholm 1798).

SÖRLIN 1988: S. SÖRLIN, Framtidslandet: debatten om Norrland och naturresurserna under det industriella genombrottet (Umeå 1988).

SÖRLIN 1992: S. SÖRLIN, Sveriges moderna miljöhistoria. In: K. V. Abrahamsson et al. (eds.), Humanekologi (Stockholm 1992) 381–440.

STARBÄCK 1915: K. STARBÄCK, Naturskydd. In: H. Conwentz/K. Starbäck, Naturskydd och industri: Två föredrag (Stockholm 1915) 27–40.

STÅHLBERG/SVANBERG 2016: S. STÅHLBERG/I. SVANBERG, A Russian Polar Bear in Stockholm. Notes on Animal Diplomacy. Svenska Linnesällskapets tidskrift 2016, 107–116.

SVANBERG 2010: I. SVANBERG, Nyss visades i Uppsala två små nyfödda lejonungar upp: om svenska lejon. Uppland 2010, 102–133.

TEGNÉR 1895: A. TEGNÉR, Sjung med oss, mamma! 3, 18 småvisor (Stockholm 1895).

TEGNÉR 1903: A. TEGNER, Mors lilla Olle och andra visor (Stockholm 1903).

ULLÉN 2020: M. ULLÉN, En medeltida bildvärld. Dädesjö gamla kyrka (Stockholm 2020).

WALLANDER/WETTERBERGH 1864/1865: W. WALLANDER/C. A. WETTERBERGH, Svenska folket: sådant det ännu lefver vid elfvom, på berg och i dalom (Stockholm 1864/1865).

WERKMÄSTER 1996: B. WERKMÄSTER, Jenny Nyström och illustrationen. In: M. Gynning (red.), Jenny Nyström: målaren och illustratören (Stockholm 1996).

Associate Prof. Dr. Åsa Ahrland
Department of Urban and Rural Development
Swedish University of Agricultural Sciences
Uppsala
Sweden
asa.ahrland@slu.se

Associate Prof. Dr. Gert Magnusson
Jernkontoret
Stockholm
Sweden
gert.y.magnusson@gmail.com

Fig. 1. Bear hunting in winter, painting by Wilhelm Wallander, 1858. Many depictions of bear hunting show a struggle between bear and man. In Swedish folk tradition, the bear has the wits of one man and the strength of seven men. Nordiska museet, Stockholm (photo: Nordiska museet).

Fig. 2. Illustrations from Fritz von Dardel's album on the travels of Mr Black & Mr Smith to Scandinavia. The humorous approach conveys von Dardel's disinterest in hunting as a sport. a: A bear suddenly appears when the party is having breakfast in the Norwegian wilderness; b: They track down the bear to his den. The animal is only wounded by Mr Smith's shot, who subsequently applies the trick of pretending to be dead. In this drawing, the bear blames himself that he was so easily tricked, and swears that he will never be so gullible again; c: "The best moment in Mr Smith's life" (after VON DARDEL 1991, 35–43; photos G. Magnusson); d: Mr Black & Mr Smith visiting Mr Lloyd. The wealthy Englishman, Llewellyn Lloyd (1792–1876), established himself in Sweden in the 1820s. A keen fisherman and hunter, he particularly enjoyed the thrills of bear hunting. Lloyd published several books, for instance "Fields Sports of the North of Europe" in 1827/1828 (cf. DIRKE, this volume; GARNERT/RUNDQUIST 1991, 169–170, 210–212; ill. after VON DARDEL 1991, 67; photo G. Magnusson).

Fig. 3. The Sickelsjö bear, painting by David Klöcker Ehrenstrahl. Nationalmuseum, Stockholm (photo H. Thorwid, Nationalmuseum; DelaLika 4.0 Internationell CC BY-SA 4.0).

Fig. 4. Drawings of animals by the young prince and later King Charles XII (Kungliga handskriftsamlingen: Karl XII, Riksarkivet, Stockholm).

Fig. 5. Commemorative stone at Dödmansudden, erected in 1935, with the text: "In a fight with a hamnbjörn [a human bear or werebear, respectively], on 22/2/1691, the farmer Nils Persson and the dragoon Anders Slaghöök, both from Hålland [a small village in the vicinity], fell here". It was moved to its current location in 1965 in connection with the construction of the Håkren dam and power plant (photo B. Åkerström).

Fig. 6. The bear and the beauty. This woodcut from Olaus Magnus' Historia de gentibus septentrionalibus, *1555, illustrates the tradition of the origin of the Danish royal family in a relationship between a beautiful maiden and a bear. Instead of eating the maiden, the bear falls in love (photo G. Magnusson).*

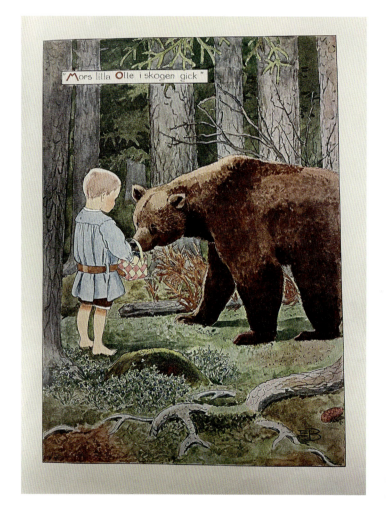

Fig. 7. Mors lilla Olle, *by Elsa Beskow. Kungliga biblioteket, Stockholm (photo Å. Ahrland).*

Fig. 8. Barnkretong, goache by Astrid Sampe, 1939, for NK:s Textilkammare. Original for a printed textile depicting well-known Swedish nursery songs. In the centre there is Mors Lilla Olle by Alice Tegnér (photo Designarkivet).

Fig. 9. Illustrations by Jenny Nyström. a: Cover of a Christmas periodical for children (LILLEPUTT 1916). A small boy with a red-orange hood sits on top of a pile of Christmas presents, while two little bear cubs are pulling his sled through the snow. Kungliga biblioteket, Stockholm (photo Å. Ahrland); b: Image illustrating a story in Läsebok för småskolan *(Reader for primary school; AMBROSIUS 1906), with a bear cub and a young boy walking to school together. Kungliga biblioteket, Stockholm (photo Å. Ahrland).*

Fig. 10. Illustration of a brown bear in Läsebok för Folkskolan, *showing striking similarities with the Swedish artist Johan Wilhelm Palmstruch's scientific illustration of a brown bear in* Svensk zoologi *(photo G. Magnusson).*

Fig. 11. Palmstruch, in turn, refers to Nicolas Maréchal's depiction of a brown bear in the catalogue of the zoo in Paris, Le ménagerie du Muséum d'histoire naturelle, *from 1801 (photo G. Magnusson).*

Fig. 12. The author Selma Lagerlöf (see LINDÉN, this volume) seemed to take inspiration from the true story of a young boy's experience with a bear family in Särsjön in Dalarna in 1850. In Nils Hogerssons underbara resa *(Nils' adventures), the elf-sized boy Nils' encounter is more scary. He is tossed around by the cubs and is afraid of being eaten. In the end, he saves the large male bear from being shot and they part as friends. Illustration by Bertil Lybeck (1887–1945) in the 1931 edition (photo G. Magnusson).*

Fig. 13. The Swedish Museum of Natural History in Stockholm was inaugurated in 1916. The pillars by the main gate are adorned with a female bear with cubs on one side (a) and with a male brown bear on the other (b). Riksmuseet (photos Å. Ahrland).

Fig. 14. Carl Milles' two sculptures "Playing bears" were made for Berzelii park in central Stockholm in 1906 (photo G. Magnusson).

Fig. 15. By the turn of the century in 1900, school posters had been established as a pedagogical tool. One of the more frequent artists in this genre was the landscape and animal painter Nils Tirén (1885–1935). In this poster from 1919, he has depicted a bear family in their natural environment. P. A. Norstedt & Söner, Stockholm (photo Malmö museer, Attribution-NonCommercial-NoDerivs 2.5 Generic CC BY-NC-ND 2.5).

Fig. 16. Bear scene on the eaves in Dädesjö church, dated to the 1180s. There are several symbolic motifs, including a bear threatening a peaceful ram. During the Middle Ages, the bear was seen as a symbol of wrath, while the domesticated ram was a docile, sacrificial animal serving as a symbol of Christ's suffering (after Ullén 2020, 54).

Fig. 17. The scene in Härkeberga church shows the moment when, according to the narrative in the Bible, God realises David's qualities as a leader of the children of Israel. The bear symbolises rage and evil, while David represents the good whilst defending his herd from the wild beast. Samuel 17:34–35 (English Standard version): "But David said to Saul, Your servant used to keep sheep for his father. And when there came a lion, or a bear, and took a lamb from the flock, I went after him and struck him and delivered it out of his mouth. And if he arose against me, I caught him by his beard and struck him and killed him." The paintings were made by the well-known church painter Albertus Pictor in the 1480s (photo G. Magnusson).

Fig. 18. Man and standing bear, by Erik Eliasson, 1781, interiors from Backhans farm, Svärdsjö, Dalarna, now in Dalarnas museum, Falun. The painting shows Paul's conversion on the road to Damascus, a narrative from the Acts of the Apostles. The motif appears in other traditional Dalecarlia paintings, however, only Erik Eliasson has depicted the scene with a bear. The bear probably symbolises Paul's anger against Christians prior to his salvation through the encounter on the road with the Holy Spirit, after which he became one of the most important apostles of Christianity (photo G. Magnusson).

Fig. 19. Bear crushing a hedgehog. Illustration from Olaus Magnus' Historia de gentibus septentrionalibus *from 1555 (photo G. Magnusson).*

Fig. 20. Olaus Magnus devoted several chapters to the varied diet of bears. In this illustration, he shows the techniques they used in their hunt for honey in trees and in the nests of ground bees. Woodcut from Olaus Magnus' Historia de gentibus septentrionalibus *from 1555 (photo G. Magnusson).*

Fig. 21. Cover of the Bamse comics by Rune Andréasson from 1978. When Bamse eats his grandmother's special honey, he becomes the strongest bear in the world and can put all things right (photo Å. Ahrland).

Fig. 22. Tamed bear collecting money from spectators after his performance. Woodcut from Olaus Magnus' Historia de gentibus septentrionalibus *from 1555 (photo G. Magnusson).*

Fig. 23. Queen Ulrika Eleonora the elder's painting from 1682 shows children and their attendants playing with a young, tame bear. Standing on its hind legs, the bear seems to scare some of them. The man is prepared to throw a stone if needed, while the pet dog looks calm and inquisitive (photo Nationalmuseum [Erkännande-DelaLika 4.0 Internationell, CC BY-SA 4.0]).

Fig. 24. A brown bear in the frieze of the bed chamber of Queen Christina the elder in Gripsholm, painted by Sigfrid Henriksson in the 1590s (cf. VON MALMBORG 1971, 28; LINDGREN 1996, 235–236; photo Å. Ahrland).

Fig. 25. In 1686, David Klöcker Ehrenstrahl painted a polar bear in a mountainous landscape with two bears swimming in the background. On the stone in the left corner is the inscription: "Ursúm húnc aquatica ū(m)e Nova Zemblas", which translates to "Aquatic bear from Novaya Zemlya" (a group of islands in today's northwest Russia). Nationalmuseum, Stockholm (photo Nationalmuseum [Erkännande-DelaLika, CC BY-SA]).

Fig. 26. Funeral coat of arms for Mattias Björnklou from 1671, Ösmo church (photo G. Magnusson).

Fig. 27. The coat of arms of the duchy of the north of Finland is depicted at the 1562 sarcophagus of Gustavus I in Uppsala cathedral. It shows a crowned bear with a raised sword against the night sky (photo Å. Ahrland).

Chapter 11

Bears in Classical Antiquity

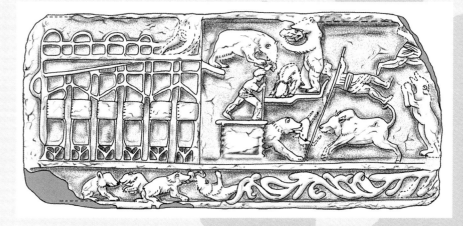

Bears in the arena, limestone relief from Cibyra in Turkey, 2nd or 3rd century AD (see HORN, this volume; drawing C. Golze, after a template).

Bear und human in Greco-Roman antiquity

By Florian Hurka

Keywords: Greco- Roman antiquity, art, literature, cult, stellar constellations, hunt, capture, animal training, circus, science, medicine

Abstract: The bear played a not insignificant role in the ancient Greco-Roman world – as a pest, it occasionally disrupted human life and sometimes even killed human beings. Humans, however, were the greater foe to bears by far; in huge numbers, bears were hunted, captured, and – particularly in Rome – killed or set on humans and other animals for the entertainment of the audience. Humans not only fear, hunt, and torment bears, but they also eat them, heal, clothe, and adorn themselves with parts of them. They invoke the power attributed to the bear in ceremony and cult as well as for magical rituals. They name stellar constellations after it, train it, imitate it in their art, tell stories about it, and turn the bear into an object of their scientific curiosity.

Both literary and archaeological sources of Greco-Roman antiquity provide evidence of the existence of bears in almost all regions in Europe, Asia Minor, and North Africa.[1] Occasional errors in them cannot be ruled out,[2] and some of the ancient writers, particularly when writing fictional texts, probably rather have let their imagination run wild than to refer to verified knowledge.[3] It must be considered a certainty that human behaviour influenced the animals' population, and bears even were completely driven out of their habitats in many places.[4] The question whether the ancient world

---

1   KELLER (1887, 115), after a detailed presentation of the evidence (translation from German): "So we would have regarded the animal as living all around the Mediterranean: from the Barbary via Spain, Gaul, Germania, Italy, Greece, Asia Minor, and Syria back down to the land of the pyramids; even in the large islands of Great Britain, Sicily, and Crete, it appears to have been widespread once". TOYNBEE (1973, 94) is slightly more cautious in his interpretation of the literary evidence and the (zoo)archaeological finds: "throughout the Roman world – east, west, north, south, and centre – bears were to be obtained".
2   A frequently-quoted statement by Pliny the Elder (1st century AD): "it is known that there are no bears in Africa" (Pliny's *Naturalis Historia* 8, 131; for more details on Pliny, see below). Numerous references to bears from or in Northern Africa contradict this claim (cf. TOYNBEE 1973, 94); MACKINNON (2014, 163) takes a conciliatory position: "Pliny [...] stating that, at least in his time, it was apparently well-known that bears do not occur in Africa".
3   In his heroic epic *Aeneid*, for example, the poet Virgil (70–19 BC), working under Emperor Augustus, has the hero Acestes, mythical founder of the Sicilian city Segesta, appear in the skin of a Libyan she-bear (Virgil's *Aeneid* 5, 37). The plot is purely fictional; thus, it cannot be inferred from instances such as this that after the destruction of Troy there were men in Sicily wearing the skin of Libyan bears (regarding the occurrence of bears in Northern Africa, see above), but at best how at the time of writing author and audience would have imagined a well-travelled hero.
4   MACKINNON 2014, 163, with reference to zooarchaeological findings (also already KELLER 1887, 106, 108, albeit without reliable methodology).

knew not only the brown bear (*Ursus arctos*) but also the polar bear (*Ursus maritimus*) is commonly answered in the negative.⁵

The earliest literary evidence for bears is found in Homer's *Iliad*, which in 24 books tells the story of an episode of the Trojan War (Latacz 1985, 23).⁶ In the epic (which, based on long oral mythical tradition, dates from the second half of the 8th century BC), bears do not make an appearance in its actual narrative at all – to be hunted and slain by one of the heroes, for example – but at least the animal features in the context of the so-called *ékphrasis* – that is, it is mentioned in the poetic description of a work of art, which is considered particularly sophisticated. On the shield made for Achilles by the god Hephaestus, the stellar constellation of *Ursa Major* (the "greater she-bear"; *árktos*) is depicted, as part of a cosmological framework (Homer's *Iliad* 18, 487–489): On it, the god fashions the earth, the sea, and the sky as well as the Seven Sisters of the Pleiades in it, the hunter Orion, raised to the sky, and "the She-Bear that by name also 'the wain' is called, / that rotates on the spot and always stalks Orion, / and that alone is barred from the bath in Oceanus' tides" (see Künzl, this volume).

In his account of the she-bear, Homer takes for granted that his readers know that a circumpolar heavenly body never leaves the northern night sky.⁷ Instead, he rather prioritises the juxtaposition of human and animal; the she-bear stalks the hunter in its desire for water (almost humanised, it wants to bathe), which is denied to it by Orion.

Homer does not relate how the she-bear came to be in the starry sky in the first place. As in other ancient myths, tradition is inconsistent on this point; in the elaborate didactic poem by the Hellenistic poet Aratus (about 310–245 BC), there is the tale about how Zeus, in appreciation of their one-year-long services as his wet nurses, transfers the two nymphs Helice and Cynosura into the starry sky, as *Ursa Major* and *Ursa Minor*, respectively.⁸ For Aratus, the recourse to remote variants is characteristic, so it cannot be surprising that the myth of Helice and Cynosura by no means is the most common version of the origin story of the stellar constellation of the (she-)bear.

Much more frequently, Callisto is mentioned in this context, an Arcadian nymph, who was raped by Zeus while in the retinue of the virginal Artemis, the goddess of the hunt. When Artemis discovered Callisto's pregnancy, the nymph not only was sent away by her, but after she had given birth to

---

5   Taking a different position, Toynbee (1973, 94, and, following her, MacKinnon 2014, 163), draws her conclusion from a claim by the poet Calpurnius Siculus (1st century AD) that in Rome he saw seals fight bears (*Ecloga* 7, 65–66): "for according to Calpurnius Siculus bears, which must certainly have been polar bears, could be seen in Rome chasing seals". Firstly, however, Calpurnius' *vitulus aequoreus* ("sea-calf") cannot positively be translated as "seal", secondly and most importantly, the games did not take "natural realism" (here presumably Atlantic seals versus polar bears instead of brown bears) too seriously in other matters, too (see below). The white bears in Thrace mentioned by Pausanias in the 2nd century AD in his description of Greece (*Helládos Periēgēsis* 8, 17, 3) cannot have been polar bears (Llewellyn-Jones/Lewis 2018, 317). And the suggestion by Eichinger (2005, 76) that Oppian (also 2nd century AD) "surely" mentions polar bears (*Halieutica* 5, 38–40) is not correct: *phókai* are seals (Fajen 1999, 381, considers *Monachus monachus* Hermann, the Mediterranean monk seal), which in Oppian's text scare "shaggy bears" and are able to best them in a fight even on dry land.

6   That is, the events concerning Achilles' temporary withdrawal from the fighting in the tenth and last year of the war (Latacz 1985, 91–93).

7   This circumstance gave the she-bear (*Ursa Major*) a significance for ancient seafaring that also is apparent in literary texts: Homer (*Odyssey* 5, 273–274) has his Odysseus be led by the constellation on his homeward journey, and so does Ovid (*Metamorphoses* 13, 293), who in his *Metamorphoses* (dating to about the time of the birth of Christ) likes to mention the constellation anyway (2, 171–172; 4, 625; in 13, 726, Aeneas, having lost his home after the fall of Troy, also gets his bearings from the bear on his journey to Italy).

8   Aratus' *Phainomena* (30–32). Perhaps it was only through Thales – in the 6th century BC and originating in the Orient – that the Little Bear (*Ursa Minor*) came to be in the sky of classic antiquity in the first place; before that, the respective stars were part of *Draco* (Fasching 1998, 126). Repaying the nurses by turning them into bears might reflect a special motherly aspect also attributed to the she-bear on other issues; in their infancy, the two mythical figures Atalanta and Paris were nursed by she-bears (Apollodorus' of Athens *Bibliotheca* 3, 9, 2; 3, 12, 5). Perhaps these tales are linked to the ancient notion that after their birth, bear cubs have to be given their shape by the licking of their mother (for this, see below).

a son was turned into a she-bear by the jealous Hera, Zeus' wife. Years later, she meets her son Arcas, the progenitor of the Arcadians, who fatefully has grown up to be a great hunter. Saving them both, Zeus transfers mother and son (the latter now called "Boötes") into the starry sky. The resentful Hera, however, arranges that the she-bear is not allowed to leave the night sky, but in all eternity is pursued by Boötes and his two hunting dogs.[9]

The myth of Callisto sufficiently explains that bears play a role in the cult of Artemis (they even were sacrificed to her[10]). Even without reference to Callisto, however, bears are important to the cult of Artemis – after all, as goddess of hunt and nature, she is linked to animals in general, including large beasts such as boars and lions (ROSCHER 1890–1894, II, 1, 564–566). This also applies to bears: In atonement for the mythical killing of a bear by Athenian youths, the festival of the *arkteía* was celebrated in Athens every five years, when young girls would dress up as she-bears (*árktoi*) to placate Artemis.[11] Priests of either sex wore bear masks during the festivities, too, and a small black-figure krater that depicts girls and priests disguised in this manner at the ceremony of the *arkteía* also features the image of a real bear.[12]

The Romans not only equated the Greek Artemis with the Italic Diana, but in the same process of appropriation (*interpretatio Romana*), they also adopted Artio, the Celtic goddess of the hunt and/or fertility, who was venerated by the Gallic-Celtic tribes of the Helvetii and Treveri in the shape of a bear (KAUFMANN-HEINIMANN 2002, 48). For the cults of other deities, such as Cybele, Mercury, Zeus, and others, bears are attested as well.[13] Furthermore, small bear figurines were found, especially in late antique burials of Romano-British children, that might have been protective talismans (?).[14]

But the bear had a place not only in cult and religion, but it also occurs in several varieties of magic, including transition magic. In the *Odyssey* (which is slightly younger than the *Iliad*), the baldric of Hercules in the underworld is adorned not only with lions and boars, but also with bears (Homer's *Odyssey* 11, 610–612): "Terrifying was the belt about his breast, a baldric of gold, on which wondrous things were fashioned, bears and wild boars, and lions with flashing eyes, and conflicts, and battles, and murders, and slaying of men". This is not mere decoration; rather, it is meant to place the strength and ferocity of the animals, which in ancient literature often are characterised as violent and cruel,[15]

---

9   Aside from the main version sketched out here, which is told in particular detail in Ovid's *Metamorphoses* (2, 401–530), there are numerous variants (cf. ROSCHER 1890–1894, II, 1, 931–935) that played their role in the cult, too. In the Arcadian town Scias, for example, the sanctuary of Artemis Kalliste was located on the burial place of Callisto, who, according to this variant of the myth, was not put into the sky, but – in her bear shape – killed by Artemis (Pausanias' *Hellados Periēgēsis* 8, 3,7; 8, 35, 8). Archaeological evidence also substantiates the story of Callisto, most impressively on a 4th-century red-figure *pelike* that depicts the moment of the nymph's transformation – in terror, Callisto looks at her left hand that already has become a paw (see the figure published in EICHINGER 2005, 327).

10   In Patras, Pausanias reports, bears and other animals were sacrificed alive to Artemis (*Hellados Periēgēsis* 7, 18, 13). He asserts that he watched with his own eyes as a bear tried to escape being burned to death, but was brought back to the pyre by the priests attending to the sacrifice. "No one has been injured by the animals in the process yet", he claims. Votive offerings to Artemis in the shape of bears (statuettes, reliefs, bear's teeth) also occur elsewhere (EICHINGER 2005, 63).

11   Regarding the *arkteía*, cf. PERLMAN 1989, 116–121. SCANLON (2002, 164) interprets the act of dressing up as she-bears as a sort of initiation ritual; the bear, perceived as half-human (see below) symbolises the transition from the girl (wild, maidenly) to the woman (tamed, wife/mother).

12   Cf. figure in EICHINGER 2005, 326. KAHIL (1977, 91) concludes that real, living bears took part in the *arkteía*, too (regarding tame animals, see below in the main text).

13   EICHINGER 2005, 62, 97. Zeus himself in one of his numerous love affairs is said to have approached the coveted woman in the guise of a bear (Clement's of Rome *Homiliae* 5, 13, 2).

14   E.g. in York, Malton, and Colchester (TOYNBEE 1973, 99), as well as in Trier and Cologne (EICHINGER 2005, 95). It remains unclear why the bear was invoked here – possibly for its motherly qualities or in reference to Christian symbolism (hibernation as a symbol of Resurrection: COMSTOCK/VERMEULE 1971, 347); possibly these objects merely were toys (TOYNBEE 1973, 100)? Bear's teeth were used as amulets, too (KELLER 1887, 121; see O'REGAN, this volume).

15   Cf. CHERUBINI's (2009, 82) "eloquent dossier" (with omissions) – *furor* ("rage, madness": Horace's *Ars Poetica* 472; cf. Martial's *Epigrammata* 6.64.27–32), *rabies* ("anger": Lucan's *Pharsalia* 6.220, 222; Pliny's *Naturalis historia* 8.130), *fenestus* ("deadly": Apuleius' *Metamorphoses* 7.24), *truculentus* ("cruel": Ovid's *Metamorphoses* 13.803), *ferus* ("wild":

in an expressive relation to the human violence depicted in the imagery – both traits defined Hercules' life (and, according to Homer, continued to do so even in Hades, where Hercules is portrayed as fierce and combative).[16]

Similar appropriations of the bear's aura of a powerful and wild beast also occur elsewhere; in legends, demigods are begot on the skins of bears or lions, or in their youth are fed on the offal of lions, wild boar, or bears.[17] Also, participants in historic battles are described in fictitious texts (without historical validity) as wearing bear costumes.[18] But in reality also, the military drew on the symbolic power of the bear; in Rome, distinguished ranks such as the horn players (*cornicines*) or the bearers of standards and effigies (*aquiliferi* and *imaginiferi*) wore the skins and/or the heads of bears.[19]

The transition from magical appropriation to art and decoration is fluid: Arcadian coins use the bear motif in order to refer to the mythical progenitor of Arcadia, the aforementioned Arkas, son of Callisto.[20] Sculpture, reliefs, and pottery show the animal, frequently in connection with human cultural practices (e.g. hunt, arena, or cult),[21] but there also are entirely non-propagandistic commodities, such as a bear statuette that was part of a grater, or the handle of a pestle decorated with bear motifs (EICHINGER 2005, 69, 106). Both objects could have been used in the preparation of cosmetics, one ingredient of which was bear's grease (ZAHLHAAS 1996, 103). Corresponding storage boxes occasionally were decorated with bear ornaments, too. There are Roman oil lamps depicting bears in their *discus*, as well as vessels in the shape of bears (*rhyta*) that perhaps were used for the storage of bear's grease, which was a valued item in the Roman medicine chest (EICHINGER 2005, 89, 106). Obtained primarily from the animal's kidneys, it promised to alleviate a wide spectrum of ailments,

---

Seneca's *Oedipus* 151), *trux* ("fierce": Gaius Valerius Flaccus' *Argonautica* 2.73), gifted with an *ingenium immansuetum* ("uncontrollable nature": Ovid's *Metamorphoses* 15.85). The bear's cry is described by the verb *saevire* ("to rage": Virgil's *Aeneid* 7.17), and even when it is drinking, the animal *aquammor suvorat* ("devours water with bites": Pliny's *Naturalis historia* 10.201). These are the fierce beasts that Oppian will call a "bloody race", gifted with "a deadly mouth with sharp teeth" and a "fierce heart" (Oppian's *Cynegetica* 3, 139–145).

16 At the same time, the ornament might symbolise the victory of humankind over nature's threats to it: Hercules is a mythical conqueror of threats, and among his twelve labours he defeated the Nemean Lion and the Erymanthian Boar; according to Diodorus Siculus' *Bibliotheca historica* 4, 17 (1st century BC/AD), he is also said to have eradicated the Cretan bears.

17 Aeneas, mythical hero of the Trojan War and later Proto-Roman, is supposed to have been sired by Aphrodite and the Trojan Anchises on a bed decorated with the skins of bears or lions, according to stanza 159 of the pseudo-Homeric *Hymn to Aphrodite* (date unclear, but certainly post-Homeric and pre-Hellenistic: LESKY 1971, 110–112). Regarding offal, cf. Achilles (Apollodorus' of Athens *Bibliotheca* 3, 172). In contrast, bears appear in the family trees of mythical heroes only in exceptional cases: Odysseus' grandfather Arkeisios is said to have had a she-bear for his mother (*Etymologicum Magnum* 144, 23–32 [according to Aristotle]), cf. FRANCO 2017, 52–53. The liaison between the Thracian virgin Polyphonte and a bear (brought about by the offended love goddess Aphrodite) resulted in the birth of two savage giants (Antoninus Liberalis' *Metamorphoses* 21; cf. CHERUBINI 2009, 79).

18 Cf. in addition to the passage by Virgil mentioned above (note 3) also Silius Italicus (c. AD 25–100), who in his epic *Punica* (4, 558) features an Italic warrior wearing the skin of a Samnite she-bear; also Statius (AD 40/50–pre-96) in his *Thebaid* (4, 304); cf. other examples in KELLER 1887, 121.

19 Vegetius' *De re militari* 2, 16 (4th century AD), cf. DEVIJER 1992, 146. According to Pausanias' *Hellados Periēgēsis*, the Arcadians also wore the skins of wolves and bears during the fighting in the First Messenian War (8th century BC; Pausanias 4, 11, 3). This information was provided to Pausanias by Myron of Priene (probably 3rd century BC), a source that is now lost (Pausanias 4, 6, 1–2) and can hardly be regarded as trustworthy. A civilian use of bear clothes is reported by the Greek geographer Strabo (c. 64 BC – AD 19), who claims that in Mauretania, people not only sleep on bear skins, but also dress in them (*Geographica* 17, 3, 7). Despite some travels he undertook himself, most of Strabo's information on regional geography is based on literary sources (cf. LESKY 1971, 993).

20 LLEWELLYN-JONES/LEWIS 2018, 320. KELLER (1887, 109–111) mentions bear coins from Celtic Gaul, Spain, and Thrace as well.

21 For archaeological evidence, see TOYNBEE 1973, 99, and EICHINGER 2005, *passim*. In addition, there are a few literary references, such as Martial's *Epigrammata* 3, 19 (1st century AD), which mention a bronze bear sculpture in Rome (now lost).

including chilblains,[22] gout, hair loss, dandruff, pains of the neck or loins, pains of the feet, burns, and ulcers (Pliny's *Naturalis historia* 28, 29, 163, 192, 198, 221, 235, 251). Bear bile (respiratory tract, eye complaints: ibid., 28, 167, 193), external application of bear's blood (abscesses: ibid., 28, 217) or bear's lung (excoriated feet: Dioscorides' *De materia medica* 2, 38), and the consumption of testicles (epilepsy[23]) were recommended as well. In farming, the blood and fat of bears also served as pesticides and antifreeze agents.[24]

The influence of bears on cult, magic, decoration, and medicine probably was based on respect for, possibly also the fear of the bear's strength. It is a fact that there were encounters that were fatal for humans.[25] More often, the bear acted as a pest for agriculture – a number of archaeological objects show a bear plundering fruit trees,[26] and in poetry, the bear occasionally is described as a threat to farm life as well.[27] Even if such poetic evidence cannot readily be transferred to contemporaneous agriculture (it was written by urban poets who in their compositions do not betray any profound agrarian knowledge), it nevertheless testifies to the image of the bear in the minds of the poets and their sophisticated audience.

This image may by all means be founded on personal experience: The earliest documentation of performances of bears in Rome (so-called *venationes*, "hunts") is provided by the historian Livius, who worked under Emperor Augustus, for the year 169 BC.[28] Archaeological and literary evidence of bears in Roman circus events attest to this practice right into late antiquity (ancient literary sources frequently mention high numbers of up to 1,000 animals).[29] At the conclusion of the spectacle, the animals mostly lost their lives (the meat might have been eaten afterwards[30]), even if their death

---

22 Dioscorides' *De materia medica* 2, 76. Pedanius Dioscorides is the author of *On Medical Material*, the most important pharmacological book in antiquity (second half of the 1st century AD); cf. LESKY 1971, 999.
23 Pliny's *Naturalis historia* 28, 224. Bear bile also is supposed to be beneficial against this disorder, if it is allowed to melt in the mouth (Dioscorides' *De materia medica* 2, 78).
24 Pliny's *Naturalis historia* 17, 265. A mass utilisation is hardly conceivable, of course.
25 From Thyrion (Acarnania), the funerary inscription of a woman who was killed by a bear is known from the 1st century BC (PERLMAN 1989, 115).
26 E.g. central discs of oil lamps as well as mosaics and reliefs (EICHINGER 2005, 89).
27 The Roman poets Horace (65–8 BC) and Ovid (43 BC – AD 17/18) tell of bears prowling around sheep shelters and being considered a danger even to larger livestock (Horace's *Epodi* 16, 51; Ovid's *Metamorphoses* 7, 546). Both passages, however, do not reflect the poets' living environments: Horace conjures up a fanciful counter-world to a Rome ravaged by civil war – a utopian new beginning in the Islands of the Blessed, where "does not growl the bear around the sheep pen at night", while Ovid has the mythical Aeacus report how during the spread of an epidemic in Aegina, wild animals suffer as well – the boar loses its fierceness, the weakened bear does not dare attack strong cattle anymore.
28 Livy's *Ab urbe condita* 44, 18, 8 (forty bears in the Circus Maximus). In his monumental book *Ab urbe condita*, Livy draws on numerous unnamed accounts (in doing so, he frequently lacks criticism and diligence: FUHRMANN 2005, 346). Only an impersonal *notatum est* ("it is noted") is added to the information about the bear at the games of the Aediles Publius Lentulus and Scipio Nasica in 169 BC.
29 For extensive archaeological and literary evidence, see EICHINGER 2005, 79–81. One thousand bears are mentioned in the so-called *Historia Augusta* (SHA), a collection of the biographies of thirty emperors that is of unclear authorship and date (VON ALBRECHT 1994, 1102–1104), in regard to the games of the later Emperor Gordian I (c. AD 159–238; *SHA Gordiani Tres* 3, 6, 7). The historian Cassius Dio (2nd/3rd century AD) is sceptical about these numbers, however (*Historia Romana* 61, 9, 1); cf. MACKINNON (2014, 466): "The figures given in our ancient texts may be exaggerated, but there is no question that the number of animals that perished during several centuries of spectacles was enormous".
30 PRELL 1997, 86. The gruesome information provided by Tertullian (*Apologeticum* 9, 11; around AD 198) that the Roman audience demanded bears' stomachs filled with human flesh, probably should be regarded as the polemic hyperbole of a Christian author attacking pagan practices, but it nonetheless confirms the custom of eating the flesh of the animals killed in the Circus (at least at the end of the 2nd century AD). This fits in with a scene from the comical-whimsical novel by Apuleius (also 2nd century AD); in it, all the bears bought for a spectacle die before the games, and the peasantry devours the cadavers (the episode is given its macabre climax by the protagonists' idea to dress up one of them as a bear and sell him to the unlucky organiser of the games, which ends in tragedy when the false bear is hunted and killed: Apuleius' *Metamorphoses* 4, 13–21). In the parody of a dinner party in the picaresque novel by Petronius (probably 1st century AD), the taste of bear meat is compared to that of roasted boar (Petronius' *Satyrica* 66, 5–6); at least to the dinner guests in the novel, the taste appears to be unfamiliar (from this, KELLER [1887, 122] draws the conclusion that the

instance, he is hoping for the speedy delivery of Italic bears (the more, the better; cf. Symmachus' *Epistulae* 7, 121), in another, for bears of Dalmatian origin (ibid., 9, 142), then he is afraid that the merchandise he ordered might clandestinely be replaced with inferior animals (ibid., 7, 121), he writes about bears from overseas (ibid., 9, 135), and finally mentions the professional bear traders (*ursorum negotiatores*; ibid., 5, 62), who, in this case, prove to be not particularly professional – only one bear cub was delivered to Rome in time, and that was almost starved.[46]

In addition to capturing bears alive, there was the actual hunt aimed to kill the animal. Here, it is necessary to differentiate between pest control, which, among other methods, used poisoned bait,[47] and the hunt for pleasure or prestige – the "gentleman hunt" (MACKINNON 2014, 207–208). Many instances are known of this latter in connection with the mythical and real grandees of antiquity; the epicist Statius, for example, has the hero Achilles (grown strong from infancy on a diet of the animal's offal; cf. Apollodor's of Athens *Bibliotheca* 3, 172) hunt bears, and Plutarch attests to the bear hunt in his biography of Alexander the Great, which contains all sorts of anecdotes (cf. Statius' *Achilleid* 2, 96–156; Plutarch's *Alexandros* 41, 2). There also is impressive archaeological evidence. A Cretan bronze shield (probably from the 8th century BC) portrays the hunt for bears (and lions) as a heroic feat by mounted and un-mounted warriors and archers (EICHINGER 2005, 68), and the royal tomb of the Macedonian Argead dynasty (to which belonged Alexander the Great) in Vergina (4th century BC) features a hunting frieze that depicts a royal hunt for lions and bears – on rocky terrain, a bear is being attacked with lances, spears, and nets.[48]

In Rome, it was especially the emperors who hunted bears, perhaps inspired by the Hellenistic royal courts (LORENZ 2000, 92); after the successful hunt for a she-bear, Hadrian (reign: AD 117–138), for example, founded the town Hadrianotherae in Mysia (cf. *Historia Augusta [SHA] Hadrianus* 20, 13; Cassius Dio, *Historia Romana* 69, 10, 2) and had himself portrayed prestigiously in a relief hunting the bear;[49] in the Boeotian city of Thespiae, he dedicated the trophy of a killed she-bear to the god Eros in a self-authored inscription (*Epigrammata graeca* [Kaibel] 329, no. 811).

The perhaps least pragmatic approach to bears is displayed by the ancient natural sciences under the domineering influence of Aristotle (384–322 BC). Just as in regard to other animals, the great empiricist in his zoological studies takes interest particularly in the bears' mating habits – heightened aggressiveness during the mating season (Aristotle's *Historia animalium* 571b, 26–28), copulation primarily in a lying position, the animals' winter rest lasting at least forty days (ibid., 579a, 18–20), and the process of giving birth during winter rest (ibid., 579a, 18–20). Their wide-ranging menu also is familiar to him (a bear has a very varied diet, such as fruit, honey, crustaceans, ants, or meat).[50] In several of these aspects, Aristotle sees parallels to humans.[51]

---

46 Symmachus' *Epistulae* 2, 76. In view of the literary and artistic features of the epistles (VON ALBRECHT 1994, 1146–1148), it is doubtful whether this amusing turn of events can be taken at face value.
47 Xenophon's *Cynegeticus* 11, 2. Author, soldier, and politician Xenophon (*c.* 430 – *c.* 354 BC) perhaps wrote his book on hunting skills during his youth; it might, however, have come from someone else's pen (LESKY 1971, 695).
48 EICHINGER 2005, 69, 337. Other Greek burial monuments also depict the bear hunt (TOYNBEE 1973, 94–95; EICHINGER 2005, 67).
49 Preserved as one of the four Hadrianic *tondi* of the Arch of Constantine in Rome (TOYNBEE 1973, 94).
50 Aristotle's *Historia animalium* 594b, 5–7. Aristotle teaches, however, that bears only eat rotting meat (594b, 16–17). Aelianus (*De natura animalium* 5, 49) claims the opposite and thence recommends dropping on the ground in the case of a bear attack and holding one's breath. This behaviour is also deemed an effective procedure in a bear attack in a fable by Aesop – while an unfaithful friend escapes up a tree, the other one plays dead, and after a long examination is spared by the bear (*Corpus Fabularum Aesopicarum* 66).
51 Apart from being omnivorous and (wrongly assumed) mating in a lying position, the bear, according to Aristotle (*Historia animalium* 507b, 594b), is walking upright just like humans (594b), has only one stomach, and its paw is similar to the human hand (five fingers, three phalanges: 498a). An explicit connection between the limbs of humans and bears is drawn in Oppian's *Cynegetica* 3, 144: "hands like human hands, and feet like human feet".

A large portion of the information provided by Aristotle is correct and may, at least in part, be founded on the practical knowledge of a hunter (EPSTEIN 2019, 217). There are, however, also misconceptions such as the claim that bears (just like foxes or lions) not only give birth to tiny (of a size between a weasel and a mouse), but also physically largely undeveloped young.[52] False assumptions as this often even are embellished by the ancient natural sciences after Aristotle;[53] the account, for example, of the almost shapeless bear cubs is supplemented by the information that they are only given their shape by the extensive licking of their mother,[54] and Aristotle's remark that at "about the period" of hibernation, bears grew "especially fat, so that they hardly are able to move" (Aristotle's *Historia animalium* 600a, 29), is turned by Pliny the Elder into the notion that bears grew fat "miraculously" *during* hibernation (in caves and thickets) and subsisted after a particularly deep first period of rest by sucking on their front paws.[55]

At least in part based on personal experience, Aelian and Pliny the Elder provide some further information regarding anatomy and behaviour of bears; they claim that the head was the weakest part of their body (in the arena, they frequently are killed by a single blow to the head[56]), that they attacked oxen by attaching all four limbs to their head and horns to wear them out,[57] and that they customarily climbed down from trees backwards (Aelian's *De natura animalium* 6, 9; Pliny's *Naturalis historia* 8, 130). "No other animal", concludes Pliny his descriptions on a harsh and moralising note, "acts more cleverly in its misdeeds despite its stupidity" (*Naturalis historia* 8, 131).

These Plinian "misdeeds" of bears probably stem from the animal being perceived – as touched upon above – as a pest and from its image as being wild and cruel. In contrast, the somewhat paradox "stupid cleverness" on one hand possibly refers to the bears' successes in foiling human attempts to keep them away (e.g. stop them from raiding fruit trees) and to capture and/or kill them (i.e. the challenges of the bear hunt),[58] on the other on the bears' seemingly cumbersome manner of moving; to symbolise the power of love, bears, allegedly so sluggish, occasionally are depicted with Aphrodite/

---

52 Aristotle's *Historia animalium* 540a, 1. KELLER (1887, 123): "This is one of the most widespread myths of the natural history of classic Antiquity". At least, newborn bears are, in fact, very small, naked, and largely helpless.
53 The most important known successors of Aristotle are Aelian and Pliny the Elder: Claudius Aelianus (*c.* AD 170–235), author of an extensive book on animals (*De natura animalium*), was a classic (Roman) "study room scholar", who obtained his knowledge of the world through the study of Greek books (and who also wrote in Greek); cf. LESKY 1971, 953. Gaius Plinius Secundus (AD 23 or 24–79), on the other hand, was a man of action; after a distinguished military and administrative career, he died as the commander of the western fleet in the Mediterranean in the attempt to evacuate the victims of the eruption of Mt Vesuvius. He won literary acclaim as the author of historical and scientific studies, of which only his natural history (*Naturalis historia*) in 37 books still exists today (it is the largest surviving prose work of antiquity). Like Aelian, Pliny – without doing research of his own – consulted the most significant reference books directly or at least in excerpts (Pliny lists 327 Greek and 146 Roman authors). The zoological books primarily are based in the traditions of Aristotle's school and that of his successor Theophrastos, supplemented by notes taken from Roman authors (VON ALBRECHT 1994, 1004). A special study about bears penned by Sostratus (1st century AD) is lost (cf. KELLER 1887, 123; 1909/1913, 180).
54 This concept is very common not only in ancient specialised writing (Plutarch's *Moralia* 494; Pliny's *Naturalis historia* 8, 126; Oppian's *Cynegetica* 3, 163–165; Aelian's *De natura animalium* 2, 19 and 6, 3). In his book mentioned (146–148), Oppian explains the incompleteness of bear cubs with the bears' desire to mate again as soon as possible. In a biography of Virgil written during the time of the Roman Empire, the dictum of the Roman national poet can be found that the arduous honing of his verses was similar to the licking of the bear cubs by their mother (Suetonius' *Vita Vergilii* 22).
55 Pliny's *Naturalis historia* 8, 127. Aelian (*De natura animalium* 6, 3) specifies: sucking on the right paw. KELLER (1887, 123) points out that the licking of the paws in fact been has observed on animals in captivity (confirmed by LOZZA 1998, 33; for the care of skin and claws).
56 Pliny's *Naturalis historia* 8, 130. Pliny also is aware that bears protect their heads with their paws when jumping off a rock.
57 Pliny's *Naturalis historia* 8, 131. The origin of this notion might well be an occurrence in the arena (cf. the relevant mosaic in DUNBABIN 1978, fig. 48).
58 LLEWELLYN-JONES/LEWIS 2018, 321, including the reference to Plutarch's *Moralia* 971e: She-bears withdraw into their caves backwards to trick their pursuers.

Venus or Erotes ("even the most awkward apathy must follow the magic of love"; KELLER 1909/1913, 176 [translation from German]; for further references, see EICHINGER 2005, 84).

Not explicitly as stupid, but at any rate as more stupid than the clever fox, the bear is depicted in a fable by Aesop, in which a bear and a lion fight to exhaustion over a fawn – only to watch in dismay as a fox then carries away the prize.[59]

In summary, it can be said that the bear played a not insignificant role in the Greco-Roman world. As a pest, it occasionally disrupted human life and sometimes even killed human beings. Humans, however, were the greater foe to bears by far: In huge numbers, bears were hunted, captured, and – particularly in Rome – killed or set on humans and other animals for the entertainment of the audience. Humans not only fear, hunt, and torment bears, but they also eat them, heal, clothe, and adorn themselves with parts of them. They invoke the power attributed to the bear in ceremony and cult as well as for magical rituals. They name stellar constellations after it, train it, imitate it in their art, tell stories about it, and turn the bear into an object of their scientific curiosity.

BIBLIOGRAPHY

Primary sources

Ab urbe condita: Titus Livius, Livy in fourteen volumes, vol. 13: Books XLIII–XLV, with an English translation by A. C. Schlesinger (Cambridge [Massachusetts] 1951).

Achilleid: Statius, P. Papini Stati Achilleis recensuit A. Marastoni (Leipzig 1974).

Ad martyras: Tertullian, Q. Septimi Florentis Tertulliani Opera, Pars 4: Ad martyras (et al.); recensuit V. Bulhart (Wien 1957).

Aeneid: Virgil, P. Vergilii Maronis Opera recensuit R. A. B. Mynors (Oxford 1972).

Alexandros: Plutarch, Plutarch's Lives in eleven volumes, with an English translation by B. Perrin, vol. 7 (London 1919).

Antidosis: Isocrates, Opera Omnia recensuit B. G. Mandelaras (München 2003).

Apologeticum: Tertullian, Apologeticum, with an introduction and translation by T. Georges (Darmstadt 2015 [with Latin text]).

Argonautica: Gaius Valerius Flaccus, Gai Valerii Flacci Setini Balbi Argonauti con libros octo; recensuit W. W. Ehlers (Stuttgart 1980).

Ars Poetica: Horaz, Q. Horatius Flaccus: Opera recensuit D. R. Shackleton Bailey (München ⁴2001).

Bacchylides: Bacchylides, Carmina cum fragmentis post B. Snell edidit H. Maehler (Leipzig 1970).

Bibliotheca historica: Diodorus Siculus, Diodorus of Sicily in twelve volumes, with an English translation by C. H. Oldfather, vol. 3 (Cambridge [Massachusetts] 1939).

Bibliotheca: Apollodoros von Athen, Apollodoros: Bibliotheke (Helden- und Göttersagen), edited by P. Dräger (Düsseldorf, Zürich 2005).

Codex Theodosianus: Theodosiani libros XVI […] ediderunt Th. Mommeen et P. M. Meyer (Berlin 1905).

Corpus Fabularum Aesopicarum: Äsop, Leben und Fabeln, edited by N. Holzberg (Berlin 2021).

Cynegetica: Oppian, Pseudo-Oppian: Kynegetica, edited by S. Renker (Berlin 2021).

Cynegeticus: Xenophon, Xenophon in seven volumes, vol. 7: scripta minora by E. C. Marchant (London 1968).

De ira: Seneca, Moral Essays, with an English translation by J. W. Basore (Cambridge [Massachusetts] 1928).

De materia medica: Pedanius Dioscorides of Anazarbus, De materia medica, translated by L. Y. Beck (Hildesheim 2005 [with Greek text]).

De mortibus Persecutorum: Lactantius, De mortibus persecutorum, edited and translated by J. L. Creed (Oxford 1984).

De natura animalium: Claudius Aelianus, De natura animalium, edited by M. García Valdés/L. A. Llera Fueyo/L. Rodríguez-Noriega Guillén (Berlin 2006).

De re militari: Vegetius, Epitomarei militari, edited by M. D. Reeve (Oxford 2004).

De significatione verborum: Pompeius Festus, Sexti Pompei Festi De verborum significatu quae supersunt cum Pauli epitome edidit W. M. Lindsay (Leipzig 1913).

---

59 *Corpus Fabularum Aesopicarum* 147. The masquerade of the fable conceals types of human beings and human destinies, of course, so that in fact humans are at the centre of interest instead of bears. Aesop's perhaps somewhat irritating moral is: The fable shows that those get angry with good reason who watch the profit of their labours being carried away by others. A related form of human self-reference probably is the basis of the proverbial phrase not to go looking for the bear's tracks when the bear itself is there already (*Bacchylides*, fr. 6 [after Snell's edition], 6th/5th century BC); here, the universal adage is derived from the bear hunt that in a difficult situation, the crucial facts have to be faced.

De vita Pythagorica: Iamblichos, Iamblichi de vita Pythagorica ediderunt L. Deubner et U. Klein (Stuttgart 1975).

Digesta: Paulus, Corpus iuris civilis ediderunt Th. Mommsen et P. Krueger, vol. 1 (Berlin 1872.

Ecloga: Calpurnius Siculus, Calpurini Siculi Eclogae a cura di M. A. Vinchesi (Firenze 2014).

Eidyllia: Theokrit, Theocritus, Moschus, Bion, edited and translated by N. Hopkinson (Cambridge [Massachusetts] 2015).

Epigrammata: Martial, M. Valeri Martialis Epigrammata edidit D. R. Shackleton Bailey (Stuttgart 1990).

Epigrammata graeca (Kaibel): Epigrammata graeca ex lapidibus conlecta edidit G. Kaibel (Berlin 1878).

Epistulae: Symmachus, Q. Aurelii Symmachi quae supersunt edidit O. Seeck (Berlin 1883).

Epodi: Horaz (see *Ars Poetica*).

Etymologicum Magnum: Etymologicum Magnum seu magnum Grammaticae penu edidit F. Sylburg (Leipzig 1816/1818).

Geographica: Strabo, Strabonis Geographica recognovit A. Meineke (Leipzig 1898–1903)

Halieutica: Oppianus, Halieutica. Einführung, Text, Übersetzung in die deutsche Sprache und ausführlicher Katalog der Meeresfauna von F. Fajen (Stuttgart 1999).

Historia animalium: Aristotle, Historia Animalium, edited by D. M. Balme (Cambridge 2002).

Historia animalium (Epstein): Aristotle, Historia Animalium, Buch V, eingeleitet, übersetzt und kommentiert von K. Epstein (Berlin 2019).

Historia Augusta (SHA): The Scriptores Historiae Augustae in three volumes, with an English translation by D. Magie (Cambridge [Massachusetts] 1921).

Historia Romana: Cassius Dio, Cassii Dionis Cocceiani Historiarum Romanarum quae supersunt edidit U. Ph. Boissevain (Berlin 1955).

Hymnus to Aphrodite: Pseudo-Homer, Three Homerichyms: to Apollo, Hermes and Aphrodite, edited by N. Richardson (Cambridge 2010).

Hellados Periēgēsis: Pausanias, Pausaniae Graeciae descriptio ed. M. H. Rocha-Pereira (Leipzig 1989).

Homiliae: Clemens von Rom, Die Pseudoklementinen: Homilien (lateinisch-deutsch), edited by B. Rehm (Berlin ²1969).

Iliad: Homer, Homeri Ilias recensuit M. L. West (Stuttgart 1998/2000).

Institutionum libri IV: Gaius, Institutiones, edited by U. Manthe (Darmstadt 2004).

Liber Spectaculorum: Martial (see *Epigrammata*: Martial).

Metamorphoses: Ovid, P. Ovidii Nasonis Metamorphoses edidit W. S. Anderson (Leipzig ³1985).

Metamorphoses: Antoninus Liberalis, Les métamorphoses. Texte établi par M. Papathomopoulos (Paris 1968).

Metamorphoses: Apuleius, Apulei Metamorphoseon libri XI recensuit M. Zimmerman (Oxford 2012).

Moralia: Plutarch, Moralia, edited by Ch. Weise and M. Vogel (Wiesbaden 2012).

Naturalis Historia: Pliny: Natural history in 10 volumes with an English translation by W. H. S. Jones (Cambridge [Massachusetts] 1958–1966).

Odyssey: Homer, Homerus: Odyssea recensuit M. L. West (Stuttgart 2017).

Oedipus: Seneca, L. Annaei Senecae Tragoedia recensuit F. Leo (Berlin 1878).

Onomasticon: Pollux, Iulii Pollucis Onomasticon ex recensione I. Bekkeri (Berlin 1846).

Phainomena: Aratus, Phaenomena, edited by D. Kidd (Cambridge 1997).

Pharsalia: Lucan, M. Annaei Lucani De bello civili libri X edidit D. R. Shackleton Bailey (Berlin 2009).

Punica: Silius Italicus, Sili Italici Punica edidit I. Delz (Stuttgart 1987).

Satyrica: Petronius, Petronii Arbitri Satyricon edidit K. Müller (München 1961).

SHA: see *Historia Augusta*.

Thebaid: Statius, P. Papini Stati Thebais ediderunt A. Klotz et Th. C. Klinnert (Leipzig 1973).

Vita Vergilii: Sueton, Vitae Vergilianae antiquae ed. C. Hardie (London 1966).

Secondary sources

Cherubini 2009: L. Cherubini, The Virgin, the Bear, the Upside-Down 'Strix'. Arethusa 42, 2009, 77–97.

Comstock/Vermeule 1971: M. Comstock/C. Vermeule, Greek, Etruscan and Roman Bronzes in the Museum of Fine Arts (Boston 1971).

De Garies Davies 1973: N. De Garies Davies, The Tomb of Rekh-Mi-Re at Thebes. Publications of the Metropolitan Museum of Art, Egyptian Expedition, 11 (New York 1973).

Den Boeft et al. 2013: J. Den Boeft/J. W. Drijvers/ D. Den Hengst/H. C. Teitler, Philological and Historical Commentary on Ammianus Marcellinus XXIX = vol. 10 (Leiden 2013).

Devijer 1992: H. Devijer, Equestrian Officers of the Roman Imperial Army 2 = Mavors. Roman Researches 6 (Leiden 1992).

Dunbabin 1978: K. Dunbabin, The mosaics of Roman North Africa: studies in iconography and patronage (Oxford 1978).

Eichinger 2005: W. Eichinger, Der Bär und seine Darstellung in der Antike. Schriftenreihe Antiquitates 32 (Hamburg 2005).

Epstein 2019: see *Primary sources*, *Historia animalium* (Epstein).

Fajen 1999: F. Fajen (ed.), Oppianus: Halieutica. Einführung, Text, Übersetzung in die deutsche Sprache und ausführlicher Katalog der Meeresfauna (Stuttgart 1999).

Fasching 1998: G. Fasching, Sternbilder und ihre Mythen (Wien ³1998).

FRANCO 2017: C. Franco, Greek and Latin Words for Human-Animal Bond: Metaphors and Taboos. In: T. Fögen/E. Thomas (eds), Interactions between Animals and Humans in Graeco-Roman Antiquity (Berlin, Boston 2017) 39–60.

FUHRMANN 2005: M. Fuhrmann, Geschichte der römischen Literatur. Reclams Universal-Bibliothek 17658 (Stuttgart 2005).

JENNISON 1937: G. Jennison, Animals for Show and Pleasure in Ancient Rome. Publications of the University of Manchester/Victoria University of Manchester 258 (Manchester 1937).

KAHIL 1977: L. Kahil, L'Artémis de Brauron: Rites et mystères. Antike Kunst 20(2), 1977, 86–98.

KAUFMANN-HEINIMANN 2002: A. Kaufmann-Heinimann, Dea Artio, die Bärengöttin von Muri. Römische Bronzestatuetten aus einem ländlichen Heiligtum (Bern 2002).

KELLER 1887: O. Keller, Thiere des classischen Alterthums in culturgeschichtlicher Beziehung (Innsbruck 1887).

KELLER 1909/1913: O. Keller, Die antike Tierwelt. Zwei Bände (Leipzig 1909/1913).

LATACZ 1985: J. Latacz, Homer: eine Einführung. Artemis-Einführungen 20 (München 1985).

LESKY 1971: A. Lesky, Geschichte der griechischen Literatur (München ³1971).

LLEWELLYN-JONES/LEWIS 2018: L. Llewellyn-Jones/S. Lewis, The Culture of Animals in Antiquity: a Sourcebook with Commentaries (Abingdon, New York 2018).

LORENZ 2000: G. Lorenz, Tiere im Leben der alten Kulturen. Alltag und Kultur im Altertum 5 (Wien 2000).

LOZZA 1998: H. Lozza, Auf den Spuren der Bären: zur Vergangenheit und Zukunft der Braunbären in der Schweiz. Begleitbroschüre zur Bärenausstellung des Schweizerischen Nationalparks. Konzept und Text: Hans Lozza; unter Mitwirkung von Tobias Kamer (Chur 1998).

MACKINNON 2014: M. MacKinnon, Fauna of the Ancient Mediterranean World. In: G. L. Campbell (ed.), The Oxford Handbook of Animals in Classical Thought and Life (Oxford 2014) 156–179.

PERLMAN 1989: P. Perlman, Acting the she-bear for Artemis. Arethusa 22(2), 1989, 111–133.

PRELL 1997: M. Prell, Sozialökonomische Untersuchungen zur Armut im antiken Rom. Beiträge zur Wirtschafts- und Sozialgeschichte (Stuttgart 1997).

ROSCHER 1890–1894: W. H. Roscher (ed.), Ausführliches Lexikon der griechischen und römischen Mythologie Bd. 2/Abt. I (Leipzig 1890–1894).

SCANLON 2002: T. F. Scanlon, Eros and Greek Athletics (Oxford 2002).

SHELTON 2014: J.-A. Shelton, Spectacles of Animal Abuse. In: G. L. Campbell (ed.), The Oxford Handbook of Animals in Classical Thought and Life (Oxford 2014) 461–477.

THOMAS 2017: E. Thomas, Geographies of Human-Animal Relations in Classical Antiquity. In: T. Fögen/E. Thomas (eds.), Interactions between Animals and Humans in Graeco-Roman Antiquity (Berlin, Boston 2017) 339–368.

TOYNBEE 1973: J. M. C. Toynbee, Animals in Roman life and art (Ithaca 1973).

VON ALBRECHT 1994: M. von Albrecht, Geschichte der römischen Literatur: von Andronicus bis Boethius mit Berücksichtigung ihrer Bedeutung für die Neuzeit 2 (München ²1994).

ZAHLHAAS 1996: G. Zahlhaas, Aus Noahs Arche: Tierbilder der Sammlung Mildenberg aus fünf Jahrtausenden. Ausstellungskatalog (München 1996).

PD Dr. habil. Florian Hurka
Christian-Albrechts-Universität zu Kiel
Institut für Klassische Altertumskunde
Kiel
Germany
fhurka@email.uni-kiel.de

Bears in Early and Middle Byzantine art (330–1204)

By Martina Horn

Keywords: Byzantine art, mosaic pavement, bear hunting, Consular Diptychs, circus games

Abstract: Most Byzantine artefacts that depict bears derive from the Early Byzantine era (c. 330–641), especially the 6th century, and follow Roman pictorial formulas and traditions. The objects largely originate from the regions of the Levant and the capital, Constantinople (present-day Istanbul, Turkey). There are only single, isolated artistic references from the Middle Byzantine period. A manuscript from the 11th century illustrates a cycle concerning the life of a bear. The few visual images of a later date are hardly innovative but instead based on ancient patterns. In general, the natural strength of the bear is often portrayed. Many church pavements in Jordan give an impression of its primeval ferocity and its exoticism as part of the magnificent God-created cosmos, and as a symbol of the Messianic Kingdom of peace. This is in contrast to the creation-theological arguments that usually justify the mastery of men over animals. Men have the know-how to hunt and kill bears, to protect themselves against bears, or to domesticate wild bears. Triumph over the strong bear symbolises the human virtue of fortitude but, even more so, also the heroic power of the emperor, as can be seen from the example of King David. There is pictorial and written evidence that bears were valued both for their excitability and their skill in performing tricks. They were raised in parks and were trained by bear keepers for acrobatic shows to entertain the audiences in Constantinople. The function of bears as guardians against evil is not, or only rarely, present. Byzantine art deals with bears as a dichotomous phenomenon. The imagery reflects admiration for the bear's untamed natural behaviour and at the same time the idea of a human-dominated cultural sphere, where bears are subordinated to the rule of men.

Introduction

The Early and Middle Byzantine epochs include artworks created in the period from the reign of Emperor Constantine I (306–337)[1] to the Crusader conquest of Constantinople (present-day Istanbul, Turkey) in 1204. In the 6th century, under the reign of Emperor Justinian I (527–565), the Byzantine Empire had extensively expanded its territory and penetrated geographic regions far from its capital in Constantinople, encompassing parts of the Italian peninsula, the Levant,[2] the Middle East, and North Africa. This period is distinguished by an exceptional record of artistic patronage and production. Despite numerous beautiful examples from this Early Byzantine period, images of

---

1 More precisely, from the re-foundation of Constantinople (the former Byzantium) in 330 and the transfer of the imperial seat of the Roman Empire to the new capital.
2 The Levant refers to the Eastern Mediterranean region of Western Asia including present-day Syria, Israel, Lebanon, Palestine, Jordan, and parts of Turkey.

bears are only of marginal importance in the later periods of visual art. The bear is classified with the species of wild and dangerous beasts and often not emphasised with clarity. Regarding the political zoology, greater significance is assigned to the representation of lions or eagles as symbols of the imperial power of the Byzantine emperors than the imagery of bears.[3]

Maybe this is why there are no modern scholarly works that present a comprehensive overview of the role of bears in Byzantine art, but only studies of single artistic objects, such as the research on the Consular Diptychs or Jordan mosaic pavements. According to the archaeozoological monograph by H. Kroll on animals in Byzantium, the bear did not play a significant role in daily life in Byzantine settlements (KROLL 2010). In T. Schmidt's study on the instrumentalisation of animals in political discourse, there is almost no perception of bears (SCHMIDT 2020). The following article provides a brief overview of the diverse pictorial aspects that reflect the role of bears, with the main focus on Early Byzantine artefacts (Fig. 1).

THE BEAR IN DEPICTIONS OF BIBLICAL EPISODES

There are some images of bears in Byzantine art which are based on Old Testament texts. First, we have various depictions of the creation of the land animals, including the beasts of the earth, on the fifth day of the creation of the world (Gen 1:23–25); for example, the 13th-century mosaics on the Genesis Cupola at the Basilica of San Marco, Venice, interpreted as a revised version of the Late Antique illuminated manuscript, the Cotton Genesis (Ms. Cotton Otho B. VI),[4] illustrate Christ blessing the land animals that are approaching two by two, among them a pair of threatening-looking bears (for an image, see NIERO 2001, 258).

Another mosaic on the Genesis Cupola shows Adam giving names to all the livestock and to every beast of the field (Gen 2:19–20), as God has authorised him to do. The two playful and friendly bears among the other wild animals reflect the peaceful nature of the terrestrial Paradise (Fig. 2; see NIERO 2001, 263). The 5th-century panel from the ivory Carrand Diptych,[5] showing Adam in Paradise, demonstrates his role as dominator of the paradisiac wildlife. Exotic and wild animals are frolicking around the large, naked figure of Adam, and a little bear is running about his feet and turning its head to the lions next to it. The bear, defined as a creature of God, lives in peaceful paradisiac coexistence with the other animals and at the same time is under the rule of man. The domination and authority of humans over beasts can be explained by their creation in God's image and likeness (Gen 1:26–28).

One of the nine Byzantine silver plates from the Cypriot Lambousa Treasure (dated to 629/630),[6] with scenes from the life of King David, according to the Old Testament narrative in the book of Samuel (Septuagint = 1 Kings), is embossed with the episode of David killing a bear (Fig. 3). The biblical story characterises David as a shepherd who protects his flock against wild beasts, among them lions and bears.[7] The plate represents David wearing a flying cape, his body energetically twisted in

---

3 For lions and eagles as metaphors for Byzantine rulership, *porphyrogénnētos* (purple-born), imperial authority, and martial virtues, see SCHMIDT 2020.
4 Generally dated to the 6th century; the fragments of the manuscript, which was largely destroyed in a fire, are today in the British Library, London, UK.
5 Museo nazionale del Bargello, Florence, Italy; for an image see https://www.akg-images.com/C.aspx?VP3=SearchResult&ITEMID=2UMDHURDHJ8W (accessed 29.06.2020).
6 Dated by control stamps; today in the Museum of Antiquities, Nicosia, Cyprus; for the plate, see LEADER 2000, fig. 7.
7 1 Sam 17:34: And David said to Saul, Thy servant was tending the flock for his father; and when a lion came and a she-bear, and took a sheep out of the flock, 35 then I went forth after him, and smote him, and drew the spoil out of his mouth: and as he rose up against me, then I caught hold of his throat, and smote him, and slew him. 36 Thy servant smote both the lion and the bear, and the uncircumcised Philistine shall be as one of them: shall I not go and smite him, and remove this day a reproach from Israel? For who is this uncircumcised one, who has defied the army of the living God? 37 The Lord

the direction of the huge bear, seizing the fur between its ears, pressing his knee into its back and swinging out with his right hand to kill it with a tapered knife. The bear is turning its head to David and baring its fearsome teeth. David looks like a mythological hero in an ancient drama. In case of the plate discussed here, his heroic victory over the bear can possibly be associated with the military triumphs of the emperor, Heraclius (610–641), based on his identification as a second King David.[8] The focus lies on the imperial virtue of strength and glory, and the overwhelmed bear symbolises the conquered mighty and dangerous enemy. A marginal miniature in the 9th-century Cludow Psalter (Moscow, Hist. Mus. Ms. Gr. 129d)[9] from Constantinople depicts David as the composer of the psalms (fol. 147v). As a biblical reference to his life, David is shown killing a lion and a bear, grabbing the bear by its ear and hitting it with a stick.

The first page (fol. 1r) of the Christian Topography of Kosmas Indikopleustes (Laur. Plut. 9.28)[10] illustrates David as a shepherd, guarding his flock and looking up to three dogs that are mauling and killing a huge brown bear (Fig. 4; see ANDERSON 2013, plate I). The bear is lying stretched out on its back, its fore paws are in the air, it is bleeding from deep wounds, and blood is running out of its nose and open mouth. One dog is biting its stomach, another its hindquarters and a third its throat. The inscription reads: "David tending his father's sheep", which is probably reminiscent of Ps 151:1. A direct single combat between David and the bear is not to be seen here. This role as an observer seems unusual for the image of David as a hero. The iconographic formula of this miniature, which has no written accompaniment, perhaps originates from illustrations of classical hunting scenes, which often depict dogs slaying a bear or another wild animal.

The bear in Christian martyrdom episodes

In the collection of the RGZM Mainz are two Roman terra sigillata bowls (dated to 350–430), categorised as African Red Slip Ware (ARS) from the province of Africa Proconsularis (North Africa). Each is appliquéd with two relief figures, a female prisoner bound to a stake behind her and a huge upright standing bear facing backwards.[11] On the first bowl, the bear leaps at the woman, digs its claws into the victim's body and shows its teeth ferociously, but then it turns its head away from the woman (Fig. 5).[12] On the other one, the bear approaches the hoisted female victim from the left with outstretched paws and its head turned backwards (Fig. 6; see VAN DEN HOEK 2013, 428–434; VAN DEN HOEK/HERRMANN 2013, 94–105 and figs. 15–16; STIEGEMANN 1996).[13] Possibly, the two plates depict an ordinary *damnatio-ad-bestias*-episode – the execution of convicts or captives, barbarian prisoners, and condemned criminals by exposing them to wild animals, such as lions or bears, as shown in vivid examples on Roman pottery from the 2th and 3th centuries (for examples, see VAN DEN HOEK 2013, 420–429; VAN DEN HOEK/HERRMANN 2013, 94–101). But the hesitancy on the part of the bear does not correspond to the *damnatio* in the traditional sense. A clear change of attitude is notable,

---

who delivered me out of the paw of the lion and out the paw of the bear, he will deliver me out of the hand of this uncircumcised Philistine. Text from the Septuagint, after https://biblehub.com/sep/1_samuel/17.htm (accessed 29.06.2020).
8 See the discussion in LEADER 2000, 413–418: We have no textual evidence in Byzantine sources about the identification of David and Heraclius, but David could generally be used as the ideal type for the Byzantine ruler.
9 State Historical Museum, Moscow, Russia; for image see https://upload.wikimedia.org/wikipedia/commons/a/ad/Chludov_david.jpg (accessed 29.06.2020).
10 Dated to the 10th/11th century, in the Biblioteca Medicea Laurenziana, Florence, Italy; for the codex, see ANDERSON 2013.
11 Another bowl with an identical motif is in the Römisch-Germanisches Museum, Cologne, Germany; for an image see VAN DEN HOEK/HERRMANN 2013, fig. 15.
12 Römisch-Germanisches Zentralmuseum (RGZM), Mainz, Germany, Inv.-Nr. O.41911.
13 Römisch-Germanisches Zentralmuseum (RGZM), Mainz, Germany, Inv.-Nr. O.41962.

cruelty has transformed into sympathy for the victim. Can this episode be interpreted as a special kind of Christian martyrdom tradition?

The bear-and-woman-motif and the bear's distraction could be literarily associated with episodes in the *passiones* of female Christian martyrs in North Africa. We know from written sources that Maxima, Donatilla, and Secunda, the three virgins of Tuburga, were executed during Valerian's persecution. When four she-bears were let loose to kill the girls in the amphitheatre, the beasts showed no inclination to harm them; instead they refused to touch them, rolled in the sand and laid down at the martyrs' feet. Saint Marciana Caesariensis of Mauretania, who died in 303, was thrown to some wild bears in the arena. The beasts dug their paws into the girl's breast, but then they meekly retreated.[14] Perhaps, the bowls illustrate how a virgin martyr had been protected from wild bears by divine intervention. One striking argument against such an interpretation might be that only very few Late Antique images of Christian martyrdom by wild beasts are known.[15] The hesitant attitude could also point to the unpredictability of bears that has been documented in literature; they sometimes simply have no desire to kill their victims (for examples see VAN DEN HOEK 2013, 430).

In the Middle Ages, the divine miracle-motif of the bear's sudden tameness is much more popular in depictions of martyrdom episodes. The Church of Hagia Euphemia at the hippodrome of Constantinople has a cycle with fourteen episodes portraying the martyrdom of Euphemia from Chalcedon.[16] Symeon Metaphrastes' *Menologion* (10th century) tells us about the tortures she endured under the reign of emperor Diocletian: Euphemia is led to the arena, but the four lions and three bears there do not touch her, they only lick her feet (for the Greek text, see HALKIN 1965, 161). After she has prayed to God for letting her suffer the martyrdom, a she-bear bites her and she dies. This episode is included in the painted *vita*-cycle with *tituli* from her *passio* in the Church of Hagia Euphemia at Constantinople, dating to the late 13th century (NAUMANN/BELTING 1966, 136–139 fig. 44).

THE BEAR ON BYZANTINE CHURCH PAVEMENTS

From the 6th century, the age of Justinian I, there are numerous mosaic floors in Byzantine churches and chapels in the provinces of Jordan, Libya, Palestine, Syria, Arabia, and Greece with representations of daily life scenes, with hunters or farmers, and with exotic and fierce animals.[17] The naves of the one or three-aisled basilicas depict birds, different types of animals and beasts, such as bears or lions, as well as men in combat with or chasing wild animals. The forms and pattern formulae are adapted from the Roman repertoire with similar images, borrowed especially from the pagan floor mosaics in North African Roman villas. The exotic and rural motifs have been transmitted to Christian churches.

In the Church of the Martyrs at Al-Kahdir, Madaba, Jordan (dated to the 6th century), the mosaic pavement of the nave depicts panels with animals under trees laden with fruit. Two bears romp below a date-laden palm tree (PICCIRILLO 1993, 129–131 fig. 142). The Church of Cosmas and Damian at Gerasa, Jordan (dated to 533), has a mosaic floor in the nave with squares containing different species of animals, and also a row with wild bears shown leaping around (MAGUIRE 1987, 34–36 fig. 41). In the three-aisled basilica at Petra, Jordan (dated to c. 550), the north side is paved with mosaics,[18]

---

14 For the martyr acts and legends of the North African female saints, see TILLEY 1996, 13–24.
15 For the discussion about the interpretation, see STIEGEMANN 1996, 141; VAN DEN HOEK/HERRMANN 2013, 101–106.
16 For the iconographic analysis and reconstruction of the frescoes, see NAUMANN/BELTING 1966.
17 For detailed studies on the pavements, see HACHLILI 2009; cf. KITZINGER 1951; MAGUIRE 1987; PICCIRILLO 1993; ZOHAR 2017.
18 For the entire mosaic pavement, see HACHLILI 2009, figs. VI.6–8, for the row of bears fig. VI.7a.

which are composed of the inhabited vine scroll pattern, in which vine tendrils emerge from a vase and form frames for medallions which enclose pairs of animals as well as six human figures. In this central-axial symmetrical composition, the rows alternate with pairs of birds and animals, each in different postures. A basket of grapes is flanked by two bears in dissimilar poses (Fig. 7). One is depicted as ferocious, with an open mouth and bared teeth, whereas its counterpart appears calm in a posture of docility, with an inclined head that represents the ambivalence between the bear's natural savageness and paradisiac peaceful coexistence. The nave of the East Church at Qars-El-Lebia, Libya (dated to 539–540), shows a mosaic floor divided into fifty square panels with pagan and religious motifs, including a multitude of animals, half-human creatures, buildings, and Nilotic scenes.[19] This extensive representation of earth and sea as part of God's creation is composed of various land and water animals. In one field, a huge jumping bear is baring its teeth (Fig. 8). A similar depiction of a huge bear, standing on his hind paws, appears in an inhabited vine scroll on the mosaic pavement in the nave of the Church at Qabr Hiram, Lebanon, dated to the 6th century (see HACHLILI 2009, 136 fig. VI.18).

The mosaics portray a representation of the natural world as an image of the whole prosperous earth with its varied inhabitants, with humans and animals, and all things that do creep and fly. In a religious church context, the floor can be understood as an image of God's creation; as a symbol of the richness and variety of His living creatures, including wild beasts, such as bears (for the different means of interpretation, see HACHLILI 2009, 286–288). The church donor-inscription in the Theotokos chapel attached to the south aisle of the basilica in Mount Nebo, Jordan (dated to the 7th century), interprets the world as God's donation to man and praises the creator with the acclamation: "O, Creator and Maker of all things, Christ our Lord ..." (see MAGUIRE 1987, 49). All creatures, even dangerous animals, such as bears, are under the protection and blessing of God, according to the biblical passage in Gen 1:24–25 concerning the creation and benediction of land-living animals on the fifth day. More than just picturing the terrestrial created world, the mosaics convey the Messianic vision of the celestial Paradise. The prophecy of Jes 11:6–8,[20] which we can read in a Greek quotation in the Acropolis Church in Ma'in, Jordan, dated to 719/720 (see PICCIRILLO 1993, fig. 312), describes the amicable arrangement between predator and prey animals in future times of salvation in the celestial Paradise, the *filia* (friendship) among all animals in the Garden of Eden.

THE BEAR IN HUNTING EPISODES (*VENATIONES*)

Many church floor mosaics include the theme of hunters on horseback, or on foot, who are in combat with a bear or another wild beast.

The large, central floor in the old Diakonikon Baptistery at the Memorial of Moses on Mt. Nebo, Jordan (dated to 530), shows four separate rows of hunting and pastoral scenes.[21] In the second register, two men on horses, accompanied by their dogs, are hunting a bear and a wild boar with spears. The huntsman on the left has hit the brown bear in the back with his long spear. The bear is turning its head to the hunter (Fig. 9; see PICCIRILLO 1993, fig. 169). On Mt. Nebo, in the Church of SS. Procopius and Lot at Mukkayyat, Jordan (dated to *c.* 557), there survives a floor mosaic in the nave with inhabited scrolls, including agricultural scenes with human figures in various rural activities.[22] The scene,

---

19 Dated by an inscription, see MAGUIRE 1987, 44–48; for more information on the pavement, see HACHLILI 2009, 101–109.
20 V. 7: καὶ βοῦς καὶ ἄρκος ἅμα βοσκηθήσονται ...
21 Dated by two inscriptions of Bishop Elias; see PICCIRILLO 1993, 146 and figs. 166, 169, 182; HACHLILI 2009, fig. VII.13c.
22 There are three Christian dedication and commemoration inscriptions, see PICCIRILLO 1993, 164–165 and figs. 201–215, esp. 202; for a further depiction, see HACHLILI 2009, fig. VI.11.5; ZOHAR 2017, fig. 2.

consisting of two medallions, is similar to that in the aforementioned Diakonikon, with the common iconographical formula: A man is striking a huge leaping brown bear with a spear (Fig. 10; see Piccirillo 1993, fig. 202). The north wing floor mosaic of the northern transept basilica at Nikopolis, Epirus, Greece, donated by Bishop Dometios (dated after 550), represents a chasing scene in a frieze with a chain of medallions, alternately filled with hunters and wild beasts (see Kitzinger 1951, figs. 21–22; Maguire 1987, 21–24). Both are each in a separate circular scroll formed by a vine rinceau. One of the hunters is attacking a bear with a long lance. The mortally wounded bear is sticking out its tongue.

In the north aisle of the Church of St. Elias at Kissufim, Negev, Israel (dated to 578), we find a modified pictorial pattern of a hostile encounter between man and bear: A hunter-soldier armed with a huge round shield and a sword is confronting a bear (Fig. 11; see Hachlili 2009, 160 and figs. VII.4; VII.12a).[23] The impression of defense and protection, rather than of attack or combat, predominates this image. The upper chapel of the Church of the Priest John at Khirbat Al-Makhayyat, Mt. Nebo, Jordan (6th century), illustrates a man armed, just as at Kissufim, with a sword and a shield, who defends himself against a jumping bear (see Maguire 1987, 69–72 fig. 79; Piccirillo 1993, 174). Both are depicted in their own inhabited scroll medallions, separated by a rolled acanthus leaf. The hunter does not attack or spear the bear, nor has he overpowered it. Another example of this theme is found in the Church of Deacon Thomas at ʿUyun Musa Valley, Jordan (dated to the 6th century; see Piccirillo 1993, 186–188; Hachlili 2009, figs. VI.10; VII.12d). In the panel above the vine rinceau, a barefooted hunter with a shield and a sword is facing a dangerous wild bear that has bared its teeth (Fig. 12; see Hachlili 2009, fig. VII.12d).

The hunting motif focuses on man's terrestrial rule over creation. Despite their natural weakness, men have the power to protect themselves and have the intelligence to arm themselves against fierce and savage beasts. By these means, hunters cause the bears to fear them. The mosaics demonstrate the cultural superiority of man over the force of the wild animal, the triumph of civilisation over nature. The pictures also emphasise the heroic and glorious nature of hunting. Hunting events were very popular among the upper classes, not only in Roman times, but also in the Early Byzantine area. On the one hand, the episodes are purely decorative and glorify the activities of the ecclesiastical donors, on the other hand they have Christian significance; they symbolise man's dominion over the animals, the biblical theme of the exploitation of nature by humans based on the divine image in man (Gen 1:26–28).

In the Church of the Rivers at Umm Al-Rasas, Jordan (dated to 579 or 594), on the frame of the central nave mosaic, a bear trainer with a sword on his back is holding a brown bear with bowed head by a long rope (Fig. 13; see Piccirillo 1993, 240–214 and fig. 389). His exotic dress, blonde hair and long moustache may indicate that this man is not a native, and perhaps comes from an inner Asiatic country. The portrayal of a man leading exotic and savage animals, such as ostriches, zebras, or bears, on a leash like camels could be interpreted as a demonstration of man's power to dominate and domesticate exotic and wild animals.

The bear depicted in the pavement mosaic discussed above is a local Syrian bear (*Ursus arctos syriacus*), which lived in the higher mountains and forests of Syria, Jordan, Lebanon, and Palestine. The fur of this relatively small species is light brown or straw-coloured, the hair of the whiskers is grey-brown, and it is recognisable by a dark stripe running across its back[24] (for images of this species, see Figs. 7–13).

---

23 Today exhibited at the Israel Museum, Jerusalem.
24 For further information, see https://en.wikipedia.org/wiki/Syrian_brown_bear (accessed 29.06.2020).

An Early Byzantine brass jug from Budakálasz, Hungary (dated to the beginning of the 6th century),[25] depicts various isolated hunting scenes in bas-reliefs representing combats as a duel between man and animal in a purely dramatic and realistic manner. Four of the friezes show hunters killing bears in different ways (Fig. 14; see VIDA 2017, figs. 15–16, 19, 21). On the first relief, a hunter on foot is ramming his lance deeply into a huge bear's throat, and the mortally wounded bear is falling forwards (Fig. 14.1). The next scene is really uncommon: An undressed man is hitting a bear on its head with a cudgel.[26] The bear is sitting in front of him and exhibits a behaviour that is passive and not at all combative (Fig. 14.2). On the third segment, a mounted hunter is thrusting a lance into the back of a powerful bear that is just biting into the neck of a wild boar (Fig. 14.3). On the last relief, an enormous bear, wounded by a deadly lance thrust of a hunter, is standing on its hind legs and lowering its head as sign of defeat. A dog is licking up the blood that is flowing from a big abdominal wound (Fig. 14.4). All bears are depicted in a naturalistic and plastic manner with precisely crafted fur. The jug-scenes demonstrate the *virtus* of the heroic hunter and his glorious triumph over bestial bears as a symbol of imperial power, which also implies the triumph over dangerous enemies.

THE BEAR AND ITS LIVING CONDITIONS IN CONSTANTINOPLE

The mosaic that once decorated the floors of the Great Palace in Constantinople mostly consists of bucolic, rural, hunting and mythological motifs, further genre scenes and a variety of exotic animals (for the mosaic, dated to the 6th century, now in a preservation hall, see JOBST et al. 1997). One panel shows a group of bears, one of them climbing the branches of an apple tree, another eating an apple from the ground, and a third walking along with bowed head, perhaps also searching for an apple (Fig. 15). The monumental floor mosaic evokes an ideal landscape to please the imperial elite of the royal court. The peaceful picture of feeding and climbing bears refers to a complex symbolism of the peace and harmony that imperial rule brings. But it also can be read as a reflection of the celestial Garden of Eden. The same motif of a bear eating fruit appears on the ivory throne of Archbishop Maximinian in Ravenna (dated to the 6th century), here again being a symbol of paradisiac peace (Fig. 16; see CECCHELLI 1936, plate XII).

The family of bears on the Great Palace mosaic looks as if it has been taken from nature's template and inspired by a realistic setting. Literary sources, such as numerous *ekphrasis* or chronicles, give various descriptions of menageries, animal parks and animal farms in the palace area of Constantinople, where wild beasts were kept before performances in the arena (SEVCENKO 2002, 75–81). Furthermore, we can read about game parks outside the city where wild animals were fed and trained for entertainment performances and emperors' hunting events (SEVCENKO 2002, 69–75). The most frequently mentioned garden spaces are the two suburban parks, the Philopation and the Aretai, and two in the area of the Great Palace, the Mesokepion and the Mangana.[27] We also know the term beast-rearer (θηριοτρόφος) from written sources that describe keepers in the parks who feed the animals, especially in the wintertime (SEVCENKO 2002, 72 and note 11). Archaeological finds of animal bones in the Theodosian harbour of Yenikapi in Constantinople document the possible transport by sea of bears for the imperial menagerie (for the archaeological excavations, see KROLL 2010, 68–69).

---

25 The jug was found in an Avar grave (no. 740) during the 1989 excavations; it had been given to a deceased as a burial gift. Today in the Ferenczy Museum in Szentendre, Hungary; for the jug, see VIDA 2017.
26 The iconographic model of a naked hero with a cudgel can be deduced from mythologic scenes, such as of Heracles.
27 Today, there are few, if any, surviving remains of these parks. For the literary sources on the parks, see MAGUIRE 2000.

Procopius tells us in his *Anecdota* IX,2 about Acacius, the father of the Byzantine empress, Theodora (*c.* 500–548). He worked as a keeper of animals used in the circus.[28] He was "an adherent of the Green Circus faction,[29] a man whom they called Master of the bears" (ἀρκοτρόφος; for Greek text and English translation, see DEWING/DOWNEY 1935, 102–103). Besides the mosaic in the Church of the Rivers at Umm Al-Rasas, Jordan (cf. Fig. 13), we have no Byzantine pictorial testimonies of beast-rearer activities, but we can get the idea from Roman monuments. In the necropolis at the ancient town of Cibyra in Asia Minor, limestone reliefs from the 2nd or 3rd century show a wild bear's life in captivity and how bears were treated by caretakers and *venatores* (BERN/EKINCI 2019, figs. 26–27; 29–30).[30] On blocks C1 and 3a–b, different stages of a *venatio* are depicted – the release of the bears from their cages, various artistic and fighting activities, and carrying the dead animals out of the arena (Fig. 17; see BERN/EKINCI 2019, figs. 26–27, 30). The drawing of the lost piece (block C1) shows a huge cage with wooden slats, probably a sliding fence, and a gabled roof with six compartments. In front of the cage, four bears are playing and scrapping. A fifth upright bear is snapping at two men who are trying to escape, while a third man is hiding under the roof. Block C2 also shows a cage with six boxes and wooden sliding gates from which bears are released to fight in the arena. Another bear is coming out of a kennel, the trapdoor of which an animal-keeper is opening. The bear has the body or parts of an animal in its mouth. There are three other bears; one is standing on all fours while, in front of him, a bear in an upright standing position has a small animal trapped between its forefeet and is biting into it. A third bear is rising up against a *venator* who is trying to escape. In the lower relief-frame, five little bear cubs are play-fighting and frolicking. On blocks C3a–b, the cage is mounted on wheels like a caged wagon. In one of the compartments, a bear is putting its head through the bars of the cage. On the right, there are hunting scenes and animal-keepers carrying dead animals away, among them two bears. One man is pulling a dead bear away by its hind paws, the second bear is lying on its back and two servants are tying its paws together. In the lower border, four bears are romping. Blocks C4–8 depict many *venationes* and different combat-scenes between wild beasts and men; a lot of bears are struggling with *venatores*, biting and attacking the fighters and, furthermore, the combatants are performing tricks, such as turning somersaults[31] over the bears (see BERN/EKINCI 2019, figs. 34, 36, 38, 40, 42). The relief vividly bears witness to the tradition of bloody and dangerous circus and arena games in Roman, but also in Late Antique times. The bear represented strength and power and was one of the main animal figures in the combats. In the age of Justinian and Theodora, the tradition of bear-fights was still alive, as we can read in the *Anecdota*.

The bloody animal-baiting spectacles were first banned by Constantine I (325) and later by Anastasius (499) in the eastern part of the Byzantine empire, but the prohibitions and restrictions were not properly put into effect (PUK 2014, 229, 269–271). Justinian I once again authorised fights with wild animals to be performed on the anniversary celebrations of the inauguration of the consuls (LEHMANN 1990, 173; PUK 2014, 271). The Council of Trullo (691/692) determined the penalty of excommunication for bloody animal combats. The huge popularity of the *venationes* is evident in numerous Consular Diptychs (LEHMANN 1990, figs. 13–18; PUK 2014, figs. 95–97). A pictorial reflec-

---

28 He probably had the same function as the *usarii*, who cared for the bears in the Roman provinces of Gaul and Germania – bears that were utilised to fight in the circus plays in Rome, see WAMERS 2009, 33.
29 Blues and Greens were political and religious factions in the Byzantine Empire in the 6th century. They took their names from the colours worn by the circus charioteers.
30 Today in the Burdur Archaeological Museum (Pisidia, Turkey).
31 The dangerous activity in which men execute somersaults over the backs of wild animals has an old pictorial tradition. The famous Minoan Toreador fresco from Knossos (today in the Heraklion Archaeology Museum, Crete, *c.* 1450 BC) shows acrobats performing somersaults and handstands across the back of a huge and ferocious bull. Here the bull-leaping could be part of an ancient magic-religious cult. For image and description, see https://www.historytoday.com/miscellanies/inside-ancient-bull-cult (accessed 29.06.2020).

tion of the Byzantine hippodrome spectacle, inspired by the performances in Constantinople, can be seen in the 11th-century fresco in the Hagia Sofia in Kiev.[32] The painting portrays the imperial box (*kathisma*), a horse race from start to finish, circus life, musicians and dancers, and a *venatio* with a bear, where a mounted hunter is attacking it with a spear (Fig. 18; see VELMANS 1999, fig. 42). Using bears from personal menageries for fighting or hunting spectacles may have continued until the 12th century in the hippodrome in Constantinople (SEVCENKO 2002, 76).

However, due to Christian moral concerns and changing social concepts, the focus of the games in the amphitheatre shifted from bloody to bloodless and acrobatic attractions with bears. The Serdica-relief and the late Consular Diptychs[33] from the 6th century document this process. On the marble Serdica-plate in Sofia,[34] we see the common *venatio* depictions with *venatores*, equipped with body belts (*balteus*), puttees (*fasciae*) and bandages on their arms and shoulders. The combatants have mechanical instruments for provoking the bears, such as ropes and brass knuckles (*caestus*). The relief also presents a bear's attack on a *venator* and a bear who is biting a bull in its neck. In addition, innovative acrobatic scenes, performed by bears and showmen are depicted (Fig. 19). Two actors with animal masks use acoustic stimuli, e.g. hammer, anvil and hand drum, to anger and exhaust a small bear sitting on a chair and licking its paws. One bear is jumping up against a huge four-winged wooden bogie, fixed in the ground. Behind the bogie, a man is swinging a rope. These images on the Serdica-relief probably reflect the spectacles in Constantinople. A rare motif on this relief shows a man's fistfight with a bear.[35] An athlete with boxing gloves is confronting an upright, standing bear. A relief on a marble slab from Malkara (Turkey) also depicts a boxing match between a gladiator and a bear. Another fighter is wrestling with a huge bear[36] (Fig. 20). This confrontation is like a man-to-man combat and can be understood as "a code for defeating prominent enemies heroically" (OEHRL 2013, 307).

THE BEAR IN ACROBATIC GAMES

Five Byzantine ivory Consular Diptychs illustrate the innovative and spectacular preferences in the entertainment genre, the performance of acrobatic tricks and bloodless, but also dangerous games with exotic animals, especially bears. Bears were the preferred animals to carry out these tricks using varied equipment. Three diptychs, produced for the consulate of Areobindus in Constantinople (dated 506), portray pantomimic plays featuring bears on the bottoms of the panels. On the left plate of the piece in Zurich,[37] one athlete is stimulating a huge bear with raised arms and another is attempting to escape from an enraged bear. On the left side a human dummy, carried into the arena, is

---

32 The fresco was made by Byzantine artists in the 11th century, it decorates the walls of a private turret in the church and the hallway leading up to it, where the prince would host noble guests, see VELMANS 1999, 123–124 and fig. 42; DAUTERMANN MAGUIRE/MAGUIRE 2007, 30.

33 The Consular Diptychs originated in Rome as gifts of high-ranking civil elites, commissioned for those who had supported their candidacy. In 476, the imperial power was transferred to Constantinople, and from this point we have diptychs of the East Roman consuls. The bottoms of the panels often depict the circus games which the consuls had to arrange at their inauguration ceremony. The diptychs were used until the end of the consulate under Justinian I in 542.

34 Today in the National Archaeological Museum of Sofia; for the relief, see LEHMANN 1990; PUK 2014, fig. 109; normally dated to the 3th–5th centuries; concerning the question of dating see the stylistic and thematic analysis of LEHMANN 1990, 146–172; he dates it to the first half of the 5th century (page 172).

35 Plinius reported that an athlete had a real chance of winning a fistfight with a bear (Nat. Hist. 8,130); see LEHMANN 1990, 143 und note 13; PUK 2014, VI 252 and note 134.

36 Parts of a grave stele, today in the Archaeological Museum, Istanbul, Inv. 1219 T, dated to the 3rd century. For this subject, see OEHRL 2013; as a further example of a man wrestling with a bear, he mentions a relief on the Sens Cathedral (12th century), Burgundy, France: OEHRL 2013, 306 fig. 15.

37 Schweizer Nationalmuseum-Landesmuseum Zürich, Inv. no. A 3564; for the diptych, see DELBRUECK 1929, 110–111 no. 9 (plate 9); VOLBACH 1976, 32–33 no. 8 (plate 4.8); EASTMOND 2010, fig. 1.

used to increase the anger of the bear. Depicted below this, one performer is trying to catch a bear with a swinging lasso. Another actor is hiding behind one of the panels of a partitioned wooden turnstile (*cochlea*), while a bear is unhurriedly pushing against it (Fig. 21; see DELBRUECK 1929, plate 9). The Diptych in Paris[38] shows an egg-shaped cage made of wooden sticks protecting the man inside against a bear's angry bites and sharp claws. Underneath, three actors, equipped with rods and slings, are provoking two bears with excited gestures. A human dummy is also used here to provoke the wild beasts (Fig. 22). On the St. Petersburg Diptych plate,[39] there is again a wooden turnstile, the rotation of which excites the bear while, at the same time, the continued motion irritates it and protects the performer against the bear's bites. Next to this scene, an acrobat is attempting to jump, using a bar, over a fiercely upright bear with bared teeth. In the lower register, an actor, high up on a railed bridge-like structure, is provoking a bear who is angrily running up and down under the bar of the construction and trying to attack another performer. Two more men are sitting in the rotating baskets of a carousel, each basket is attached to a stick that can be pivoted back and forth around the axis of the central pole. A bear is jumping furiously against the middle bar (Fig. 23).

Two of the Diptychs of consul Anastasius (dated 517) illustrate similar acrobatic feats with bears. On the now-lost Diptych plate once kept in Berlin,[40] a huge carousel-construction is depicted. Two actors are sitting in large wickerwork baskets, fixed to a big pole by a rope, which allows them to circulate and to swing up and down in order to provoke the towering bear. Another performer is beating it with a stick in order to increase its rage. Below this scene, there is again a bear-jumper with a cane (Fig. 24; see DELBRUECK 1929, plate 20). The ivory plate in Paris[41] shows two wooden turnstiles with bears angrily sneaking around them, and performers who are annoying and persecuting the bears with looped ropes (Fig. 25). A Byzantine marble relief now kept in St. Petersburg (dated to *c.* 500)[42] illustrates circus scenes with two acrobats performing somersaults over bears who are leaping up to grab them, and a mounted fighter struggling with two feral bears (Fig. 26). A fragment of an illuminated papyrus,[43] kept in the British Library, depicts a performance in the arena comparable to those shown on the diptychs. A heavily muscled, infuriated brown bear is leaping at a pair of human legs above him. The cause of its incitement is the piece of violet cloth in the upper right corner (Fig. 27). The best trick for the actor to escape the exasperated bear was a summersault over its back by means of a pole (*contomonobolon*),[44] as seen on the Berlin and St. Petersburg Diptychs. The two plates of the Diptych in the Louvre (dated to *c.* 400)[45] with depictions of Roman *ludi* also demonstrate how wild bears were stimulated by swinging cloths to attack the *venator* (Fig. 28).

The ferocity and savagery of the huge and strong bears awaken fear and respect among the spectators, and at the same time the skill and mastery as well as the risky and dangerous performance of the protagonists evoke their admiration. Based on Roman tradition, in Early Byzantine times bears were mainly used for the exotic amusement of the audience.

---

38 For the diptych, see DELBRUECK 1929, 112–113 no. 11 (plate 11); VOLBACH 1976, 33 no. 10 (plate 5.10).
39 For the diptych see DELBRUECK 1929, 114–115 no. 12 (plate 12); VOLBACH 1976, 33–34 no. 11 (plate 5.11).
40 For the diptych, see DELBRUECK 1929, 127 no. 20 (plate 20); VOLBACH 1976, 35–36 no. 17 (plate 8.17).
41 For the diptych see DELBRUECK 1929, 131 no. 21 (plate 21); VOLBACH 1976, 36–37 no. 21 (plate 9.21).
42 For the relief, see SEVCENKO 2001, 122 note 10.
43 The dating of papyrus 3035 (3rd to 6th century) is uncertain. It may have originated in Egypt. For the manuscript, see WEITZMANN 1979, 95–98 no. 86.
44 See DELBRUECK 1929, 76; LEHMANN 1990, 143 with reference to the Anthologia Graeca 9, 533.
45 Possibly of Gallic origin; for the diptych, see DELBRUECK 1929, 221 no. 57 (plate 57); VOLBACH 1976, 53 no. 58 (plate 31.58).

Taming a bear

The existence of tamed bears, which were raised and trained in the parks in Constantinople, is documented by a written source that mentions a gift sent by the Byzantine emperor, Michael VI (1056–1057), to Fatimid Caliph al-Mustansir bi-Allah (see Sevcenko 2002, 82 and note 62). Among other exotic animals, bears that played musical instruments were presented to the Caliph. Such a bear, sitting on a stool and playing a string instrument, is shown on a fresco in the audience hall of the bath complex in the Umayyad Palace in Qusayr Amra, Jordan (8th century; Fig. 29; see Piccirillo 1993, 353).

At the imperial court in Constantinople, in the age of Justinian, exhibitions of domesticated animals (θηρίων θέαι or θεατροκονύγιον), parades and shows with bears, performed with specially trained animal coaches, are described in the literature (see Puk 2014, 281). Concerning the presence of dancing bears in Byzantium, we have no pictorial sources that show bears performing this action, only some oil ampullas in form of a bear, which may be Roman or Early Byzantine (3rd or 4th century). There are three bronze vessels, each shaped like a dancing bear, probably used for oils or ointments for bathing (Fig. 30). At one of them, kept in the Cleveland Museum of Art, the bear wears a massive collar with a round opening, probably to fasten an iron chain (Fig. 30.1). In the Museum of Fine Arts, Boston, the bear that forms the ampulla has straps around its stomach and between its forelegs to indicate a harness, which could be a reference to a domesticated dancing bear (Fig. 30.2). The third vessel, kept in the Metropolitan Museum, New York, shows a heavy twisted rope around the neck of the bear and an iron chain attached to the collar's circular hole (Fig. 30.3). The church acts of the Council of Trullo (691/692) emphasise the problem of the captive bears' treatment by men in *canones* 61: "[…] be subjected to the canon for six years. And to this penalty they also should be subjected who carry about she-bears or animals of the kind for the diversion and injury of the simple […]".[46]

The bear's life

The illuminated manuscript of Pseudo-Oppian's *Cynegetica* in Venice (Cod. Gr. Z. 479, dated to *c*. 1060), includes several animal and hunting depictions that were possibly copied from an earlier illustrated codex. The Venice manuscript is based on the text of Pseudo-Oppian's *Cynegetica*, dedicated to Caracalla (211–217).[47] The 11th-century illustrations show man's triumph over bears by chasing and killing them, but they also portray hunters who are mortally wounded by the bears they have pursued. Successful bear hunting can be achieved by active or passive techniques, with bow and arrows (fols. 4v; 44r), with hunting dogs and sticks (fol. 64v), with spears, or with net traps (fols. 20r; 56v; 40r; 65v; Spatharakis 2004, figs. 41, 117, 132–133; see Fig. 31.2–4). On fol. 20v, an overturned hunter is lying on the ground and a huge, furious bear is pouncing on him. Unable to move, the victim is trapped and the bear is mauling him by biting and scratching (Fig. 31.1). On fol. 56v, a hunter is pinned down on his stomach by the enormous weight of a bear that has already been caught in a bag-net hanging from a stick between two trees[48] (Fig. 31.2). These illuminations demonstrate the potential risk of injury or death for the bear-hunters.

---

46 See Troianos 1990, 45 with the Greek text: τοὺς τὰς ἄρκτους ἐπισυρομένους […] πρὸς παίγιον καὶ βλάβην τῶν ἁπλουστέρων […].
47 Biblioteca Marciana, Venice, Italy. It has been suggested that the illustrations are copied from an illuminated Late Antique codex of the *Cynegetica* or another, perhaps Aristotelian, source; for the manuscript, see Sevcenko 2001; Spatharakis 2004; the poem *Cynegetica*, dedicated to Caracalla, was written in the 3rd century; see also Williams 2018, 474 note 1.
48 The same subject is seen on an ivory casket from the 11th century in the State Hermitage Museum, St. Petersburg, Russia – a warrior has fallen down and is besieged by an attacking bear; for a depiction see Sevcenko 2001, fig. 9. For the bear as a dangerous enemy, see Oehrl 2013, 304–305.

Other themes are the bear's *eros*,[49] the cubs' birth, and the bear's natural superiority over humans (SPATHARAKIS 2004, figs. 92–95; see Fig. 32). The verses III, 139–158 of the *Cynegetica* describe the lives of wild bears (*ἄρκτοι*): Bears are "a deadly race with crafty wit" (SPATHARAKIS 2004, 129). Their roar is terrible, and their hearts are fierce. This is followed by some negative comments about the lechery of female bears: The she-bear incessantly desires mating and the pleasures of union, even while still pregnant. When the day of birth arrives, "she puts pressure on her womb and does violence on the goddesses of birth" (SPATHARAKIS 2004, 129), so that she can further indulge in lust. On the miniature fol. 44r top, there is an illustration based on this text – a she-bear is sitting in a tree and, with her left paw, she is pressing her womb from which blood is flowing down. A hunter is aiming at her with his bow and arrow. Next to her, three little bears are depicted, one is walking and the others are climbing in the trees and eating leaves and grapes (Fig. 32.1). On the left side of fol. 44r bottom, there is a bear in an upright stance next to a tree, his open mouth may refer to the "terrible roar" mentioned in the text. To his right, two bears are lying on top of each other and copulating to express their passion for sexual activity (Fig. 32.2). The verses 159–169 tell us about the rearing of the she-bear's cubs, which are half-formed and not yet fully developed after birth. She has to lick her whining offspring with her tongue, in order to shape and to finish it.[50] On fol. 44v, a female bear is standing on all fours in a cave and licking one of her partially shapeless cubs (Fig. 32.3). Verses 170–182 describe the survival of a female bear in wintertime. Her hair is shaggy, and she is hiding in a cave. Instead of eating she is licking her paws. On fol. 45r, the hibernating of a bear, lying in a cave, is depicted (Fig. 32.4).

Basil of Caesarea (4th century), in his homily 9 of the *Hexaemeron*, describes the bear's nature as sluggish, but also sly and very secretive, and he emphasises the bear's ability to hibernate in the freezing cold.[51] Basil interprets the bear's natural instinct to survive as a wonder that reveals the wisdom of the divine Creator. The *Cynegetica*-codex studies the bear's natural animalistic strength, its power to survive and its advantages over humans: The bear's sexual instinct is not burdened with the complexities of the human experience of love and psychological suffering, and childbirth takes place without assistance, without need of midwifery. The manuscript points out the reciprocal phenomenon of human and animal power and weaknesses, the strong interdependency between man's cultural hegemony and the bear's natural force.

THE BEAR'S PROTECTIVE FUNCTION

Bone finds of wild animals document hunting practices in Byzantine villages (cf. KROLL 2010). In the settlements of Iatrus Krivina, Nicopolis ad Istrum (present-day Bulgaria), Pontes (present-day Serbia), and Sagalassos (present-day Turkey), sporadic skeletal elements from brown bears (*Ursus arctos*) from the 6th century have been excavated (KROLL 2010, 53–54, 77, 198). Bears were valuable; they were kept for hunting in parks and the circus *venationes*. But, in the case of Sagalassos, for example, the bear bones might be explained in the context of a fur trade (KROLL 2010, 85). We know about the use and the high reputation of bear skins from the historian Ioannis Skylitzes (THURN 1973, 280). According to his *Synopsis Historion* 22 (dated to the late 11th century), Emperor Nikephoros II

---

49 For the theme of animal love in the *Cynegetica*, see WILLIAMS 2018.
50 This idea that bear cubs were born as shapeless lumps of flesh and had to be licked into shape by their mothers is found in many ancient sources, including Aristotle and Pliny the Elder; see SPATHARAKIS 2004, 132.
51 For the text, see https://www.newadvent.org/fathers/32019.htm (accessed 29.06.2020): "The bear has a sluggish nature, ways of its own, a sly character, and is very secretive; therefore, it has an analogous body, heavy, thick, without articulations such as are necessary for a cold dweller in dens."

Phokas (963–969) used to sleep ascetically, wrapped only in a bear skin, on the floor of his palace, and he took the bear skin with him on his campaigns. He got the fur as a gift from his uncle, the Holy Monk, Michael Maleinos, who had worn it himself. The bear skin is considered as protective and as a lucky charm for the owner and, in addition, it is thought to transmit the bear's strength to the wearer.

The depictions of fierce bears and their prey were also used to visualise the bear's natural force as a magical symbol of superiority over enemies and protection against evil. In the carving on the wooden door panel of the Church of St. Nicholas Bolnichki at Ohrid, Macedonia, Bulgaria,[52] two bears, raptors, and warrior saints defend the entryway as gatekeepers (Fig. 33). The bears, devouring their prey, symbolise the annihilation of evil. The wild creatures' power can thus adopt an apotropaic function which makes them guardians of church entrances.

Summary

The depictions of bears in Early and Middle Byzantine art can be put into sacred as well as into secular contexts. The Old Testament stories describe the bear as a creature blessed by God, but also justify the God-given power of man to domesticate it. In the episodes of King David, the bear functions as a symbol of strength, and David's victory over it constitutes his mighty power and his martial virtues. These heroic qualities serve as an ideal for the Byzantine emperors. Visual sources of Christian martyrdom underline the bear's ferocity and dangerous nature for humans, and at the same time its wayward character. The Byzantine 6th-century church pavements in Syria and Jordan portray the bear as a part of God's creation and as a symbol of the Messianic kingdom and the paradisiac peace. Hunting scenes emphasise man's dominion and mastery of wild beasts, such as bears, and further underline the heroic aspect of killing dangerous animals. In Constantinople, bears were kept in animal parks for aristocratic hunting events and performances in the arena. The Consular Diptychs from the 6th century show, apart from the *venationes*, also bloodless spectacles that consist of acrobatic tricks using equipment such as wooden turnstiles or rotating baskets. It is possible to train bears for these exiting games because they are easy to provoke by acoustic and mechanical stimulation. Furthermore, the savage nature of the huge and powerful bears awakens fear among the spectators, and at the same time admiration for the actors' skill and courage. The circus games with bears, on the occasion of the inauguration of the consuls, were aimed at enhancing the imperial reputation of the Byzantine elite. The aspect of taming bears for dancing performances may be testified to by some oil ampullas in the form of a bear with a collar, harness, and rope. In Pseudo-Oppian's *Cynegetica*, there is a pictorial cycle about the bear's life. Besides various practical instructions for hunting bears, the manuscript accentuates the bear's natural strength and its instincts that, being better than those of man, help him to survive. Some examples indicate the apotropaic and protective function of the wild bear's power against evil, which can be explained by man's respect for its natural aura of strength. All pictorial and written sources demonstrate a strong interdependency between man's hegemony and the bear's ferocity, between nature and culture, between the bear's awesome power and man's intelligent but sometimes abusive superiority.

---

52 Dated to between the 12th and 14th centuries, today in the National History Museum at Sofia; for an image see DAUTERMANN MAGUIRE/MAGUIRE 2007, fig. 64.

Bibliography

Anderson 2013: J. C. Anderson, The Christian Topography of Kosmas Indikopleustes. Firenze, Biblioteca Medicea Laurenziana, Plut. 9.28. The Map of the Universe redrawn in the sixth Century, with a Contribution on the Slavic Recension (Rome 2013).

Bern/Ekinci 2019: C. Bern/H. A. Ekinci, Gladiatorial Games in the Greek East: A Complex of Reliefs from Cibyra. Anatolian Studies 65, 2019, 143–179.

Boeck 2009: E. Boeck, Simulating the Hippodrome: The Performance of Power in Kiev's St. Sophia. Art Bulletin XCI(3), 2009, 283–301.

Cecchelli 1936: C. Cecchelli, La cattedra di Massimiano ed altri avorii romano-orientali (Roma 1936).

Cohen 1993: R. Cohen, A Byzantine Church and its Mosaic Floors at Kissufim. In: Y. Tsafrir (ed.), Ancient Churches Revealed (Jerusalem 1993) 277–282.

Dautermann Maguire/Maguire 2007: E. Dautermann Maguire/H. Maguire, Other Icons: Art and Power in Byzantine secular Culture (Princeton, Oxford 2007).

Delbrueck 1929: R. Delbrueck, Die Consulardiptychen und verwandte Denkmäler. Studien zur spätantiken Kunstgeschichte 1–2 (Leipzig 1929).

Dewing/Downey 1935: H. B. Dewing/G. Downey (eds.), Procopius. Buildings, History of the Wars, and Secret History. Loeb Classical Library, 7 vol., vol. VI: The Anecdota or secret History. Greek Text with English Translation (Cambridge [MA] 1935).

Eastmond 2010: A. Eastmond, Consular Diptychs, Rhetoric and the Languages of Art in sixth-century Constantinople. Art History 33, 2010, 742–765.

Fiema et al. 2001: Z. T. Fiema/Ch. Kanellopoulos/T. Waliszewski/R. Schick, The Petra Church. American Center of Oriental Research Publications 3 (Amman 2001).

Hachlili 2009: R. Hachlili, Ancient Mosaic Pavements. Themes, Issues and Trends (Leiden, Boston 2009).

Halkin 1965: F. Halkin, Euphémie de Chalcédoine. Légendes Byzantines. Subsidia Hagiographica 41 (Bruxelles 1965).

Van den Hoek 2013: A. Van den Hoek, Execution as Entertainment. The Roman Context of Martyrdom. In: A. Van den Hoek/J. J. Herrmann (eds.), Pottery, Pavements and Paradise. Iconographic and Textual Studies on Late Antiquity. Supplements to Vigiliae Christianae 122 (Leiden, Boston 2013) 405–434.

Van den Hoek/Herrmann 2013: A. Van den Hoek/J. J. Herrmann, Thecla the Beast Fighter: A female Emblem of Deliverance in Early Christian popular Art. In: A. van den Hoek/J. J. Herrmann (eds.), Pottery, Pavements and Paradise. Iconographic and textual Studies on Late Antiquity. Supplements to Vigiliae Christianae 122 (Leiden, Boston 2013) 65–106.

Jobst et al. 1997: W. Jobst/B. Erdal/C. Gurtner, Istanbul. The Great Palace Mosaic. The Story of its Exploration, Preservation and Exhibition 1983–1997 (Istanbul 1997).

Kitzinger 1951: E. Kitzinger, Studies in Late Antique and Early Byzantine Floor Mosaics. I. Mosaics at Nikopolis. Dumbarton Oaks Papers 6, 1951, 83–122.

Kroll 2010: H. Kroll, Tiere im byzantinischen Reich: archäozoologische Forschungen im Überblick. Monographien des RGZM 87 (Mainz 2010).

Leader 2000: R. Leader, The David Plates Revisited: Transforming the Secular in Early Byzantium. The Art Bulletin 82(3), 2000, 407–427.

Lehmann 1990: S. Lehmann, Ein spätantikes Relief mit Zirkusspielen aus Serdica in Thrakien. Bonner Jahrbücher 190, 1990, 139–174.

Maguire 1987: H. Maguire, Earth and Ocean: The terrestrial World in Early Byzantine Art (Pennsylvania 1987).

Maguire 2000: H. Maguire, Garden and Parks in Constantinople. Dumbarton Oaks Papers 54, 2000, 251–264.

Naumann/Belting 1966: R. Naumann/H. Belting, Die Euphemia-Kirche am Hippodrom zu Istanbul und ihre Fresken. Istanbuler Forschungen 25 (Berlin 1966).

Niero 2001: A. Niero, Die Genesiskuppel. Ein Beispiel ikonographischer Lesart. In: E. Vio (ed.), San Marco. Geschichte, Kunst, Kultur (München 2001) 248–281.

Oehrl 2013: S. Oehrl, Svá beitum vér björnuna á mörkinni norðr – Bear hunting and its ideological context (as a background for the interpretation of bear claws and other remains of bears in Germanic graves of the 1st millennium AD). In: O. Grimm/U. Schmölcke (eds.), Hunting in northern Europe until 1500 AD. Old traditions and regional developments, continental sources and continental influences. Papers presented at a workshop organized by the Centre for Baltic and Scandinavian Archaeology (ZBSA), Schleswig, June 16th and 17th, 2011. Schriften des Archäologischen Landesmuseums, Ergänzungsreihe 7 (Neumünster 2013) 297–332.

Piccirillo 1993: M. Piccirillo, The Mosaics of Jordan. American Center of Oriental Research I (Amman 1993).

Puk 2014: A. Puk, Das römische Spielewesen in der Spätantike. Millennium-Studien 48 (Berlin 2014).

Schmidt 2020: T. Schmidt, Politische Tierbildlichkeit in Byzanz. Vom späten 11. bis zum beginnenden 13. Jahrhundert. Mainzer Veröffentlichungen zur Byzantinistik 16 (Wiesbaden 2020).

Sevcenko 2001: N. Sevcenko, Eaten Alive: Animal Attacks in the Venice Cynegetica. In: I. Anagnostakis/T. G. Kolias/E. Papadopoulou (eds.), Animals and Environment in Byzantium (7th–12th c.) (Athens 2001) 115–135.

Sevcenko 2002: N. Sevcenko, Wild Animals in the Byzantine Park. In: A. Littlewood/H. Maguire/J. Wolschke-Bulmahn (eds.), Byzantine Garden Culture (Dumbarton Oaks 2002) 69–86.

SPATHARAKIS 2004: I. SPATHARAKIS, The Illustrations of the Cynegetica in Venice, Codex Marcianus Graecus Z 139 (Leiden 2004).

STIEGEMANN 1996: CH. STIEGEMANN (ed.), Frühchristliche Kunst in Rom und Konstantinopel. Schätze aus dem Museum für Spätantike und Byzantinische Kunst Berlin (Paderborn 1996).

THURN 1973: H. THURN (ed.), Ioannis Scylitzae Synopsis historiarum. Corpus Fontium Historiae Byzantinae 5 (Berlin 1973).

TILLEY 1996: M. A. Tilley (ed.), Donatist Martyr Stories: The Church in Conflict in Roman North Africa (Liverpool 1996).

TROIANOS 1990: S. N. TROIANOS, Zauberei und Giftmischerei in mittelbyzantinischer Zeit. In: G. Prinzing/D. Simon (eds.), Fest und Alltag in Byzanz (München 1990) 37–51.

VELMANS 1999: T. VELMANS, Byzanz. Fresken und Mosaike (Zürich, Düsseldorf 1999).

VIDA 2017: T. VIDA, Die frühbyzantinische Messingkanne mit Jagdszenen von Budakalász (Ungarn) (Budapest 2017).

VOLBACH 1976: W. F. VOLBACH, Elfenbeinarbeiten der Spätantike und des frühen Mittelalters (Mainz 1976).

WAMERS 2009: E. WAMERS, Von Bären und Männern. Berserker, Bärenkämpfer und Bärenführer im frühen Mittelalter. Zeitschrift für Archäologie des Mittelalters 37, 2009, 1–46.

WEITZMANN 1979: K. WEITZMANN (ed.), Age of Spirituality. Late Antique and Early Christian Art, third to seventh Century. Catalogue of the Exhibition at the Metropolitan Museum of Art New York (New York 1979).

WILLIAMS 2018: C. WILLIAMS, The Poetry of Animals in Love. A Reading of Oppian's *Halieutica* and *Cynegetica*. In: S. Finkmann/A. Behrendt/A. Walter (eds.), Antike Erzähl- und Deutungsmuster: Zwischen Exemplarität und Transformation (Berlin 2018) 473–500.

ZOHAR 2017: D. ZOHAR, Sanctifying the Daily: Continuation and Innovation in Early Church Mosaic Floors. Proceedings Workshop Anchoring Sanctity, OIKOS, Radboud University (Nijmegen 2017).

Dr. Martina Horn
Institut für Kunstgeschichte und Musikwissenschaft,
Abteilung Christliche Archäologie und Byzantinische Kunstgeschichte
Johannes Gutenberg-Universität Mainz
Mainz
Germany
marthorn@uni-mainz.de

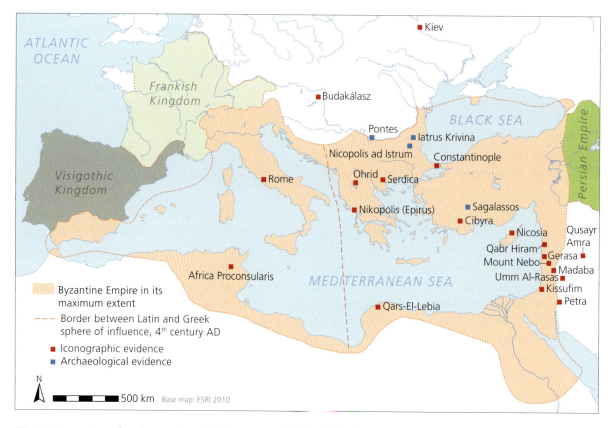

Fig. 1. Sites mentioned in the text (map GIS department, ZBSA, Schleswig).

Fig. 2. Adam naming the animals. Mosaic, Genesis Cupola, Basilica of San Marco, Venice, Italy, 13th century (© Mondadori Portfolio/Electa/S. Anelli / Bridgeman Images).

Fig. 3. *David killing a bear. Silver plate, Lambousa Treasure, Museum of Antiquities, Nicosia, Cyprus, 629/630 (© akg images, A. Held).*

Fig. 4. *David and the bear. Manuscript of Christian Topography by Kosmas Indikopleustes (Laur. Plut. 9.28), fol.1r, 10th–11th century (© Biblioteca Medicea Laurenziana, Florence, Italy).*

Fig. 5. *Bear and martyrdom. Terra sigillata bowl from North Africa, Römisch-Germanisches Zentralmuseum Mainz, Germany, 350–430 (© Römisch-Germanisches Zentralmuseum, photo Lübke & Wiedemann, Leonberg [ehemals Stuttgart]).*

Fig. 6. *Bear and martyrdom. Terra sigillata bowl from North Africa, Römisch-Germanisches Zentralmuseum Mainz, Germany, 350–430 (© Römisch-Germanisches Zentralmuseum, photo Lübke & Wiedemann, Leonberg [ehemals Stuttgart]).*

Fig. 7. Bears. Floor mosaic, Early Byzantine basilica, Petra, Jordan, c. 550 (after FIEMA et al. 2001, 312, Section 12; © The American Center of Research [ACOR], Amman, Jordan).

Fig. 8. Bear. Floor mosaic, East Church, Qars-El-Lebia, Libya, 539–540 (© temehu.com).

Fig. 9. Bear and hunter. Floor mosaic, Diakonikon-Baptistery, Memorial of Moses, Mt. Nebo, Jordan, 530 (after PICCIRILLO 1993, fig. 169; © The American Center of Research [ACOR], Amman, Jordan).

Fig. 10. Bear and hunter. Floor mosaic, Church of SS. Procopius and Lot, Mukkayyat, Mt. Nebo, Jordan, c. 557 (after Piccirillo *1993, fig. 202; © The American Center of Research [ACOR], Amman, Jordan).*

Fig. 11. Bear and hunter. Floor mosaic, Church of St. Elias, Kissufim, Northern Negev, Israel, 578 (© The Israel Museum, Jerusalem, Israel / Bridgeman Images).

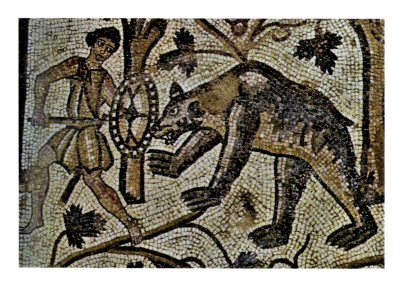

Fig. 12. Bear and hunter. Floor mosaic, Church of Deacon Thomas, ʿUyun Musa Valley, Jordan, 6th century (after HACHLILI 2009, col. pl. VII 1d; photo M. Picirillo; © Studium Biblicum Franciscanum [Israel], Photographic Archive).

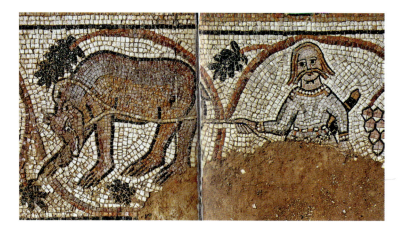

Fig. 13. Bear and bear trainer. Floor mosaic, Church of the Rivers, Umm Al-Rasas, Jordan, 579 or 594 (after PICCIRILLO 1993, fig. 389; © The American Center of Research [ACOR], Amman, Jordan).

Fig. 14. Hunting episodes. Brass jug from a grave in Budakálasz, Hungary, 6th century (© Ferenczy Museum, Szentendre, Hungary; photo B. Deim).

Fig. 15. Bears. Floor mosaic, Great Palace Mosaics Museum, Istanbul, Turkey, 6th century (© Universal Images Group / Bridgeman Images).

Fig. 16. Bear. Ivory throne of Archbishop Maximinian, Museo arcivescovile, Ravenna, Italy, 6th century. The bear is depicted on the back of the throne. 1: Full view; 2: Detail (© Archiepiscopal Museum of Ravenna, Italy; Opera di Religione della Diocesi di Ravenna).

Fig. 17. Bears in the arena. Limestone reliefs from Cibyra, blocks C 1, C 2 and C 3a–b, 2nd or 3rd century (drawing C. Golze, after templates; Burdur Archaeological Museum, Pisidia, Turkey).

Fig. 18. Venatio. Fresco, Church of Agia Sofia, Kiev, Ukraine, 11th century (© Saint Sophia of Kyiv Cathedral, Ukraine).

Fig. 19. Venationes. Marble relief-plate from Serdica, Bulgaria, 6th century (National Archaeological Institute with Museum at the Bulgarian Academy of Science [NAIM – BAS]; photo V. Petrov).

Fig. 20. Boxing and wrestling with a bear. Marble relief-slab from Malkara, Turkey, 3rd century (Archaeological Museum, Istanbul, Turkey; © akg-images, photo E. Lessing).

Fig. 21. Acrobatic games with bears. Ivory Consular Diptych of Areobindus, 506 (© Schweizer Nationalmuseum-Landesmuseum, Zürich, Switzerland, Inv.-Nr. A-3564).

Fig. 22. Acrobatic games with bears. Ivory Consular Diptych of Areobindus, 506 (Musée national du Moyen Âge [Musée de Cluny], Paris, France; © bpk / RMN / Grand Palais / photo Th. Ollivier).

Fig. 23. Acrobatic games with bears. Ivory Consular Diptych of Areobindus, 506 (The State Hermitage Museum, St. Petersburg, Russia; © The State Hermitage Museum, photo V. Terebenin, D. Sirotkin).

Fig. 24. Acrobatic games with bears. Ivory Consular Diptych of Anastasius, 517. Lost, formerly Antiquarium, Berlin, Germany (after DELBRUECK 1929, plate 20).

Fig. 25. Acrobatic games with bears. Ivory Consular Diptych of Anastasius, 517 (© Bibliothèque nationale de France, Paris, France).

Fig. 26. Circus scenes with bears. Marble relief, c. 500 (© The State Hermitage Museum, St. Petersburg, Russia; photo V. Terebenin, D. Sirotkin).

Fig. 27. Circus scene with a bear. Papyrus 3035, 3rd–6th centuries (© British Library Board / Bridgeman Images).

Fig. 28. Games with bears. Ivory Diptych, c. 400 (© RMN-Grand Palais [musée du Louvre], Paris, France; photo J.-G. Berizzi).

Fig. 29. Musician bear. Fresco, Umayyad Palace, Qusayr Amra, Jordan, 8th century (© Museum With No Frontiers [MWNF], Discover Islamic Art).

Fig. 30. Bronze bear-shaped vessels. 3rd or 4th century. (1: © Cleveland Museum of Art, Ohio, USA, purchase from the J. H. Wade Fund / Bridgeman Images; 2: © Museum of Fine Arts, Boston, USA, gift of Burlon Y. Berry 62.1203; 3: © Metropolitan Museum, New York, USA, Edith Perry Chapman Fund, 1966).

Fig. 31. A bear's life. Four folios from the manuscript of Pseudo-Oppian's Cynegetica, *c. 1060, from above: fol. 20v bottom; fol. 56v; fol. 64v; fol. 65v (© Biblioteca Marciana, Venice, Italy; Marc. Gr. 479).*

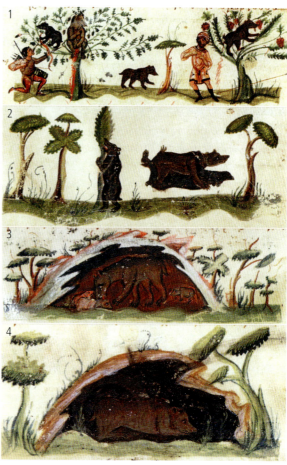

Fig. 32. A bear's life. Four folios from the manuscript of Pseudo-Oppian's Cynegetica, *c. 1060), from above: fol. 44r top; fol. 44r bottom; fol. 44v; fol. 45r (© Biblioteca Marciana, Venice, Italy; Marc. Gr. 479).*

Fig. 33. Wooden door with bears. Church of St. Nicholas Bolnichki, Ohrid, Bulgaria (12th–14th century 1: Full view; 2: Detail (© National Museum of History, Sofia, Bulgaria, inv. no. НИМ29339).

Chapter 12

Further reading: Bears in a broader perspective

The figurine from Wray, possibly a shaman who wears a bear costume. Hopewell (100 BC – AD 500), Newark Earthworks, Licking Col, Ohio, USA (see HULL, this volume; drawing by A. C. Lange, after a template).

The role of bears in the Late Bronze and Early Iron Ages in southern Germany, with a focus on the Hallstatt period

By Melanie Augstein

Keywords: Urnfield period, Hallstatt period, bear remains, hunting trophies, amulets/apotropaia, status indicators

Abstract: In this article, bear remains are examined particularly in the context of Late Bronze and Early Iron Age settlements and especially in burial contexts. Although these periods represent "domestic animal periods", evidence of bears is found sporadically but regularly in the archaeological record. This evidence consists of both unworked and worked bear teeth and claws, which raises the question why these particular body parts were important to humans. The field of interpretation ranges from hunting trophies to amulets/apotropaia and status indicators. However, the few finds from contexts that can provide information on the meaning and significance of bear remains in relation to age and gender of the people equipped with them are very varied and (too) few in number.

Introduction[1]

The bear is undoubtedly an impressive predator, whose face everyone can visualise – even though most bear species in the wild in central European latitudes have long ceased to exist, or exist only as relic populations. The (second) last bear in Germany was shot in the Upper Bavarian town of Ruhpolding in 1835. Only twice since then – in 2006 and 2019 – has the brown bear (*Ursus arctos*) tried to reinhabit his old home in Bavaria. In 2006, a so-called "problem bear" named Bruno achieved sad fame as his visit ended fatally; he was pursued and finally shot dead. As a consequence of this difficult human-animal relationship, real bears are more or less only known from zoos – but our everyday lives are permeated by bears. Almost everyone has had a teddy bear in their childhood, has read "Winnie-the-Pooh", eaten Haribo Gold Bears, or knows bears from fairy tales and legends.[2] Ursa Major and Ursa Minor are amongst the most well-known constellations in the northern hemisphere (cf. Künzl, this volume), and also in heraldry there are numerous images of bears (cf. e.g. the coat of

---

1 I am indebted to Oliver Grimm for inviting me to contribute to this publication. I had my first "encounter" with bears at the Lower Saxony State Museum recording old finds from the Late Pre-Roman and Early Roman Iron Age burial ground of Nienbüttel. While perusing one of the few recorded cremation remains from a bronze urn, which was said to contain no grave goods except for a piece of so-called urn resin, seventeen calcinated bear phalanges unexpectedly emerged (Augstein/Karlsen 2019, 232 fig. 9).
2 The bear plays a more or less central role in various fairy tales by the Brothers Grimm, such as in "Bearskin", "The Willow-Wren and the Bear", "The Two Brothers", "Snow White and Rose Red", or "The Clever Little Tailor" (https://www.grimmstories.com/en/grimm_fairy-tales/index; see also Hirsch, this volume).

arms of Berlin or Bern). And finally, fans of the fantasy series "Game of Thrones" are sure to know House Mormont of Bear Island.

The human-bear relationship is in fact as old as humankind. Bears were (and are) hunted for their fur and sometimes for their meat,[3] but they also have a highly symbolic connotation (cf. BRUNNER 2010). Probably because of their size and strength, their physiognomy and the way they move – especially the ability to stand upright and thus appear human-like – bears play an important role in mythology and in the bear cults of various cultural groups.[4] In the following, the remarks on the cultural significance of the bear refer to a narrowly defined spatial context – southern Germany in the sense of the Alpine foothills, especially today's Free State of Bavaria, the setting of Bruno's last showdown. Despite the fact that this area has been involved in numerous contact and communication networks through the ages, it has its own cultural character and is therefore suitable for such an overview. In addition, the state of research and publication can be described as positive. At the same time, this focus, which is also oriented towards modern borders, should not be interpreted too radically. Whenever appropriate, finds from the neighbouring regions are included (Fig. 1). The temporal focus of this study is on the Hallstatt period, but there will be perspectives on the previous Late Bronze Age/Urnfield period as well as on the following Latène period (cf. Table 1).

The attempt to reconstruct human-animal relations

Animals have always played an important role for humans. This is also true for prehistoric times. This importance has been proven archaeologically in all areas of life, in settlement contexts as well as in sepulchral and non-sepulchral ritual contexts. They are thus culturally connoted – and this applies not only to animals that have been bred and modified in appearance and attributes, but more generally to animals that are given a certain meaning by humans. Here, prehistoric archaeology is faced with the problem of having to build a bridge between the "material" and the "non-material" worlds. Actions and attributions of meanings that were once extremely complex can only be grasped at from material traces alone. The question arises as to what extent conclusions can be drawn about complex phenomena such as social structure, political organisation, or worlds of belief and symbols on this basis. This also applies to the reconstruction of human-animal relationships as historically constructed, spatio-temporal and socially specific phenomena. In the context of Human-Animal Studies, animals are understood as historical actors with their own agency (cf. ROSCHER 2012; PETRUS 2015). Human-Animal Studies see themselves as explicitly interdisciplinary, and for this reason they can also be of importance for archaeological research. One question is how animal history should be understood theoretically and methodologically. This debate includes the question of sources that can be used to describe animals historically (ROSCHER 2012, 8) – but, so far, Human-Animal Studies have only dealt sporadically with antiquity (ibid., 5; cf. PETRUS 2015, 158). Archaeozoology brings together both Human-Animal Studies and Prehistoric Archaeology, as it relates to questions of the materiality of animals, of where and how their physical remains (such as bones, feathers, fur, etc.) can be found, and what conclusions can be formed about human societies and their actions and practices on this

---

3 Despite its nutritional value, bear meat was rarely consumed in many periods and cultures (SCHMÖLCKE et al. 2017, 910).
4 Callisto, an attendant nymph of Artemis, the goddess of the hunt, was seduced and made pregnant by Zeus, whereupon Artemis turned her into a she-bear. In order to prevent Callisto from being killed in the hunt, she was transferred to the sky as the constellation of Ursa Major, the Great Bear (BRUNNER 2010, 22). The bear also plays a role in Celtic mythology and in Celtic traditions – the (supposed) evidence, however, can be found in particular in the southern and eastern Gaulish and Alpine regions or in the Insular Celtic area and is based in part only on linguistic considerations (cf. SJÖBLOM 2012). Therefore, "the actual significance of the bear and the beliefs connected with it are not easy to trace" (SJÖBLOM 2012, 77; cf. SCHMÖLCKE et al. 2017, 902).

foundation (ROSCHER 2012, 4). So, how can one reconstruct the role of the bear in relation to humans in prehistoric contexts, meaning contexts that are characterised by the absence of written sources? In these cases, bears can only be identified if they have left material traces. These can consist of themselves, i.e. their remains, in the form of bones, teeth, claws, or fur, but they also include images.

Bears and the world of the living

Animal bones from settlement contexts should not only be seen from a functional perspective,[5] but on the whole they can mostly be regarded as being representative of the nutritional practice of a group of people (VON DEN DRIESCH 1993, 126; MÜLLER-SCHEESSEL/TREBSCHE 2007, 84). Against this background, let us now take a look at the established record in our chosen area, which starts with the Urnfield Culture (c. 1200–800 BC). If one looks at the range of animal bones, agricultural animals such as cattle, sheep, goats and pigs, which have been recorded since the Neolithic, also dominate the spectrum in the Bronze Age; additionally, the horse increased the range of livestock (FALKENSTEIN 2009, 153, 156), while wild animals only played a subordinate role. The proportion and composition of wild animal bones in settlements can also be traced back to the landscape and the environmental impact of humans. In any case, hunting seems to have lost much of its importance in central Europe during the Bronze Age (JOCKENHÖVEL 1994, 35; FALKENSTEIN 2009, 152). According to Jörg Schibler and Jacqueline Studer, the rare but regular finds of bear skulls and foot bones in the Swiss wetland settlements point to bear skins inside the houses (SCHIBLER/STUDER 1998, 188 fig. 77, cited after FALKENSTEIN 2009, 152). For the southern German foothills of the Alps, there also is isolated evidence for the presence of the bear from settlement contexts. In the lower occupation layers of the Achalm, a hilltop settlement in the northern foreland of the central Swabian Alb near Reutlingen (Baden-Wuerttemberg), which dates to the Late Urnfield period (WILK 2019, 135), two brown bear bones (metacarpus IX and phalanx I ante.) were found (ibid., 151, 174 tab. 49). Hunting in general was of minor importance – at least for the supply of the residents (ibid., 157). Wild animals also played a very insignificant role in the Bronze and Iron Age settlement on the Kiabichl in Faggen (Tyrol; TECCHIATI 2012, 21), and this applies once again to the bear – only a brown bear tibia was found here (ibid., 62). Pierced bear claws, as they are also occasionally known from settlement contexts from the Urnfield period, however, have a distinct artefact character. A pendant made from a bear's claw comes from the settlement near Graben in Swabia (DIETRICH/SORGE 1992, 76), and pierced bear teeth were found in Mörigen on the southern bank of Lake Biel in Switzerland (BERNATZKY-GOETZE 1987).[6] One of the most interesting new finds is certainly the one from Erding (Upper Bavaria), where an extraordinary ensemble of objects came to light in pit 1733. In the southern half of the pit there were a small bronze ring, a bronze utensil with flattened ends, a badly-preserved bronze wire mesh and a dovetail-shaped pendant (BIERMEIER/KOWALSKI 2010, 26 with fig. 54), while at the western edge of the pit a vessel was found in which, among other things, a Neolithic stone axe, a stone ball and a bronze hook had been deposited; further, fragments of several miniature vessels were found in the filling (ibid., with figs. 55 and 56). Among the animal bones, a left rear rabbit paw and a left rear wolf paw came to light; also interesting is a perforated pig tooth and the same from a bear (ibid., 26, 27 fig. 57). It is, in particular, the combination of the objects that places the Urnfield period

---

5   BEECH (1995, 169) points towards the fact that archaeozoologists in central Europe tend to analyse animal bones in terms of subsistence, consumption and economic practices, while the symbolic and ritual component is often faded out.
6   BERNATZKY-GOETZE (1987, 19) mentions two brown bear teeth in addition to a radius. Presumably, the teeth are the two objects with the numbers 9 and 10 on plate 116. The species are not specifically mentioned in the table references (ibid., 77, 79 tab. 4) nor in the table directory (ibid., 172).

ensemble in a ritual context, and they have thus been interpreted as the paraphernalia of a shaman (ibid., 27).

If one wanted to summarise the diffuse impression that the few finds are able to convey, one could say that there are isolated, individual finds of bear bones that are difficult to interpret; it cannot be judged with certainty whether they indicate bears as a source of meat or bear skins as a part of the furnishing of buildings (ultimately, both kinds of use can hardly be separated from each other).

The fact that wild animal bones are rare, and also that hunting in the area of research did not play an essential role in the subsequent Hallstatt period (e.g. TORBRÜGGE 1979, 189), has been confirmed by more recent studies. Juliane Stadler dealt primarily with the question of food goods, in particular meat goods in graves from that period (c. 800–450 BC), but she also included animal bones from settlement contexts as a reference (STADLER 2010). The picture is dominated by domestic animals such as cattle, sheep/goats and pigs (cf. MÜLLER-SCHEESSEL/TREBSCHE 2007); wild animals are again only an exception (STADLER 2010, 78).[7] Among the wild animal bones from the excavations in 2004 and 2005 at the Late Hallstatt hilltop settlement on the Göllersreuther Platte on the southern Franconian Jura, only 0.1 % of the animal bones – that is, one single artefact and, interestingly, the only worked bone artefact in the settlement, namely a tooth that could have been worn as a pendant (SCHATZ 2006, 6, 12) – could be assigned to the brown bear (ibid., 7 fig. 2; SCHUSSMANN 2009, 99 fig. 13).[8] This suggests an open landscape with noticeably decreasing forests, which will not have offered an ideal habitat for the bear (SCHATZ 2006, 1, 14; SCHUSSMANN 2009, 98). The following Early Latène period is culturally closely interlinked with the Hallstatt period; burial grounds as well as settlements are often occupied across the "epoch borders". Bones from at least three and possibly nine bear individuals – if the occupation layers are separated – come from the Hallstatt and Early Latène hilltop settlement of Eiersberg (Lower Franconia).[9] Klaus Kerth and Anke Müller-Depreux have dealt with the quantity of wild animal bones within certain types of settlement (KERTH/MÜLLER-DEPREUX 2004). They noted that only a few settlements had a significantly increased proportion of wild animals. Reference should be made here explicitly to Niedererlbach (Lower Bavaria). In the settlement of Erdwerk I – one of the so-called *Herrenhöfe* (manors) – the proportion of game is extremely high with a total bone weight of almost 25 % (ibid., 224) – this is explained by the physiographic conditions on site (ibid., 226). The bear catches the eye, as it is represented with a minimum of eight individuals (ibid., 223, 224 tab. 1).[10] In this context, it is interesting that the nearby barrow cemetery furnished proof of a bear skin or a bear's paw as a grave good (see below).

As previously mentioned, the Early Latène period is still culturally closely related to the Hallstatt period, but there is a break in the following Middle Latène period. Not only were burial customs changing – it was also a phase of upheaval in other respects as well. The time of the Celtic migrations,

---

7 The faunal assemblage from the rectangular enclosures near Osterholz (Baden-Wuerttemberg) consists of 95 % domestic animals. Individual bear remains were found, but there is no information on the amount (SCHATZ/STEPHAN 2005, 5). The archaeozoological analyses of other settlements that were also carried out as part of the DFG priority programme 1171 "Early centralization and urbanization. Genesis and development of early Celtic princely sites and their territorial surroundings" draw a very similar picture. All sites have a strong dominance of domestic animals (BIEL et al. 2006, 1–2 fig. 1), while wild animals – although documented in different quantities – play a clearly subordinate role (cf. Fig. 2).

8 This tooth, a right upper canine, comes from a young adult or, at most, middle-aged probably male bear. It has a perforation below the root tip, but this has been torn. A little below, there is another small hole, possibly indicating an attempt to repair the pendant (SCHATZ 2006, 13).

9 However, the researchers are sceptical about the high number of individuals (KERTH/WACHTER 1993, 73–74). For comparison; for the Late Hallstatt period "princely seat" of Heuneburg (Baden-Wuerttemberg) with its extensive assemblage of animal bones, "only" (this is stressed by KERTH/WACHTER 1993, 74) thirteen bear individuals – including evidence of an exceptionally large individual (GRAF 1967, 17) – are documented (BRAUN-SCHMIDT 1983, 125, cited after KERTH/WACHTER 1993, 74).

10 For comparison; only two single bear bones originate from the Early Latène period settlement of "Reps", 400 m west of the "princely grave" of Hochdorf (Baden-Wuerttemberg; SCHATZ 2009, 25 tab. 4, 45).

and the reverse flow of settlers who were in contact with the Mediterranean civilisations, brought a change in settlement patterns and economic systems. Large, open settlements and oppida emerged in which trade and craftsmanship flourished. Among the animal remains, wild fauna is occasionally represented, but, on the scale of things, hunting was of very low significance, as numerous archaeozoological analyses illustrate (MÉNIEL 2002, 226–227, cited after TREBSCHE 2013, 216). One of the largest osteologically evaluated find complexes of this period stems from the Middle and Late Latène period oppidum of Manching near Ingolstadt in Upper Bavaria, as it is one of the most substantially explored oppida in central Europe. The archaeozoological record enables us to draw a differentiated picture of the role of animals in nutrition and cult in a proto-urban settlement. If one focuses on the bear, the relevant information can be quickly assessed as only three brown bear bones are reported from the occupation layer (BOESSNECK et al. 1971, 100).[11]

EXCURSUS: THE ROLE OF HUNTING IN GENERAL AND BEAR HUNTING IN PARTICULAR IN THE HALLSTATT PERIOD – A CASE STUDY FROM THE *OSTHALLSTATTKREIS*

As stated earlier, animal bones from settlement contexts are generally regarded as representative of the nutritional practice of a group of people. As we have seen, in the case of the bear it is usually a question of individual bones, which are difficult to interpret. In addition, however, one can also find clearly defined artefacts, such as pierced claws and teeth. Since bears are not domestic animals, hunting would be a plausible reason for the presence of bear remains in settlements. The hunting of fur-bearing animals is difficult to prove in archaeological contexts as they were probably skinned on the spot and the meat of carnivores does not seem to have been consumed on a large scale.[12] Thus the carcasses were often not brought into the settlements and hence their bones did not end up in the settlement waste (VON DEN DRIESCH 1993, 127; TREBSCHE 2018, 237). However, the scarcity, yet ubiquity, of bones from fur-bearing animals like the bear indicates that they were hunted (TREBSCHE 2013, 227; cf. JACOMET et al. 1999, 133 fig. 61). Peter Trebsche investigated the role of hunting – in general – in the Hallstatt and Early Latène periods, among others, on the basis of 74 archaeozoological studies from settlements of these periods (TREBSCHE 2013, 227–228). He was able to demonstrate that, in most settlements, hunting was 1) a regular activity, 2) seemed not to be the privilege of a certain social group, and 3) was not restricted to different settlement types or to certain game species, as hunting opportunities depended on available game resources in a site's catchment area (ibid., 225, 228; cf. Fig. 2). Due to their height, sheer mass and aggressiveness, bears are undoubtedly a huntsman's challenge. So, who was hunting bears?

If one is looking for depictions of hunting in general, the path leads to the *Osthallstattkreis*, which is at least partially close to the area of study.[13] Alexandrine Eibner has identified three groups of images based on representation conventions (EIBNER 2004, 621–623). One is represented by the

---

11 Only two bones of *Ursus arctos* were found within the Late Latène period layer in a ditch at the enclosure (*Viereckschanze*) of Riedlingen (Baden-Wuerttemberg); these are said to be an indication of the use of the fur (DOLL 2009, 323, 324 fig. 11, 329 tab. 3).
12 The fact that bears were most likely consumed is evidenced by bear bones from grave 250 of the Dürrnberg "Eislfeld", which are interpreted as the remains of meat goods, but which, however, cannot be clearly assigned to any buried individual (WENDLING 2018, 286 tab. 2, 288). From the same period, there is evidence for the butchery of bears in the Early Latène settlement of Jenišův Újezd in Bohemia (BEECH 1995). The consumption of bear meat is also assumed for Heuneburg (cf. VON DEN DRIESCH/BOESSNECK 1989, 150). The traces on the bones from Jenišův Újezd point to dismembering activities, but it is quite unclear if the animal was butchered for its meat (and associated bone working or other craft activities) or for ritual purposes (cf. BEECH 1995, 167–168). Apart from that, the boundaries between the two can be fluid.
13 Occasionally, at least for the older Hallstatt period, the Bavarian eastern settler groups were already related to the *Osthallstattkreis* (e.g. ETTEL 2006, 152 fig. 3, 154).

so-called *Situlenkunst* (situla art) that spread from northern Italy to the Alpine region, Carniola and Istria. This refers to toreutic works, where the figures were plastically driven out of the sheet metal from the underside. In part, the environment in which the hunt takes place – the forest – is present through the depiction of plants (ibid., 623). Bears do not appear at all in situla art (EGG/KRAMER 2013, 454); in particular, stag hunting and, on a smaller scale, hare hunting are significant here (EIBNER 2004, 623, 635 tab. 1). The second one leads us to the so-called Hallstatt period Kalenderberg group, situated in the east of Austria, Slovakia and western Hungary. From this context, nine images interpreted as hunting scenes are currently known (TREBSCHE 2018, 230 tab. 1). They were applied on ceramic vessels. Most of them are from graves, only one case comes from a hilltop settlement (ibid., 229). As far as the prey is concerned, in three images deer and once an aurochs could be clearly identified by their antlers or horns (ibid., 229). It is interesting to compare pictorial representations with the archaeozoological remains from settlement contexts, which provide additional information about the frequency, intensity and importance of hunting for meat supply, as well as about preferred hunting prey. According to the percentage of animal bones, the bear only plays a subordinate role in Hallstatt period settlements of the Kalenderberg group (ibid., 236) – the animal is found only in every fifth settlement (ibid., 237). The relationship between the *hunted* species (as shown in the form of animal bones from settlements) and the *depicted* species is remarkable – of the, at least eleven, animal species hunted in the Kalenderberg group, only red deer and aurochs can be identified in the images (ibid., 237), while hares, wild boars, foxes, badgers, wolves, beavers – or bears – are not depicted.[14] The images are therefore not a representative selection of individual hunting experiences, but a stereotypical pictorial formula that may be based on real hunting events, but offers a highly selective or completely imagined perspective (ibid., 238; cf. KOCH 2003 concerning the *Situlenkunst*).[15]

The third group, however, finally leads us to the bear (cf. EIBNER 2004, 635 tab. 1). On cistas (bucket-shaped vessels made of sheet bronze) that were found among the grave goods of the Kröllkogel and the Pommerkogel in Kleinklein in Styria, we finally find images of this animal or, more precisely, of bear hunting (Fig. 3). These vessels are made using hallmark techniques and show a group of warriors in military equipment – evidenced by shields and panaches – on a bear hunt (EGG/KRAMER 2013, 454, 458–459; 2016, 91, 233; cf. the collage of the respective images from both graves in EGG/KRAMER 2013, 459 fig. 201). According to Markus Egg and Diether Kramer, in the Mediterranean world militarily equipped hunters only appear when it comes to the battle of a mythical hero against the beastly monster, and the authors see such content also represented in Kleinklein. Perhaps this hunting scene portraying glorious founding heroes from mythical past times was of fundamental importance for the identity construction of societies in this region during the Hallstatt period (EGG/KRAMER 2013, 459; cf. BERNHARD/GUŠTIN 2019, 43). Holger Wendling, on the other hand, points out that narratives from ancient Greece illustrate the responsibility of armed sections of the population to hunt dangerous and harmful animals like wild boar, bears and wolves (WENDLING 2018, 298). Trophy hunting is closely related to such a "mandate of protection", which puts the hunter in danger and requires bravery and strength. Interestingly enough, evidence of trophy hunting, like claws or teeth, is often found in women's or children's graves. We shall return to that later.

---

14 Adrienne Frie summarises in her study on animal representation in Early Iron Age southeastern Slovenia that, amongst the number of depictions of wild animals, prey animals are more the focus of most of these depictions than the predators; bears for example are nearly absent (FRIE 2017, 356).

15 Something very similar can be observed for the Bavarian region during the Hallstatt period; here, too, the various levels of selection and mechanisms of attribution are observable. Almost exclusively horses and birds are pictured – precisely those animal species that not only played a marginal role in the human diet, as reflected in the archaeozoological analyses of animal bones from settlement contexts, but which are also significantly not observable in graves in the form of meat goods (cf. AUGSTEIN 2017).

In any case, it is remarkable that on the one hand depictions of bear hunting were found in the area of Kröllkogel and Pommerkogel, both east of the hilltop settlement of Burgstallkogel, while in the nearby Höchschusterwald barrow group, west of the Burgstallkogel, one of the few Hallstatt period graves with bear remains came to light (see below).

Bears and the world of the dead

Animal remains are not isolated phenomena in the spatio-temporal context that is our focus here. In specialised literature, animal bones in graves are usually interpreted as evidence of meat goods. However, this is only one possible explanation for the existence of animal remains in this special social space, grave or burial ground. Analytically, a distinction can be made between an animal grave, an animal sacrifice, and meat goods. However, this classification is only partially applicable to archaeological findings – not least because the boundaries between the categories are possibly permeable (Augstein 2014, 86–87).[16]

Again, let us first take a look at the Urnfield period. In the context of the study and publication of the urnfields from the Regensburg area (Hennig 1993), where the first systematic and broadly based animal bone analyses from larger find complexes from that period in Bavaria were carried out (ibid., 36),[17] bears were not found among the animal remains. In my opinion, little has changed in the basic results of this study in general; other burial grounds of this period that have been published in the meantime confirm this picture.[18] However, now there are occasionally bear remains in some other Urnfield period graves (cf. Table 2),[19] which reveal that brown bears were hunted from time to time. Remains such as animal teeth, certain stones, or astragals, are usually interpreted as amulets (see below). In Regensburg-Burgweinting "Nordwest" in the Upper Palatinate, for example, grave goods with a presumed amulet function were found in some of the cremation graves, including a bear tooth (Zuber 2011, 289 fig. 24; cf. Buckel 2008, 135 pl. 1.10). It was found within the cremation remains of a 40-to-60-year-old person, presumably a man,[20] and was associated with other enigmatic objects, like a stone ball, a petrified shell with an unfinished drill-hole, and a flat

---

16 The material reflection of actions in the context of an animal burial does not necessarily differ from the material reflection in the context of an animal grave good or an animal sacrifice. Only the reconstruction of the context, the motivation of the animal deposit, combined with the knowledge of the circumstances of death of the animal can provide information about its character (Augstein 2014, 87).

17 The domestic animal species clearly dominate – sheep/goats, pigs, cattle, and dogs have been identified, while wild animals are only represented in four out of 44 graves by deer and stag (Hennig 1993, 36). Hennig (ibid.) points out that these remains consist mainly of the muscle-/meat-bearing bones of young animals, which are almost always the front and rear limbs. The graves with animal bones or meat goods often also stand out because of their burial equipment, with weapons, bronzes, bronze vessels, eye-catching ceramic vessels, or even gold (ibid., 38), found in graves of men, women and children (ibid., 39).

18 The *status quo* formulated by Hennig (1993) is confirmed, for example, by two recently excavated and extensively investigated burial grounds in the area of research. In Straubing-Sand (Lower Bavaria; see Frisch 2018), as well as in Zuchering (Upper Bavaria; see Schütz 2006), sheep/goats, pigs, and cattle dominate; canids, amphibians, and fish are documented in Straubing on a very small scale, but for both large and well-excavated burials grounds, the presence of bear remains has not been proven.

19 In fact, animal teeth found in graves of this period are often not specifically addressed in terms of the species (cf. the respective category for which often only the information "animal" or "predator" tooth is provided: Buckel 2008, 133–136).

20 For this and further information I am indebted to Joachim Zuber (Kelheim/Regensburg).

stone disk (Fig. 4).²¹ For more findings, one has to switch to Austria. One of the objects from the cremation burial of an adult to mature man from grave 20 in the cemetery at Horn (Lower Austria), which dates to the Early Urnfield period, was a calcined, perforated claw from a brown bear (LOCHNER 1991, 151) and in the richly furnished cremation grave 68 of Innsbruck-Wilten (Tyrol) two centrally perforated fangs of a bear were found (WAGNER 1943, 38, 132 pl. 34.15).

As far as the Hallstatt period is concerned, meat goods appear regularly in graves of the northern pre-alpine region (cf. MÜLLER-SCHEESSEL/TREBSCHE 2007; STADLER 2010).²² Juliane Stadler has analysed the proportion of animal bones in eight cemeteries from that era from Baden-Wuerttemberg and Bavaria. In three cases, no bones of wild animals were found at all (STADLER 2010, 49 fig. 14, 221 tab. 1), so, on the scale of things, wild animals only played a subordinate role, at least as meat goods (ibid., 49).²³ Any evidence for wild animals is dominated by wild boar and red deer, followed by hare, birds, foxes, and fish (ibid., 50 fig. 15, 221 tab. 2), while bears are not even mentioned. Recently, Sebastian Beermann has dealt with bear claws and bear skins from Pre-Roman Iron Age grave contexts (BEERMANN 2016); previously, the focus concerning this topic was more on the 1st millennium AD (cf. WAMERS 2009; GRIMM 2013). The terminology used is not without its problems. "Bear skin graves" are those with unprocessed bear phalanges – even if there is only a single find (BEERMANN 2016, 15).²⁴ To reconstruct an entire coat from this is methodologically improper in my opinion. First, to prove a bear *skin*, remains of hair and skin/tissue are needed. And second, since (single) paws can have a different meaning than the entire fur, I would – at least analytically – plead for a terminological differentiation between "graves with bear skins" and "graves with bear paws".²⁵ I am aware that this will often not be possible due to the source-related limitations. Nevertheless, Beermann highlighted some interesting points: Processed and unprocessed bear claws do not appear together in one grave context; graves with processed and unprocessed bear claws are rarely found together, even in the same burial ground (BEERMANN 2016, 14). Furthermore, bear phalanges appear significantly more often in cremation than in inhumation graves – this will also be due to the better preservation conditions for calcinated compared to uncalcinated bones (ibid., 14), but basically, both burial rites are integrated into specific cultural, ritual or religious contexts that also include the specific handling of living beings as well as things.

Let us now look at the archaeological record. As mentioned at the beginning, the focus of this study lies on the Early Iron Age. Only a handful of examples from Hallstatt period graves with

---

21 Michal Přichystal also points to the association of pendants made from animal teeth or animal claws with other conspicuous objects, such as stones with natural or artificial openings, archaica, amber or coral beads, a bronze hand, or a bulla (PŘICHYSTAL 2007, 350).

22 It should be noted that certain animal species appear in grave contexts that do not (necessarily) match the composition of the animal bones in the settlements, and there are also significant differences with regard to the slaughter age of the animals (MÜLLER-SCHEESSEL/TREBSCHE 2007, 76, 78 fig. 7; STADLER 2010, 78). Apparently there are different levels of meaning of animal species in different areas of past realities (AUGSTEIN 2017, 152).

23 It is a little confusing that Untereggersberg is one of the burial grounds which, according to some of Stadler's diagrams and tables, did not yield any wild animal bones (STADLER 2010, 49 fig. 14, 221 tab. 1), but appears at the same time in the list of graves with artefacts made from wild animal bones (ibid., 48 list 1).

24 The actual number of phalanges differs (BEERMANN 2016, 39 fig. 30). In terms of complete bear skins, twenty phalanges would be expected (ibid., 14, 39) – if there are fewer, diverse reasons can be imagined. On the one hand, the claws of an outstretched bear skin could have been overlooked due to their peripheral location, or, concerning cremations, could have been destroyed by the high temperatures of the pyre. On the other hand, and perhaps more interesting, there are culturally determined actions, such as an offering of a "cut-up", which means an intentionally incomplete fur – or just individual paws. Egon Wamers points out the special importance of bear paws in the context of the bear festival celebrated by the Sámi and Niwchen in northern Scandinavia and eastern Siberia, respectively; they were cooked and eaten separately, but were also used in fortune-telling and in oracles. Replicated as amulets on the shaman's robes, they could have had an apotropaic effect or help the shaman on his journey to the Otherworld (WAMERS 2015, 47).

25 Oliver Grimm as well only speaks of "bear-skin burials" or "bear-skin graves" if the presence of an actual bear skin is proven. Otherwise, in the case of bear claws, he speaks of "bear-related furnishings" or "bear-related findings" (GRIMM 2013).

bear remains are to be mentioned here.[26] The list starts with two examples from Austria. BEERMANN (2016, 34; translation by the author) cites grave 99 from the eponymous site of Hallstatt in the Salzkammergut in Upper Austria as a "very early example of the appearance of the bear skin custom". In the context of the inhumation of a mature to senile man from the Late Hallstatt period, ten bear phalanges, among other things, were found that point to the presence of a bear skin, or (methodically more correct) several bear paws.[27] According to Beermann (ibid.; translation by the author), this is the "oldest direct proof for furnishing with a hide" and the oldest combination of armament and bear, which on the continent is actually more a phenomenon of the 1st century BC to the 2nd century AD (ibid., 36). This has to be put into perspective. On the one hand, with the above-mentioned mound 13 from the Höchschusterwald grave group of Kleinklein in Styria, we probably have an even older grave with a bear skin or bear paws, in which, at the same time, a combination of bear and armament is proven. In this grave, dating to the Early Hallstatt period (Ha C1b), calcinated unworked bear claws[28] were found within the cremation remains of a male individual (BERNHARD/GUŠTIN 2019, 36 note 10, 39). They have been interpreted as an indication that a bear skin was put on the pyre (ibid., 38; 39). The deceased is identified as an archer because of the high number of arrowheads (ibid., 39).[29]

In fact, in addition to the two Austrian sites,[30] there are other early examples to be mentioned (cf. Table 2). One of these is the cemetery of Werbach (Baden-Wuerttemberg).[31] While wild animals again do not appear as meat goods (WEHRBERGER 1984, 135), a broken bear phalange with two holes was found in grave 14 dating to Ha C/Ha D1 (Fig. 5a), the inhumation of a possibly female adult (ibid., 187). Discolouration by bronze oxide could be observed at the breaking point (ibid., 189). Apart from the bear phalange, the grave is remarkable in a number of ways. The rich burial equipment includes two eye-catching so-called *Hohlwulstringe* found in the pelvic area as well as an enigmatic hollow ball with a bronze grommet or socket,[32] and the arms of the deceased were strongly flexed towards the shoulders so that the hands came to rest in the collarbone area (cf. AUGSTEIN 2009, 13 fig. 1.3). Such a staging of the body can be found repeatedly in the burial grounds of the Altmühl valley, the Isar valley, and the Tauber valley (see MÜLLER-SCHEESSEL 2008; NIKULKA 2008; AUGSTEIN 2009;

---

26 Leif Hansen sees the rarity of bear skins in graves as being due to preservation conditions (HANSEN 2013, 253): however, the phalanges of a bear can be easily identified due to their characteristic morphology (BEERMANN 2016, 12, 14 fig. 3). There is an exceptional new find of actually a bear skin from the central burial of tumulus 17 in the Hohmichele barrow group near the Heuneburg hillfort. For further information and relevant literature I am indebted to Bettina Arnold (Milwaukee). The record has not yet been published conclusively (but cf. ARNOLD 2019, esp. 224; RAST-EICHER in print). Apart from that, there are several bear remains to be found on the Dürrnberg near Hallein, but this site also points to the rarity of brown bears in Early Iron Age burial contexts, because in the inventory of the c. 400 burial sites on the Dürrnberg, bear teeth and a bear's claw were only found in six graves (PAULI 1975, 130; WENDLING 2018, 287–288, cf. 286 tab. 2). It is thus surprising that Andreas Bernhard and Mitja Guštin postulate that bear claws, especially as amulets, are well known from Iron Age graves (BERNHARD/GUŠTIN 2019, 38–39).
27 In Hallstatt, pierced bear teeth were also found in graves 139 and 535 (KROMER 1959, pl. 11.14, 78.16).
28 Sometimes there is talk of "several", sometimes of "numerous" claws (BERNHARD/GUŠTIN 2019, 36, 39) – unfortunately the finds are no longer available (ibid., 36–37 note 10), and there is also no illustration.
29 However, no archers hunting bears are pictured on the cistas from Kröllkogel and Pommerkogel.
30 A perforated lower canine of a brown bear from the Hallstatt burial ground of Statzendorf in Lower Austria (REBAY 2006, (I), 184) is also to be mentioned, which, however, cannot be matched to a certain grave, but is regarded as complementary to a perforated stone, a perforated boar tooth, and a miniature handle bowl (Inventory GD03; REBAY 2006, (II), 243; 469 PA56109b).
31 Due to the specific topography of the burial ground with a "honeycomb structure" the Taubertal group is culturally closely related to the Hallstatt period groups of the Altmühl and Naab valleys (AUGSTEIN 2015, 1, 19 with note 31).
32 Such an item was also found in grave 12 (WEHRBERGER 1984, 183, 205 fig. B3), the burial of an adult non-gendered person (ibid., 182). According to WEHRBERGER (ibid., 140–141), objects like these belong to a hip dress or hip jewellery. In the case of Werbach, where they were found in the head area, he assumes that they must have been displaced in the course of grave disturbances. This is understandable in the case of the heavily disturbed grave 12, but I am sceptical about the less disturbed grave 14 and have considered a kind of "ceremonial device" or "sceptre" instead of an item of dress in another publication (AUGSTEIN 2009, 19–20).

2011; 2013a; b), always in women's or – less often – children's graves. The first ones are often the most well-equipped women of the respective burial grounds, but something other than "wealth" seems to have determined the status of these women (cf. AUGSTEIN 2013b, 370). Grave 21 from Kelheim-"Am Urnenfeld" (Lower Bavaria), however, is a child burial. The infant, who died at the age of about six months, was given a small bronze bracelet, small bronze spiral roulettes indicating a chain, and a pierced brown bear's tooth (MEIBORG/MÜLLER 1997, 162 pl. 130C).[33]

But there are some more examples. The already-mentioned grave 4/1990 at Niedererlbach, which should be dated as generally Late Hallstatt (Ha D), is said to have been robbed and unfortunately heavily disturbed (KOCH 1992, 56). It is a collective burial[34] of one man and two women; the objects still in the grave and the remains of meat goods (parts of a young pig and a young cow) cannot be assigned to any of the individuals.[35] The same is true for the four phalanges of a brown bear as "indications of an added fetish or bear fur" (ibid., 56; translation by the author). In grave 30 of Riedenburg-Untereggersberg (Lower Bavaria; Fig. 5b), which also dates to the Late Hallstatt period (Ha D1; NIKULKA 1998, 117; 137), two bear claws were found in the northern part of the chamber of "burial horizon II", which were only identified during the analysis of the animal bones; therefore, their exact position cannot be determined (ibid., 235–236). They belong to the disturbed primary burial, the inhumation of a (probably young adult) man (ibid., 235), whose other grave goods include two iron spear heads (ibid., 237, 236 fig. 78).[36]

The low importance of hunting during the following Latène period, as already mentioned above, is also reflected in the funeral system. Beermann's so-called "bear skin graves" of the Late Pre-Roman and Early Roman Iron Age seem to be "an almost exclusively Germanic phenomenon" (BEERMANN 2016, 19; translation by the author). In the area of research, the bear is virtually invisible.[37] The only exceptions in the quasi "Celtic" milieu are a few "elite graves" in England, Luxembourg, and France, "which can be assigned to a Gallo-Roman context" (ibid., 19; cf. 34–36; translation by the author), and here the offering of bear skins "may have been adopted as a foreign custom" (ibid., 36; translation by the author).[38]

So, what remains to be noted? First of all, in comparison to those of the Pre-Roman Iron Age in the Germanic area, the Hallstatt graves observed here are all inhumation graves except for the ones in Kleinklein and Landersdorf.[39] Furthermore, in the graves presented here there are four cases of

---

33 In this burial ground the offering of pendants is in fact limited to infant graves. In the five examples, in addition to the bear's tooth, a clay bead, a wheel-shaped object, and an amber bead were found. Due to the materials used, they are interpreted as amulets (MEIBORG/MÜLLER 1997, 96).

34 Hubert Koch speaks of a "multiple burial" (KOCH 1992, 56). The term, however, implies the simultaneous deposition of the deceased. Since the time lag between the burials cannot be determined and subsequent burials were a regular practice during the Hallstatt period, I prefer to refer to them as "collective burials" (cf. EGGERT 2012, 59, 62).

35 Bear bones from Hallstatt period graves are rare – all the more reason to have a close look at the contexts and connecting factors. Although grave 4/1990 from Niedererlbach is severely disturbed and the bear claws cannot be assigned to a distinct burial, two women found their last rest here. All women buried in this necropolis – and thus probably also the two from grave 4/1990 – show the same angulation of the arms (KOCH 1992, 60 note 13) as the deceased from grave 14 at Werbach.

36 Unfortunately, there is no image of the bear claws, so that it is not certain whether these are worked or unworked objects. The latter would be the criterion for adding grave 30 to the group of early – in the sense of Hallstatt period – graves with bear skin or bear paws, and it would also provide another example of the combination of bear and armament.

37 From the Early Latène period, there is a grave excavated early at Ingelfingen-Criesbach (Baden-Wuerttemberg) from which, besides some rings and an iron needle with an amber head, a neck ring with ten glass beads and a bear tooth pendant was brought to light. There is no anthropological analysis, there are just said to be about "apparently three inhumation graves" (KLEIN 2004, 341 no. 475).

38 In fact, Beermann's mapping of the "bear skin graves" shows only one single site in the area of study (BEERMANN 2016, 18 map 1) – this is the Merovingian cemetery of Großprüfening, city of Regensburg, where a single bear phalange was found in a cremation grave (EICHINGER/LOSERT 2004, 98).

39 In cremation grave 1 at Landersdorf, a calcinated and perforated bear claw was found within the cremation remains of an adult, presumably a woman, and a child (HOPPE 1986, 166).

unworked bear claws, which point to a bear's fur or paw (Table 2). If it really was a bear skin, the question of the meaning of its presence in the grave arises. Pragmatically speaking, a fur is first and foremost a soft underlay. But the potential emotional charge behind the offering must also be considered. In the belligerent-martial societies of the central European Iron Age, victory over dangerous and powerful animals such as wolf, bear, boar, aurochs, or bison and the ostentatious display of the fur promised prestige (WENDLING 2018, 299). This may have been true in life as well as in death. Looking at the specific figures (BEERMANN 2016, 39 fig. 30), there are only very few graves where the number of phalanges makes the offering of a complete bear skin possible.[40] Among the four graves with unworked bear claws, only grave 99 from Hallstatt with ten phalanges indicates the mandatory presence of more than one bear paw.

In five cases, on the other hand, we are dealing with artificially worked objects (Table 2; Fig. 6). The meaning will differ from that of the unworked bear remains. Interestingly, animal teeth or claws as pendants in graves are always only present as individual pieces (SCHÖNFELDER 1994, 217). According to the general consensus, they have been interpreted as amulets.[41] Objects with a potential amulet function resemble each other throughout the ages and across geographical and cultural areas, in most cases up to the present – often, they are in fact animal teeth or animal claws. The offering of an amulet reflects less complex ideas of gods and myths and more of a kind of "folk belief", which was based on the power of individual objects as a means of protection and magic for individual persons (BUCKEL 2008, 112).[42] Ludwig Pauli noted in his study of Iron Age amulets that bear canines were the second most common animal teeth after boar tusks, found primarily with women and children (PAULI 1975, 129–130), who were considered to be typically in need of protection.[43] While Leif Hansen also claims that these items had an apotropaic function and were not predominantly hunting trophies (HANSEN 2013, 252), Adrienne Frie stresses the fact that the "most dangerous parts of the bear were harvested and retained", which indicates "that these were quite literally hunting trophies" (FRIE 2017, 335). Another facet is that amulets made from animal bone or teeth were supposed to have "transfer qualities of the animal to the human, be it vital force, strength, or power" (SCHMÖLCKE et al. 2017, 906).[44] If one takes this further, one comes across evidence for transformations of bears into humans and vice versa (BRUNNER 2010, 20–33).

---

40 Beermann notes grave 1909/11 from Großromstedt (Thuringia) with 17 phalanges (BEERMANN 2016, 40, 87), but Gustav Eichhorn only mentions five specimens (EICHHORN 1927, 233, 234, 273).

41 For a definition and the history of research see BUCKEL 2008, 72–73. Amulets are often pendants that were worn around the neck (ibid., 72). An "amulet function" will, however, often be hardly distinguishable from a "jewellery function" (WARNEKE 1999, 195; cf. BUCKEL 2008, 121; SCHMÖLCKE et al. 2017, 907). Further, pendants can basically have manifold and overlapping functions. "Only the wearer's belief in magical powers or the demonstration of jewellery as an expression of wealth or the like gives the object its meaning" (WARNEKE 1999, 195; translation by the author).

42 In the funeral context, this can be both protection *for* the deceased and protection *against* the deceased. PAULI (1975, 140–144) lists examples of amulets in the context of "irregular" or "deviant burials", which are interpreted as a means of defence against dangerous or restless dead.

43 Cf. PŘICHYSTAL 2007, 350. From the Roman provinces in Britain and Germany, some small bear figurines are known that also mainly come from infant and younger child burials (SCHMÖLCKE et al. 2017, 904, 905 fig. 4). They can be "connected with the complex of motherhood, protection and escort. The animals might have functioned as guides as well as companions and protectors, ensuring that the small child does not travel to the afterworld alone and unguarded" (ibid., 904). Bear tooth pendants have also been found in the Merovingian period graves of children and women (ibid., 907).

44 Alice Choyke stresses that using specific skeletal remains of particular animals attributes particular powers to them or, in other words, specific apotropaic beliefs are associated with specific body parts (CHOYKE 2010, 200). On the one hand, skeletal elements can represent the whole animal (ibid., 202), on the other hand it is noticeable that just certain elements of the head (especially the teeth) and the feet – those body parts associated with defence and movement – were used for making amulets (ibid., 208). She (ibid., 201) calls them "bones of power". The same body parts seem to have been relevant in other periods. SCHMÖLCKE et al. (2017, 903) quote examples both from the Palaeolithic and modern Nordic societies.

The end of the Bronze Age is an epoch of explicit cultic expression – just think of the numerous deposits or hoards, including those of "cult equipment" such as the gold cones/gold hats, or symbolically charged objects in graves, such as metal vessels with a depiction of the *Vogel-Sonnen-Barke* ("bird-sun-boat"), or cult wagons. While the mythological-cosmological pictorial language – apart from isolated reminiscences, e.g. in the Hochdorf "princely grave" – was largely abandoned at the end of the Urnfield period, so-called burnt offering sites have been documented in the Bavarian Alpine foothills to show a continuity of cult practices from the Urnfield to the Hallstatt period. Some of these are massive, hill-like accumulations of shattered pottery, heavily burnt animal bones and ashes in often exposed topographical locations. Here, too, domestic animal species of cattle and sheep/goat dominate; pigs are not important, wild animals are negligible in quantitative terms (LANG 2005, 32; STADLER 2010, 78; cf. AUGSTEIN 2017, 153), and bears play no role at all (cf. WEISS 1997).

Finally, concerning cult practices in the following Latène period, we will first take a look at ancient Teurnia, near Spittal an der Drau (Carinthia). Andreas Lippert interprets the ash layer from the Middle and Late Latène period settlement phases and the objects found therein – especially weapons – also as relics of a burnt offering site (LIPPERT 1992). The animal teeth found there – two mandibular canines of a brown bear in addition to two of young domestic pigs – are thus interpreted as the remains of burnt offerings.[45] The Latène period indeed revealed further differentiated evidence for complex cult events – e.g. the sanctuaries in northern France, such as Gournay-sur-Aronde or Ribemont-sur-Ancre, the complexes of Roseldorf in Austria, or Mormont in Switzerland. In Gournay, it was domestic animals that played a central role in cult events over the entire use phase (SCHERR 2013a, 227), while in Ribemont no large animal sacrifices took place during its most prominent use phase (first half of the 3rd century BC) as a *tropaion* with decapitated, armed warriors (SCHERR 2013b, 235).[46] Domestic animals also dominate the find spectrum of the large central place of Roseldorf, for which differentiated offerings in districts known as "sanctuaries" are evident, mainly in the Middle Latène period (HOLZER 2009, 77). Wild animals are poorly or not at all represented and obviously were of no importance in cult practice. Finally, in 2006, an important ritual place from the 2nd and 1st centuries BC was discovered on Mormont Hill in the Swiss canton of Vaud. Here, over two hundred and fifty structures, pits and shafts, up to 5 m deep, were found, in which vessels, tools, jewellery, coins, grindstones, and the remains of people and animals had been deposited (DIETRICH et al. 2009, 3). Only 0.1 % of the animal remains were from game (MÉNIEL 2009, 10). The skull of an adult bear was found in one of the pits (ibid., 8 with fig.) – his right canine, on the other hand, was found in a pit about 30 m away (ibid. 8). Together with the skull of a wolf, they represent the only predators documented there and have been interpreted as prestigious hunting trophies (ibid., 10).

However, there are no comparable structures in the area in question. It should therefore be noted that the bear appears only to a limited extent in cult practices. Thus, explanations for the "invisibility" of an animal that is at least physiognomically extremely visible are finally yet to be found.

---

45 This would thus be unusual in two respects – on the one hand, wild animals are not usually found in the context of burnt offering sites while, on the other hand, Lippert himself points out that burnt offering sites are unfamiliar in the Celtic region (LIPPERT 1992, 297; sceptical about this functional approach: GLASER 1993, esp. 295).

46 From the middle of the 2nd century BC, however, animal sacrifices have finally been documented in large numbers – again, mainly domestic animals were concerned (SCHERR 2013b, 237).

Conclusion: Why did the bear not matter?

While the bear played a prominent role in the 1ˢᵗ millennium AD, especially in the Scandinavian area and in northern central Europe (cf. SCHÖNFELDER 1994; WAMERS 2009; GRIMM 2013), it played only a minor one in the Pre-Roman metal ages of southern and southwestern Germany. Although wild animals were obviously hunted then, the bear makes only a brief appearance. Rather, the periods seem to represent "domestic animal periods". Even if there are signs of social differentiation in settlement patterns and burial practices, the Urnfield, Hallstatt and Early Latène periods were primarily characterised by a rural way of life and economy. This is even true for the later part of the Latène period, in which social and structural changes occurred that were accompanied by differentiation in settlement, trade and craftsmanship – but here, too, the dominance of domestic animal species in the archaeological evidence remains.

The reason that the bear played such a small role in the hunting of animals could perhaps be that it had already retired into the large forests – human intervention in its environment certainly reached regionally noticeable scales in the 1ˢᵗ millennium BC, and there is little doubt that natural conditions have a strong influence on the presence or absence of wildlife. At the same time, however, it is precisely this rarity and dangerousness that can lead to an accumulation of significance. After all, the quasi-absence of the bear does not necessarily mean that this animal has not played a role. It is also conceivable that a taboo was created, as has been proven for bears from numerous ethnographic and historical contexts (SCHMÖLCKE et al. 2017, 901–902; cf. BRUNNER 2010, 9). The interpretation of bear remains in the literature ranges between hunting trophy, apotropaion, and status indicator. Let us take a last look at the Hallstatt period burials with bear relics – how are the definite finds of bear remains to be interpreted?

First of all, the infant grave at Kelheim, the only case of a subadult individual with bear remains, stands out. According to the "classical" interpretation, the bear tooth would have had a protective, apotropaic function. For the other graves, however, it is more difficult to find coherence in terms of age, gender, and potential social role. Based on sources from the Middle Ages, the hunt for big game is assumed *a priori* to have been a gender-specific domain of men, but, especially for 1ˢᵗ-millennium Scandinavia, one has to ask how the frequency of bear teeth and bear claws in women's graves can be explained. Against this background, the portrayal of a successful bear hunter and his trophy that ends up as a grave good gets a little shaky. On the one hand, Siegmund Oehrl refers to "hunting women" as they are documented in various sources – strong, courageous women who stand up to the dangerous predator (OEHRL 2013, 312–313). At the same time, he tries to depolarise this with an alternative narrative that lets the husband act as a successful bear hunter, who gives his deceased lover his most valuable hunting trophy to take with her on her final journey. In the afterlife "everyone [...] should see that the deceased was nothing less than a hero's wife".[47]

It is important to avoid androcentric views which, *a priori* and undifferentiated, assign passive roles to women (for the female "princely grave" phenomenon in the Early Iron Age cf. ARNOLD 1995), not only because predator remains are thoroughly associated with status-bearing women from various contexts.[48] But, while in some proto- and early historic circumstances there is a recurring

---

47 Martin Schönfelder, who clearly understands bear claws in graves as a "male attribute", argues similarly (SCHÖNFELDER 1994, 220). He explains the few exceptions thus: "the grave goods of the woman include the prestige goods of the husband who arranged the funeral".
48 Torun Zachrisson and Maja Krzewińska introduce an example from the Migration and Merovingian periods in present-day Sweden – the "Lynx Ladies", women who are richly furnished with grave goods (among others lynx skins) and who, in addition to their function as head of the household, have assumed an important role in cult and ritual practices (ZACHRISSON/KRZEWIŃSKA 2019).

connection between bear remains and women's and children's graves, these do not dominate in the spatio-temporal context that is at issue here. Among the few burials with bear remains from the area in question, the presence of women has been proven in only exceptional cases – in one case (Werbach) almost certainly, in another (Niedererlbach) only possibly. In one case, we have a pierced bear phalange, in the other we have four apparently unworked phalanges, possibly a bear paw. The Werbach find, if one considers its entire context, could be seen as the representation of a person experienced in the subject of roles in cult and ritual. Here, the bear phalange was found together with features such as the functionally ambiguous bronze hollow ball with grommet and the symbolic staging of the corpse. The buried person belongs to a group of women with a special status that was not just based on "wealth". It should be asked whether the bear remains are to be understood as status indicators.[49] In Kleinklein, for example, we have an archer who is included in the "circle of better equipped warriors in Kleinklein society", with a "position between the princes and the simple warriors" (BERNHARD/GUŠTIN 2019, 39, 41; translations by the author). NIKULKA (1998, 174) in turn refers to grave 30 from Untereggersberg with its bear remains, located in a separated group of graves, to which, in addition to the two potential "primary or founder graves" (the burials of particularly old men with particularly large grave monuments and rich grave furnishings), further graves of inhumated men of advanced age belong, "most of whom have also been provided with special grave goods" (translations by the author).

While in the Urnfield and Hallstatt periods – and this is apparently not due to the state of research – only a small group of people was attributed with bear remains,[50] the bear appears significantly more frequently in burials at the end of the Latène period or at the transition from the Latène period to the Early Roman Iron Age. Initially, this applies mainly to the central and northern German regions; however, from the 2nd century AD onwards, the number of finds increased in Scandinavia (WAMERS 2009, 9 fig. 5). The remains involve bear claws as well as bear skins (WAMERS 2009; GRIMM 2013). Since around four hundred of the graves with bear remains are known from northern Europe and at least a hundred from central Europe (GRIMM 2013, 290, 291), there is talk of a "mass phenomenon" (ibid., 291; SCHMÖLCKE et al. 2017, 905) – a phenomenon that even had an impact on the local bear populations (SCHMÖLCKE et al. 2017, 905; cf. LINDHOLM/LJUNGKVIST 2016).

In many cultures, both past and present, bears are what Ulrich Schmölcke, Daniel Groß and Elena Nikulina call "dominant symbols" (SCHMÖLCKE et al. 2017, 902), which is rooted in their ambivalence and in their representation of many different actions and meanings. Animals always held a special significance in mythology, religious beliefs and practices. They were far more than passive foils of human action; they had agency. Animals and their representations can therefore be understood as part of a complex system of symbolic communication. When it comes to bears, the archaeological record indicates that they were rarely hunted and they were also rarely part of the iconographic repertoire. But it is perhaps that scarcity that creates their importance. Bears are actually highlighted in the archaeological record and thus allow insights into the world of ideas and the construction of the reality of past people.[51]

---

49 I understand "status" as the determination of the social position of an individual in relation to other individuals in the community, and this relationship does not necessarily have to be hierarchical (cf. AUGSTEIN 2015, 193 with note 52).
50 Concerning the Hallstatt period, HANSEN (2013, 253) has pointed out that the teeth and claws of predators such as the bear are documented "in rich burials as well as in 'normal' graves". A clear "elite context" does not seem to be tangible here.
51 Postscriptum: Only after typesetting and layout I came across another bear claw – as there is talk of an "amulet", it is probably perforated. It originates from the richly furnished burial of an adult male in the central chamber of the large Hallstatt period tumulus at Repperndorf, district of Kitzingen (Lower Franconia, Bavaria). That burial, however, had already been robbed in antique times (WAMSER 1982).

Bibliography

Arnold 1995: B. Arnold, 'Honorary males' or women of substance? Gender, status, and power in Iron-Age Europe. Journal of European Archaeology 3(2), 1995, 153–168.

Arnold 2019: B. Arnold, Expect the unexpected: implications of recent analyses of mortuary vessels for early Iron Age social configurations and commensality in southwest Germany. In: Ph. Stockhammer/J. Fries-Knoblach (eds.), Was tranken die frühen Kelten? Bedeutungen und Funktionen mediterraner Importgefäße im früheisenzeitlichen Mitteleuropa (Leiden 2019) 213–230.

Augstein 2009: M. Augstein, Der Körper als Zeichen? Deutungsmöglichkeiten von Körperinszenierungen im hallstattzeitlichen Bestattungsritual. In: R. Karl/J. Leskovar (eds.), Interpretierte Eisenzeiten – Fallstudien, Methoden, Theorie. Tagungsbeiträge der 3. Linzer Gespräche zur interpretativen Eisenzeitarchäologie. Studien zur Kulturgeschichte von Oberösterreich 22 (Linz 2009) 11–26.

Augstein 2011: M. Augstein, Ein Grab mit Halbmondfibeln aus Dietfurt a. d. Altmühl, Lkr. Neumarkt i. d. Opf. – Aspekte der Distinktion im Rahmen hallstattzeitlicher Bestattungssitten. Germania 87, 2009 (2011), 41–74.

Augstein 2013a: M. Augstein, Gräber – Orte der Lebenden und der Toten, Medien der Kommunikation. In: R. Karl/J. Leskovar (eds.), Interpretierte Eisenzeiten – Fallstudien, Methoden, Theorie. Tagungsbeiträge der 5. Linzer Gespräche zur interpretativen Eisenzeitarchäologie. Studien zur Kulturgeschichte von Oberösterreich 37 (Linz 2013) 107–122.

Augstein 2013b: M. Augstein, ‚Reguläre' und ‚irreguläre' Bestattungen der Hallstattzeit Nordostbayerns. In: N. Müller-Scheeßel (ed.), ‚Irreguläre' Bestattungen in der Urgeschichte: Norm, Ritual, Strafe …? Kolloquien zur Vor- und Frühgeschichte 19 (Bonn 2013) 357–368.

Augstein 2014: M. Augstein, Gefährte, Opfer, Statussymbol? Tierdeponierungen im Kontext prähistorischer Bestattungsplätze. In: J. Ullrich/A. Ulrich (eds.), Tiere und Tod. Tierstudien 5, 2014, 75–88.

Augstein 2015: M. Augstein, Das Gräberfeld der Hallstatt- und Frühlatènezeit von Dietfurt an der Altmühl (‚Tankstelle'). Ein Beitrag zur Analyse einer Mikroregion. Universitätsforschungen zur Prähistorischen Archäologie 262 (Bonn 2015).

Augstein 2017: M. Augstein, Das Tier und sein Kontext. Tierdarstellungen und Tierfunde der Hallstattzeit Nordostbayerns. In: V. Brieske/A. Dickers/M. M. Rind (eds.), Tiere und Tierdarstellungen in der Archäologie. Beiträge zum Kolloquium in Gedenken an Torsten Capelle, 30.–31. Oktober 2015 in Herne. Veröffentlichungen der Altertumskommission für Westfalen 22 (Münster 2017) 143–157.

Augstein/Karlsen 2019: M. Augstein/H.-J. Karlsen, Nienbüttel – New Research on Old Graves. In: M. Augstein/M. Hardt (eds.), Sächsische Leute und Länder – Benennung und Lokalisierung von Gruppenidentitäten im ersten Jahrtausend. Neue Studien zur Sachsenforschung 10 (Braunschweig 2019) 227–235.

Beech 1995: M. Beech, A Matter of Taste? Some Evidence for the Butchery of Horses, Dogs and Bears at the La Tène Settlement of Jenišův Újezd. In: J. Blažek/P. Meduna (eds.), Archeologické výzkumy v severozápadních Čechách v letech 1983–1992 (Most 1995) 165–170.

Beermann 2016: S. Beermann, Bärenkrallen und Bärenfelle in Brand- und Körpergräbern der vorrömischen Eisenzeit bis Völkerwanderungszeit in Mittel- und Nordeuropa. Universitätsforschungen zur Prähistorischen Archäologie 279 (Bonn 2016).

Bernatzky-Goetze 1987: M. Bernatzky-Goetze, Mörigen – Die spätbronzezeitlichen Funde. Antiqua 16 (Basel 1987).

Bernhard/Guštin 2019: A. Bernhard/M. Guštin, Zu einem frühhallstattzeitlichen Bogenschützen aus Kleinklein/ O zgodnjehalštatskem lokostrelcu iz Kleinkleina. Arheološki vestnik 70, 2019, 31–47.

Biel et al. 2006: J. Biel/E. Stephan/K. Schatz, Archäozoologische Untersuchung der Faunenfunde aus hallstatt- und frühlatènezeitlichen Siedlungen und Gräbern – Studien zur Wirtschaftsgeschichte im Umfeld frühkeltischer Fürstensitze. http://www.fuerstensitze.de/dna_media/www3-Biel+445f0383a228c.pdf.

Biermeier/Kowalski 2010: St. Biermeier/A. Kowalski, Grabungsbericht Erding-Kletthamer Feld. http://www.singularch.de/referenzen/erding_bericht.pdf.

Boessneck et al. 1971: J. Boessneck/A. von den Driesch/U. Meyer-Lemppenau/E. Wechsler-von Ohlen, Die Tierknochenfunde aus dem Oppidum von Manching. Die Ausgrabungen in Manching 6 (Wiesbaden 1971).

Braun-Schmidt 1983: A. Braun-Schmidt, Tierknochenfunde von der Heuneburg, einem frühkeltischen Herrensitz bei Hundersingen an der Donau (Grabungen 1966 bis 1979) (München 1983).

Brunner 2010: B. Brunner, Bär und Mensch. Die Geschichte einer Beziehung (Darmstadt 2010).

Buckel 2008: I. Buckel, Beigaben mit Kult- und Amulettcharakter in bronze- und urnenfelderzeitlichen Grabfunden Bayerns. In: Festgabe 40 Jahre Lehrstuhl für Vor- und Frühgeschichte der Universität Regensburg 1968–2008. Regensburger Beiträge zur Prähistorischen Archäologie 20 (Regensburg 2008) 71–293.

Choyke 2010: A. Choyke, The Bone is the Beast: Animal Amulets and Ornaments in Power and Magic. In: D. Campana/P. Crabtree/S. D. deFrance/J. Lev-Tov/A. Choyke (eds.), Anthropological Approaches to Zooarchaeology: Complexity, Colonialism, and Animal Transformations (Oxford 2010) 197–209.

Dietrich/Sorge 1992: H. Dietrich/G. Sorge, Eine urnenfelderzeitliche Siedlung bei Graben, Landkreis Augsburg, Schwaben. Das Archäologische Jahr in Bayern 1991 (1992) 75–76.

Dietrich et al. 2009: E. Dietrich/C. Nitu/C. Brunetti, Les fouilles de 2006 à 2009 et les premières études. In: Archeodunum (ed.), Le Mormont – Un sanctuaire des Helvètes en terre vaudoise vers 100 avant J.-C. 3–4. http://www.archeodunum.ch/FILES/mc9/124_tmp_197.pdf.

Doll 2009: M. Doll, Die Tierknochenfunde aus der Viereckschanze „Klinge" bei Riedlingen. In: Ch. Bollacher, Die keltische Viereckschanze „Auf der Klinge" bei

Riedlingen. Materialhefte zur Archäologie in Baden-Württemberg 88 (Stuttgart 2009) 287–371.

VON DEN DRIESCH 1993: A. VON DEN DRIESCH, Haustierhaltung und Jagd bei den Kelten in Süddeutschland. In: H. Dannheimer/R. Gebhard (eds.), Das keltische Jahrtausend (München, Mainz ³1993) 126–133.

VON DEN DRIESCH/BOESSNECK 1989: A. VON DEN DRIESCH/ J. BOESSNECK, Abschlußbericht über die zooarchäologischen Untersuchungen an Tierknochenfunden von der Heuneburg. In: E. Gersbach (ed.), Ausgrabungsmethodik und Stratigraphie der Heuneburg. Heuneburgstudien VI. Römisch-Germanische Forschungen 45 (Mainz 1989) 131–157.

EGG/KRAMER 2005: M. EGG/D. KRAMER, Krieger – Feste – Totenopfer. Der letzte Hallstattfürst von Kleinklein in der Steiermark. Mosaiksteine. Forschungen am Römisch-Germanischen Zentralmuseum 1 (Mainz 2005).

EGG/KRAMER 2013: M. EGG/D. KRAMER, Die hallstattzeitlichen Fürstengräber von Kleinklein in der Steiermark: Der Kröllkogel. Monographien des Römisch-Germanischen Zentralmuseums 110 (Mainz 2013).

EGG/KRAMER 2016: M. EGG/D. KRAMER, Die hallstattzeitlichen Fürstengräber von Kleinklein in der Steiermark: Die beiden Hartnermichelkogel und der Pommerkogel. Monographien des Römisch-Germanischen Zentralmuseums 125 (Mainz 2016).

EGGERT 2012: M. K. H. EGGERT, Prähistorische Archäologie – Konzepte und Methoden (Tübingen, Basel ⁴2012).

EIBNER 2004: A. EIBNER, Die Bedeutung der Jagd im Leben der eisenzeitlichen Gesellschaft – dargestellt anhand der Bildüberlieferungen. In: H. Heftner/K. Tomaschitz (eds.), Ad fontes! Festschrift für Gerhard Dobesch zum 65. Geburtstag am 15. September 2004, dargebracht von Kollegen, Schülern und Freunden (Wien 2004) 621–644.

EICHHORN 1927: G. EICHHORN, Der Urnenfriedhof auf der Schanze bei Großromstedt. Mannus-Bibliothek 41 (Leipzig 1927).

EICHINGER/LOSERT 2004: W. EICHINGER/H. LOSERT, Ein merowingerzeitliches Brandgräberfeld östlich-donauländischer Prägung bei Großprüfening. Das Archäologische Jahr in Bayern 2003 (2004) 98–101.

ETTEL 2006: P. ETTEL, Frühe Kelten am Schnittpunkt zwischen Ost und West. In: Gesellschaft für Archäologie in Bayern and Bayerisches Landesamt für Denkmalpflege (eds.), Archäologie in Bayern – Fenster zur Vergangenheit (Regensburg 2006) 152–156.

FALKENSTEIN 2009: F. FALKENSTEIN, Zur Subsistenzwirtschaft der Bronzezeit in Mittel- und Südosteuropa. In: M. Bartelheim/H. Stäuble (eds.), Die wirtschaftlichen Grundlagen der Bronzezeit Europas / The Economic Foundations of the European Bronze Age. Forschungen zur Archäometrie und Altertumswissenschaft 4 (Rahden/ Westf. 2009) 147–176.

FRIE 2017: A. C. FRIE, Cultural Constructions of Nature: Animal Representation and Use in Early Iron Age Southeastern Slovenia. University of Wisconsin, Theses and Dissertations 1472 (Madison 2017). https://dc.uwm.edu/ cgi/viewcontent.cgi?article=2477&context=etd.

FRISCH 2018: A. FRISCH, Die urnenfelderzeitliche Nekropole von Straubing-Sand. Beiträge zur Archäologie in Niederbayern 6 (Büchenbach 2018).

GLASER 1993: F. GLASER, Der behauptete Brandopferplatz und der tatsächliche Fundort eiserner Waffen in Teurnia. Carinthia 183, 1993, 289–295.

GRAF 1967: G. GRAF, Tierknochenfunde von der Heuneburg, einem frühkeltischen Fürstensitz bei Hundersingen an der Donau (Grabungen von 1959 und 1963) – Nichtpaarhufer (Stuttgart 1967).

GRIMM 2013: O. GRIMM, Bear-skins in northern European burials and some remarks on other bear-related furnishings in the north and middle of Europe in the 1st millennium AD. In: GRIMM/SCHMÖLCKE 2013, 277–296.

GRIMM/SCHMÖLCKE 2013: O. GRIMM/U. SCHMÖLCKE (eds.), Hunting in northern Europe until 1500 AD. Old traditions and regional developments, continental sources and continental influences. Papers presented at a workshop organized by the Centre for Baltic and Scandinavian Archaeology (ZBSA) Schleswig, June 16th and 17th, 2011. Schriften des Archäologischen Landesmuseums, Ergänzungsreihe 7 (Neumünster 2013).

HANSEN 2013: L. HANSEN, Hunting in the Hallstatt period – The example of the Eberdingen-Hochdorf 'princely grave'. In: GRIMM/SCHMÖLCKE 2013, 239–258.

HENNIG 1993: H. HENNIG, Urnenfelder aus dem Regensburger Raum. Materialhefte zur Bayerischen Vorgeschichte A 65 (Kallmünz/Opf. 1993).

HOLZER 2009: V. HOLZER, Roseldorf – Interdisziplinäre Forschungen zur größten keltischen Zentralsiedlung Österreichs. Schriftenreihe der Forschung im Verbund 102 (Wien 2009).

HOPPE 1986: M. HOPPE, Die Grabfunde der Hallstattzeit in Mittelfranken. Materialhefte zur Bayerischen Vorgeschichte A 55 (Kallmünz/Opf. 1986).

JACOMET et al. 1999: ST. JACOMET/CH. JACQUAT/CH. MAISE/ J. SCHIBLER/B. STOPP/J. STUDER/L. WICK/M. WINTER, Klima, Umwelt, Landwirtschaft und Ernährung. In: F. Müller/G. Kaenel/G. Lüscher (eds.), Die Schweiz vom Paläolithikum bis zum frühen Mittelalter IV: Eisenzeit (Basel 1999) 93–136.

JOCKENHÖVEL 1994: A. JOCKENHÖVEL, Umwelt – Landwirtschaft – Ernährung. In: A. Jockenhövel/W. Kubach (eds.), Bronzezeit in Deutschland. Archäologie in Deutschland, special issue 1994 (Stuttgart 1994) 31–35.

KERTH/MÜLLER-DEPREUX 2004: K. KERTH/A. MÜLLER-DEPREUX, Die tierischen Nahrungsressourcen der späthallstatt-/frühlatènezeitlichen Siedlung „Erdwerk I" bei Niedererlbach (Lkr. Landshut, Niederbayern). Germania 82(1), 2004, 219–234.

KERTH/WACHTER 1993: K. KERTH/N. WACHTER, Die Tierknochenfunde aus drei nordbayerischen Siedlungen der Hallstatt- und Frühlatènezeit. Bayerische Vorgeschichtsblätter 53, 1993, 61–77.

KLEIN 2004: F. KLEIN, Siedlungsfunde der ausgehenden Späthallstatt und frühen Latènezeit aus Württemberg. https://publikationen.uni-tuebingen.de/xmlui/handle/10900/46237.

KOCH 1992: H. KOCH, Grabfunde der Hallstattzeit aus dem Isartal bei Niedererlbach, Lkr. Landshut. Bayerische Vorgeschichtsblätter 57, 1992, 49–75.

KOCH 2003: L. C. KOCH, Zu den Deutungsmöglichkeiten der Situlenkunst. In: U. Veit/T. L. Kienlin/Ch. Kümmel/ S. Schmidt (eds.), Spuren und Botschaften: Interpretationen materieller Kultur. Tübinger Archäologische Taschenbücher 4 (Münster, New York, München, Berlin 2003) 347–367.

Kromer 1959: K. Kromer, Das Gräberfeld von Hallstatt (Firenze 1959).

Lang 2005: A. Lang, Alpine Brandopferplätze. TÜVA-Mitteilungen 6/7, 2003/2004 (2005) 27–62.

Lindholm/Ljungkvist 2016: K.-J. Lindholm/J. Ljungkvist, The Bear in the Grave: Exploitation of Top Predator and Herbivore Resources in First Millennium Sweden – First Trends from a Long-Term Research Project. European Journal of Archaeology 19(1), 2016, 3–27.

Lippert 1992: A. Lippert, Ein latènezeitlicher Opferplatz in Teurnia bei Spittal an der Drau. In: Festschrift zum 50jährigen Bestehen des Instituts für Ur- und Frühgeschichte der Leopold-Franzens-Universität Innsbruck. Universitätsforschungen zur Prähistorischen Archäologie 8 (Bonn 1992) 285–304.

Lochner 1991: M. Lochner, Ein Gräberfeld der älteren Urnenfelderzeit aus Horn, Niederösterreich. Archaeologia Austriaca 74, 1991, 137–220.

Meiborg/Müller 1997: Ch. Meiborg/A. Müller, Die urnenfelder- und hallstattzeitliche Siedlung „Kanal I" und das frühhallstattzeitliche Gräberfeld „Am Urnenfeld" von Kelheim. Archäologie am Main-Donau-Kanal 12 (Espelkamp 1997).

Méniel 2002: P. Méniel, La chasse en Gaule – une activité aristocratique? In: V. Guichard/F. Perrin (eds.), L'aristocratie celte à la fin de l'âge du Fer (IIe s. avant J.-C.–Ier s. après J.-C.). Collection Bibracte 5 (Glux-en-Glenne 2002) 223–230.

Méniel 2009: P. Méniel, Veaux, vaches, cochons, chevaux ... In: Archeodunum (ed.), Le Mormont – Un sanctuaire des Helvètes en terre vaudoise vers 100 avant J.-C. [8] 9–11. http://www.archeodunum.ch/FILES/mc9/124_tmp_197.pdf.

Müller-Scheessel 2008: N. Müller-Scheessel, Auffälligkeiten bei Armhaltungen hallstattzeitlicher Körperbestattungen – postdeponale Eingriffe, funktionale Notwendigkeit oder kulturelle Zeichen? In: Ch. Kümmel/B. Schweizer/U. Veit (eds.), Körperinszenierung – Objektsammlung – Monumentalisierung: Totenritual und Grabkult in frühen Gesellschaften. Archäologische Quellen in kulturwissenschaftlicher Perspektive. Tübinger Archäologische Taschenbücher 6 (Münster, New York, München, Berlin 2008) 517–535.

Müller-Scheessel/Trebsche 2007: N. Müller-Scheessel/P. Trebsche, Das Schwein und andere Haustiere in Siedlungen und Gräbern der Hallstattzeit Mitteleuropas. Germania 85, 2007, 61–94.

Nikulka 1998: F. Nikulka, Das hallstatt- und frühlatènezeitliche Gräberfeld von Riedenburg-Untereggersberg, Lkr. Kelheim, Niederbayern. Archäologie am Main-Donau-Kanal 13 (Rahden/Westf. 1998).

Nikulka 2008: F. Nikulka, Bestattungsvarianten, Zeichensprache und Kommunikationslinien. In: F. Verse/B. Knoche/J. Graefe/M. Hohlbein/K. Schierhold/C. Siemann/ M. Uckelmann/G. Woltermann (eds.), Durch die Zeiten ... Festschrift für Albrecht Jockenhövel zum 65. Geburtstag. Internationale Archäologie Studia Honoraria 28 (Rahden/Westf. 2008) 373–382.

Oehrl 2013: S. Oehrl, *Svá beitum vér bjǫrnuna á mörkinni norðr* – Bear hunting and its ideological context (as a background for the interpretation of bear claws and other remains of bears in Germanic graves of the 1st millennium AD). In: Grimm/Schmölcke 2013, 297–332.

Pauli 1975: L. Pauli, Keltischer Volksglaube. Amulette und Sonderbestattungen am Dürrnberg bei Hallein und im eisenzeitlichen Mitteleuropa. Münchner Beiträge zur Vor- und Frühgeschichte 28 (München 1975).

Petrus 2015: K. Petrus, Human-Animal Studies. In: A. Ferrari/K. Petrus (eds.), Lexikon der Mensch-Tier-Beziehungen (Bielefeld 2015) 156–160.

Přichystal 2007: M. Přichystal, Dva provrtané medvědí zuby z halštatského sídliště v Kralicích na Hané (okr. Prostějov). Příspěvek k postavení medvěda v symbolice starší doby železné (Zwei durchbohrte Bärenzähne aus der hallstattzeitlichen Siedlung in Kralice na Hané [Bez. Prostějov]. Ein Beitrag zur Rolle des Bären in der Symbolik der älteren Eisenzeit). Pravěk N. Ř. 17, 2007, 329–356.

Rast-Eicher in print: A. Rast-Eicher, Textilfasern und Felle aus Altheim-Heiligkreuztal, „Wald Speckhau" Hügel 17 und 18. In: B. Arnold/M. L. Murray, with contributions by T. Kreß, A Landscape of Ancestors: Archaeological Investigations of Two Iron Age Burial Mounds in the Hohmichele Group, Baden-Württemberg. Forschungen und Berichte zur Vor- und Frühgeschichte in Baden-Württemberg (Stuttgart, in print).

Rebay 2006: K. C. Rebay, Das hallstattzeitliche Gräberfeld von Statzendorf in Niederösterreich. Möglichkeiten und Grenzen der Interpretation von Sozialindexberechnungen I und II. Universitätsforschungen zur Prähistorischen Archäologie 135 (Bonn 2006).

Roscher 2012: M. Roscher, Human-Animal Studies, Version 1.0. Docupedia-Zeitgeschichte, 25.01.2012, 1–15. http://docupedia.de/zg/Human-Animal_Studies.

Schatz 2006: K. Schatz, Die Tierknochenfunde aus der späthallstattzeitlichen Höhensiedlung „Göllersreuther Platte", südliche Frankenalb – Zwischenbericht zum Abschluss der Grabungen 2004 und 2005. https://publikationen.uni-tuebingen.de/xmlui/bitstream/handle/10900/44012/pdf/Schatz_Goellersreuth.pdf.

Schatz 2009: K. Schatz, Die Tierknochenfunde aus der frühlatènezeitlichen Siedlung Eberdingen-Hochdorf „Reps" – Archäozoologische Untersuchungen zur Wirtschaftsweise, Landnutzung und Ernährung der frühen Kelten im mittleren Neckarraum. In: K. Schatz/H.-P. Stika, Hochdorf VII. Archäologische Untersuchungen zur frühen Eisenzeit im mittleren Neckarraum. Forschungen und Berichte zur Vor- und Frühgeschichte in Baden-Württemberg 107 (Stuttgart 2009) 19–91.

Schatz/Stephan 2005: K. Schatz/E. Stephan, Die Tierknochenfunde aus den Rechteckhöfen im Gewann „Zaunäcker" bei Osterholz, Gde. Kirchheim am Ries. http://docplayer.org/130191993-Kristine-schatz-elisabeth-stephan-die-tierknochenfunde-aus-den-rechteckhoefen-im-gewann-zaunaecker-bei-osterholz-gde.html.

Scherr 2013a: J. Scherr, Gournay-sur-Aronde (Der Neue Pauly Addenda & Corrigenda). Orbis Terrarum. Internationale Zeitschrift für Historische Geographie der Alten Welt 11, 2013, 225–233.

Scherr 2013b: J. Scherr, Ribemont-sur-Ancre (Der neue Pauly Addenda & Corrigenda). Orbis Terrarum. Internationale Zeitschrift für Historische Geographie der Alten Welt 11, 2013, 234–242.

Schibler/Studer 1998: J. Schibler/J. Studer, Haustierhaltung und Jagd während der Bronzezeit in der Schweiz. In: S. Hochuli (ed.), Die Schweiz vom Paläolithikum bis zum frühen Mittelalter III. Bronzezeit (Basel 1998) 171–191.

Schmölcke et al. 2017: U. Schmölcke/D. Gross/E. A. Nikulina, Bears and beavers: 'The Browns' in daily life and spiritual world. In: B. V. Eriksen/A. Abegg-Wigg/R. Bleile/U. Ickerodt (eds.), Interaktion ohne Grenzen: Beispiele archäologischer Forschungen am Beginn des 21. Jahrhunderts / Interaction without borders: exemplary archaeological research at the beginning of the 21st century (Schleswig 2017) 901–916.

Schönfelder 1994: M. Schönfelder, Bear Claws in Germanic Graves. Oxford Journal of Archaeology 13, 1994, 217–227.

Schussmann 2009: M. Schussmann, Frühe Zentralisierungsprozesse auf der südlichen Frankenalb – Ausgrabungen und Forschungen zur Urnenfelder-, Hallstatt- und Frühlatènezeit im Schwarzach- und Thalachtal. In: Ch. Bockisch-Bräuer/B. Mühldorfer (eds.), Beiträge zur Hallstatt- und Latènezeit in Nordostbayern und Thüringen. Tagung vom 26.–28. Oktober 2007 in Nürnberg. Beiträge zur Vorgeschichte Nordostbayerns 7 (Nürnberg 2009) 87–101.

Schütz 2006: C. Schütz, Das urnenfelderzeitliche Gräberfeld von Zuchering-Ost, Stadt Ingolstadt. Materialhefte zur Bayerischen Vorgeschichte A 90 (Kallmünz/Opf. 2006).

Sjöblom 2012: T. Sjöblom, The Great Mother. The Cult of the Bear in Celtic Traditions. Studia Celtica Fennica 3, 2012, 71–78.

Stadler 2010: J. Stadler, Nahrung für die Toten? Speisebeigaben in hallstattzeitlichen Gräbern und ihre kulturhistorische Bedeutung. Universitätsforschungen zur Prähistorischen Archäologie 186 (Bonn 2010).

Tecchiati 2012: U. Tecchiati, Die Tierknochen aus der bronze- und eisenzeitlichen Siedlung auf dem Kiabichl bei Faggen (Tirol, Österreich). Annalen des Naturhistorischen Museums in Wien, Serie A 114, 2012, 21–78.

Torbrügge 1979: W. Torbrügge, Die Hallstattzeit in der Oberpfalz I. Auswertung und Gesamtkatalog. Materialhefte zur Bayerischen Vorgeschichte A 39 (Kallmünz/Opf. 1979).

Trebsche 2013: P. Trebsche, Hunting in the Hallstatt and Early La Tène Cultures: the economic and social importance. In: Grimm/Schmölcke 2013, 215–238.

Trebsche 2018: P. Trebsche, Die hallstattzeitlichen Jagddarstellungen der Kalenderberggruppe – zu einem Altfund von Rauheneck bei Baden (Niederösterreich). Annalen des Naturhistorischen Museums in Wien, Serie A 120, 2018, 211–244.

Wagner 1943: K. H. Wagner, Nordtiroler Urnenfelder. Römisch-Germanische Forschungen 15 (Berlin 1943).

Wamers 2009: E. Wamers, Von Bären und Männern: Berserker, Bärenkämpfer und Bärenführer im frühen Mittelalter. Zeitschrift für Archäologie des Mittelalters 37, 2009, 1–46.

Wamers 2015: E. Wamers, Bärenkult und Schamanenzauber: Rituale früher Jäger (Regensburg 2015).

Wamser 1982: L. Wamser, Ein hallstattzeitlicher Großgrabhügel bei Repperndorf, Landkreis Kitzingen, Unterfranken. Das Archäologische Jahr in Bayern 1981 (1982), 110–111.

Warneke 1999: Th. F. Warneke, Hallstatt- und frühlatènezeitlicher Anhängerschmuck. Studien zu Metallanhängern des 8.–5. Jahrhunderts zwischen Main und Po. Internationale Archäologie 50 (Rahden/Westf. 1999).

Wehrberger 1984: K. Wehrberger, Das hallstattzeitliche Gräberfeld von Werbach, Main-Tauber-Kreis. Fundberichte aus Baden-Württemberg 9, 1984, 81–221.

Weiss 1997: R.-M. Weiss, Prähistorische Brandopferplätze in Bayern. Internationale Archäologie 35 (Espelkamp 1997).

Wendling 2018: H. Wendling, Keine Schonzeit – Jagd und Wild am eisenzeitlichen Dürrnberg. Annalen des Naturhistorischen Museums in Wien, Serie A 120, 2018, 281–304.

Wilk 2019: A. B. Wilk, Archäozoologische Auswertung und kulturhistorischer Vergleich der Tierknochenfunde. In: M. Halle/U. Veit/A. B. Wilk, Die bronze- und eisenzeitliche Höhensiedlung auf der Achalm bei Reutlingen (Baden-Württemberg). Die archäologischen Ausgrabungen am „Rappenplatz" in den Jahren 2000 bis 2005. Leipziger Forschungen zur Ur- und Frühgeschichtlichen Archäologie 10 (Leipzig 2019) 135–174.

Zachrisson/Krzewińska 2019: T. Zachrisson/M. Krzewińska, The 'Lynx Ladies' – Burials Furnished with Lynx Skins from the Migration and Merovingian Periods found in Present-day Sweden. In: M. Augstein/M. Hardt (eds.), Sächsische Leute und Länder – Benennung und Lokalisierung von Gruppenidentitäten im ersten Jahrtausend. Neue Studien zur Sachsenforschung 10 (Braunschweig 2019) 103–119.

Zuber 2011: J. Zuber, Nicht nur Rauch und Feuer – Neues zur Urnenfelderzeit in Ostbayern. In: M. Chytráček/H. Gruber/J. Michálek/R. Sandner/K. Schmotz (eds.), Archäologische Arbeitsgemeinschaft Ostbayern/West- und Südböhmen/Oberösterreich – 20. Treffen 23. bis 26. Juni 2010 in Eschenbach i. d. OPf. Fines Transire 20, 2011 (Rahden/Westf. 2011) 267–313.

Dr. Melanie Augstein
Professur für Ur- und Frühgeschichte am Historischen Seminar
Universität Leipzig
Germany
melanie.augstein@uni-leipzig.de

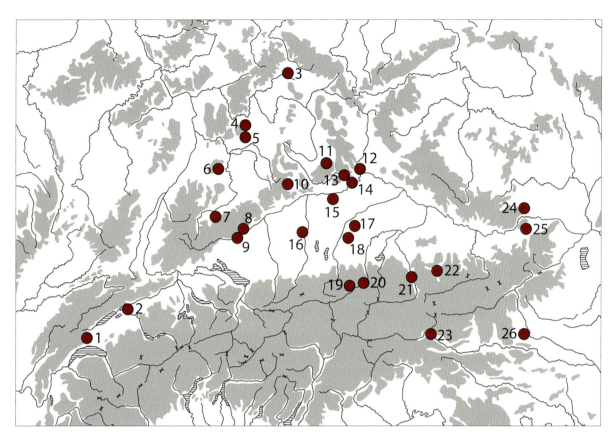

Fig. 1. Sites with bear remains mentioned in the text. 1: Mormont Hill (canton of Vaud, Switzerland); 2: Mörigen at Lake Biel (canton of Bern, Switzerland); 3: Eiersberg, Rhön-Grabfeld district (Bavaria, Germany); 4: Werbach, Main-Tauber district (Baden-Wuerttemberg, Germany); 5: Ingelfingen-Criesbach, Hohenlohe district (Baden-Wuerttemberg, Germany); 6: Hochdorf, district of Ludwigsburg (Baden-Wuerttemberg, Germany); 7: Achalm near Reutlingen, district of Reutlingen (Baden-Wuerttemberg, Germany); 8: Riedlingen, district of Tübingen (Baden-Wuerttemberg, Germany); 9: Hohmichele barrow group near the Heuneburg near Hundersingen, district of Sigmaringen (Baden-Wuerttemberg, Germany); 10: Osterholz, district of Kirchheim am Ries (Baden-Wuerttemberg, Germany); 11: Göllersreuther Platte near Landersdorf, district of Roth (Bavaria, Germany); 12: Burgweinting, city of Regensburg (Bavaria, Germany); 13: Riedenburg-Untereggersberg, district of Kelheim (Bavaria, Germany); 14: Kelheim, district of Kelheim (Bavaria, Germany); 15: Manching, district of Pfaffenhofen an der Ilm (Bavaria, Germany); 16: Graben near Augsburg, district of Augsburg (Bavaria, Germany); 17: Niedererlbach, district of Landshut (Bavaria, Germany); 18: Erding, district of Erding (Bavaria, Germany); 19: Kiabichl near Faggen, district of Landeck (Tyrol, Austria); 20: Wilten, city of Innsbruck (Tyrol, Austria); 21: Dürrnberg near Hallein, district of Hallein (Salzburg, Austria); 22: Hallstatt, district of Gmunden (Upper Austria, Austria); 23: Teurnia near Spittal an der Drau, district of Spittal an der Drau (Carinthia, Austria); 24: Horn, district of Horn (Lower Austria, Austria); 25: Statzendorf, district of Sankt Pölten-Land (Lower Austria, Austria); 26: Kleinklein, district of Leibnitz (Styria, Austria) (map M. Augstein, on the basis of "Tübinger Stumme Karte").

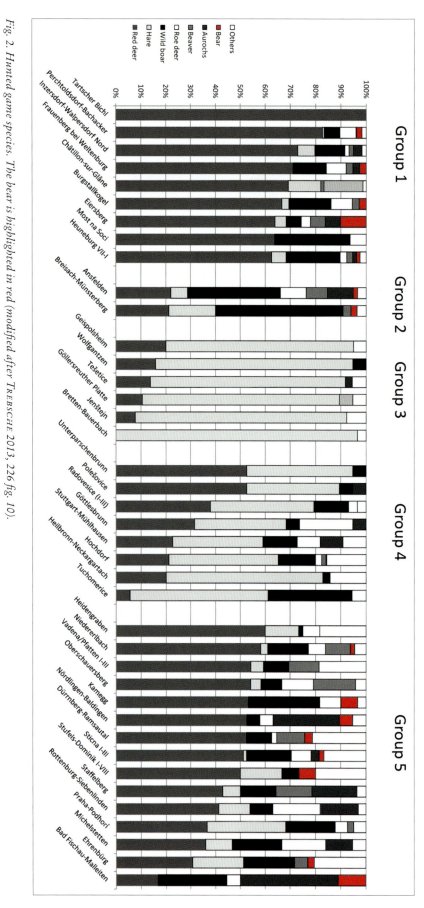

Fig. 2. Hunted game species. The bear is highlighted in red (modified after Trebsche 2013, 226 fig. 10).

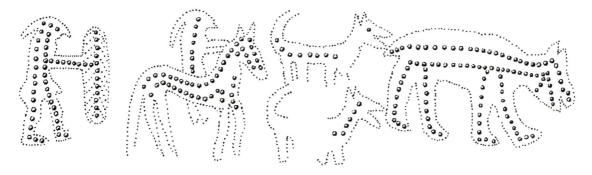

Fig. 3. Detail of the bear hunting scene on a cista from Kröllkogel, Kleinklein (Styria; after HANSEN 2013, 249 fig. 13; for a larger section see EGG/KRAMER 2005, 30–31 fig. 24).

Fig. 4. Enigmatic objects from an Urnfield period cremation grave from Burgweinting, city of Regensburg (photos P. Ferstl, city of Regensburg).

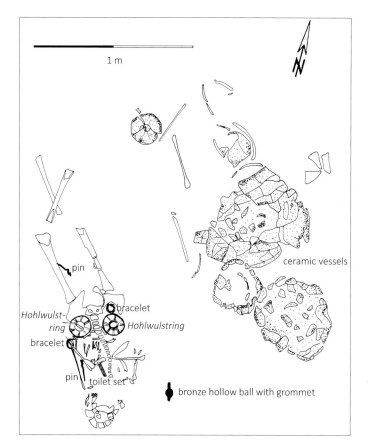

Fig. 5a. Hallstatt period inhumation grave (disturbed to varying extent) with bear remains: Werbach, grave 14 (after WEHRBERGER 1984, 186 fig. 40). The accurate position of the bear remains is not marked on the plan.

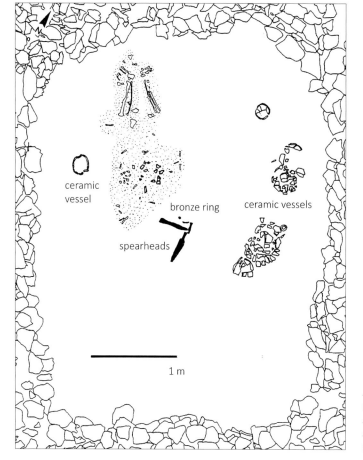

Fig. 5b. Hallstatt period inhumation grave (disturbed to varying extent) with bear remains: Riedenburg, Untereggersberg, grave 30 (after NIKULKA 1998, 236 fig. 78). The accurate position of the bear remains is not marked on the plan.

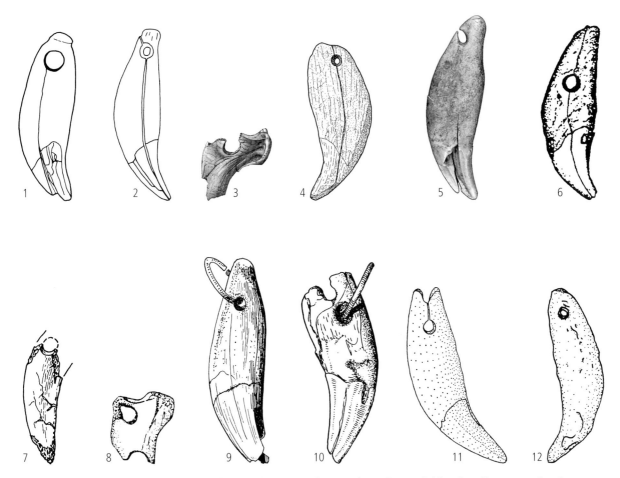

Fig. 6. Selection of worked bear remains from settlements and graves from the Urnfield and Hallstatt periods (objects not to scale). 1: Mörigen (after BERNATZKY-GOETZE 1987, pl. 116.9); 2: Mörigen (after BERNATZKY-GOETZE 1987, pl. 116.10); 3: Horn, grave 20 (after LOCHNER 1991, 197 pl. 25.7); 4: Burgweinting (after ZUBER 2011, 289 fig. 24); 5: Göllersreuther Platte (after SCHATZ 2006, 13 fig. 5a); 6: Kelheim-"Am Urnenfeld", grave 21 (after MEIBORG/MÜLLER 1997, pl. 130 C); 7: Thalmässing-Landersdorf, "Brandgrab 1" (after HOPPE 1986, pl. 110.11); 8: Werbach, grave 14 (after WEHRBERGER 1984, 211 fig. 62.16); 9: Hallstatt, grave 139 (after KROMER 1959, pl. 11.14); 10: Hallstatt grave 535 (after KROMER 1959, pl. 78.16); 11: Dürrnberg, grave 32/2 (after PAULI 1975, fig. 6.6); 12: Statzendorf, inventory GD03 (after REBAY 2006, (II) 469 fig. PA56109b).

Table 1. The archaeological periods relevant to this study.

Archaeological period	Relative chronological phases	Absolute chronology
Urnfield period	Ha A	c. 1200–c. 1050/1020 BC
	Ha B	c. 1050/1020–c. 800 BC
Hallstatt period	Ha C	c. 800–c. 620 BC
	Ha D	c. 620–c. 450 BC
Latène period	LT A – LT B	c. 450–c. 260/250 BC
	LT C	c. 260/250–c. 150/120 BC
	LT D	c. 150/120 BC–c. turn of eras

Table 2. Bear remains, associated grave goods, and – if available – anthropological classification of the deceased for selected graves of the Urnfield and Hallstatt periods.

Site / Grave	Bear remains	Dating	Anthropological analyses	Associated objects / Grave goods	References
Regensburg-Burgweinting "Nordwest"	Perforated bear tooth, length 8.3 cm	BZ D	Cremation Male by tendency, 40–60 years	Stone ball (diameter 5.4 cm), petrified shell with an unfinished drill-hole (6.1 x 5.2 cm), flat stone disk (diameter 5.2 cm); besides the urn fragments from two ceramic vessels	Buckel 2008; Zuber 2011
Horn Grave 20	Calcinated claw of a brown bear (phalanx-3-fragment), perforated at the base	Ha A1	Cremation Male, adult-mature Possibly second individual	Fragments of six ceramic vessels	Lochner 1991
Innsbruck-Wilten Grave 68	Two perforated fangs of a brown bear	Ha A1	Cremation –	Knife, belt buckle, fragments of a bracelet with square profile, two pendants (bird-shaped and wheel-shaped), fragments of a decorative disk with concentric circles, four pins with *Vasenkopf*, ceramic vessel	Wagner 1943
Kleinklein Höchsusterwald group Mound 13	"Several" / "numerous" calcinated, unworked bear claws	Ha C1b	Cremation Male	Bronze *Mehrkopfnadeln*, bronze *Hohlverteiler*, bronze bead, amber bead, bronze double hook, fittings, bone arrow heads, iron winged axe, iron socketed axe, knife, lugs, rings, chisel, ceramic vessel (urn)	Bernhard/Guštin 2019
Kelheim-"Am Urnenfeld" Grave 21	Perforated brown bear tooth	Ha C	Inhumation Infans I (c. 6 months)	Small bronze bracelet, small bronze spiral roulettes	Meiborg/Müller 1997
Werbach Grave 14	Bear phalange, double-drilled, found near the skull, without further details of the position	Ha C/Ha D1	Inhumation Female?, adult	Bronze pins, bronze bracelets, bronze *Hohlwulstringe*, bronze hollow ball with grommet, toilet set, bronze spiral rings, ceramic vessels	Wehrberger 1984
Untereggersberg Grave 30	Two bear claws	Ha D1	Inhumation Certainly male, adult, probably early adult	Two iron spearheads, small bronze ring, fragments of an iron wire, ceramic vessels	Nikulka 1998
Thalmässing-Landersdorf "Brandgrab 1"	Calcinated bear claw, perforated, fragmented	Ha D1	Cremation Multiple burial Female by tendency, adult, and child	Bronze *Melonenarmband*, bronze twisted neck ring, ceramic rattle ball, three ceramic vessels	Hoppe 1986

Cont. Tab. 2.

Cont. Tab. 2

Site / Grave	Bear remains	Dating	Anthropological analyses	Associated objects / Grave goods	References
Niedererlbach Grave 4/1990	Four phalanges of a brown bear	Ha D	Inhumation Disturbed collective burial Male, mature and female, early mature and female, early adult	Diverse bronze rings (head-/hair dress, or items of horse's harness), fragments of an iron snaffle, iron pin, fragments of amber beads, ceramic fragments, fragments of a bronze sheet and roulette; probably tooth of a sheep/goat and atlas of a dog/wolf	Koch 1992
Hallstatt Grave 99 (new excavations)	Ten phalanges, unburnt	Ha D	Inhumation Male, mature–senile	Iron weapons (axe, sword or knife, dagger), amber jewellery (ribbed amber beads)	Beermann 2016; https://www.diepresse.com/585994/ (last accessed: January 13, 2021)
Hallstatt Grave 139 (old excavations)	Perforated brown bear's tooth, provided with a bronze ring, right lower canine of *Ursus arctos*, length 7.1 cm	Hallstatt period (generally)	Inhumation – (archaeologically: male, "warrior")	Two open bronze wire rings, bronze *Kugelkopfnadel*, fragment of an iron spearhead	Kromer 1959
Hallstatt Grave 535 (old excavations)	Bear canine with two drillings at the root. One hole is broken out, in the other one there is a ring of bronze wire pulled in, length 6.5 cm	Hallstatt period (generally)	Inhumation – (children's grave because of the size of the human remains)	Three bronze wire rings	Kromer 1959
Statzendorf Inventory GD03	Perforated lower canine of a brown bear	Hallstatt period (generally)	–	Perforated stone, perforated boar tooth and miniature bowl with handle	Rebay 2006

Tracking former royal dignity:
The bear in medieval German literature

By Sabine Obermaier

Keywords: Medieval German literature, Vita Galli, Reynke de Vos, Nibelungenlied, Schrätel und Wasserbär

Abstract: In medieval German literature, the bear – compared to the horse, the dragon, the lion or the fox – leads a rather marginal existence. At first glance, Michel Pastoureau's thesis of the bear as the "fallen king" seems fully confirmed: The bear has not only been eliminated from the medieval forest, but also from German literature. This article tracks, and indeed finds, remains of the bear's "former royal dignity" in the legend of St Gallus (Vita Galli, 8th century, German translation in the 15th century), in the animal epic Reynke de Vos (1498) and in the heroic epic Nibelungenlied (around 1200). An additional sideways glance at the tale of Schrätel und Wasserbär, or Kobold und Eisbär (late 13th century) sheds light on these findings from a different perspective. Even if it cannot be proven beyond doubt that knowledge of the brown bear as the "old king of animals" was actually still present in the minds of authors and audiences, the readings referred to in this article show that the above texts gain in depth when bearing this in mind, although with an interesting result: Whether the bear is the "royal vassal of the king of kings" (Vita Galli), a "rejected alternative to the king" (Reynke de Vos), the "representative of a royal competitor to be eliminated" (Nibelungenlied), or the origin of a myth about a rural household spirit as a "royal gift with consequences" (Schrätel und Wasserbär) – in all cases showing traces of his "royal dignity" – Pastoureau's thesis of the bear as a king-no-more nevertheless finds confirmation.

Introduction

Certainly, there are more prominent animals in medieval German literature than the bear – just think of the omnipresent horses, of the many dragons to be defeated, of Iwein's lion, or of the fox protagonist in animal epics. Not even the Old and Middle High German (MHG) versions of the *Physiologus* contain a chapter on the bear.[1] One also searches for him in vain in the large fable collections by Stricker, Ulrich Boner, or Heinrich Steinhöwel.[2] Coats of arms with bears are found much

---

1   Cf. *Der altdeutsche Physiologus* (without bear!). HENKEL (1976, 42 note 113) mentions the bear among those animals that do not belong to the inventory of the *Physiologus* ("unter den Tieren […], die nicht zum Physiologus-Bestand gehören"), only to be found later on in the *Novus Physiologus*. On the bear in MHG texts of spiritual allegoresis of nature, cf. SCHMIDTKE 1968, vol. 1, 247–249.
2   The catalogue of DICKE/GRUBMÜLLER 1987, 50–52, lists only Cyrillus' fables and fables of the early modern period

less frequently than coats of arms with lions and eagles,[3] and bear coats of arms in literature are few and far between.[4] Dreams of bears in medieval German literature are quite modest in number,[5] and Konrad von Megenberg, in his *Buch der Natur*, limits himself to an uncommented replication of the bear chapter from the Latin original, an abbreviated version of the *Liber de natura rerum* by Thomas of Cantimpré (which specialists refer to as redaction III or Thomas III).[6]

These findings seem to fully confirm Michel Pastoureau's thesis of the bear as the "fallen king" (PASTOUREAU 2011 [French original edition 2007]): At first glance, it appears that the Christian Middle Ages successfully eliminated the former "king of the forest" (PASTOUREAU 2011, 2) not only from the German forests, but also from German literature. However, there are striking bear appearances in medieval German literature, which superficially speaking seem to confirm Pastoureau's thesis, but at the same time acquire a deeper meaning against the background of the "old royal dignity" of the bear.

Vita Galli: The bear as a royal vassal of God

Late antique and early medieval hagiographies constitute important ground on which Pastoureau builds his thesis. In these legends, wild bears obey the saint's will, even though this is very much against their nature. These bears are, as it were, "domesticated" (PASTOUREAU 2011, 89–90, 95–105). In principle, the *Vita Galli* (VG) in the adaptation by Walahfrid Strabo (833) and its German translation (VGdt, extant in the St Gallen Codex 602, fol. 14r–15r, originating in 1460) also follow this pattern: Gallus tells the bear to fetch wood for a fire, a task which the bear willingly fulfills (Fig. 1). He is rewarded with a loaf of bread, but is then expelled from the valley forever. The deacon Hiltibod pretends to be asleep and observes everything, recognising the holiness of Gallus in this encounter with the bear.

In addition to *ursus* ("bear"), the Latin text uses the entire array for naming wild animals when designating the bear. Thus the narrator lets the *homo Dei* ("man of God") speak *ad feram* (VG 11–12; "to the wild animal"), which the MHG text renders with *zuo dem wilden tier* (VGdt 8–9). The saint addresses the animal as a *bestia* (VG 12; "animal", "wild animal", "beast of prey", "beast"), as do both the narrator (VG 21) and Gallus's companion at later points in the text (VG 23, here "classifying" it among the *bestiae heremi*: "animals of the wilderness"), which the MHG text always renders as "animal" (VGdt 9, 21, 23–24: *tier der wüsty*). When the bear turns around to obey the saint, the narrator finally uses the name *belua* (VG 13; "wild beast/monster"), which the German text translates again as *daz wilde tier* (VGdt 11; "the wild animal"). Only in the Latin text does the bear appear anthropomorphised as *famulans* (VG 16; "servant"), when he is given bread and accepts it. However, in the German-language text, Gallus gives the bread *dem wilden tier* (VGdt 14; "to the wild animal"), and it is *der bär* (VGdt 15; "the bear") who takes it. Here, then, the bear remains a bear, a wild animal.

It is, however, striking that this bear is a rather "unbearlike" bear from the start: It picks up *gewarsamklich* (VGdt 6; "prudently, carefully"; see VG 11: *caute*) the *brösemly und die stükly* (VGdt 6–7; "crumbs and remains"; see VG 10: *micas et fragmenta*) given by the men (quite in contrast to the bears of other bear legends, who devour the saint's draught or pack animal), and willingly obeys the saint, although the saint actually demands the bear's service for no reason other than *in dem namen des her-*

(K41–K45); even the index entries for the keyword *bear* do not lead to famous fables.
3   SCHEIBELREITER 2009, 39; in a prominent position, the bear of Berne: cf. SCHMID 2000; cf. also FRANZ 2011, 53–58.
4   HARTMANN 2009, 173, mentions a bear paw coat of arms in Heinrich von dem Türlin's *Crône*.
5   Consider Daniel's dream of the bear as the representative of the Persian Empire in the *Annolied*, str. XIII, and in the *Kaiserchronik*, verses 526–594 (cf. FIEBIG 1995), as well as the bears from Charles's dreams in the *Rolandslied* by the priest (*Pfaffe*) Konrad, 3076–3082 and 7083–7127 (cf. VAN WELL 2016, chap. VI.3). Another dream about a bear occurs in *Kaiserchronik*, 14202–14211 (I'd like to thank Dr. Corinna Biesterfeldt for this reference).
6   Konrad von Megenberg, *Buch der Natur*, Kap. III.A.68; see also Thomas III.

ren (VGdt 9–10; "in the name of the Lord"; VG 12: *in nomine Domini*). In the Gallus legend, the order to collect wood for the fire (!) is not meant as a punishment or as compensation (in other legends, the bear has to replace the devoured draught or pack animal), but rather means a step in the direction of civilising the wilderness, which is also mirrored in the subsequent handing over of the bread: The service of the bear is rewarded, saint and bear have thus entered into a quasi-vasallic exchange relationship. One could even go as far as seeing in the bear receiving a *panem integrum* (VG 15–16; "an entire bread"; VGdt 14: *ain ganz brod*) an image of the Last Supper and thus the Christianisation of the bear (e.g. SCHÄR 2011, 456; in contrast to NEF 1982, 105, for whom the sharing (!) of bread would have been an impermissible and dangerous closeness to communion ("das Teilen [!] eines Brotes [...] eine unzulässige und gefährliche Annäherung an die Kommunion").

With Gallus's second order, the bear – again in the name of the Lord – is now expelled from the valley. It is in the Steinach valley, where Gallus will establish a hermitage, which will later become the Abbey of St Gall. The bear's habitat is limited to the mountains, so that it can no longer harm people or cattle (VG 19; VGdt 18–19). This again brings up the image of the "evil" and "dangerous" bear (hence there can be no question of a "domestication" of the bear in the strict sense). Max Schär interprets the scene as a division of the habitat ("Teilung des Lebensraums"): The wild animal is not chased away or exterminated, but is given a living zone, which is, however, circumscribed and delimited by humans. So man appears as master and the animal as servant (SCHÄR 2011, 456: "Das wilde Tier wird nicht verjagt oder ausgerottet, sondern erhält einen Aufenthaltsbereich, der allerdings vom Menschen umschrieben und abgegrenzt wird. So erscheint der Mensch als Herr und das Tier als Knecht."). I do not share this conclusion and instead contend that the animal remains "master" of the mountains, even if this role is assigned to him by the *man gottes* (VGdt 21; "man of god"; VG 20–21: *vir Deo carus*, "the man who is dear to God"), the representative of the "greatest Lord". However, the peaceful coexistence of humans and animals in the same habitat – as Schär has remarked correctly – anticipates Isaiah's "animal peace" (Isa. 11,6–8): By the saint and the wild animal encountering in peace, you almost want to say: in eucharistic community, the kingdom of God appears (SCHÄR 2011, 457: "Indem der Heilige und das wilde Tier sich im Frieden, man möchte fast sagen: in eucharistischer Gemeinschaft begegnen, scheint das Reich Gottes auf"). This is why Hiltibod shall remain silent about the miracle until he sees the glory of God (VG 24–25; VGdt 25–26).

In narratological terms, the bear of the *Vita Galli* is therefore more than just the antagonist of the sacred, the evil that is overcome. It is more than a random wild animal that is made submissive by virtue of the saint's holiness (OBERMAIER 2007, 46; cf. ALEXANDER 2008, ch. 3). If the bear is still considered as the old "king of the forest" (and the educated Walahfrid can certainly be trusted to think so), then something really extraordinary is happening here. The "king of the forest" now lives in peaceful coexistence with the representatives of the "king of kings", even if he is restricted in his rule and is now an obsequious (but nonetheless royal) "vassal of God". However, whether this thought was still present in the minds of the 15th-century German audience remains questionable.

REYNKE DE VOS: THE BEAR AS A REJECTED ALTERNATIVE TO THE KING

The medieval animal epic made the wolf and not the bear into the fox's antagonist. The Carolingian (!) fable of the sick lion (*Leo aeger* [La], 8th century),[7] which may even be regarded as the nucleus of the fox epic, shows that this could have turned out very differently. In this fable, it is the bear who speaks

---

[7] The authorship is uncertain, and neither Paulus Diaconus nor Notker Balbulus can be conclusively identified as the author (see ZIOLKOWSKI 1993, 62).

first and demands the death penalty for the fox (offense: lese majesty). The fox evades judgment by pretending to have sought and found an effective cure for the lion – the bear skin. Moreover, the flayed bear is mocked by the fox: *Quis dedit, urse pater, capite hanc gestare tyaram / et manicas vestris quis dedit has manibus?* (La 65–66: "Who gave you, Father Bear, this tiara [= papal crown] to wear on your head, and who gave you these gloves on your hands?").[8] The bear is no longer king here, but is instead mockingly "crowned" by the fox with the stigmata of his flaying. From the *Roman de Renart* (around 1175) onwards, Brun the bear is only one (minor) figure among many, but as a representative of the clergy he is still part of the animal society at the court of the lion. And his name gives it away: Brun is a brown bear – the prototypical bear in the European Middle Ages (Pastoureau 2011, 166–167).

In the MHG *Reinhart Fuchs* (RF) of the Alsatian poet Heinrich (c. 1192), Brun the bear also belongs to the nobility and, as a *kapelan* (RF 1511), is a member of the royal court chapel and advisor to the king (RF 1528: *des kvniges rat*). He appears at the trial against the fox as *vorspreche* ("advocate" – today we would say: lawyer) of the wolf Ysengrin (RF 1366–69) and as a court messenger. It is also he who, based on his profound legal knowledge, advises Reinhart to take the "oath on the dog's tooth" to trap him (RF 1126). By doing so, the bear arrogates the role of the cunning fox in a certain way. Perhaps this scholarly aura is the reason why the fox addresses the bear as *edeler schribere* (RF 1525; "noble scribe"), but it may also contain a perfidious play on words: a *schri-bere*, a "scream-bear" is what the bear will be when, after the honey adventure instigated by the fox, he is almost beaten to death by the farmers (RF 1555). In the honey adventure, the bear finally presents himself again with all the negative properties the Christian Middle Ages ascribe to him: stupidity, greed, voracity.[9] It is no coincidence that the fox's warning *werbet mit sinnen* (RF 1547; "act with understanding") remains unheard, and, significantly, it is the head (!) that the bear pushes *in daz bloch* (RF 1550; *bloch* is "the tree trunk" which becomes a "trap" for Brun, and finally a kind of "pillory"; cf. Widmaier 1993, 98–99). In the tradition of natural history, the bear's head is considered "weak", a belief which was initially only related to its physique, later to its mental abilities.[10]

For Pastoureau, the honey adventure – along with other episodes of the *Roman de Renart* tradition – marks the sad climax and end point of the "dethronement" of the bear by the Christian Middle Ages: The bear appears as "a stupid animal, ridiculous, humiliated" (Pastoureau 2011, 167).

In addition to the honey adventure, there is however another episode in the Middle Low German *Reynke de Vos* (RdV; 1498) in which the bear plays a central role: Reynke's conspiracy lie. Convicted and already standing under the gallows, Reynke accuses his accusers Brun the Bear, Ysegrim the Wolf, and Hyntze the Cat (and, to make it more credible, also his friend Grymbart and his own father) of plotting to install the bear as king, a coup which Reynke, however, claims to have successfully thwarted. The fox is thus "lying the bear" to where his place used to be according to Michel Pastoureau: on the throne (Obermaier 2016, 138: "Der Fuchs lügt damit den Bären dorthin, wo er nach Michel Pastoureau ursprünglich einmal war: auf den Thron."). But even in Reynke's web of lies, the scheme does not originate from the bear himself, it is rather represented as an idea of the fox's father. Reynke, however, claims that this corresponded to the bear's long-cherished wishes (RdV 2154: *Wente he des lange hadde begherd* "because this was his wish for a long time"). The text, though,

---

8 Reading it as a "political allegory" is possible, but, according to Ziolkowski 1993, 65–66, not imperative.
9 It is certainly no coincidence that in a 1463 print of Vincent of Beauvais's *Speculum historiale*, on Maître François's illustration of the seven deadly sins, it is the bear that rides on *gula*, hence gluttony (!), which is here impersonated by a clergyman.
10 Cf. Plinius, *Naturalis Historia*, lib. 8, 54, 130: *invalidissimum urso caput* ("the weakest part of the bear is the head"). Albertus Magnus, *De animalibus*, lib. 22, §107, ultimately places the weakness in the brain. An interpretation of the scene that focuses on the physicality of the bear was recently offered by Darilek 2020, chap. 3.2.

knows nothing of the bear's possible ambitions for the throne; nor is it about the reinstatement of a former king. Nevertheless, Reynke's lie makes use of the idea of the bear as the "former (!) king of the animals" – should this idea still be present at all around 1498 – insofar as Reynke represents the bear as a traditional (!) king model, namely the model of Caesarean power (REEMTSMA 2013, 165–166: "Modell caesareischer Macht"; cf. also OBERMAIER 2016, 141). Furthermore, Reynke does not leave out any of the usual negative stereotypes in order to denigrate the bear. In Reynke's narrative, the bear appears as an animal of the wilderness, not of courtly civilisation (RdV 2147); he is *schalck vnde quaet* (RdV 2219; "bad and nasty") and *vul van groter ouerdaet* (RdV 2220; "very presumptuous"). Reynke calls the bear *eynen bur* and *eynen vneddelen vrad* (RdV 2229; "a peasant" and "a common glutton") – a picture confirmed by the honey adventure (RdV, chap. 6–9) and the glosses (see OBERMAIER 2016, 140 note 12). This conspiracy was undertaken *[n]ot myt god, men des duuels macht* (RdV 2166; "not driven by God, but by the devil"). The bear is thus "demonised" by the fox in the same way, according to Pastoureau, as by the medieval clergy (PASTOUREAU 2011, 113–134). Yet it is the fox who is playing a devilish game here, which the fox perhaps even suggests himself by adding *Vnde myt mynes vaders ghewelde* (RdV 2167; "with the help of my father's power/possession"). What stands out though, is that the bear's supposed pretensions to the throne are of no interest to King Noble. He is only interested in the treasure that the fox cleverly and casually mentions (see OBERMAIER 2016).

But why is it specifically the bear that Reynke "lies onto the throne", and not the Fox father, Grymbart the Badger, Hyntze the Cat, or even his enemy, Ysegrim the Wolf? The prologue of the story recounts that within the hierarchy of Reynardian animal society, the bear belongs – like the other "*grype[n]*" ("predators", together with the wolf, the lynx, and the leopard) – to the high nobility or to the *vorsten vnd heren der werlt, de syk eddel holden* (RdV, prologue; "to the princes and secular lords who consider [!] themselves noble"). As a representative of the clergy, the badger is ruled out from a legal standpoint, as are the foxes, who as *banreheren* ("bannerets") belong to the lower nobility, and the cat who is probably meant by *vnde der ghelyken* ("and the like"). The only one left is the wolf, to whom the bear is physically superior, just as the bear – within this group of conspirators (which Reynke carefully put together) – is the largest and strongest native predator. Size, strength and nobility (OBERMAIER 2016, 140: "Größe, Stärke und Adel") predestine the bear to be a royal candidate. In other words: Who else could have been "lied onto the throne" as counter-king to the lion? Certainly, it might be difficult to prove that at the end of the 15th century one still had an inkling of the "former kingship" of the bear and knew about its displacement by the lion. On the other hand, it is against this specific background that the "lion's fall" and "bear's enthronement" (even if they only exist as manipulative suggestions of the fox) gain their humour. The fact that both lion and bear expose themselves as incompetent rulers in the fox's imagination is the icing on the cake (cf. WALTENBERGER 2013, 219).

One should not forget that this bear (just as the lion, the fox, and all other animals) is an animal of the fable/animal epic which, by definition, stands for a "type of human". But if one considers literary animals according to Roland Borgards as material-semiotic hybrids (BORGARDS 2016, 236: "materiell-semiotische Mischwesen"), then it should not be a coincidence that it is precisely the role of the strong but only moderately intelligent, yet all the greedier member of the high nobility attributed to the bear. In doing so, the text makes skilful use of medieval bear roles. As a figure of the animal epic diegesis, the bear in *Reynke de Vos* appears to be at the climax of the process that Pastoureau describes as the "fall of the bear", because here, this animal is nothing but ridiculous (admittedly, the other animals including the lion are no less ridiculous). As a (chess) figure in the metadiegesis of the fox, i.e. within Reynke's web of lies, the bear is again a potential, even if ultimately rejected (!), "aspirant to the throne", a "counter-king" born from the manipulative imagination/fantasy of the fox.

2002, 829: Why Siegfried hunts with the sword remains unclear ["Warum Siegfried mit dem Schwert jagt, bleibt unklar"]), but the use of the sword is perhaps not that strange at all (see OEHRL 2013, 302–303 and note 12; cf. the term *[bēr-/swīn-]swert* for a weapon for wild boar hunting in DALBY 1965, 19, 236). In this respect, caution is required towards interpretations which read the use of the sword for hunting as exuberant disdain for the symbols of chivalry (REICHERT 2005, 444, cf. 430: "übermütige Geringschätzung der Symbole des Rittertums"). But even my first, attractive thesis that it is the "courtly" Siegfried (here appropriately referred to as *Kriemhilde man*, NL 959.2; "Kriemhild's husband") who kills the "mighty hero" Siegfried cannot be maintained, even if it is precisely Siegfried's courtliness that supports his murder at the spring (see NL 975.3–4).[21] The sword with which Siegfried kills the bear is likely to be the double-edged sword Balmung – that is the weapon of the "splendid hero" from the Nibelungenland – which was mentioned just prior in Siegfried's description (NL 952.1). So this sword is precisely not "a symbol of the courtly world" (AUTERI 1998, 59: "simbolo del mondo cortese"). The commentary on Siegfried's murder in the bear episode is therefore to be read differently: Siegfried kills himself in the bear (WILSON 1981, 252, speaks of "self-destruction"), thus acting simultaneously as the killing subject (!) and as the killed object (!). Both aspects define Siegfried as being Siegfried (just as this inserted statuesque portrait again illustrates). In the death of the bear (= Siegfried) by Siegfried's own hand, the perfidious paradox of the "murder hunt" is represented by an impressive picture: Siegfried kills himself by taking part (as the hunter) in a hunt which actually applies to himself (as the hunted; cf. TALLY 1983, 128).[22]

Hagen's calculation works: A competitive hunt (instead of a war-campaign) for *bern unde swîn* (NL 908.2: "bears and wild boars") or *swîn / bern und wisende* (NL 913,3–4: "wild boars, bears and wisents") is entirely to Siegfried's taste, because in the wild boar hunt (WILSON 1981, 250), which is particularly dangerous due to the direct confrontation between humans and animals (WILSON 1981, 250), he can – and indeed does – show that he is ein *kreftec man* (NL 960.1; "a strong man"). Siegfried is therefore a "best hunter" by the grace of Hagen (FASBENDER 2007, 20: „Siegfried ist also ein ‚bester Jäger' von Hagens Gnaden"), because "the best hunters" will be those who bring down the best hunter (ibid.: „die „besten jegere werden die sein, die den besten Jäger zur Strecke bringen").

Let us look again at Siegfried's hunt: The bear is Siegfried's very last (and initially unintended) prey. Siegfried's first prey, which he beats to death with his bare hands, has not yet been clearly identified, and according to the conjecture by Bartsch/De Boor (NL BARTSCH/DE BOOR; following Wilhelm Grimm) it may be a vil *starkes halpswuol* (NL 932.3; "a half-grown but very strong wild boar").[23] Then he shoots with the bow an *ungefüegen lewen* (NL 932.4; "a huge lion") found by his (scent) hound (*bracke*), which only runs three more leaps (!) after the shot (NL 933.3; *drîer sprünge* [!] *lanc*). A colourful list of prehistoric, foreign, fantasy and real big game animals follows (cf. PANZER 1955, 196; DALBY 1965, xvii note 42), and this marks the climax of his being re-mythicised (FASBENDER 2007, 20: "den Höhepunkt seiner [= Siegfried's] Re-Mythisierung"). Finally, a large boar (NL 935.1) is particularly emphasised, which he, *der Kriemhilde man* (!),[24] – like the bear later on – kills *mit dem swerte* (!) (NL 936.1; "with the sword"). That the wild boar stands in such a striking position and in

---

21 See MÜLLER 2002, 73: Siegfried will always be both: magnificent hero and courtly knight, even his murder by the Wormsers will be inferred from this contradiction ("Siegfried wird immer beides sein: Heros und höfischer Ritter, und aus diesem Widerspruch wird schließlich sogar seine Ermordung durch die Wormser abgeleitet werden.").
22 SINGER 1967, 173: "Siegfried regards himself a hunter and is in reality the hunted." Or, as the text states: *von helden kunde nimmêr wirs gejaget sîn. / ein tier daz si sluogen, daz weinten edeliu kint* (NL 999.2–3; "Heroes could never have hunted in a worse manner. Noble girls mourned the game that the heroes had slain.")
23 According to MATTHIAS (1883, 493) the wording transmitted by the textual tradition does not allow for an unambiguous identification; recently, REICHERT (2019, 148, concerning 932.2–3) drew a similar conclusion.
24 Concerning the well-reflected use of this antonomasia by the poet of the *Nibelungenlied*, see SCHWEITZER 1972, 360–361.

such a clear parallel to the bear is probably no coincidence, for it corresponds too well to Kriemhild's boar dream, in which Siegfried is hunted down and killed by two wild boars (NL 918.2–3), which are commonly associated with Gunther and Hagen (VAN WELL 2016, 133–134),[25] although it is Brünhild who has advised to murder (*gerâten*, NL 1007.4), and Kriemhild who has betrayed Siegfried's vulnerable spot to Hagen (NL 900.4). Furthermore, the lion might be more than only cock-and-bull-stories or "Jägerlatein" (BARTSCH/DE BOOR 1988, 156, concerning NL 935.2; so REICHERT 2019, 148 concerning the passage altogether): Around 1200, the lion is the royal symbol *par excellence* (JÄCKEL 2006), so it is all the more remarkable that the lion does not rank first here. For Gale L. Wilson, the hunt gives an "image of Siegfried destroying his own enemies" on the one hand (WILSON 1981, 252), and on the other hand all these animals reflect different facets of Siegfried (WILSON 1981, 254–255, 260). Both interpretations seem problematic to me and, indeed, completely impossible altogether. Even if the lion's "three leaps" may vaguely call to mind the courtship deception (and the "degraded" lion may remind us of Gunther's leadership weaknesses), such clear correspondences between the hunted animals and the human characters of the *Nibelungenlied* hardly seem possible to prove in the same meaningful manner as can be done with the bear. Nevertheless, one can say that the hunt clearly demonstrates one last time the manifold threat Siegfried represents to the Burgundian kingdom. It also makes clear that in an open confrontation with the Burgundians, Siegfried would have been infinitely superior, without much of an effort.

It is hardly a coincidence that it is a bear that stands for Siegfried; the bear represents the "ancient" king of the animals and thus an "old style" royalty, which with regard to Siegfried means a kingship that – very differently from how one thinks and acts in "courtly" Worms – is still entirely based on strength.[26] Without wanting to overstretch the interpretative opposition between the lion and the bear, the death of the bear makes it clear that in Siegfried – remember Siegfried's provocation at his first appearance in Worms (NL 107.1–108.4) – a powerful competitor to the royal dignity in Worms is eliminated.

A sideways glance – *Schrätel und Wasserbär*: A polar bear as a king's gift with consequences

We encounter a very different kind of bear in the anonymous[27] tale *Schrätel und Wasserbär* (SuW; end of the 13th century, also known as *Kobold und Eisbär*, cf. WILLIAMS 1983). In this tale, the king of Norway sends a *Norman* (SuW 22; "Norseman") to bring *einen zamen wazzerbern* (SuW 15; "a tame polar bear") as a gift to the king of Denmark.[28] The MHG word *wazzerber* is only used in this text; the white colour of the bear (SuW 17), its origin from the north, and, last but not least, its value as a royal gift undoubtedly suggest a polar bear. Upon their arrival in Denmark, the bear-leader and the bear take up quarters with a farmer who has lost the rule over his house to a *schretel* (a goblin). That does not scare the Norseman, and they eat and then go to sleep in the bakehouse. But when the goblin begins to make trouble at night, the Norseman hides in the oven. Only the polar bear can successfully defend itself against the goblin, but only after the goblin has tormented it three times with his

---

25 SCHRÖDER (1960, 121) sees – rather problematically – the boar-Hagen association justified in legendary history.
26 This would make clear why Brünhild was also considered (cf. note 28): The bear, Siegfried, and Brünhild are part of an archaic "Germanic" world. VOORWINDEN (1995, 167), however, considers the bear's function as an announcer of danger or as an announcer of approaching death ("als Ankündiger von Gefahr bzw. als Ankündiger des nahenden Todes"), which the animal fulfills in literary dreams, as a possible reason for the choice of the bear.
27 Heinrich von Freiberg's authorship was disproved by FISCHER 1983, 165.
28 The beginning of the story recalls the journey of the Icelander Audun. There is controversy on whether this story belongs to the core of the narrative (cf. GRUBMÜLLER 2011, 1264; with a different approach taken by RÖHRICH 1977, 1221).

spear. The next morning the bear-leader and the bear bid farewell, but while working in the fields, the farmer meets the dishevelled goblin who asks him whether this terrible "big cat" is still with him. The farmer has the presence of mind to reply: *jâ, jâ, mîne grôze katze, / dir ze trutze und ze tratze / lebet sie, du boesez wihtel, noch. / [...] / vünf jungen si mir hînt gewan; [...] der alten katzen alle gelîch* (SuW 325–332; "Yes, my big cat is still alive, to anger and to defy you, you evil spirit; [...] she gave birth to five cubs tonight [...], all equal to the old cat"), whereupon the goblin disappears forever.[29]

The polar bear is well chosen for this role. Not only do historical facts confirm that the polar bear replaces the brown bear in his function as a gift for royal menageries (PASTOUREAU 2011, 58–59; cf. also SCHIER 1935, 172–173; PARAVICINI 2003, 591, with examples: 578–579), but its white fur also makes him an animal of great value (PARAVICINI 2003, 588) and – according to Christian colour symbolism – opposes it, as a "light" and "pure" (even more: "tame") creature, to the "dark" powers of the devilish (and "wild") goblin, which itself wears a red hooded coat (SuW 190). However, the landlord is not completely without fear when faced with the polar bear, as expressed in the question: *waz tieres vüeret ir an der hant? / ist diu selbe crêatiure / gehiure oder ungehiure? / daz eislîche kunder, / ist es ein merwunder?* (SuW 68–72; "What kind of animal are you leading by the hand? Is this creature good or bad? This terrible animal, is it a sea monster?"). Therefore, the polar bear has the greatest possible "goblin closeness" and, at the same time, the greatest possible "goblin distance". It is thus, at the same time, an adequate opponent as well as a positive counterpart of the evil goblin.

Martin Todtenhaupt is rightly surprised that the tale's prologue speaks of *hovelicher mære* (SuW 1: "a courtly story" ["einer höfischen Geschichte"]) with regard to the story to be told, although the court "just [sets] the outer framework of the plot" (TODTENHAUPT 1977, 302: "nur den äußeren Rahmen der Handlung") and two of the acting figures are essentially "non-courtly figures" (ibid., 304: "zwei von ihrem Stand her nicht-höfische Figuren"). Due to its quality as a "courtly gift" and its white colour, Todtenhaupt sees the polar bear as an excellent representative of the courtly sphere (ibid.: "zum hervorragenden Repräsentanten des Höfischen"), but one that is reduced to purely physical violence (ibid., 305: "rein physische Gewalt"), and which is also sleepy and unwilling (ibid., 304: "schläfrig und unwillig"). As a result, Todtenhaupt wants to see in the tale a program of the late courtly renaissance *en miniature*, which calls for a positive role and identity of the court, definitely in the sense of the high courtly ideal (ibid., 308–309: "eine Programmschrift der späthöfischen Renaissance *en miniature*, die [...] eine positive Rolle und Identität des Hofes, durchaus im Sinne des hochhöfischen Ideals, anmahnt"). It is also conceivable, however, that the adjective *hovelîch* does not refer to the content and staff of the tale but only to its stylistic aspects, because this funny story (WILLIAMS 1983, 1279: "zum Lachen" [for laughing], cf. FISCHER's [1983, 104] classification under "schwankhafte Mären" [humorous tales]) is indeed elegantly designed with all the stylistic devices of courtly storytelling (DE BOOR 1997, 237: „mit allen Stilmitteln höfischer Erzählkunst elegant gestaltet").

From a narratological point of view, here the bear's status as figure changes in a very striking way. At the beginning of the tale, the polar bear, as a royal gift, is only an object, not a figure (and – in agreement with the prologue – one is tempted to consider the two kings as the protagonists of the coming narrative). The bear only becomes a (co-)figure in the travel part, whereby the relationship between the bear-leader and the bear changes the figure status of both: *dem bern dâ gegeben wart / gein Tenemarken ûf die vart / ein wegewîser villân* (SuW 19–21; "The bear was given a farmer [i.e. the Norseman] as a knowledgeable guide on the way to Denmark"). On the one hand, the bear-leader is given to the bear as a companion, on the other hand, grammatically the bear-leader is the subject. The relationship is described "the other way round" at the end of the section: *im folgte an siner hant der*

---

29 On controversially discussed parallels with the 13th-century Middle Dutch poem *Van bere Wisselau* and with the Thidreksaga (in which a man disguised as a bear defeats the giants), see TAYLOR 1919, 309–311; RÖHRICH 1977, 1221–1222.

ber (SuW 42; "the bear was following him, being led by the hand"). Here the bear is indeed the grammatical subject, but he appears as a companion to the human figure. The *wirt* (i.e. the "landlord"), at whose property they stop for refreshments, also perceives the bear as a companion to the bear-leader; and in this function the bear also becomes the object of speech (!) for the two human figures. However, at the request of the bear-leader (v. 144–145) the bear is also "catered for". Thus, in this part, the bear is one (animal) figure among other (human) figures. Only in the nocturnal fight with the goblin does the bear become an actor, indeed the main character, the protagonist of the tale.[30] In contrast, the goblin is and constantly remains in the role of the antagonist (of the host and his visitors and thus of the bear). The real highlight of this little story, however, is that in the conversation between the landlord and the goblin on the morning following the night of the fight, the bear (as the "big cat" that has had five young ones) becomes the character of a metadiegetic story. This turns the peasant landlord into the narrator/inventor of a (fictional) story that finally drives the goblin away (whereby the point is about more than just a scare, cf. LINHART 1995, 428–429).

The physical presence and the corporeal superiority of the polar bear can only solve the landlord's goblin problem for a short time. But the physical presence of the bear and, even more, the goblin's memory of the physical violence it has experienced are essential prerequisites for the peasant's narrative of the polar bear, as the "big cat" with five kittens, to be invented at all and ultimately have a lasting effect. The bear teaches the goblin fear, and the bear's success as well as the goblin's fear allow the farmer to recognise his own fear of the goblin as surmountable. He can thus regain power over events and above all over his own homestead. The bear is therefore not only a "gift among kings", its temporary presence is also a "gift to the farmer" – a gift that the farmer knows how to use for himself, with cunning compensation for his lack of strength (cf. TODTENHAUPT 1977, 304).

For the goblin, the polar bear is not a ghost (as the goblin is for the peasant), but rather it "rationalises" the animal that is unknown to him, into a "cat" with gigantic dimensions. This "translation" into the peasant's imaginary world appropriates the narrative of the farmer, as does the enlargement process by inventing a whole "giant cat clan" in addition to the "giant cat". One is tempted to see the lion in the "big cat", who, according to Pastoureau, replaces the bear, albeit the brown bear (!), as king of the animals (PASTOUREAU 2011, 3; cf. also 135–155). The lion was only classified as a (predatory) cat after the Middle Ages; such classification is alien to medieval natural history, just as the polar bear is also not the former "king of the forest". Nevertheless, it should be more than a witty contrast that here the polar bear appears as a "big cat" because in the Middle Ages the cat is an important farm animal as a mouse catcher (WALKER-MEIKLE 2012, 10), and so in its function of freeing the house from annoying pests, the polar bear is actually a "cat".

The peasant's cunning turns the polar bear from a "king's gift" into the "founding father" of a new (fictitious) family genealogy for the long-term security of his own rural power over house and farm. Understandably, the creature cannot take over this function as a "royal" polar bear, but only if modified into an animal fitting his new role as household spirit and his new rural environment, precisely as an imaginary fabulous cat. Thereby, the tale is not only a nice story with an amusing punchline (DE BOOR 1997, 236: "eine hübsche Geschichte mit einer amüsanten Pointe"), but also a lesson on the power of myths.

---

30 JANNIDIS (2004, 103) mentions the following as possible criteria for a main character: The main character is involved more actively in the continuation of the plot than the secondary character; a main character is a character who takes up more space in the presentation than a secondary character; a major character is [...] any character who participates in significantly more events than the other characters ("Die Hauptfigur ist aktiver an der Fortführung der Handlung beteiligt als die Nebenfigur." / "Hauptfigur ist eine Figur, die einen größeren Raum in der Darstellung einnimmt als eine Nebenfigur." / "Eine Hauptfigur ist [...] jede Figur, die an bedeutend mehr Ereignissen partizipiert als die anderen Figuren.").

Conclusion

This short overview on bears in medieval German literature makes it clear that reflections of the former royal dignity of the (brown) bear cannot be proven beyond doubt, but interpretations gain in depth upon inclusion of this dimension. In this regard, the bear of *Vita Galli* does not just appear as the wild animal obeying the saint. When the "king of the forest" as "vassal of God" willingly serves the representative of the "king of kings" and accepts the restrictions placed on his habitat, the bear is indeed only a "king with limited royal power", but he is still a king. In Reynke's conspiracy lie in *Reynke de Vos*, however, the bear becomes the "rejected alternative to the king". He is "conceivable as a king", but basically no longer worthy of royal dignity – and not just in the eyes of Reynke, who makes the bear the representative of an outdated rule model. As a "sign of a past form of royalty" based entirely on strength, the bear in the *Nibelungenlied* is the ideal embodiment of Siegfried. In the image of the bear it appears clearly that with Siegfried a powerful competitor for the royal dignity is eliminated. The fact that Siegfried kills himself in the bear is less a comment on the question of his guilt than an impressive picture of this paradoxical "murder hunt", which turns the hunter into the hunted. The tale *Schrätel und Wasserbär* should be seen slightly differently. Through his very nature, the polar bear cannot contain reflections of the old "king of the forest". His only affinity to royalty is that he is handled as a "gift among kings". His successful nocturnal fight against the goblin also turns the bear into a gift for the peasant landlord. The latter reinvents the bear – in the shape of the "big cat" who gave birth to five young ones – as the "prime ancestor" of a new (fictitious) family genealogy of household spirits and thus regains permanent power over his farm and his lands.

As a "royal vassal of the king of kings", as a "rejected alternative to a king", or as a "representative of a king's competitor to be eliminated", the bear is no longer a king, even where his old royal dignity can still be meaningfully included. As a "king's gift with consequences", he even becomes the origin of a rural household spirit myth. Even where traces of the "old royal dignity" of the bear can still be sensed, it seems confirmed that the bear is indeed a "fallen" king in the sense of Pastoureau.

(Translation: Larissa Birrer)

Bibliography

Primary sources

Albertus Magnus: *De animalibus*. Ed. H. Stadler. 2 vols. (Münster 1916–1920).

Altdeutscher Physiologus: Altdeutscher Physiologus. Ed. F. Maurer (Tübingen 1967).

Annolied: Annolied. Ed. M. Rödiger. Monumenta Germaniae Historica, Deutsche Chroniken I,2 (Hannover 1895). https://www.dmgh.de/mgh_dt_chron_1_2/index.htm#page/(63).

Buch der Natur: see Konrad von Megenberg.

Kaiserchronik: Kaiserchronik. Ed. E. Schröder. Monumenta Germaniae Historica, Deutsche Chroniken 1,1 (Hannover 1895). https://www.dmgh.de/mgh_dt_chron_1_1/index.htm#page/(1).

Konrad von Megenberg: *Das Buch der Natur*. Eds. R. Luff/ G. Steer (Tübingen 2003).

(Pfaffe) Konrad: *Rolandslied*. Ed. C. Wesle. 3. Aufl. bes. von P. Wapnewski (Tübingen 1985).

La: *Leo aeger, vulpis et vrvs*. In: Lateinische Fabeln des Mittelalters. Ed. H. C. Schnur (Berlin 1979) 26–31.

Liber de natura rerum: see Thomas von Cantimpratensis/ Thomas III.

NL: *Nibelungenlied*. Nach der St. Galler Hs. Ed. H. Reichert (Berlin, New York 2005).

NL (Bartsch/De Boor): Das Nibelungenlied. Ed. K. Bartsch/H. De Boor (Mannheim 1988).

NL (Brackert): Das Nibelungenlied. Ed./transl. H. Brackert (Frankfurt a. Main 1970).

NL (Grosse): Das Nibelungenlied. Ed./transl. S. Grosse. Reclams Universalbibliothek 644 (Stuttgart 2002).

NL (Heinzle): Das Nibelungenlied und die Klage. Ed. J. Heinzle. Bibliothek des Mittelalters 12 (Berlin 2013).

Physiologus: see Altdeutscher Physiologus

Plinius: *Naturalis Historia*, lib. 8. Ed./transl. R. König/G. Winkler (Düsseldorf ²2007).

RdV: *Reinke* [sic!] *de Vos*. Eds. F. Prien/A. Leitzmann. Altdeutsche Textbibliothek 8 (Tübingen 1960).

RF: Der Elsässer Heinrich, *Reinhart Fuchs*. Ed. K. Düwel. Altdeutsche Textbibliothek 96 (Tübingen 1984).

Rolandslied: see (Pfaffe) Konrad.

Roman de Renart: *Le Roman de Renart*. Transl./intr. H. Jauss-Meyer. Klassische Texte des Romanischen Mittelalters in zweisprachigen Ausgaben 5 (München 1965).

SuW: *Schrätel und Wasserbär*. In: Novellistik des Mittelalters. Ed. K. Grubmüller. Bibliothek des Mittelalters 23 (Berlin 2011) 698–717.

Thomas Cantimpratensis (= Thomas of Cantimpré), *Liber de natura rerum*. Ed. H. Boese (Berlin 1973).

Thomas III (= Thomas von Cantimpré), *Liber de naturis rerum*. Redaktion III (Thomas III), Ed. project group B2 of the SFB [Collaborative Research Centre/Special Research Area] 226 Würzburg-Eichstätt under the direction of B. K. Vollmann (typed manuscr., 1992).

VG: *Vita sancti Galli*, see Walahfrid Strabo.

VGdt: *Vita Galli deutsch*. In: E. Rüsch, Gallus und der Bär. Geschichte und Legende (St. Gallen 1950) 13–14.

Wahlafrid Strabo: Walahfrid Strabo, *Vita sancti Galli*. Das Leben des heiligen Gallus. Lateinisch/Deutsch. Transl. F. Schnoor (Stuttgart 2012 [ch. 11, here p. 50]).

Secondary sources

ALEXANDER 2008: D. ALEXANDER, Saints and Animals in the Middle Ages (Woodbridge 2008).

AUTERI 1998: L. AUTERI, *ein tier daz si sluogen*. Simbologia animale e destino di morte nel *Nibelungenlied*. AION. Annali dell'Università degli stui di Napoli „L'Orientale"/Sezione Germanica N. F. 8, 1998, 41–60.

BAARTSCH/DE BOOR 1988: K. BARTSCH/H. DE BOOR, Commentary on NL (BARTSCH/DE BOOR), see: Primary sources.

BINDSCHEDLER 1985: M. BINDSCHEDLER, Vom Bärenfang im Mittelalter. Zu einem strîtelîn zwischen Philologie und Volkskunde. In: A. Schnyder (ed.), Mittelalter und Moderne. Gesammelte Schriften zur Literatur (Bern, Stuttgart 1985) 203–206.

DE BOOR 1997: H. DE BOOR, Die deutsche Literatur im späten Mittelalter. Teil 1: 1250–1350. Geschichte der deutschen Literatur von den Anfängen bis zur Gegenwart 3 (München ⁵1997 [originally published 1962; 1997 version revised by J. Janota]).

BORGARDS 2016: R. BORGARDS, Tiere und Literatur. In: R. Borgards (ed.), Tiere. Kulturwissenschaftliches Handbuch (Stuttgart 2016) 225–244.

DALBY 1965: D. DALBY, Lexicon of Medieval German Hunt. A Lexicon of Middle High German Terms (1050–1500) associated with the chase, hunting with bows, falconry, trapping and fowling (Berlin 1965).

DARILEK 2020: M. DARILEK, Füchsische Desintegration. Studien zum *Reinhart Fuchs* im Vergleich zum *Roman de Renart* (Heidelberg 2020).

DICKE/GRUBMÜLLER 1987: G. DICKE/K. GRUBMÜLLER, Die Fabeln des Mittelalters und der frühen Neuzeit. Ein Katalog der deutschen Versionen und ihrer lateinischen Entsprechungen (Munich 1987). https://digi20.digitale-sammlungen.de//de/fs1/object/display/bsb00050026_00001.html.

FASBENDER 2007: C. FASBENDER, Siegfrieds Wald-Tod. Versuch über die Semantik von Räumen im Nibelungenlied. In: N. Staubach/V. Johanterwage (eds.), Außen und Innen. Räume und ihre Symbolik im Mittelalter (Frankfurt a. Main u.a. 2007) 13–24.

FIEBIG 1995: A. FIEBIG, *vier tier wilde*. Weltdeutung nach Daniel in der ,Kaiserchronik'. In: A. Fiebig/H.-J. Schiewer (eds.), Deutsche Literatur und Sprache von 1050–1200. Festschrift für Ursula Hennig zum 65. Geburtstag (Berlin 1995) 27–49.

FISCHER 1983: H. FISCHER, Studien zur deutschen Märendichtung (Tübingen ²1983).

FRANZ 2011: L. FRANZ, Wahre Wunder. Tiere als Funktions- und Bedeutungsträger in mittelalterlichen Gründungslegenden (Heidelberg 2011).

FRIEDRICH 2006: J. FRIEDRICH, Phraseologisches Wörterbuch des Mittelhochdeutschen. Redensarten, Sprichwörter und andere feste Wortverbindungen in Texten von 1050–1350 (Tübingen 2006).

GLOGAU 1993: D. R. GLOGAU, Untersuchungen zu einer konstruktivistischen Mediävistik. Tiere und Pflanzen im „Tristan" Gottfrieds von Straßburg und im „Nibelungenlied" (Essen 1993).

GRIEBEL 2020: J. GRIEBEL, „das thier friszt, der mensch iszt". Zur Diachronie der lexikalischen Mensch-Tier-Grenze im Deutschen (Heidelberg 2020).

GROSSE 2002: S. GROSSE, Commentary on NL [GROSSE], see: Primary sources, 719–1039.

GRUBMÜLLER 2011: K. GRUBMÜLLER, Commentary on SuW (see: Primary sources), 1261–1268.

HARTMANN 2009: H. HARTMANN, Tiere in der historischen und literarischen Heraldik des Mittelalters. Ein Aufriss. In: S. Obermaier (ed.), Tiere und Fabelwesen im Mittelalter (Berlin 2009) 147–179.

HEINZLE 2013: J. HEINZLE, Commentary on NL (HEINZLE), see: Primary sources, 987–1748.

HENKEL 1976: N. HENKEL, Studien zum Physiologus im Mittelalter (Tübingen 1976).

JÄCKEL 2006: D. JÄCKEL, Der Herrscher als Löwe. Ursprung und Gebrauch eines politischen Symbols im Früh- und Hochmittelalter (Cologne, Weimar, Vienna 2006).

JANNIDIS 2004: F. JANNIDIS, Figur und Person. Beitrag zu einer historischen Narratologie (Berlin, New York 2004).

KRAUSE 1996: B. KRAUSE, Die Jagd als Lebensform und höfisches ,spil'. Mit einer Interpretation des ,bast' in Gottfrieds von Straßburg Tristan (Stuttgart 1996).

Lindner 1940: K. Lindner, Die Jagd im frühen Mittelalter (Berlin 1940).

Linhart 1995: D. Linhart, Hausgeister in Franken. Zur Phänomenologie, Überlieferungsgeschichte und gelehrten Deutung bestimmter hilfreicher oder schädlicher Sagengestalten (Dettelbach 1995).

Matthias 1883: E. Matthias, Die Jagd im Nibelungenliede. Zeitschrift für deutsche Philologie 15, 1883, 471–501.

Mowatt/Sacker 1967: D. G. Mowatt/H. Sacker, The Nibelungenlied. An Interpretative Commentary (Toronto 1967).

Müller 1998: J.-D. Müller, Spielregeln für den Untergang. Die Welt des Nibelungenliedes (Tübingen 1998).

Müller 2002: J.-D. Müller, Das Nibelungenlied (Berlin 2002).

Nef 1982: R. Nef, Gallus und der Bär. In: Gallusstadt. Jahrbuch der Stadt St. Gallen 1, 1982, 101–109.

Obermaier 2007: S. Obermaier, Der Heilige und sein Tier, das Tier und sein Heiliger – Ein Problemaufriss. In: W. G. Rohr/T. Honegger (eds.), Tier und Religion, Das Mittelalter 12(2), 2007, 46–63.

Obermaier 2016: S. Obermaier, Der Bär auf dem Thron. Reflexionen des Politischen in Reynkes Verschwörungslüge. In: J. Glück/K. Lukuaschek/M. Waltenberger (eds.), Reflexionen des Politischen in der europäischen Tierepik (Berlin, New York 2016) 138–155.

Oehrl 2013: S. Oehrl, Bear hunting and its ideological context (as a background for the interpretation of bear claws and other remains of bears in Germanic graves of the 1st millenium AD). In: O. Grimm/U. Schmölke (eds.), Hunting in northern Europe until 1500 AD. Old traditions and regional developements, continental sources and continental influences. Schriften des Archäologischen Landesmuseums, Ergänzungsreihe 7 (Neumünster 2013) 297–332.

Panzer 1955: F. Panzer, Das Nibelungenlied. Entstehung und Gestalt (Stuttgart 1955).

Paravicini 2003: W. Paravicini, Tiere aus dem Norden. Deutsches Archiv für die Erforschung des Mittelalters 59(2), 2003, 559–591.

Pastoureau 2011: M. Pastoureau, The Bear: History of a Fallen King. Transl. G. Holoch (Cambridge [Mass.] 2011). French original edition: L'ours. Histoire d'un roi déchu (Paris 2007).

Reemtsma 2013: J. P. Reemtsma, Vertrauen und Gewalt. Versuch über eine besondere Konstellation der Moderne (Hamburg 2013).

Reichert 2005: H. Reichert, Interpretation. Commentary on NL (Reichert), see: Primary sources, 317–539.

Reichert 2019: H. Reichert, Nibelungenlied-Lehrwerk. Sprachlicher Kommentar, mittelhochdeutsche Grammatik, Wörterbuch (Vienna ²2019).

Röhrich 1977: L. Röhrich, Bärenführer. In: Enzyklopädie des Märchens 1, 1977, 1217–1225.

Schär 2011: M. Schär, Gallus. Der Heilige in seiner Zeit (Basel 2011).

Scheibelreiter 2009: G. Scheibelreiter, Tiersymbolik und Wappen im Mittelalter. Grundsätzliche Überlegungen. In: G. Scheibelreiter, Wappenbild und Verwandtschaftsgeflecht. Kultur- und mentalitätsgeschichtliche Forschungen zu Heraldik und Genealogie (Vienna, Munich 2009) 43–56.

Schier 1935: B. Schier, Die Sage vom Schrätel und Wasserbären. Mitteldeutsche Blätter für Volkskunde 10(6), 1935, 163–180.

Schmid 2000: R. Schmid, Nu dar, du edels müetzelin, dar! Bern und der Bär im 14. bis 16. Jahrhundert. In: P. Michel (ed.), Symbole im Dienste der Darstellung von Identität (Bern u.a. 2000) 159–179.

Schmidtke 1968: D. Schmidtke, Geistliche Tierinterpretation in der deutschsprachigen Literatur des Mittelalters (1100–1500) (Berlin 1968).

Schröder 1960: F. R. Schröder, Sigfrids Tod. Germanisch-Romanische Monatsschrift N. F. 10, 1960, 111–122.

Schweitzer 1972: E. C. Schweitzer, Tradition and Originality in the Narrative of Siegfried's Death in the Nibelungenlied. Euphorion 66, 1972, 355–364.

Singer 1967: C. S. Singer, The Hunting Contest: An Interpretation of the Sixteenth Aventiure of the Nibelungenlied. The Germanic Review 42, 1967, 163–183.

Tally 1983: J. A. Tally, The dragon's progress (Ann Arbor [Michigan] 1983).

Taylor 1919: A. Taylor, Schrätel und Wasserbär. Modern Philology 17(6), 1991, 57–76.

Thiébaux 1974: M. Thiébaux, The Stag of Love. The Chase in Medieval Literature (Ithaca, London 1974).

Todtenhaupt 1997: M. Todtenhaupt, Kobold und Eisbär. Ein höfisches Märe ohne den höfischen Adel. In: B. Andersson/G. M. Müller (eds.), Kleine Beiträge zur Germanistik. Festschrift für John Evert Härd (Uppsala 1997) 301–309.

Voorwinden 1995: N. Voorwinden, Das mittelniederländische Fragment ‚Van bere Wisselau' und sein Verhältnis zur deutschen Heldensage. Amsterdamer Beiträge zur älteren Germanistik 41, 1995, 161–174.

Walker-Meikle 2012: K. Walker-Meikle, Medieval Pets (Woodbridge 2012).

Waltenberger 2013: M. Waltenberger, Die Legitimität der Löwen. Zum politischen Diskurs der frühneuzeitlichen Tierfabel. In: A. Höfele/J.-D. Müller/W. Oesterreicher (eds.), Die Frühe Neuzeit. Revisionen einer Epoche (Berlin, Boston 2013) 203–228.

Van Well 2016: B. Van Well, Mir troumt hînaht ein troum. Untersuchung zur Erzählweise von Träumen in mittelhochdeutscher Epik (Göttingen 2016).

Widmaier 1993: S. Widmaier, Das Recht im „Reinhart Fuchs" (Berlin, New York 1993).

Williams 1983: U. Williams, ‚Kobold und Eisbär' (‚Schrätel und Wasserbär'). In: Die deutsche Literatur des Mittelalters. Ein Verfasserlexikon (VL²) 4, 1983, 1279–1280.

Wilson 1981: G. L. Wilson, Epic and Symbolic Functions of the Hunt in Five Medieval German Epics (Chapel Hill 1981).

Ziolkowski 1993: J. M. Ziolkowski, Talking Animals. Medieval Latin Beast Poetry, 750–1150 (Pennsylvania 1993).

Prof. Dr. Sabine Obermaier
Deutsches Institut
Johannes Gutenberg-Universität Mainz
Mainz
Germany
sabine.obermaier@uni-mainz.de

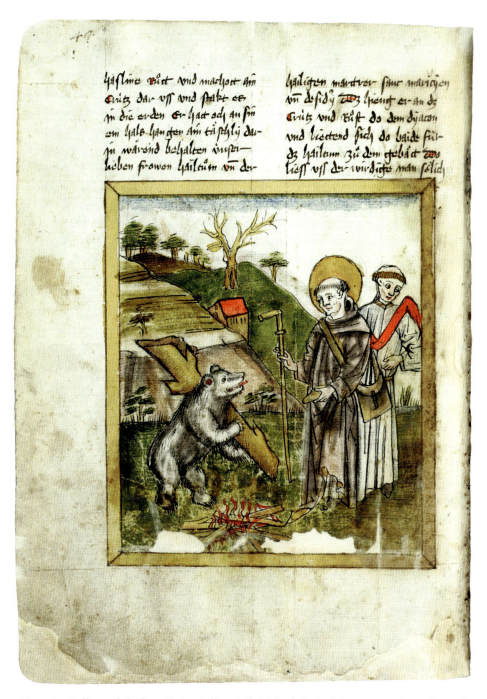

Fig. 1. St. Gallus and the bear (© St. Gallen, Stiftsbibliothek, Cod. Sang. 602 – Deutsche Heiligenleben, pag. 44; AD 1451–1460; NonCommercial 4.0 International [CC BY-NC 4.0]).

"The Bear's Son Tale": Traces of an ursine genealogy and bear ceremonialism in a pan-European oral tradition

By Roslyn M. Frank

Keywords: Bear ceremonialism, animism, relational ontology, Basque, Ursus arctos

Abstract: The most widely disseminated set of European folktales is the "Bear's Son Tale" (ATU 301), also known as "John the Bear", whose main character is the offspring of a bear and a human female. The possible implications of the ursine genealogy central to the structure of these stories is explored, starting with a discussion of the way the tales have been treated by folklorists up until now, including how they have been classified in ways that leave aside the ethnographic evidence for real world manifestations of the ursine genealogy. In contrast, I have attempted to identify the intangible remains of this animist ontology and how it is embedded in well-documented beliefs, traditions, rituals and performance art across much of Europe and most particularly in the Pyrenean region. Central to the endeavor has been the work I have carried out for many years on the Basque culture and language which allowed me to discover that the Basques used to believe humans descended from bears, a belief in consonance with the tenets of circumboreal bear ceremonialism. Two aspects of this widely disseminated set of European folktales will be highlighted. The first is how the folktales have acted to transmit the belief in an ursine ancestry across time, while the second is the way the animist relational ontology embedded in the tales can provide a means of accessing the extra-textual imprint of the belief system in the real world.

Introduction

For the past forty years I have explored the ramifications of an archaic belief that I encountered while doing fieldwork among the Basque people back in the early 1980s, namely, that Basques used to believe humans descended from bears. Although my informants had alluded to aspects of this belief indirectly, it was not until the late 1980s that a report documenting the belief was published (Peillen 1986). Up until that time it had been passed orally from one generation of Basque speakers to the next, who were always careful not to share the information with non-Basque speakers. Soon after I had discovered the existence of this ursine genealogy, other bits and pieces of ethnographic evidence began to fall into place, among them folktales that speak of a young woman who mates with a bear and gives birth to a half-human, half-bear offspring. Once the ursine origin of humans was plugged into the interpretive frame of these stories, the adventures of the hybrid being took on a new significance. As a result, I began to process other European ethnographic data through a different lens, one that was no longer purely anthropocentric in nature, but rather animist, informed more by what has been referred to in recent years as a relational ontology (Bird-David 1999; Harvey 2006; Harrison-Buck 2018).

Early on, he has an encounter which allows him to acquire his Animal Helpers. Walking along a path in the woods, he spies four animals ahead of him standing next to the carcass of a deer. They are a lion, a dog, an eagle, and an ant. Lion calls out to him: "We're hungry and have been arguing about how to divide up the meat. Can you help us?"[3] Little Bear responds saying that he will try. "Lion, I'll give you the haunch which is what you like best." And to Dog, he gives the ribs. Addressing Eagle, he says: "To you I'll give the innards and intestines because you don't have any teeth, and this is what you like best." Finally, to the tiny Ant, Little Bear says, "To you I'll give the skin and bones and when you've eaten the marrow from the bones you can use them for your house when it rains." With that, Lion responds: "You've done so well with the division that we want to reward you." And each of them gives him a talisman, telling him that when he requires their help all he needs to do is touch it and call out the animal's name. That way he will gain the animal's innate abilities and take on the shape of the animal in question. Lion gives him a tuft of fur, Dog another tuft, the Eagle a feather, and little Ant a leg because she has several.

Time passes, and Little Bear finds himself at a farmstead where he meets the young woman who lives there with her old father. Naturally, since all good stories need a romantic twist, Little Bear falls in love and wants to run off with the young woman. But she explains that she cannot leave because she must care for her old father who happens to be immortal. Little Bear insists that there must be a way to get the old man to die.

At this point the first example of shape-shifting takes place, an element typical of an animist mindset. The young woman tells him to come back the next day and enter the garden where she will be combing the old man's hair and removing his lice. Little Bear is to climb up into the tree located next to them and hide in its branches while she asks the old fellow what will make him die. So Little Bear shows up, shape-shifted into an ant, and climbs silently up into the tree from where he overhears the old man's response: "For me to die, the challenger will have to do battle with my brother who is a shape-shifter, too. He will appear as a porcupine and the challenger must show up as a lion and engage in battle with him. If he triumphs, a hare will appear, and the challenger must turn into a dog and catch it." The old man continues explaining: "Once the hare is caught, a pigeon will fly up and my opponent must turn into an eagle, snatch the pigeon, open it, remove the egg inside, and then take the egg and break it on my forehead. When that happens the egg inside my head will break and I will become mortal and die."

Since Little Bear has been pre-equipped with his talismans, he is well able to follow these instructions successfully, shape-shifting into one animal after another, while his opponent does the same. In the end the Little Bear succeeds, and his opponent, having received the fatal blow to his forehead, is no longer immortal. From one point of view, the identity of the antagonist appears to be clear – he is the father of the young woman. However, other versions of the tale point directly to his identification with the serpent or dragon who is killed by a blow to his forehead with the magical egg (SATRÚSTEGUI 1975, 18–21; FRANK 2019).

A closer analysis of the plot brings into view other cultural conceptualizations that informed the interpretive framework for the tale, revealing it to be an animist ontology typical of hunter-gatherers who lived (or live) in close contact with bears and other wild animals, along with the associated belief in shape-shifting that is firmly embedded in the narrative itself (HALLOWELL 1926; BRIGHTMAN 2002; BERRES et al. 2004; BRIGHTMAN et al. 2012). On this view, the backdrop of the story becomes Nature itself, upon which the actions are projected. A child, seeing an eagle swooping down to catch its prey, could have interpreted the scene, symbolically, as an exteriorization of a familiar episode from the traditional narrative.

---

3   Unless indicated otherwise, the translations are the author's.

When interpreted on this deeper level, what we find in the tale is a series of purely ritual battles between two shape-shifters, one who is already half-bear, and his older adversary. From this perspective, the role of the four Animal Helpers is of fundamental importance to the hero, beginning with the smallest one, Ant. Moreover, there is a pattern to the ritual confrontations: They are encounters between a predator and its prey (Table 1).

In the end, it is the magic pigeon egg that makes the old man become mortal like the rest of Nature, subject to life and death, rather than standing apart as a transcendent immortal being.

Ecocentric coding

There are other lessons that in times past a child might have learned from this story. It is a tale that goes counter to the so-called "law of the jungle", the Social Darwinian view of Nature that interpreted the "survival of the fittest" scenario as the superiority of brute strength and/or self-interest in the so-called "struggle for survival" (Weiss 2010). This is a view that featured prominently in 19th-century thought and is still alive today in some quarters. In contrast to that view of Nature, the equitable division of the dead deer can be read as a parable of sharing and reciprocity in which Little Bear restores the natural order of things. It speaks of the harmony and balance of Nature and the interlocking networks of dependencies that act to maintain that natural order. Large carnivores bring down the prey. Smaller carnivores then approach to eat the scraps. Next in line come the scavenger birds, eagles and vultures. And, finally, the insects arrive to pick clean the skin and bones. Viewed from the perspective of modern conservation biology, the food web described implicitly in the narrative suggests an understanding of the dynamics of "trophic cascades" and the concept of "keystone species," for example, where the actions of large carnivores impact the complex food-web dynamics in positive ways (Paine 1969; 1980; Polis et al. 2000). At the same time, it speaks to us of the eternal cycle of life and death.

As we have seen, the ursine ancestry of the main character, his hybrid nature, is highlighted in these European tales and appears to be grounded in an archaic belief that held that humans descended from bears. This can be compared to the many Native American origin myths that claim bears as ancestors, too. And even if direct descent is not claimed, bears often share family relations through marriage and sexual relations, beliefs frequently exteriorized in narratives and legends. As a result, in these Native American oral traditions bears often appear as both ancestors and kin (Kassabaum/Peles 2020, 111).

The plot of the European tale unfolds in a landscape infused with trophic relations, a metaphysics characterized by an awareness of the intricate reciprocal relations inherent in Nature. The complex food-chain network of predator-prey interactions is emphasized, rather than the triumph of "man over beast". Animals are collaborators and function as active participants, not passive by-standers. Overall, the plot is framed by elements typical of an animist worldview. When compared to Native American spirituality, we find evidence of shape-shifting and can identify what appears to be the European counterpart of a Native American vision quest by means of which the seeker obtains spirit animal helpers and creates a medicine bundle composed of talismans symbolizing each guardian animal (McGaa 1990, 75–83; Woodhead 1992, 60–63, 121–128; Waugh 1996, 56–60; Encyclopaedia Britannica 2015; Posthumus 2018). The plot itself revolves around ritual combats between two shape-shifters, each aided by their respective Animal Helpers (Shepard 1999; 2007; Frank 2016a; 2019).

However, these aspects of the tales did not attract the attention of scholars, probably because researchers were unfamiliar with the tenets of an animist ontology. Instead of pursuing how evidence for the ursine genealogy implicit in these tales might be explained or otherwise documented ethnographically, it has been ignored and treated as a mere curiosity. Indeed, by the end of the 19th century folklorists were busy concentrating their efforts on a very different task, that of classifying the narratives according to motif and tale type (COSQUIN 1887, 1–27; DUNDES 1997; FRANK 2015). By 1910 Panzer had documented 221 European variants of ATU 301, the descent of the Bear's Son hero to the Underworld (PANZER 1910). In a study published in 1959, 57 Hungarian versions of the tale were mentioned (KISS 1959), and in 1992, Stitt, in his study of *Beowulf and the Bear's Son: Epic Saga, and Fairytale in Northern Germanic Tradition*, recorded 120 variants of the Bear's Son story for Scandinavia alone (STITT 1992, 25–27, 209–217). The cycle of oral tales is present in all the Indo-European language groups of Europe as well as in Basque and Finno-Ugric languages, i.e. in Finnish, Saami and Magyar (Hungarian), and there is even one example of the tale found among the Mansi (Voguls) of Siberia (VON SADOVSZKY/HOPPÁL 1995, 118–120, 152).

In addition to the preoccupation with classification by motif and tale type, what one also observes is that when the tale has been analyzed, that analysis has been geared primarily to finding counterparts to the tale in epic poems and sagas, that is, in literary works (PANZER 1910; GLOSECKI 1988; 1989; STITT 1992; ANDERSON 2016). What has not been explored properly is whether the ursine genealogy intrinsic to the tales was part of larger animist worldview that left traces in the ethnographic and ethnohistoric record, that is, in the real world.

We need to remember that the label frequently used by folklorists, namely "The Bear's Son", is a broad one.[4] It encompasses ATU 301 and all its variants. In addition, this tale type has been linked to ATU 650A "Strong John." A key element in the plotline of ATU 301 is the fact that the protagonist acquires two or three unusually strong and fully anthropomorphic companions who help him. What has not been recognized, however, is that the fully anthropomorphic helpers have replaced the Animal Helpers found in the older variants. This caused the tale of the four animals to end up being relegated to a totally separate and supposedly unrelated tale type, ATU 554, called "The Grateful Animals". And, to complicate matters even more, as will be demonstrated, the Animal Helpers resurface in a tale type referred to as ATU 302, namely "The Ogre's (Devil's) Heart in the Egg".

To summarize, over time the plot of the tale broke into pieces. The variants that resulted were reshaped, adjusting themselves in different ways to reflect the changing cultural norms of Europeans. Hence, we can see that narratives associated with the expression "The Bear's Son" focus on ATU 301 (plus ATU 301A, B, C, and D, variants referred to globally by folklorists as "The Three Princesses"). At the same time there are other tale types that form part of the same narrative tradition: ATU 650A "Strong John", ATU 554 "The Grateful Animals", and ATU 302, now shortened to "Soul in an Egg". Over time the storyline and episodes associated with earlier versions of the tale ended up becoming fragmented and, as a result, became classified as different tale types (ATU 650A, ATU 554, ATU 302, and ATU 301, plus at least four subtypes of ATU 301).

THE DISAPPEARANCE OF THE ANIMAL HELPERS

In the tale known as "The Bear's Son" (ATU 301), the Animal Helpers disappear entirely from view, replaced by figures more in consonance with the later worldview. Instead of animal helpers, the hero acquires two or three anthropomorphic companions, huge male figures with extraordinary strength.

---

4   ATU 301, referred to globally as "The Bear's Son", is the term used to refer particularly to versions of the tales that have been compared to northern sagas, such as *Beowulf*. In other instances, the same set of tales carries a title that highlights the name of the protagonist in that language: *John the Bear* in English, *Juan el Osito* in Spanish, *Jan de l'Os* in Catalan, *Jan l'Ourset* in Gascon, *Jean de l'Ours* in French, *Giovanni l'Orso* in Italian, *Hans Bär* in German, and *Ivanuska* as well as *Ivanko Medvedko* in Slavic languages.

Although the main protagonist is still often portrayed as half-bear and half-human, born of a human female and a bear, his helper companions are no longer animals as is the case in the earlier versions. Although it is not clear exactly at what point in time the human helpers were substituted for the four animals, DE BLÉCOURT (2012, 179–181) discusses a literary tale dating from 1634 in which the animals morph into helping brothers-in-law or brothers, a tale type classed as ATU 552.

With the passage of time a kind of bifurcation takes place, giving rise to two types of variants. One type is represented by the set of stories that continue to retain the animal helpers (ATU 302 and ATU 554) and, to a certain extent, imprints from the animist coding typical of the earlier storyline. But in these tales, the ursine identity of the hero is often lost. The other type consists of the set of stories (ATU 301) known as the tales of "The Bear's Son" in which the ursine ancestry of the main character is retained, but the animal helpers have been transformed into human companions.

Not uncommonly, the hero known as the Bear's Son was elevated even further, as elements from a hierarchically organized society were introduced into the frame, a backdrop typical of the genre of literary "magic" or "wonder tales". These became popular and highly influential. They circulated in various print collections published initially in the late 17th century and were followed up by other extensive compilations of tales in the 18th and 19th centuries. These changes allowed the main character to be reconfigured as a hero who descends to the Underworld (often represented as simply a deep hole in the ground) where he rescues a princess (or three princesses). In this instance, each of the princesses gives him a talisman that later on allows him to be recognized as their savior. In the end he is rewarded at court and marries the daughter of the king. Another dominant theme is that of betrayal. The hero is abandoned by his companions who refuse to haul him up out of the hole, an act of betrayal that in the end the hero punishes. Clearly, the backdrop has been transformed and a new interpretive template is now functioning. However, a few key elements from the older storyline still show up.

Factors leading to changes in the text and the fragmentation of its narrative structure

Over time, the cultural schemas supporting the earlier animist ontology lost their hold and were replaced by new ones. This contributed to the loss of awareness on the part of subsequent generations of storytellers and translators of the tale concerning the ursine genealogy, the deeper meaning of the four Animal Helpers and the significance of the shape-shifting that takes place. In the Middle Ages, Christianity contributes to this erosion, bringing about a strict dichotomy, a conceptual divide which attempted to set humans totally apart from animals. And, as this portrayal of animal otherness becomes increasingly entrenched in the worldview utilized by both the tellers of the tale and their audiences, it becomes harder to understand the earlier interpretive framework.

In the eyes of the Church, the European bear cult was perceived as a threat and impediment to the conversion of the popular classes. For example, as Pastoureau notes, the Germanic veneration for the bear shows that the animal was a being apart, an intermediary creature between the animal and human worlds and considered even an ancestor or relative of humans (PASTOUREAU 2011, 2). During the Middle Ages, worship of the bear, however, was not confined to the Germanic world for it was also deeply engrained among the Slavs, who admired the bear as much as the Germans did. Further proof of this veneration lies in the fact that both Germanic and Slavic languages use *noa* terms – euphemisms – to refer to the animal. This recalls the wide-spread pattern of semantic avoidance documented among the indigenous peoples of North America and Eurasia (BLACK 1998; SOKOLOVA 2000; PASTOUREAU 2011; NAGY 2017, 46–52). As is well known, the Germanic as well as the Slavic words for "bear" are not the same as those used in the other Indo-European languages (see NEDOMA and UDOLPH, this volume; Pokorny 1959, III, 875; Buck 1988; Praneuf 1989). The words for "bear" in other Indo-European languages derive from a common Proto-Indo-European root, namely, *h_2rtkos.[5] It is logical to assume that the semantic avoidance evidenced in Germanic and Slavic languages was brought about because earlier speakers of these languages were operating from a mindset similar to that of indigenous hunter-gatherer populations who also show deference to bears, often viewing them as kin or ancestors and believing that bears hear everything that is said, especially what is said about them.

Wherever bear ceremonialism is practiced, such patterns of semantic avoidance go hand in hand with the belief that the bear is omniscient and, therefore, has the power to hear all that is said about him. Addressing the bear with its real name is considered dangerous, so hunters avoid mentioning it, choosing rather to speak about the animal using euphemisms. In times past, the common term utilized by speakers of Slavic languages was *medved* "honey-eater", which today is the word used to refer generically to a bear, while Germanic tribes preferred to call him the "brown one", an expression that gave rise eventually to the English word *bear*, linked etymologically to the English words *brown* and *bruin* (cf. Nedoma and Udolph, this volume). Consequently, the Slavic and Germanic words for "bear" can be considered semantic residue left over from this older ursine belief system.

In times past, evidence for the belief in the supernatural powers of the bear, especially its omniscience, was not limited to these parts of Europe. Indeed, throughout much of Europe "in the Carolingian period, the bear continued to be seen as a divine figure, an ancestral god whose worship took on various forms but remained solidly rooted, impeding the conversion of pagan peoples" (Pastoureau 2011, 3). Almost everywhere, from the Pyrenees to the Baltic, "the bear stood as a rival to Christ. The Church thought it appropriate to declare war on the bear, to fight him by all means possible" (Pastoureau 2011, 3).

Whereas the struggle of the medieval Church against the bear was ultimately successful, that is, in terms of eliminating most overt traces of the ancient ursine cults, what it was not able to do was to stop storytellers from passing on tenets of the ursine belief system, albeit covertly, in the form of traditional folktales. These stories insured that the hero's life and times would be transmitted orally, under the radar, across generations even as the storytellers and the members of their audiences were increasingly unaware of the message implicit in the plot itself. With the passage of time, the narratives would be translated from one language to another, a process that would introduce modifications and significant fragmentation of the narrative structure along with an increasing loss of awareness of the underlying animist cosmovision. As noted, one of the significant changes that can be detected in the tales is the bifurcation that took place with respect to the identity of the Animal Helpers. In the older version there are four animals who give the hero the power to shape-shift, a characteristic typical of animist ontologies. In what are more recently anthropocentrically coded versions the hero's helpers are transformed into two or three men endowed with immense strength as is the hero himself.

Kinship affirmation

The ursine genealogy which portrays bears as ancestors and therefore kin is a central tenet of bear ceremonialism. Therefore, it is not surprising to find stories in which a human female mates with or marries a bear in locations where bear ceremonialism was or still is practiced (Deans 1889; Edsman 1956; McClellan 1970; Henderson 2020). The presence of such stories is one of the constants of

---

5   For discussion of the Basque term for "bear" as well as further discussion of the Indo-European forms, cf. Frank 2017, 93–109.

bear ceremonialism (Rockwell 1991; Shepard/Sanders 1992; Frank 2016b; Wiget 2021). In Europe this trait was accompanied by the belief that viable offspring could result from such a union. Surprising as it might seem, for centuries Europeans continued to believe not only that bears often mated with human females but that a mixed blood offspring could result. Allegedly true stories about a bear abducting a young woman and the two of them living together continued to circulate in the Pyrenean region well into the 19th century.

Just how widespread this belief really was is underlined by the work of William of Auvergne (1228–1249) who was one of the most remarkable intellectuals of the first half of the 13th century. The question of bear-human mating was taken up in one of William of Auvergne's works, *De universe creaturarum*. In that work Auvergne reports an *exemplum notissimum*, i.e. a well-known exemplary story: One day in Saxony, a bear of enormous strength abducted the wife of a knight and imprisoned her in a cave where he usually hibernated. She was a woman of great beauty, so that after a time her body awakened the bear's concupiscence. He raped her, and in his infernal den had sexual relations with her for several years. Three sons were born of this union. Happily, one day a woodsman freed the woman and her sons, she returned to her husband and lodged her sons near the château in the sight of everyone; later, they were even knighted in the presence of the great Saxon barons. They did not differ from other knights except for their abundant hairiness and their habit of inclining their head slightly to the left, in the manner of bears. They were, besides, given the name of their father and were called *Ursini*, sons of the bear (Pastoureau 2011, 77).

Auvergne goes on to argue in favor of the actual existence of these unions: "Unlike mules, the offspring of cross-species mating, children born from a bear and a woman can procreate and have descendants" (Pastoureau 2011, 78). It is in light of these beliefs, namely, that creatures born from the union of a woman and a bear could be fertile and have descendants, that we can understand genealogies that portray a bear as the ancestor of a king. For example, we have the chronicle *Gesta Danorum* from around 1200 by Saxo Grammaticus in which this Danish scholar recounts that the great-grandfather of King Sven II Estridsen of Denmark (1047–1076) was the son of a bear, a dynastic legend that goes back further in time. The story told held that a bear had abducted a young woman and "married" her in his cave where she later gave birth to a son. "In the official genealogies compiled at the Danish court throughout the thirteenth century – the great century of genealogies in Scandinavia – the bear as ancestor of a king held an acknowledged place. No one doubted that a bear was one of the founders of the Danish dynasty [...]" (Pastoureau 2011, 79). Rather than damaging the royal name, the animal ancestry seems to have given a mythical prestige to the Danish dynasty. Indeed, it appears to have aroused the envy of the kings of Sweden and Norway, to the point that from 1260 on the kings of Norway also claimed descent from that same founding bear.

As Pastoureau notes, nothing was invented by Saxo Grammaticus himself for as early as the 11th century, literary and narrative texts were circulating that had a bear among the ancestors of various prestigious figures. Among them was Earl Siward of Northumberland, who died in 1055 and whose father was said to have had "bear's ears". That latter feature was considered the remaining bodily inheritance of the bear ancestor that had procreated with a woman (Brunner 2007, 26; Pastoureau 2011, 80). Indeed, in many of the folktales what set the bear's son apart from other children was that he was extraordinarily hairy or that he had "bear's ears".

Finally, with respect to other ways the ursine genealogy manifested itself in social practice, we find evidence that integral to the veneration shown for the slain bear were ritual acts of affirmation of the kinship between humans and bears. The theme of a ritual wedding involving the slain bear is well documented in Finland (see Piludu, this volume). For example, according to a 18th-century account from Viitasaari in central Finland, upon the arrival of the hunters carrying the slain bear, a ritual "wedding" was performed. That ritual repeated many of the customs of ordinary weddings but in this case was celebrated between the bear and one of the girls of the village. In this way, the kinship bond between bear and humans was symbolically renewed (SARMELA 2006, 2–3, 16).[6] And, once again, a feigned wedding or coupling of a human and the slain bear finds its counterpart in other locations where bear ceremonialism has been documented.

The larger picture emerges when different threads of the overarching animist belief system are woven together. At first, they might appear to be random, unrelated ethnographic bits and pieces. But when examined using the interpretive frame of bear ceremonialism and its associated ursine genealogy, they take on new importance. For instance, we have certain scenic elements that are an integral part of the performances known as *fêtes de l'ours,* celebrated each year in the village of Arles-sur-Tech in the Pyrenean region (Fig. 2; see BAKELS/BOER, this volume). The scenes represented could have been viewed as reenactments of the initial encounter of the young woman with the bear and their subsequent mating as narrated in the folktales. This interpretation is quite possible since this zone is one where in times past the events portrayed in the tale of "Jean l'Ours" would have been familiar to everyone in attendance.

The scene in question has a young woman called Rosetta – actually a man dressed as a woman – being grabbed by the Bear who attempts to drag her into his den which has been constructed in the middle of the plaza. The Bear also goes after one or more women in attendance and brings them into his lair, a practice first described nearly a hundred years ago by an eyewitness, the British folklorist Violet Alford (ALFORD 1930, 173–174). The feigned sexual coupling of the Bear with Rosetta as well as with actual women in the audience is commonplace still today in these annual performances.

CONCLUDING COMMENTARY

To summarize, the stories classified collectively as "The Bear's Son" represent one of the most widespread motifs found in European folklore. Although the narrative has been analyzed from many angles, it is only recently that the larger implications of this set of tales have been taken into consideration, especially the way that they resonate with the tenets and practices of bear ceremonialism. This rethinking has allowed the importance of the stories to be better appreciated. When projected against a backdrop that was once informed by the belief that humans descended from bears, the narratives become a key to understanding how a central tenet of this much earlier animist cosmology came to be transmitted across time, albeit with modifications.

That humans descended from bears meant that human animals could be viewed as having a mixed ancestry. And that genealogy is one that can be traced back to the intermediary figure represented by the half-bear, half-human main character of these folktales. However, evidence for the belief in that genealogy is not limited to the storyline of these tales. Rather there is ample reason to believe that the ursine ancestry of humans also left an indelible imprint in the ethnographic record and even in language. Even though the belief that humans descended from bears was already falling out of favor during the Middle Ages, the conviction that humans and bears could mate and produce fertile offspring was accepted as factual in much of Europe throughout the Middle Ages and beyond (PASTOUREAU 2011, 68–85). Indeed, it is not unusual to find folk heroes and even actual kings tracing their own lineage back to such a bear-human mating (GLOSECKI 1988; 1989). Hence, until that older cosmology faded completely from view, the Bear's Son Tale along with its variants was probably interpreted by audiences in a very different and far more realistic fashion than it is today.

---

6 For further discussion of the ritual wedding, cf. PENTIKÄINEN 2007, 63–76.

Bibliography

Aarne/Thompson 1961: A. Aarne/S. Thompson, The Types of the Folktale: A Classification and Bibliography. Folklore Fellow Communications 184 (Helsinki ²1961).

Alford 1930: V. Alford, The Springtime Bear in the Pyrenees. Folklore XLI (Sept.), 1930, 266–279.

Anderson 2016: C. E. Anderson, Transformations of the hero? The frustrated connection between Bǫðvarr bjarki and Beowulf. Paper presented at the 26th Annual Conference of the Texas Medieval Association (TEMA), 23–25 September 2016, College Station, TX. https://tinyurl.com/Bjarki-and-Beowulf.

Arratibel 1980: J. Arratibel, Kontu zaarrak (Bilbao 1980).

Barandiaran 1973–1983: J. M. Barandiaran, Obras completas (Bilbao 1973–1983)

Barbeau 1945: M. Barbeau, Bear Mother. Journal of American Folklore LIX (231), 1945, 1–12.

Barbier 1991: J. Barbier, Legendes Basques (Donostia, Baiona 1991; originally published 1931).

Berres et al. 2004: T. E. Berrres/ D. M. Stothers/ D. Mather, Bear imagery and ritual in Northeast North America: An Update and Assessment of A. Irving Hallowell's work. Midcontinental Journal of Archaeology 28(1), 2004, 5–42.

Bertolotti 1992: M. Bertolotti, Carnevale di Massa 1950 (Torino 1992).

Bertolotti 1994: M. Bertolotti, La fiaba del figlio dell'orso e le culture siberiane dell'orso. Quaderni di Semantica XV(1), 1994, 39–56.

Bidart 1978: P. Bidart, Récits & Contes Populaires du Pays Basque 1. Recuellis par Pierre-Bidart en Basse-Navarre (Paris 1978).

Bidart 1979: P. Bidart, Récits & Contes Populaires du Pays Basque 2. Recuillis par Pierre Bidart dans le Labourd (Paris 1979).

Bird-David 1999: N. Bird-David, Animism revisited: Personhood, environment, and relational epistemology. Current Anthropology 40, 1999, 67–91.

Black 1998: L. T. Black, Bear in imagination and in ritual. Ursus 10, 1998, 323–347.

De Blécourt 2012: W. de Blécourt, Tales of Magic, Tales in Print: On the Genealogy of Fairy Tales and the Brothers Grimm (Manchester 2012).

Brightman 2002: R. A. Brightman, Grateful Prey: Rock Cree Human-Animal Relationships (Regina [Saskatchewan] 2002).

Brightman et al. 2012: M. Brightman/V. E. Grotti/ O. Ulturgasheva, Animism and invisible worlds: The place of non-humans in indigenous ontologies. In: M. Brightman/V. E. Grotti/O. Ulturgasheva (eds.), Animism in Rainforest and Tundra: Personhood, Animals, Plants and Things in Contemporary Amazonia and Siberia 2 (New York 2012) 1–27.

Brunner 2007: B. Brunner, Bears: A Brief History. Transl. from German by L. Lantz (New Haven, London 2007).

Buck 1988: C. D. Buck, A Dictionary of Selected Synonyms in the Principal Indo-European Languages (Chicago, London 1988; originally published 1949).

Cerquand 1986: J. F. Cerquand, Ipar Euskal Herriko Legenda eta Ipuinak. Basajauna, Laminak, Tartaroa. Anuntxi Aranaran Transkripzioa (San Sebastián 1986; originally published 1875–1882).

Corvino 2013: C. Corvino, Orso: Biografia di un animale: Della prehistoria allo sciamanesimo (Bologna 2013).

Cosquin 1887: E. Cosquin, Les Contes populaires de Lorraine comparés avec les Contes des autres Provinces de France et des Pays étrangers (Paris 1887). http://www.gutenberg.org/files/57892/57892-h/57892-h.htm.

Deans 1889: J. Deans, The story of the bear and his Indian wife. Journal of American Folklore 2, 1889, 255–260.

Dundes 1997: A. Dundes, The Motif-Index and the Tale Type Index: A Critique. Journal of Folklore Research 24(3), 1997, 195–202.

Edsman 1956: C.-M. Edsman, The story of the Bear Wife in Nordic tradition. Ethnos 21(1/2), 1956, 3–56.

Encyclopaedia Britannica 2015: Encyclopaedia Britannica, Vision quest: Native American Religion (Encyclopaedia Britannica 2015). http://www.britannica.com/topic/vision-quest.

Frank 2008a: R. M. Frank, Evidence in Favor of the Palaeolithic Continuity Refugium Theory (PCRT): Hamalau and its linguistic and cultural relatives. Part 1. Insula: Quaderno di Cultura Sarda 4, 2008, 91–131.

Frank 2008b: R. M. Frank, Recovering European ritual bear hunts: A comparative study of Basque and Sardinian ursine carnival performances. Insula: Quaderno di Cultura Sarda 3, 2008, 41–97.

Frank 2009: R. M. Frank, Evidence in Favor of the Palaeolithic Continuity Refugium Theory (PCRT): Hamalau and its linguistic and cultural relatives. Part 2. Insula: Quaderno di Cultura Sarda 5, 2009, 89–133.

Frank 2015: R. M. Frank, Bear Ceremonialism in relation to three ritual healers: The Basque *salutariyua*, the French *marcou* and the Italian *maramao*. In: E. Comba/D. Ormezzano (eds.), Uomini e Orsi: Morfologia del Selvaggio (Torino 2015) 41–122.

Frank 2016a: R. M. Frank, Paul Shepard's "Bear Essay": On Environmental Ethnics, Deep Ecology and Our Need for the Other-than-Human Animals. Creative Commons License. http://www.tinyurl.com/paul-shepard-bear-essay.

Frank 2016b: R. M. Frank, A status report: A review of research on the origins and diffusion of the belief in a Sky Bear. In: F. Silva/K. Malville/T. Lomsdalen/F. Ventura (eds.), The Materiality of the Sky: Proceedings of the 22nd Conference of the European Society for Astronomy in Culture (Lampeter 2016) 79–87.

Frank 2017: R. M. Frank, Shamanism in Europe? Part 2. An Essay in Collective Memory and Cognition: Bears and Badgers, Basque and Celtic. Creative Commons License (CC BY 4.0). http://tinyurl.com/essay-on-collective-memory.

Frank 2019: R. M. Frank, Translating a worldview in the longue durée: The tale of "The Bear's Son". In: A. Głaz (ed.), Languages – Cultures – Worldviews: Focus on Translation (London 2019) 53–80.

Frog 2013: Frog, Revisiting the historical-geographic methods(s). RMN: Retrospective Methods Newsletter 7, 2013, 18–34.

Gastou 1987: F. R. Gastou, Sur les Traces des Montreurs d'Ours des Pyrénées et d'Ailleurs (Toulouse 1987).

Glosecki 1988: S. O. Glosecki, The Wolf of the Bees: Germanic shamanism and the Bear Hero. Journal of Ritual Studies 2(1), 1988, 31–53.

Glosecki 1989: S. O. Glosecki, Shamanism and Old English Poetry (New York, London 1989).

Goldberg 1984: C. Goldberg, The historic-geographic method: Past and future. Journal of Folklore Research 21(1), 1984, 1–18.

Gual 2017: O. L. Gual, Les derniers Ours: Une Histoire des Fêtes de l'Ours. Quaderns del Costumari de Catalunya Nord 1, 2017, 1–495.

Hallowell 1926: A. I. Hallowell, Bear Ceremonialism in the Northern Hemisphere. American Anthropologist 28(1), 1926, 1–175.

Harrison-Buck 2018: E. Harrison-Buck, Relational matters of being: Personhood and agency in archaeology. In: E. Harrison-Buck/J. A. Hendon (eds.), Relational Identities and Other-than-human Agency in Archaeology (Boulder [Colorado] 2018) 188–197.

Harvey 2006: G. Harvey, Animism: Respecting the Living World (New York 2006).

Henderson 2020: L. Henderson, Bear tales: Ways of seeing polar bears in mythology, traditional folktales and modern-day children's literature. In: A. Llompart/L. Brugué (eds.), Contemporary Fairy-Tale Magic: Subverting Gender and Genre (Leiden, Boston 2020) 250–261.

Jason 2006: H. Jason, Review: The Types of International Folktales. A Classification and Bibliography. Part 1: Animal Tales, Tales of Magic, Religious Tales, and Realistic Tales, with an Introduction; Part 2: Tales of the Stupid Ogre, Anecdotes and Jokes, and Formula Tales; Part 3: Appendices (Folklore Fellows Communications 284–286). Fabula 47(1/2), 2006, 172.

Kassabaum/Peles 2020: M. C. Kassabaum/A. Peles, Bears as both family and food: Tracing the changing contexts of Bear Ceremonialism at the Feltus Mounds. In: S. B. Carmody/C. R. Barrier (eds.), Shaman, Priest, Practice, Belief: Materials of Ritual and Religion in Eastern Northern America (Tuscaloosa 2020) 108–126.

Kiss 1959: G. Kiss, A 301-es mesetípus magyar redakciói. Ethnographia (Budapest) LXX, 1959, 253–268.

Lajoux 1996: J.-D. Lajoux, L'homme et l'ours (Grenoble 1996).

McClellan 1970: C. McClellan, The Girl who Married the Bear (Ottawa 1970).

McGaa 1990: E. McGaa, Mother Earth Spirituality (New York 1990).

Nagy 2017: Z. Nagy, "Everybody's a bit scared of the bear": Fear of the bear along the Vasyugan river in Siberia. In: K. Hoffmann-Schickel/P. Le Roux/É. Navet (eds.), Sous la peau de l'ours: L'humanité et les ursides: Approche interdisciplinaire (Paris 2017) 453–491. https://tinyurl.com/fear-of-the-bear.

Paine 1969: R. T. Paine, A note on trophic complexity and community stability. The American Naturalist 103, 1969, 91–93.

Paine 1980: R. T. Paine, Food webs: Linkage, interaction strength and community infrastructure. Journal of Animal Ecology 49, 1980, 667–685.

Panzer 1910: F. Panzer, Studien zur germanischen Sagengeschichte. I. Beowulf (München 1910).

Pastoureau 2011: M. Pastoureau, The Bear: History of a Fallen King (Cambridge 2011).

Pauvert 2014: D. Pauvert, Le rituel de l'ours des Pyrénées aux steppes. In: Société des Études euro-asiatiques (ed.), Traditions en devenir (coutumes et croyances d'Europe et d'Asie face au monde moderne. EURASIE N° 2 (Paris 2014) 17–51.

Peillen 1986: T. Peillen, Le culte de l'ours chez les anciens basques. In: C. Dendaletche (ed.), L'ours brun: Pyrénées, Agruzzes, Mts. Cantabriques, Alpes du Trentin (Pau 1986) 171–173.

Pentikäinen 2007: J. Pentikäinen, Golden King of the Forest: The Lore of the Northern Bear (Helsinki 2007).

Polis et al. 2000: G. A. Polis/A. L. W. Sears/G. R. Huxel/D. R. Strong/J. Maron, When is a trophic cascade a trophic cascade? Tree 15(11), 2000, 473–475.

Pokorny 1959: J. Pokorny, Indogermanisches etymologisches Wörterbuch (Bern 1959).

Posthumus 2018: D. C. Posthumus, All My Relatives: Exploring Lakota Ontology, Belief and Ritual (Lincoln 2018).

Praneuf 1989: M. Praneuf, L'Ours et les Hommes dans les Traditions européennes (Paris 1989).

Rockwell 1991: D. Rockwell, Giving Voice to Bear: North American Indian Myths, Rituals and Images of the Bear (Niwot [Colorado] 1991).

Von Sadovszky/Hoppál 1995: O. J. von Sadovszky/M. Hoppál (eds.), Vogul Folklore. Collected by Bernát Munkácsi. Transl. B. Sebestyén (Budapest, Los Angeles 1995).

Sarmela 2006: M. Sarmela, The Bear in the Finnish Environment: Discontinuity of Cultural Existence. Transl. A. Silver; Appendix: R. Boom (Helsinki 2006). https://sarmela.net/_files/200000206-8a5af8b549/bear-cult.pdf.

Satrústegui 1975: J. M. Satrústegui, La leyenda del dragón en las tradiciones de Urdain. Cuadernos de Etnografía y Etnografía de Navarra 19, 1975, 13–30.

Shepard 1999: P. Shepard, The significance of bears. In: F. R. Shepard (ed.), Encounters with Nature: Essays by Paul Shepard (Washington [DC], Covelo [California] 1999) 92–97.

Shepard 2007: P. Shepard, The biological bases of bear mythology and ceremonialism. The Trumpeter: Journal of Ecosophy 23(2), 2007, 74–79.

Shepard/Sanders 1992: P. Shepard/B. Sanders, The Sacred Paw: The Bear in Nature, Myth and Literature (New York 1992).

Sokolova 2000: Z. Sokolova, The bear cult. Archaeology, Ethnology & Anthropology of Eurasia 2(2), 2000, 121–130.

Stitt 1992: J. M. Stitt, Beowulf and the Bear's Son: Epic, Saga and Fairytale in Northern Germanic Tradition (New York, London 1992).

Truffaut 1988: T. Truffaut, Apports des carnavals ruraux en Pays Basque pour l'étude de la mythologie: Le cas du 'Basa-Jaun'. Eusko-Ikaskuntza. Sociedad de Estudios Vascos, Cuadernos de Sección, Antropología-Etnología 6, 1988, 71–81.

Truffaut 2010: T. Truffaut, Contribution à la connaissance de la permanence ursine dans les diverses manifestations culturelles, culturelles et festives dans le périmètre de l'ancienne province romain de Novempopulanie. Munibe (Antropologia-Arkeologia) 61, 2010, 407–434. www.aranzadi.eus/fileadmin/docs/Munibe/2010407434AA.pdf.

Uther 2004: H.-J. Uther, The Types of International Folktales: A Classification and Bibliography. Part 1: Animal Tales, Tales of Magic, Religious Tales, and Realistic Tales, with an Introduction; Part 2: Tales of the Stupid Ogre, Anecdotes and Jokes, and Formula Tales; Part 3: Appendices (Helsinki 2004).

Vinson 1883: J. Vinson, Le Folk-lore du Pays Basque (Paris 1883).

Wallace 1949: A. Wallace, The role of the bear in Delaware Society. Pennsylvania Archaeologist 19(1/2), 1949, 37–46.

Waugh 1996: E. Waugh, Dissonant Worlds: Roger Vandersteene among the Cree (Waterloo [Canada] 1996).

Weiss 2010: K. M. Weiss, "Nature, red in tooth and claw", so what? Evolutionary Anthropology 19, 2010, 41–45.

Wiget 2021: A. Wiget, Circumpolar bear ceremonialism: Reviewing the world through ritual. A seminar organized in conjunction with the Institute of Anthropology, Russian Academy of Science, Moscow, and Nizhny Novgorod State University, April 28, 2021. https://www.youtube.com/watch?v=Y-5P6lVTHyI.

Woodhead 1992: H. Woodhead, The Spirit World (Alexandria [Virginia] 1992).

Prof. emer. Roslyn M. Frank
University of Iowa
Iowa
USA
roz-frank@uiowa.edu

Fig. 1. The seven provinces of Euskal Herria, the historical Basque Country, span France (light yellow) and Spain (rest of the map). Names in this map are in Basque (map GIS department, ZBSA, after http://en.wikipedia.org/wiki/Basque_Country_(historical_territory)).

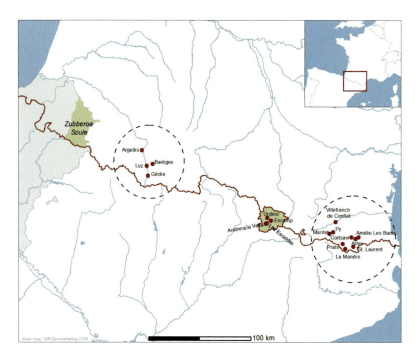

Fig. 2. Pyrenean Fêtes de l'Ours *(map GIS department, ZBSA, after* GASTOU *1987, 20).*

Table 1. The predator-prey pattern in the Bear's Son Tale.

Predator	Prey
Lion	Porcupine
Dog	Hare
Eagle	Pigeon
[Pigeon Egg]	Snake

The bear in European folktales – with a special focus on Scandinavian variants

By Angelika Hirsch

Keywords: Son of a bear, animal bridegroom, bear-skin, Grimmelshausen, berserker

Abstract: Bears can be minor characters in European folktales as well as main ones. In animal tales they are always characterised as simple-minded. In other types of folktales they can be opponents as well as helpers. In regions where the coexistence of humans and bears is commonplace, bears appear more often in functions that are occupied by other animals in similar European versions of the type. The bear is a main character in the folktale type The Three Stolen Princesses *(ATU 301),* The Animal as Bridegroom / Beauty and the Beast *(ATU 425A and C) and* Bear-Skin *(ATU 361). For the latter, however, the findings are disillusioning – traces of the berserker for example can hardly be found in Scandinavian folktales.*

State of research

The time of monographs and substantial fairy tale related publications on the bear lies decades back. Friedrich Panzer's work "Das Märchen vom Bärensohn" was published in 1910 (Panzer 1910, 3–245) and Nai-tung Ting's study on the type AT 301[1] in 1970 (Ting 1970), as a supplement to Panzer's study. The historian of religion Carl-Martin Edsman presented another work on the subject in 1956 (Edsman 1956).

The lemmas for *Bärenführer* (bear trainer) and *Bärenhäuter* (bear-skin) in the *Enzyklopädie des Märchens* (1999) were written by the fairy tale researcher Heinz Rölleke, and for *Bärensohn* (son of a bear) by the American folklorist Donald Ward. Hans-Joachim Paproth, an ethnologist, wrote the introductory article *Bär* (bear) in the mentioned *Enzyklopädie* (Paproth 1999). He had been closely associated with Scandinavia since his youth, studied in Sweden and submitted his PhD thesis on the bear ceremony among the Tungusic peoples (Paproth 1976).

This seems worth mentioning to me as the interest in bear motifs is greater among anthropologists, ethnologists, and religious studies scholars than among the folktale research community. This is not due to a lack of interest but due to the relevance of the subject. The fairy tale takes its motifs from everywhere, from the mundane environment, from legends, mythology, and popular beliefs. It fuses them into something entirely new. For one thing many magical, enchanting, cultic elements can be found, but they "vanish" in the wonder tale. Everything happens as if by itself; rituals, ceremonies

---

1 AT = type index after Aarne/Thompson, see also text below.

and the like are not necessary. The otherworld is just as natural as the real world, the hero does not know any numinous wondering (Lüthi 2005, 8–12, 63–75). Drawing conclusions from folktales about individual elements in cultures and their evolutionary processes remains rather a marginal subject of folktale research.[2]

The fierce dispute in the 1990s over the archaic character of bear cults has led to a certain reluctance in dealing with the bear as a research subject in many disciplines (see contributions by Rydving and Sundqvist, this volume).

In recent years the narrative researcher Hans-Jörg Uther has made a significant contribution to an indispensable tool for the research of folktales with his thorough revision of the type index by A. Aarne and S. Thompson (AT), which has therefore been called the Aarne-Thompson-Uther Index (ATU) since 2004 (Uther 2004). In 2015 the *Enzyklopädie des Märchens* (founded by Kurt Ranke, edited by Rolf Wilhelm Brednich et al.) was completed, a monumental work consisting of 15 volumes, which provides detailed information on all questions of narrative and folktale research and will be referred to here several times.

Without a doubt, it is impossible to imagine the European folktales without the bear motif. When it comes to systematising the individual topoi, however, it quickly becomes confusing. Since only a general overview can be given in the context of this anthology, I will do just that: In a first part I will give a general overview of the less interesting types of folktales. The second part will deal with those types in which the bear plays – or can play – a major part.

The bear as a minor character

Animal tales
The animal tales (ATU 1–99) tell about the relationship between animals in countless versions across the continent. Usually they are short, often related to animal fables and they tend to be aetiological. Typical beginnings of these stories are "The bear, the pig and the fox once made a covenant together" (Serbia: Karadžić 1854, 261) or "In the old days, when the animals could still speak, the bear and the fox were very good friends, and they sowed, reaped, threshed and ate together (Lapland in Sweden: Poestion 1886, 15–18) or "Once in summertime the bear and the wolf were walking in the forest, when the bear heard such beautiful singing from a bird, and said: 'Brother wolf, what kind of bird is it, that sings so beautifully?'" (Germany: *Der Zaunkönig und der Bär [The Wren and the Bear]*; Grimms' Kinder- und Hausmärchen [KHM] 102; cf. Grimm 1857, 87). Reliably they end with the bear being outwitted. Even though it is feared for its size and strength, it is never the brightest one. The storytellers of these popular tales agree on this throughout Europe. As popular as the bear is in this type of folktale, it is also interchangeable. The punch line usually works with another animal, too. Hardly anything specific is said about the bear as an animal in these tales.

The bear as an interchangeable element
The bear is used again and again in folktales throughout Europe when a dangerous animal is needed – e.g. by a master to get rid of a troublesome farm hand (Bukovina: *Vom Zigeuner und dem Bären [About the Gypsy and the Bear]*; cf. Waldburg 1853, 358–362). This usually ends well for the clever farm hand but fatally for the bear. Variants of "The Wolf and the Seven Young Kids" work just as well with the bear instead of the wolf (Saxony/Transylvania: *Der Wolf und die sieben jungen Geißlein [The Wolf and the Seven Young Kids]*; cf. Haltrich 1882, 82–83).

---

2   Having studied the history of religion myself, however, I personally think it is extremely worthwhile.

However, the bear is also often used in a positive function, e.g. in folktales of the type ATU 552 *The Girls Who Married Animals*: Three sisters have each been given to an animal as wives, their brother goes in search of them, wins himself a wife with the help of his brothers-in-law (bear, eagle, fish), and in the end he redeems his enchanted brothers-in-law (*Die Wunschdose [The Wish Box]*, a Danish fairy tale; cf. GRUNDTVIG 1879, 44–46).

In type ATU 567 *The Magic Bird-Heart* bears also appear frequently in tales across Europe among the animals that the heroes gain as companions and helpers (*Wattuman und Wattusin [Wattuman and Wattusin]* from Sweden; cf. HYLTÉN-CAVALLIUS/STEPHENS 1848, 94–123). The animals' help is specific: Bears give hairs from their fur as magic gifts for the journey, fishes a scale, birds a feather. But the animals in all these types remain interchangeable elements and can be replaced by another type of animal without changing anything in the meaning of the narrative.

The bear as an ecotype
The variation of the so-called interchangeable elements is very common in folktales. The material is so condensed and general that it can be adapted to one's own culture and environment with only small changes without losing the core of the plot. Bigger changes are also possible – thus interesting ecotypes[3] can arise such as within the type ATU 156, which, according to the ancient source, is titled *Androcles and the Lion* (HIRSCH 2020). With great delight in spinning the yarn it tells of a runaway slave who encounters a lion in the wasteland and, despite his initial fright, pulls a thorn out of the lion's paw, thus winning its friendship and even love. From then on, the lion takes care of him by hunting for him. At some point the slave returns to the people, is captured and sentenced to fight wild animals in the circus. The lion he is to fight does not show itself to be a wild beast but lies down in awe of the slave. So in the meantime the lion had also become a captive. When the audience understands this, all are moved and delighted. Slave and lion are given their freedom and walk the streets in harmony; everybody can admire this unusual friendship.

A particularly condensed and archaic version, in which a bear appears instead of a lion, was recorded by Y. Wichmann among the Volga Chermisses (after WICHMANN 1931, 158–159):

The Bear
A Chermissian woman was bringing noodle soup for lunch to the harvest field, and [there] a bear came towards her. It stretches out his paw and comes growling at the woman. The woman looks at the bear's paw: a large branch has gone into it. The woman pulls out the branch, and the bear is delighted, it takes the woman's soup bucket, pours out the soup, but goes into the forest with the soup bucket. The woman remains were she is in amazement. Some time passes and the bear returns: it has filled the bucket [with honey] and brings and gives the honey to the woman.

The setting of this tale is not the urban ancient world, it is the rural world of a people on the Volga. The aspect of a lifelong friendship between human and animal is missing.[4]

Just as in the ancient version outlined above the lion remains a lion – there is no magic involved, everything could actually have happened that way – the bear is and remains a bear. It says "thank you" with a generous but typical bear gift. Mutual respect is implied. It seems unlikely to me that Wichmann's text is to be read as a condensed Chermissian derivative of the ancient one. Rather, we encounter here the core of type ATU 156 which wandered through Europe and the Near East long before the famous and elaborate story of Androcles and the lion arose.

---

3   Special versions of folktales in ethnic groups or specific regions.
4   Nevertheless, I assign this folktale (and the next one) to type ATU 156, because there is no more suitable place in the type catalogue. Besides, this type is often closer to legends than to fairy tales.

A similarly impressive, very specific and singular piece, without classification in the ATU index, comes from Iceland: *Der Mann von Grimsö und der Bär* (*The man from Grimsö and the Bear;* RAUCH [no year], 76–77). On Grimsö the fire has gone out, three men walk across the frozen sound to the mainland to get fire. On the way, one of them does not dare to jump over an open spot, turns back and gets into great trouble in the changing weather. On an iceberg he encounters a female polar bear lying on her cubs. She indicates that he should lie down with her cubs, warms and nurses him and finally brings him back home. The man slaughters his best ram for the bear and her cubs. Meanwhile, the other two messengers return on a ship with fire.

The aspect of the lifelong friendship is missing here, too. Instead we have the image of a female polar bear suckling a human. The relationship between human and animal cannot be told more intimately. Here, too, the bear is and remains an animal and helps in a way that suits its capabilities, does it not? Could it not be that …? This question arises inevitably.

The bear trainer and his bear
The type ATU 1161, *The Bear Trainer and His Bear,* is known across Europe and popular because of its easiness (UTHER 2015, 291). A fiend is outwitted by a bear trainer more or less by chance. A famous Norwegian example is *Die Katze auf Dovre (The Cat on Dovre)*: Halvor's house is regularly invaded by ravenous trolls on Christmas Eve. The bear trainer who happens to pass by stays for the night, a troll child discovers the white bear, mistakes it for a cat, burns its nose wantonly, whereupon the white bear puts all the trolls to flight. The following year, the trolls ask Halvor beforehand if the *cat* is back again. Halvor answers in the affirmative and adds that she now has seven young ones – and he is rid of the trolls forever.

THE BEAR AS A MAIN CHARACTER

Son of a bear and bear as bridegroom
In two of the most well-known types of wonder tales ever the bear plays a prominent role: ATU 301, *The Three Stolen Princesses*, and ATU 425A, *The Animal As Bridegroom* / ATU 425C, *Beauty and the Beast* (see FRANK, this volume).

Both types of folktales are known throughout the Northern hemisphere and are represented with countless versions. And both tend to combine different episodes, which means that there are obvious differences in the length of these tales. Nevertheless the narrative cores of the stories are stable and the added episodes are by no means arbitrary but are either likewise common European folktale types or widespread ecotypes.

In 1910 F. Panzer published the only essential monograph (and, as far as I know, the only existing one) on ATU 301 (at that time still AaTh 301; PANZER 1910, 3–245). In *Das Märchen vom Bärensohn* (Bear's son tale) he realises a detailed study of more than 200 versions of the folktale. These are by far not all existing texts. And although the hero is actually a bear's son in less than a quarter of the versions examined, Panzer considers this "most widespread and probably also most original" (ibid., 16: "verbreiteteste und wohl auch ursprünglichste") representation to be the leading version – and consequently gives it the title *Son of a Bear*. The supernatural origin of the hero shines through in all versions. This applies not only to ATU 301 but also to ATU 650, *Strong John*, which is now listed as a separate type. The semi-animal origin of the hero gives a plausible explanation for his enormous strength (HIRSCH 2021).

The core of ATU 301 variants is as follows: A woman lives more or less willingly with a bear and has a son. When the son grows up, he leaves the bear's den for the human world. This step can be accompanied by patricide or an escape, sometimes the father even prepares the son for the human world. In any case there are considerable problems with the young man's animal strength. Often it is told with relish how the hero loads whole trees onto his back, sinks anvils into the earth with one blow, and so on. Finally he goes out into the world, often gaining two companions on the way who also have special talents. In any event, the hero hears of stolen princesses, imprisoned in a cave by a dragon or another monster. The men find the cave, but only the hero dares to go down, frees the king's daughters, has them pulled up by his comrades, and is himself abandoned by both of them and remains at the bottom. He finds helpers who carry him up, arrives at the royal court, is recognised and honoured as the real liberator.

The versions of ATU 425 are even better known to the general public (Disney has its share in this): *Beauty and the Beast* has even made it into the type designations. It is about an animal bridegroom – worldwide, marriage between human and animal is an extremely common motif (FINDEISEN 1956; RÖHRICH 1973). Apuleius ennobled the tale of *Cupid and Psyche* literarily in his *Metamorphoses* but certainly had recourse to an existing narrative (HIRSCH 2006, 171). The plot of this story is as follows: A father must promise his youngest daughter to a threatening animal, because he has unwittingly violated a taboo that usually appears to be trivial. The animal picks up the girl, brings her to his castle and takes on human form at night. After some time the girl visits her parents (often three times), and is persuaded to take a look at the mysterious bridegroom. The girl lights a candle and, moved by the man's beauty, lets a drop of wax fall on him; he then wakes up and disappears. In some versions it is only a ray of sunshine that strikes him through a tiny crack in the door. In the full versions, the heroine now undertakes a painful quest until she finally delivers her husband from the clutches of a female demon. In the short versions the quest is missing; in them, either the ray of light or the burning/beating down of the animal skin becomes the moment of redemption.

Although a completely different story is told here than in ATU 301, both types have in common that a human woman and a bear (or another animal) live together as husband and wife more or less amicably for a shorter or longer period. Both tales also have in common that it is always a large and impressive animal. In the course of the ecotypical adaptions, the son of a bear may be the son of a mare as well (PANZER 1910, 16–29; this is more common in southern Europe and the Near East) and the animal bridegroom a lion or a wolf, for example (KAWAN 2008, col. 556).

It is irrelevant whether the relationship between woman and bear is based on unfortunate circumstances (poverty, escape), coercion (bear abducts woman/holds her captive) or love – there is no doubt about the possibility of a marriage between human woman and bear man. However, it is interpreted differently: In the folktale of the *Son of a Bear*, father bear remains an animal. EDSMANN (1956) concludes, especially for the Nordic countries, that the bear's son motif is connected to the bear cult, i.e. a deep connection between human and bear. Folktales reflect this connection in their usual matter of course style. By the way, there are not only sons of a bear but daughters of a bear as well! In an archaic East Yak folktale, for example, a woman is reborn first as a flower, then as a bear's daughter, until she finally stands in the sky as a great bear with her two cubs as a constellation.[5]

In the variants of ATU 425, however, the animal bridegroom is an enchanted human being – i.e. marriages between humans and animals are not classified as normal (anymore). They are based on a misfortune, an enchantment that is resolved as the plot progresses. The famous Norwegian variants *East of the Sun and West of the Moon* and *White-Bear-King Valemon* are great examples for the full form of this type (ASBJORNSEN/MOE 2003, vol. 1, 160–169; vol. 2, 196–204). It becomes evident very quickly that the bear is an enchanted human being, the animal form can be discarded at night. Here, no animal is mating with a human. It just appears to the outside world like this for a while. As the tale progresses the bride's quest and her efforts to break the spell of a female troll are told in detail.

---

5 *Die Mosfrau (The Mos-Woman)*, cf. GULYA 1968, 26–36. Another daughter of a bear can be found in the strange Bosnian variant of ATU 310 *Die Bärenprinzessin (The Bear Princess)*, cf. PREINDLSBERGER-MRAZOVIC 1905, 81–94.

The Swedish version *Das Mädchen und der Bär* (*The Girl and the Bear*), however, works without the quest of the heroine. But she does not know anything about the enchantment. She actually believes for a long time that she has to marry a bear. Only when she has fallen in love with the animal and is ready to marry a bear, is it revealed that he is an enchanted human being (LIUNGMANN 1965, 76–80). In this story the possibility of a marriage between a human and an animal does not yet seem too far away.

Bear-skin

The type ATU 361, *Bear-Skin*, should also be given greater attention as it raises the question about a connection to the berserker. In fairy tale research it is assumed that this type is mainly distributed across northern, northeastern, central and eastern Europe (RÖTH 1998, 51: "wohl vor allem in Nord-, Nordost, Mittel- und Osteuropa").

This is the content of the Grimms' fairy tale: A war has ended, a king sends a loyal soldier away without pay. He must starve because he has learned nothing but war. The devil offers him a deal: For seven years no washing, no combing, no clipping of hair and nails, no saying of the Lord's Prayer, wearing a bear-skin and sleeping on it, but in return his pocket will be full of gold at all times. If he keeps it up, he will be free after seven years, otherwise his soul will fall to the devil. After one year he looks terrible, people avoid him. Once he helps a man out of financial need who promises him one of his daughters in return. The eldest two refuse and mock him, the youngest believes that under the repulsive exterior there is a good man and becomes engaged to him. He continues to wander the world until the seven years are up. In the end, the devil has to wash him, cut his hair and nails, and he returns to his bride's house as a distinguished man. No one recognises him until he reveals himself. The older sisters commit suicide out of envy – the devil thus gets even two souls …

The eponymous Grimms' Kinder- und Hausmärchen (KHM) 101 was called *Der Teufel Grünrock* (*The Devil Greencoat*) until the 5th edition of the KHM in 1843 and is very close to KHM 100 *Des Teufels rußiger Bruder* (*The Devil's Sooty Brother;* a crude humorous version) and also to the later version of KHM 101 *Bear-Skin*. Wilhelm Grimm added the bear-skin motif in particular. This change is not arbitrary. For this type has a significant version in a story that was published almost 200 years earlier by Hans Jacob Christoffel von Grimmelshausen under the title *Der erste Bärnhäuter* (*The first bear-skin*) in 1670. This story is similar to the Grimms' version, not only in its content and structure, but also in its wording. Grimmelhausen's text is not only valuable as a clearly older record of the story, but also because of Grimmelshausen himself. Not only is he still considered one of the most important early German writers, but as a pressed soldier he experienced the horrors of the Thirty Years' War first-hand. Grimmelshausen thus is no bookman; when he talks about being a soldier, he knows what he is doing:

On the origin of the name Bärnhäuter
Those who want to find out the origin of the German disgraceful name Bernheuter per ethymologiam have assumed that in ancient times, when the old Germans still slept on all kinds of skins, those were called by this name in mockery who remained lying on their Bärnhaut out of laziness and never wanted to do anything brave. It may be that I do not remember so far out that I could give news of it; but an ancient painting has been found at Hohen-Roht Castle, from which the enclosed portrait has also been copied, with the following account of where this name originated (translation by S. Lutkat, after TITTMANN 1877, 247).

However, the image actually printed by Grimmelshausen does not correspond to his description, of all things the bear skin is missing. He chose the following verse for the image:

So I looked, I first bear-skin
The name I got from the bear's skin
That I shot, that I do not even dread
Whether at that time I was the object of much envy.
As high as my fame had risen before,
So low must it now lie in the highest disgrace.
You can see from this: what is highly esteemed today,
That envy will overthrow in all too short a time (translation S. Lutkat, after TITTMANN 1877, 246).

Grimmelshausen's intention is to correct the meaning of the proverb *Auf der faulen (Bären)Haut liegen* (meaning "to laze about" – a literal translation would be "to lay on the lazy [bear] skin") – and to tell the true background story of the bear-skin, which he sets in 1396 (i.e. about 250 years earlier) because of the "ancient painting" and reproduces it in the following – with great similarities to our bear-skin fairy tale.

Even in his time "bear-skin" is a term of abuse – an interpretation that can also be found, for example, in the Grimm's fairy tale *Die Goldkinder* (*The Golden Children*; KHM 85): "Let him go, he is a bear-skin and as poor and bald as a church mouse, what are we to do with him!"

There are many motifs and narrative topoi that are virtually immortal and appear in ever new contexts and variants. The son of a bear, the human being transformed into a bear and the bear-skin are among them in the Northern hemisphere. The reinterpretation of the brave bear-skin hero as a lazybones and an outlaw, under whose fur (e.g. in KHM 85), however, there may be a true king, is obviously the logical consequence of a gradual process of degradation and profanation: real bear – enchanted bear – disguised bear, or respected/venerated animal – animal transformation as misfortune – animal form as an image of a despised person.

Such reinterpretations are not the result of sheer arbitrariness, probably not even conscious processes; they are connected with the changing of thought patterns and values. If one takes a closer look and examines many versions of one type of folktales, however, lines of descent can be discovered. That is why it seems logical to bring up another interpretation of the bear-skin – the berserker, the warrior in a bear's skin. "His men went without mail shirts, and they were fierce, and they were fierce as dogs or wolves. They bit their shields and were strong as bears or bulls. They slew the people and neither fire nor steel could harm them. This was called 'Walk of the Berserker'"(PEUKERT 1988, 88, here in English translation; see also SUNDQVIST, this volume). This is how Snorri writes in Heimskringla, about 400 years before Grimmelshausen and about 150 years before the time in which Grimmelshausen sets the story. In Scandinavia there are repeated reports of men who are both human and bear. Many names testify to that: Hallbjörn – Bear of the Hall, Bödvar Bjarki – Little Bear, Björn – Bear. Even if some aspects of the classification of the berserker are disputed and their interpretation, as here by Snorri, is sometimes rather positive or, as by Saxo Grammaticus, clearly negative (HUBE 2004, 295–296), these warriors in bear's clothing are a component of northern European myths and heroic epics. Astonishingly, Snorri mentions not only bears but also dogs, wolves and bulls which increases the general confusion about his famous quote. It seems that not the bear is his actual focus but the most threatening and bloodthirsty animal possible.

The proximity, assimilation or even transformation of a fighter into an animal is probably an elementary idea, i.e. an idea that has developed independently all around the world. Animal masks, amulets, or animal names, the designation of age groups of fighters according to animals such as wolves or bears as well as the attribution of other animalistic traits are encountered around the globe (BECKER 2008, col. 541–543; see different contributions, this volume).

Back to the Brothers Grimm: It is particularly striking that the soldier in the fairy tale only becomes a berserker – a bear skin bearer – when he is no longer a soldier. Berserker and soldier go together without much thought. But in the connection of berserker/no-longer-soldier the interpretation of the bear-skin image is shifted once again. Does that make sense? I think so. The soldier – so the tale goes – becomes an animal not in battle, but in peace. What he really did on the battlefield, or what outsiders ascribed to him, becomes visible only now in a repulsive way. And no one wants to have anything to do with it after the war. In short: The repulsive dirty bear-beast of a soldier and the ideal, clean world do not come together. Grimmelshausen (and then the Brothers Grimm in their adaption) tell a highly political fairy tale.

So what about the Scandinavian variants? They do not exist – at least not with the decisive name-giving element, the bear skin. Vladimir Propp, for example, writes about this type of folktale under the heading: *The Unwashed* (PROPP 1987, 164–167; see also RÖTH 1998, 50–51). This part of the devil's pact is, of course, a fixed component in all versions, in order to grasp the meaning of the story, this element is quite sufficient. The bear skin is expendable.

Last but not least I would like to mention one Norwegian folktale in this context: *The Blue Belt*, ATU 590, *The Faithless Mother*.[6] It is a long tale with many episodes: A boy is on a begging trip with his mother. On the way back he sees a blue belt lying on the road, and his mother forbids him to pick it up. Secretly, he does so anyway and thereby gains unconquerable strength. His mother turns out to be the partner of an evil troll; the two try to trick the boy into giving up the belt. In the course of this lengthy episode, the hero wins the help of the 12 lions who were supposed to be his undoing and frees a princess from the grip of a troll. Only now does his unfaithful mother succeed in robbing him of the blue belt and thus of his strength. He is abandoned at sea, the lions follow him faithfully. He succeeds in regaining the blue belt. But now he has to go to Arabia, where his princess is with her father. I quote the passage that matters (after WEBBE DASENT 1859):

Now as the lad went along he met a man who had white bear skins for sale, so he bought one of the hides and put it on; and one of the captains was to take an iron chain and lead him about, and so he went into the town and began to play pranks. At last the news came to the king's ears, that there never had been such fun in the town before, for here was a white bear that danced and cut capers just as it was bid. So a messenger came to say the bear must come to the castle at once, for the king wanted to see its tricks. So when it got to the castle every one was afraid, for such a beast they had never seen before; but the captain said there was no danger unless they laughed at it. They mustn't do that, else it would tear them to pieces. When the king heard that, he warned all the court not to laugh. But while the fun was going on, in came one of the king's maids, and began to laugh and make game of the bear, and the bear flew at her and tore her, so that there was scarce a rag of her left. Then all the court began to bewail, and the captain most of all.

"Stuff and nonsense," said the king; "she's only a maid, besides, it's more my affairs than yours."

When the show was over, it was late at night. "It's no good your going away, when it's so late," said the king. "The bear had best sleep here."

"Perhaps it might sleep in the ingle by the kitchen fire," said the captain.

"Nay," said the king, "it shall sleep up here, and it shall have pillows and cushions to sleep on." So a whole heap of pillows and cushions were brought, and the captain had a bed in a side-room.

---

6   STROEBE 1922, 263–277. Dr. Klara Stroebe has translated and edited two volumes *Nordische Volksmärchen* (Nordic Folktales) in the series *Die Märchen der Weltliteratur* (The Folktales of the World). Unfortunately, little is known about her. In the epilogue of the edition published in 1967 under the title *Norwegische Märchen* (Norwegian Folktales), revised by Reidar Th. Christiansen, it is stated: „Sie war offenbar mit der norwegischen Sprache eng vertraut, fand auch für eigentümliche, speziell norwegische Begriffe einen treffenden deutschen Ausdruck, und sie verstand es, den Ton und die Eigenart norwegischer Erzähler im deutschen anschaulich wiederzugeben." (She was obviously intimately familiar with the Norwegian language, found an apt German expression even for peculiar specific Norwegian words and knew how to vividly reproduce in German the tone and idiosyncrasy of Norwegian storytellers.)

But at midnight the king came with a lamp in his hand, and a big bunch of keys, and carried off the white bear. He passed along gallery after gallery, through doors and rooms, up-stairs and down-stairs, till at last he came to a pier which ran out into the sea. Then the king began to pull and haul at posts and pins, this one up and that one down, till at last a little house floated up to the water's edge. There he kept his daughter, for she was so dear to him that he had hid her, so that no one could find her out. He left the white bear outside while he went in and told her how it had danced and played its pranks. She said she was afraid, and dared not look at it; but he talked her over, saying there was no danger, if she only wouldn't laugh. So they brought the bear in, and locked the door, and it danced and played its tricks; but just when the fun was at its height, the Princess's maid began to laugh. Then the lad flew at her and tore her to bits, and the Princess began to cry and sob.

"Stuff and nonsense," cried the king; "all this fuss about a maid! I'll get you just as good a one again. But now I think the bear had best stay here till morning, for I don't care to have to go and lead it along all those galleries and stairs at this time of night."

"Well!" said the Princess, "if it sleeps here, I'm sure I won't."

But just then the bear curled himself up and lay down by the stove; and it was settled at last that the Princess should sleep there, too, with a light burning. But as soon as the king was well gone, the white bear came and begged her to undo his collar. The Princess was so scared she almost swooned away; but she felt about till she found the collar, and she had scarce undone it before the bear pulled his head off. Then she knew him again, and was so glad there was no end to her joy, and she wanted to tell her father at once that her deliverer had come.

This is a surprising turn in the course of the fairy tale, but one that is presented in a tone of the greatest matter-of-fact. The hero voluntarily becomes a bear-skin for a (short) time. It remains strangely unclear how far the transformation goes. The way the tearing of the maids is told sounds more like a real bear than acting "as if".

In this fairy tale we come reasonably close to the mystery of the warlike bear transformation. And so we would have found a berserker in a fairy tale after all.

Link collection to examples of fairy tale variants (in German)

Animal tales
Die Geiß mit ihren zehn Zicklein und der Bär / The Goat with her ten Kids and the Bear (Transylvania):
https://t1p.de/xw29t

Die Katze, der Fuchs, der Wolf und der Bär / The Cat, the Fox, the Wolf and the Bear (Zyrian folk poetry):
https://t1p.de/r4xaf

Der Fuchs und der Bär / The Fox and the Bear (Lapland, aetiological):
https://t1p.de/uoby0

The Bear Trainer and his Bear ATU 1161
Vom Zigeuner und dem Bären / About the Gipsy and the Bear (Bukovina):
https://t1p.de/7tbej

Das Kätzchen auf Dovre / The little cat on Dovre (Norway):
https://t1p.de/r3na9

Bear as animal helper
Von den drei Brüdern und ihren Tieren / About the three brothers and their animals (Lithuania):
https://t1p.de/1kcg9

Son of a Bear ATU 301
Bärensohn / Son of a Bear (beautiful Serbian variant, ending in a tall tale à la Münchhausen):
https://t1p.de/0y2ia

Der Bärensohn / The Son of a Bear (Pomerania):
https://t1p.de/fgy7n

Beauty and the Beast ATU 425
Der Bärenprinz / The Bear Prince (short version from Switzerland):
https://t1p.de/dwh1x

Der Bär / The Bear (short version from Austria):
https://t1p.de/s5yca

Der Bär und das Mädchen / The Bear and the Girl (Zyrian folk poetry; short and interesting; the girl frees herself alone from the evil bear):
https://t1p.de/f86wg

Zar Bär / Tsar Bear (very interesting variant from Russia, the bear is on the one hand the monster, but is on the other hand treated very respectfully and shown to be friendly to the children):
https://t1p.de/h8ysv

Östlich von der Sonne und westlich vom Mond / East of the Sun and West of the Moon (Norway):
https://t1p.de/mcl9e

Bear-Skin ATU 361
Der Ungewaschene / The Unwashed (Russia):
https://t1p.de/8u6be

Der Bärenhäuter / Bear-Skin (Germany, Grimm, on Wikisource you can also read all Grimm versions from 1812 onwards):
https://t1p.de/lsyuf

Bibliography

Asbjørnsen/Moe 2003: P. C. Asbjørnsen/J. Moe, Sämtliche Volksmärchen und Erzählungen aus Norwegen. Zwei Bände (Bad Karlshafen 2003).

Becker 2008: S. Becker, Tier, Tiere. In: Enzyklopädie des Märchens 13 (Berlin 2008) col. 541–543.

Edsman 1956: C.-M. Edsman, The story of the Bear Wife in Nordic Tradition. Ethnos 21, 1956, 38–56.

Findeisen 1956: H. Findeisen, Mensch und Tier als Liebespartner in der volksliterarischen Überlieferung Nordeurasiens und in der amerikanischen Arktis (Augsburg 1956).

Grimm 1857: Brüder Grimm, Kinder- und Haus-Märchen, Band 2 (Göttingen 1857).

Grundtvig 1879: S. Grundtvig, Dänische Volksmärchen 2 (Leipzig 1879).

Gulya 1968: J. Gulya (ed.), Sibirische Märchen, Band 1 (Düsseldorf, Köln 1968).

Haltrich 1882: J. Haltrich, Deutsche Volksmärchen aus dem Sachsenlande in Siebenbürgen (Wien 1882).

Hirsch 2006: A. Hirsch, Das Märchen von der Seele. In: H. Lox/W. Schmidt/T. Bücksteeg (eds.), Stimme des Nordens in Märchen und Mythen. Märchen und Seele (Krummwisch 2006) 171–188.

Hirsch 2020: A. Hirsch, Tiermärchen und Märchen mit Tieren. In: H. Lox/R. Lukas (eds.), Verwandlung in Märchen und Mythen. Die Bremer Stadtmusikanten (Krummwisch 2020) 229–232.

Hirsch 2021: A. Hirsch, Original, abgekupfert oder geklaut? Märchentypen und Motive auf der Wanderschaft. In: H. Lox/A. Martin/S. Lutkat (eds.), Der Wanderer im Märchen – Das wandernde Märchen. Märchen – Kunst (Königsfurt, Krummwisch 2021) 77–81.

Hube 2004: H.-J. Hube (ed.), Saxo Grammaticus (Wiesbaden 2004).

Hyltén-Cavallius/Stephens 1848: G. O. Hyltén-Cavallius/G. Stephens, Schwedische Volkssagen und Märchen. Nach mündlicher Überlieferung gesammelt und herausgegeben von Gunnar Olof Hyltén-Cavallius und George Stephens. Mit Varianten und kritischen Anmerkungen. Deutsch von Carl Oberleitner (Wien 1848).

Karadžić 1854: V. S. Karadžić, Volksmärchen der Serben. Gesammelt und aufgezeichnet von Wuk Stephanowitsch Karadžić. Ins Deutsche übersetzt von Wilhelmine Karadžić (Berlin 1854).

Kawan 2008: C. S. Kawan, Tierbraut, Tierbräutigam, Tierehe. In: Enzyklopädie des Märchens 13 (Berlin 2008) col. 556.

Liungmann 1965: W. Liungmann (ed.), Weißbär am See. Schwedische Volksmärchen von Bohuslän bis Gotland (Kassel 1965).

Lüthi 2005: M. Lüthi, Das europäische Volksmärchen – Form und Wesen (Tübingen ^{11}2005).

Panzer 1910: F. Panzer, Das Märchen vom Bärensohn. In: F. Panzer, Studien zur germanischen Sagengeschichte. Teil 1: Beowulf (München 1910) 3–245.

Paproth 1976: H.-J. Paproth, Studien über das Bärenzeremoniell. Bärenjagdriten und Bärenfeste bei den tungusischen Völkern. Skrifter utgivna av Religionshistoriska institutionen i Uppsala (Uppsala 1976).

Paproth 1999: H.-J. Paproth, Bär. In: Enzyklopädie des Märchens 1 (Berlin 1999) 1194–1203.

Peukert 1988: W.-E. Peukert, Geheimkulte (Hamburg ^{2}1988).

Poestion 1886: J. C. Poestion, Lappländische Märchen, Volkssagen, Räthsel und Sprichwörter (Wien 1886).

Preindlsberger-Mrazovic 1905: M. Preindlsberger-Mrazovic, Bosnische Volksmärchen (Innsbruck 1905).

Propp 1987: V. Propp, Die historischen Wurzeln des Zaubermärchens (München, Wien 1987).

Rauch [no year]: K. Rauch, Isländische Märchen. Märchen europäischer Völker (Gütersloh [no year]).

Röhrich 1973: L. Röhrich, Mensch und Tier im Märchen. [Reprint] In: F. Karlinger (ed.), Wege der Märchenforschung (Darmstadt 1973) 220–253 (originally published in Schweizerisches Archiv für Volkskunde 49, 1953, 165–193).

Röth 1998: D. Röth, Kleines Typenverzeichnis der europäischen Zauber- und Novellenmärchen (Hohengehren 1998).

Stroebe 1922: K. Stroebe, Nordische Volksmärchen 2: Norwegen (Jena 1922).

Ting 1970: N.-T. Ting, AT Type 301 in China and some Countries adjacent to China: A Study of a Regional Group and its Significance in World Tradition. Fabula 11, 1970, 54–125.

Tittmann 1877: J. Tittman (ed.), Der erste Bärnhauter. In: Simplicianische Schriften von Hans Jacob Christoph [sic!] von Grimmelshausen. Erster Teil (Leipzig 1877) 246–253.

Uther 2004: H.-J. Uther, The types of international folktales. A classification and bibliography. Based on the system of Antti Aarne and Stith Thompson (Helsinki 2004).

Uther 2015: H.-J. Uther, Deutscher Märchenkatalog – Ein Typenverzeichnis (Münster, New York 2015).

Waldburg 1853: R. O. Waldburg, Zwei Märchen aus der Bukowina. Zeitschrift für deutsche Mythologie und Sittenkunde 1, 1853, 358–362.

Webbe Dasent 1859: G. Webbe Dasent, Popular tales from the Norse (Edinburg ^{2}1859). Translation of: P. C. Asbjørnsen/J. Moe, Norske Folkeeventyr (Christiania 1843). https://books.google.de/books?id=nW8AAAAAMAAJ&printsec=frontcover&hl=de#v=onepage&q&f=false.

Wichmann 1931: Y. Wichmann, Volksdichtung und Volksbräuche der Tscheremissen (Helsinki 1931).

Dr. Angelika B. Hirsch
Germany
hirsch@grenzgaenge.de

The role of the bear in the Russian folk tale: Personage, plot type, and behavioural scenarios

By Inna Veselova

Keywords: Russian folklore, bear character, fairy tales, genre conventions, behavioural scenarios

Abstract: Taking four genres as examples, this paper tracks the realisation of the bear image in the system of Russian folk tale characters. The bear as a tale character is portrayed through a set of capacities defined by the genre convention of the Russian folk tale and the communicative aims of the narrators. So, the bear character is composed of a vast range of conventions. It seems evident that the folk tale as an oral activity and as a special interaction of the narrator with the audience is a presentation of methods of acting in various life circumstances and the very means of influencing the listeners. Different folk tale genres demonstrate the various scenarios of such actions. Forty-six plots with a bear as a character (with multiple records in Russian) have been categorised into plot types. All plots were reviewed from the point of view of the role performed by the bear in each genre.

The characters of the Russian folk tale form a "pantheon" in which the same animals and human beings dwell in different plots. According to the frequency of word usage as recorded in the general glossary of the COMPARATIVE INDEX OF PLOTS OF EAST-SLAVIC FOLK TALES (referred to as CIP hereafter), this pantheon embraces the bear, the fox, the rooster, the cat and the wolf, together with the *muzhik* (the Russian peasant) and the *baba* (the Russian peasant woman), along with the *tsar* and his daughter *tsarevna* (the bride), the soldier, the *barin* (landlord) and the fool (cf. CIP). The multiple reference frequency of these actors in folk tales makes the task of their general characterisation rather complicated. The point is that in various folk tale genres, and even in the tales performed by one narrator, the characters appear in various incarnations depending on the situation – as a trickster or the victim of tricksters, as a protagonist and as an antagonist and donor.

In my article, I would like to set the objective of describing the bear as a tale character through a set of capacities defined by the genre convention of the Russian folk tale and the communicative aims of the narrators. Such an approach represents each folk tale genre as a selection of poetic modes and communicative capabilities, which each narrator has learned and is accustomed to using in his/her way.

First of all, the telling of folk tales is still one of the most common and straightforward folklore performances in Russian cultural practice. Tales are typical bedtime reading for children by their parents, tale plots are staged as plays by both professional and amateur companies, and they are shot as feature and animation films. But most important is the fact that folk tales are still being orally narrated. For many years, during our field studies in the villages of the Russian North (mainly in the Vologda and Arkhangelsk regions) as well as in the northwestern Russian cities and towns, we have been recording oral folk tales and traditional anecdotes of a similar nature told by children and adults with different life experiences (education, career, family status). By storing the audio and video recordings of multiple folk tale versions as produced by our informants, we are trying to find out who was performing folk tales and when and to whom they were (or are) being performed.

I believe that tale narration remains the current symbolic practice in modern Russia, not because it relates to the storage and transfer of "worthy" archaic knowledge from one generation to another, but because I think that through the performance of folk tale plots the narrators (both male and female) continue to solve their actual communication problems. Learning narrative skills goes hand in hand with mastering a range of behavioural tactics and a wider *ethos*. Following G. Bateson, I understand *ethos* as a style of life of a separate culture defined by the value system, temperament, and personality of individuals acting as culture-bearers (BATESON 2000). In folklore studies, it is generally accepted that folk tale (fictional) prose differs from non-fiction prose by the narrator's and the audience's convention of non-reliability of the narrated events (PROPP 1984, 54). However, it seems evident that folk tale content is not about the imaginary events of an imaginary life of a protagonist who has never existed, but the proclaiming and assertion of the tactics of dwelling in real life, as well as the ethical basis of such tactics and the choice of them.

The storyteller shares with her/his audience modes of action in various circumstances known to her/him and also provides justification for selecting this or that mode. The listeners of the folk tale react to the offered modes of action by either sharing or condemning them, which means they identify proposed tactics and make them the matter of their discussion. In 2017, at Ust-Kyma village in the Arkhangelsk region, we recorded a one-hour story-telling performance by Nina Fyodorovna Klimova. She had heard these tales from her grandmother (born approximately in 1894), who told stories in the evenings to her small granddaughters before they fell asleep: "She knew a lot of tales. Like, we read a tale at first, and then she re-told it to us. But she painted them in another colour, if you see what I mean"[1] (DURAKOVA 2017, 2). Nina Fyodorovna perceived such tales as horror stories: "She knew how to scare us …"; however, it was probably one of the reasons that young Nina remembered the tales so well "because I was petrified".

Following Michel de Certeau's concept, I assume that narrating serves the task of verbalising the *modi operandi* and the way of thinking in real life: "Tales and legends seem to have the same role. They are deployed, like games, in a space outside of and isolated from daily competition, that of the past, the marvelous, the original. In that space can thus be revealed, dressed as gods or heroes, the models of good or bad ruses that can be used every day. Actions, not truths are recounted" (DE CERTEAU 1984, 23).

Folk tales as an oral activity and as a special interaction of the narrator with his/her audience are not only a description of events happening in an imaginary world. They are also a presentation of methods of acting in various life circumstances and the means of influencing the listeners. Different folk tale genres or, to be more exact, folk tale plot patterns, demonstrate various methods of such actions.

In Russian folklore studies, folk tale genres are generally classified as follows: *cumulative tales*, *animal tales*, *trickster tales*, *tall tales* (false stories), and *fairy tales* (PROPP 1976, 47). Folk tale genres are based on key composition patterns:

---

[1] All translations from Russian into English in this text by the author.

- the cumulative pattern[2] in cumulative tales;
- the trickster's tricks pattern in trickster and animal tales;[3]
- the pattern with a sequence of functions, from deficit/damage to their elimination, as defined by V. Ya. Propp (1928), is a composition unit with an act or deed of the tale character; a specific pattern for fairy tales;[4]
- the pattern which is based on the formula of the impossible constitutes the composition of tall tales.

Reflecting on composition/plot pattern typology from the point of view of their origin and importance, not only for narrative practices but for the culture in its entirety, Yu. M. Lotman (1992, 242) comes to the following conclusion: "The plot is a powerful weapon for life comprehension. Only due to the creation of the narrative forms of art has a man learned to identify the plot-like aspect of the reality, to dismember a non-discreet stream of events into several discreet units, linking them with certain meanings (i.e. semantic interpretation) and organizing them in consecutive orderly chains (i.e. syntagmatic interpretation). The accentuation of events as discrete units in the plot and attributing certain meanings to them, on the one hand, as well as providing them with definitive timely, causative-consecutive, and any other sequences, on the other hand, form the essence of a plot. [...] The more a man's behavior acquires features of liberty as compared to the automaticity of genetic programs, the more important it is for the man to build the plots of events and behavior. [...] When people create text with plots, they learn to identify plots in real life and thus interpret life for themselves". For me, the most important provisions are the following: First, the plots of imaginative narratives are directly linked to an understanding of the plots of life, second, the identification of the plots in life is both a creative and a reflexive process, and third, people learn to divide and identify. It is also significant that each person has to undergo the whole process of learning. Folk tales are the training that enables the comprehension of the plots in one's own life and the awareness of behavioural scenarios that are acceptable to the culture.

The earning of competencies presented in folk tales takes place step-by-step. A schematic description of this process assumes an almost unconscious familiarisation with the very early childhood tales of the cumulative type. Our interlocutors often inform us that they used to tell their children cumulative tales about animals. The next are the tales about already-known animal protagonists with trickster-type plots. The comprehension of fairy tale plots requires developed competencies of the listeners as, in any fairy tale plot, the journey from wrecking to elimination embraces both cumulative and trickster elements. Our next step will be the analysis of roles attributed by narrators to the bear protagonist in Russian folk tales. Also, we will try to find out which modes of acting in a weak or a strong position this character embodies.

The materials for my analysis come from various sources. First, statistics for each type of Russian folk tale with a bear-protagonist or bear character can be found in the CIP, which comprises all published folk tales in Russian up until the beginning of the 1970s. Based on this material, we will define the plot types in which we can most frequently meet the bear.

---

2   Folk tales in which the plot is built on the creation of a consecutive chain of bodies, deeds, meetings and acquisitions and which end with the destruction of the created chain. The hierarchies, orders and relations are established in these plot patterns, while the performance versions may vary with a good or bad ending for the protagonist.
3   In Russian trickster tales, peasants always deceive the landlord, gypsies very often deceive peasants, fools usually deceive normal people, and wives often deceive husbands.
4   Thus, a fairy tale is "a genre of tales, which begins with inflicting some harm or damage (kidnapping, exile, etc.) or in which a deficit or desire to possess something (a tsar sends his son to fetch a firebird) is developed through exiling the protagonist away from home with a mission, him meeting a donor who gives him a magic device or a helper to find and acquire the target of the search. Further, the tale sequence embraces a combat with an enemy (snake-fighting as the most important one), return and pursuit" (Propp 1928, 22).

of folk tales recorded over the last decades and stored in the Folklore Archive of St. Petersburg State University, it is one of the most popular. In the Vologda collection of the archive, there are 139 tales (35 plots or contaminations), and nine of them are the versions of the "Bear on Lime Leg" pattern. Eight tales have been performed by women and one by a man. Two out of eight female narrators informed us that they had learned the tale from their grandfather or father,[13] who often went to the forest hunting and told their children and grandchildren the stories based on his experience, mixing truth and lies. So, in the middle of the 20th century, this tale was narrated both by men and women, while at the end of the 20th century – when tales were recorded by folklore scientists – the narration of tales had become mostly the women's practice.

The tale's storyline is as follows: a *baba* / a *starukha* (a middle aged /an old peasant woman) sends her *muzhik* / *starik* (a peasant man / an old man) to the forest to get meat from the bear's body. The hunter axes away one of the paws of a sleeping male bear. The bear wakes up and makes a leg out of lime wood instead of the lost paw (as a result the bear has three paws and one lime leg). Then the bear goes to the village looking for his paw singing a threatening song (accusing the old woman of using his paw in her household routine). The old man and his wife hide from the bear in their *izba* (cottage) or start to bargain with him. The performers say that the tale can have a bad (the bear eats the old man and his wife) or a happy end. The happy end is considered preferable when the story is narrated to the smallest children.

The bear's threatening song consists of three parts and a final exclamation: "Om nom, I'll eat you!". In the first part of the song, the bear addresses his lime leg: "Screech, my leg, screech, screech my lime (birch) leg". In the second part of his song, the bear recites various types of dwelling places (*selo* [a large village with a church] and *derevnja* [a small one with only a chapel]), where "normal" inhabitants sleep, as opposed to the guilty (old) woman who is not sleeping: "All villages sleep, both big and small, only one woman is awake". The song's third part accuses the old lady of her use of the bear's paw: "She sits on my skin, she sucks my flesh. I'll eat her the moment I see her! Om nom!".[14] The Vologda versions of the plot have one more peculiar feature; in them, the old man and his wife play "hide-and-seek" with the bear. In the majority of variants, the old couple hides in the stove leaving a trap for the bear by opening the cellar's trapdoor on the floor. The Russian stove (*russkaja pech*) occupies the central place in the Russian village house – it heats the inside space, food is cooked in it, water is heated and boiled there, people sleep on it, get warm, and dry their garments and food stores. However, an adult can hide inside the Russian stove only if it has been specially built with a high arch and sufficient inner space. This type of stove is found only in areas where bathing and washing inside stoves is a habit. It is worth mentioning that the area where the plot of the "Bear on Lime Leg " is most common, according to A. B. Ostrovskiy, – the Higher Volga region and the Vologda region (OSTROVSKIY 1990) – coincides with the locations where there is a custom to wash in the stove. That is how A. V. Stepanov describes this hygienic and health-preserving habit: "The practice of washing, bathing, and steam-bathing in the Russian stove with an oven was quite common in the Russian North together with the 'bathing tradition'. Washing in the stove is related to the so-called Rostov-Suzdal colonisation of northern territories (by people from Rostov Velikiy and Suzdal regions, which were the central areas of Russia in the 10th to 17th centuries and still are). The height of the stove arch in regions where it was used for bathing was particularly high. People bathed in stoves

---

13 Folklore Archive of St. Petersburg State University, Vash 16-14. Recorded from Polina Alexandrovna Gorodnichnaya (born in 1928), in the village of Bereznik at the Roskomsky village community of Vashkinsky District in Vologda Region on 14.07.2002 by O. I. Alimova and E. E. Samoylova.
14 Folklore Archive of St. Petersburg State University, Syam 16-24. Recorded from Valentina Efimovna Voronova (born in 1928 in the village of Bor) in the village of Makarovskaya at the Dvinitskiy village council of Syamzhenskiy District in the Vologda Region on 30.07.2006 by Yu. Yu. Marinicheva.

when these had become cool enough after fuelling and the stovepipe had already been closed. One would have to go into the stove and leave it with one's head first. People bathed sitting with their feet directed towards the stove opening. The damper could be closed to save the heat. Newborn children of up to 40 days of age could also have been washed in the following family tradition: The grandmother (a mother-in-law or a healer) got into the stove, the mother handed her the baby, who was placed on the straightened legs of the woman in the stove with its head to the feet of the woman. After washing, the grandmother handed the baby back to mother with its head first" (STEPANOV [no date]).

From the recordings of oral performances of this tale in circumstances close to natural ones (narration for grandchildren in the presence of the recorders, or for the recorders who were in the same age group as the performer's grandchildren), it becomes evident that the narrators did not take any pains to accurately reproduce the twists and turns in the plot, but had the primary objective of frightening the audience or maintaining its closest attention. Raisa Fyodorovna Moskovtseva told us the tale about the bear[15] several times and her story always ended unexpectedly; either by pinching the listener's side at the end of the bear's song or by discourse about the material of the bear paw prosthesis or by reminiscences of the tale being performed by her grandmother. In the latter case, the narrator added the grandmother's "horror" riddles ("Where is the bear's eye in the room?" [a knot in the wood of a floorboard], "Where is the bear paw in our house?" [the broom]). The difference in the performances of the same plot demonstrates the secondary role of the plot for a narrator. The plot is just a pre-requisite for a storytelling, and the performer is free to improvise based on the storyline, seeking her/his own communicative goals.

Anna Vasilyevna Voronicheva, when narrating the tale, performed the bear's threatening song three times.[16] However, in her performance the reminiscences of her reaction as a small girl to the singing of her grandmother in the "horrifying" bear voice ("All of us: 'Oh, oh, oh, will he really eat us?'"; "All of us: 'Oh, how terrible!!!'") are more important than any variances in the tale composition. Just like other narrators, A. V. Voronicheva recollects the skills and the tricks of performers from whom she has learned the tale, such as intonations, gestures, movements and interposition with the listeners (sitting together on a stove). When talking about her fear in the course of listening to the tale, Anna Vasilyevna also remembers with relief the sense of the audience's common joy when the old couple were victorious over the bear. In one of the archived tale versions, the theme of the old man and his wife hiding is replaced by the description of the ongoing approach of the bear – past the window, through the door, to the stove and the berth, where the old woman lies in the tale, but also where the children are listening to the performer. The narrators describe the circumstances at the time of the tale-telling as sitting on the Russian stove (the one to which "the bear is approaching") before bedtime or near the stove in the course of common labour activities in the form of spinning. For the time being, we shall not pay too much attention to the fact that the bear in the tale threatens the old woman, though his paw has been axed by the hunter. The fact that the addressee of the bear's song-threat is the woman, and that the main plot interactions take place between the bear and her, leads us to assume that the basic character relations here are developing between the woman and the bear. How these relationships are presented in fairy tales, we will see later. Now we have all the evidence to suggest that performers of the "Bear on Lime Leg" plot insisted on reproducing a suspense effect, well known to them since their childhood. The suspense is created by dwelling on the borders

---

15 Folklore Archive of St. Petersburg State University, Syam 16-24. Recorded from Raisa Fyodorovna Moskovtseva (born in 1932 in the village of Martyanikha) in the village of Goluzino at the Goluzinskiy village council of Syamzhenskiy District in Vologda Region on 15.07.2005 by S. B. Adonyeva and I. S. Veselova.
16 Folklore Archive of St. Petersburg State University, Kir 16-3. Recorded from Anna Vasilyevna Voronicheva (born in 1924 in the village of Uloma) in the village of Ivanov Bor at the Ivanovoborskiy village council in the Kirillovskiy District in the Vologda Region on 20.07.2006 by M. D. Karmanova.

between the safe and the horrible. Folk tales unlock these borders on the levels of composition, semantics, pragmatics, and perception. First, a folk tale is embedded into a chain of stories in such a way as to conceal the difference between fact and fiction (for example, the hunters-storytellers mixed the tale performance with true stories of their own forest experience). Second, the folk tale's imaginary universe provides for the mutual crossing of the borders between human beings and the bear: greed and self-interest bring the hunter to the forest for easy loot, the bear takes revenge for the harm done by coming to the man's own house (in the village). Third, the very place of the tale performance coincides with its poetic universe (village, house, stove). Fourth, the narrator partially becomes the bear with the horrible singing voice and with three paws and one lime leg, who pinches the listener's side "on behalf of the bear". The suspense accumulates and intensifies due to the narration taking place at the time when the audience is caught between wakefulness and sleep, by the travelling of the bear from the forest to the village, the non-obvious frontiers between facts and fiction, the transition from singing to prosaic utterance and back, the intonation change from the "frightening" singing voice back to normal speech. Every feature helps to create and maintain the borderline condition of the audience throughout the performance. The memory of the received impression remained intact from childhood to old age. Former listeners remember the growing fear from the approach of the singing bear on a screeching artificial leg, the joy when he is overcome, or the horror of imminent death (depending on the tale finale selected by the performer in each case).

So, the number of plots of the cumulative tales with the bear among its characters is limited but the number of records is quite impressive, confirming the fact that the bear was easily identifiable and popular. The "Bear on Lime Leg" is the most common one and helps the narrator to teach his audience how to act in the atmosphere of suspense between the threat of perishing and the chance of rescue. The bear becomes a source of fear, which is realistic for those who live in a village near the forest and know the stories about real animals.

Now we move on to the analysis of the next type of plot – with the trickster and his victim. In trickster tales, where there is a competition between a cunning person and the victims of his/her deception, the bear as a character has a profoundly other nature. The CIP contains references to 34 plots that we identified as trickster's – based on cunning tricks in which the bear plays at least some role. In twelve plots out of 34, the tricksters (usually peasants) use the bear or mimic the bear to frighten their victims and obtain from them the desired object. In these plots, the bear becomes a tool in the hands of a "smooth operator". Trickster use the habitual (according to life experience and collections of fairy tales) fear of a bear in a trick with "false danger".

In 22 plots, the bear acts as a full partner in competitions and tournaments that occur in the folk tale. However, the plots of trickster tales do not indicate the winner in advance. The trick/tale outcome depends on the balance of power and the circumstances in which the trick is played. For example, in household folk tales, where the actors are a peasant vs. his wife or a peasant vs. a fox, the deceiver and his victim often swap roles. In the combat "peasant vs. the devil", the peasant will be victorious in most cases. Let us analyse the bear's chances of victory in the Russian trickster tales. In CIP 1030 – "Harvest Sharing"[17] – the partners arrange who will get a particular part of the planted crop after harvesting. The trickster (the peasant or the bear) uses his knowledge about the edible part of each crop (the upper part for wheat and the roots for turnips). The trick outcome depends on the participants of the circle of negotiation. The bear has a chance of victory if his counterpart is the devil or the fox. If the bear is facing the peasant in these negotiations, he will lose. In three plots ("The Bear on a High Wagon" [CIP 16], "The Dog and the Wolf/Bear" [CIP 101], "The Huts Made of Bast and

---

17 CIP 1030: "Harvest Sharing". "I will get the upper part of the crop and you will get the roots", arranges the peasant (the bear) in his agreement with the devil (the fox), and they plant wheat; the next year, the upper part is promised to the devil, but, because turnips are planted, the devil is again outmaneuvered.

Ice" [CIP 43]),[18] the bear enjoys a small chance of winning: mostly through successful collaboration with the dog (CIP 101) or with the rooster/tomcat (CIP 43). In 17 trickster plots, the bear loses the combat to the peasant (nine plots), to the peasant's wife (two), to the fox (five), and to the devil (one). We can see that the peasant is the most common counterpart of the bear in the competition between them and that the bear cannot succeed. In two plots, when the peasant's wife enters the scene she also manages to outsmart the bear. One episode of the woman's victory is present even in an erotic tale[19] in which the bear acts as an unlucky sexual partner of the woman (this plot had widely circulated in the form of the *lubok* pictures). Another plot about the woman's victory looks similar to the "Bear on Lime Leg" plot (CIP 161A** = AA *160 I); it is called "The Bear and the Old Woman".[20] Here, an old woman meets the bear by accident, and he threatens to eat her. The old woman promises to give some gifts to the bear in exchange for her life, naming them Strongie, Warmlie and Further-Feather. However, these gifts are not the names of valuable animals but false promises. The old woman plans to close the door with a *strong* lock, hide in the *warm* stove and escape any *further* revenge. In contrast to the "Bear on a Lime Leg" plot, the "Bear and the Old Woman" plot is distinguished by the absence of a repetitive bear song. In the plot of "The Bear and the Old Woman", the repetition and development of the old woman's trick comes to the fore. In five more plots, the bear falls victim to the fox's tricks.

Thus, in the trickster tales, the bear is very rarely a winner. If he succeeds, he mostly acts as a tool in the hands of a more advanced actor (the peasant or the dog), or cooperates with a better player (the dog or the rooster). In cases of any open intellectual one-to-one combat (with the peasant, with his wife, or with the fox) the bear mostly loses.

In this article, we have not been able to track down who was the narrator of the tales about winning over the bear and to what audience they were performed. However, we may point out that trickster tales require developed narrative competencies from the audience – the narratees must know how to identify the actors' interests and the cunning means of the deceiver. It looks like the trickster tales were told to listeners who were already familiar with the lessons taught by the cumulative tales. These establish the accepted order and hierarchy and they perform the function of marking the "strong" and "weak" positions in the social universe. As for the trickster tales, they teach how to behave in a world where one constantly finds oneself in a weak position. This suggestion comes from the development of the concepts of "tactical behaviour" introduced by Michel de Certeau, and the "weapons of the weak" made known by James Scott. Both de Certeau and Scott think that the "weapons of the weak", being the means of the non-evident resistance of those who are being dominated, include gossiping, rumors and anecdotes about the mighty, as well as legends and tales in which a character who was originally weak wins at the end (DE CERTEAU 1984; SCOTT 1985). The trickster tales provide in their opinion both a description of the "weapons of the weak" and a vivid demonstration of utilising such weapons. Thus, any victory over the bear in tales provides a person with a chance of winning (however imaginary) over a horrible and dangerous partner. Everybody with experience of visiting the wilderness knows how dangerous the bear can be, so any success in an interaction with the "master of the forest" depends on the wits and skills of his counterpart.

---

18   CIP 16: "The Bear on a High Wagon". He is taken by mistake for a priest (a general). Only two versions of this plot are known. – CIP 101: "The Dog and the Wolf (Bear)". The wolf organises the kidnapping of a baby (a sheep) with the dog and lets the dog snatch it from him for the dog to get food from his master. – CIP 43: "The Huts Made of Bast and Ice". The fox builds an ice hut for herself, while for the wolf (the bear, the hare) she builds a hut made of bast. In spring, the fox's hut melts and she tries to get possession of the bast hut.

19   CIP 152C*: "The Bear and the Woman". They wrestle after having come to an agreement that the bear shall bring the woman a hive full of honey if he occasionally tears something which belongs to her; the deceived and disgraced bear runs away.

20   CIP 161A** = AA *160 I: "The Bear and the Old Woman". The old woman survives the encounter with the bear, promising him Strongie, Warmlie, and Further-Feather as ransom; then she interprets her promises in her favour.

The Russian trickster tales (with animal and human characters) create a variety of plots. We can see a range of cunning tricks to be used in situations when any direct stand for one's interests is not possible. Sound recordings of performances of such tales stored in the Folklore Archive of St. Petersburg State University provide the evidence that the natural situation for trickster tales (or traditional anecdotes) narration is not only the performance by the elder to the younger, but rather the storytelling in a group of people of the same age sitting around a festive table or in any other informal circumstances. The archive contains several records of continuous performances of tales told in a row in a female company with such cross-cutting plots as "Lazy Wife" and "How Muzhik Taught His Wife". Despite the didactic titles, lazy wives in these tales turn out to be wiser and luckier than their husbands. The general response to the story about each successful trick was a burst of laughter, creating the effect of women's solidarity in resistance against the established forces, laws, and social/natural order. Regardless of the fact of who the winner is in each round of the tale – the wife or her husband, the old woman or the fox, the woman or the bear – the victory is never ultimate, and a smart trick will be highly appraised by the audience.

Finally, I would like to analyse the bear as a character in fairy tales, which are tales *par excellence*, i.e. they belong to the genre that first comes to the mind of any folklore specialist or native speaker when folk tales are mentioned. Most often, this type of tale ends with the victory of a protagonist, they contain magic and the action takes place "once upon a time" in an imaginary land.

There are four fairy tale plots with bear characters known in Russian folklore.[21] The bear plays the main role – of the antagonist and the donor – in plots marked as CIP 480 = AA 480*B, *C and CIP 311. In these cases, the protagonists are girls of pre-marriageable age. These tales are considered metaphoric descriptions of the wedding trials and female initiation in Russian traditional culture. In plot CIP 311, three sisters come, one by one, to the hut in the woods; the elder sisters break the ban set by the master of the hut (and of the forest) and are punished, while the last and youngest one passes the test and in some versions becomes the bear's wife: "So she walked and walked and came to a hut. She enters the hut and sees the bear who is sitting there fuelling the stove. The bear says: 'Oh pretty lass, do some weaving for me, and if you fail I'll cut your head off.' She weaved and weaved and succeeded. So they started living together" (SIMINA 1975, text nos. 17, 91).

The girl's willingness for the marriage has been defined in the village culture not only by physiological but rather by social maturity. The latter, as T. A. Bernshtam proved, was related to the girl becoming skilled in "artful handicraft", which was the combination of spinning, weaving, sewing, embroidery, singing, and circle dancing (*khorovod*) with the girl's erotic capabilities (BERNSHTAM 1999). These particular skills are demonstrated by the brides in fairy tales. As for the studied plots, the testing is done not by the *tsar*, the *tsarina*, or the bridegroom, which are common trial arrangers

---

21 The plot titles below are cited with CIP reference numbers. Each plot title is followed by its description from the glossary and the number of versions recorded therein: CIP 480, "Stepmother and Stepdaughter". The stepdaughter is taken to the forest; Morozko (Baba Yaga the witch, the forest spirit [*leshiy*], the wolf, the bear) tests the girl and rewards her (AA 480*B); the stepdaughter plays hide-and-seek with the bear, she is assisted by a mouse. 26 versions. – CIP 311: "Bear (Leshiy, Sorcerer) and Three Sisters". The elder sisters violate the prohibition to enter a special room and are murdered; the younger sister resuscitates them, hides them, conceals her visit to the banned room, makes the killer take the sisters and then herself to their home, banning him from looking at the "gifts for parents" (flees, leaving a doll on her bed). 20 versions. – CIP 650A: "Ivan the Bear's Ear". A young man (the son of a bear in many cases) shows incredible strength (at a blacksmith's shop, in the forest), sometimes brings mischief (does damage) to his master, of which his master complains; the young man goes into exile, he accomplishes feats. 43 versions. – CIP 315: "The Beast's Milk". A sister (mother) conspires with her lover to kill a young man and sends him to fetch the milk of a female wolf, she-bear or lioness, pretending to be ill; binds him; the protagonist saves his life with the help of the beasts – wolf, bear or lion cubs; destroys his enemy, punishes the hypocrite sister (mother). 72 versions.

in fairy tales, but by the most dangerous of forest beasts (the bear, the wolf) or forest spirits (Morozko[22], *leshiy*[23]).

Folklorist Ivona Zhepnikovska, in her article "Bear in a Russian Fairy Tale", asks why a bear, a wolf, or a forest owner act in a synonymous position as antagonists in both plots (CIP 480 = AA 480*B, *C and CIP 311). She writes: "The interchangeability of a bear and a wolf is explained by similar demonological concepts, including the ability to shape-shift, marriage and erotic symbolism. […] Isomorphy of the bear and *leshiy* is predictable and can be explained by their genetic relationship. The bear is *leshiy*'s prototype […], because the cult of the forces of nature, embodied in *leshiy*, among others, has been preceded by the cult of totems" (ZHEPNIKOVSKA 2012, 63). She links the mythological sources of folk tales with a totemic cult. The bear, the *leshiy* and the wolf are, in her opinion, metaphoric "ritual specialists" in a girl's *rites de passage*. The girl who passes the test, in which her life is at stake, receives her life back and a great dowry, while the one who fails perishes or gets rubbish instead of a dowry in more-merciful-to-the-audience versions. While, in CIP 311 tales, the trial lies in the sphere of the young girl's handicraft and household skills, in the tale of the CIP 480 plot, *Morozko*, or the bear, is the trial-setter who checks the heroine's patience (test by cold in the forest) or her ability to play the blind man's buff game (*zhmurki*).[24] In I. Zhepnikovska's opinion, "the idea of death and resurrection, perpetual regeneration and the repetition of life cycles forms the basis of the game of *zhmurki* – one of the key trials, which female protagonists of tales of 'Stepmother and Stepdaughter' type undergo because […] 'the blind looking for the sighted' theme is a synonym of 'the dead looking for the living' topic. We observe another confirmation of the exclusive predetermination of the bear to fulfil the role of a creature opening the mystery of life and death to the fairy tale female protagonists and, which is also quite important, the interdependence of both phenomena" (ZHEPNIKOVSKA 2012, 70). Special relations between bears and women are captured in the Russian stories about encounters with the "master of the forest" (the name for both *leshiy* and the bear in the speech of past and modern country people living in the Russian North) as well as in Russian folk pictures such as "The Bear and the Peasant Woman" (LUBOK 1984, fig. 91), in which the plot of the above-mentioned erotic tale CIP 152 is presented. One such detailed story was recorded by Evgeniy Baranov in a Moscow tavern in the 1920s from a 60-year-old woman. It is a long story that starts with the forced concubinage of a countrywoman with a bear and the birth of an ugly boy from this affair ("with a human face but the arms and legs of a bear, and the bear's ears as well"). After returning to the village, the woman begs for pardon from her husband and her landlord for her long absence and the birth of the bear's son. Then she asks a bishop (*archierey*) and the *tsar* for permission to christen the baby. In the end, the decision is taken to make the boy an exhibit in a museum and issue a pardon for the woman ("What was the poor woman guilty of? She probably did not enjoy her life at all.")

---

22 Morozko ([Grand-]Father Frost) – this is the name of the character who, in CIP 480, tests the main girl character. A father, at the request of the stepmother, leaves his own daughter (who is of marriageable age) in the winter forest for the night. There, Grandfather Frost (Morozko) asks her if she is cold. The fairy tale does not describe Morozko's appearance, we only know that he knows how to control the cold. The girl politely answers him three times that everything is fine, although she is freezing. He leaves her alive and rewards her with a rich dowry. The second girl (the stepmother's daughter), in a similar situation, rudely responds to Morozko, for which he freezes her to death. In a fairy tale with this plot from the Folklore Archive of St. Petersburg State University, the storyteller, in the finale, reports that "grandfather" was the Master of the Forest: "What kind of grandfather was that?" > "The grandfather of the forest."; cf. Folklore Archive of St. Petersburg State University, Bel 16-1. Recorded from Anna Marovna Antonova, born in 1906, in the village of Mitino, Belozersky District, Vologda Region, July 6, 1997, by E. A. Migunova, M. M. Pirogovskaya and A. Yu. Ponomareva.

23 *Leshiy* (*lesnoj*, *leshak*) – a forest spirit or a master of a forest. According to the Russian traditional worldview, each locus has its own metaphysical master. In stories about encounters with a wild beast in the forest, the narrators often refer to the bear as "master of the forest".

24 *Zhmurki* is similar to Blind Tom or the blind man's buff game – a game leader with his/her eyes blinded by a bandage tries to catch other players in a closed area.

with payment of a pension to her from the treasury and awarding her a medal (FOLKLORE TREASURES 1998, 285–290). The tales based on plots 480 and 311, folklore pictures, erotic tales and stories about events allegedly happening in real life allow the initiation of a woman by a bear, their sexual relations and common household and all the above-mentioned in various combinations. At the same time, there are no plots in the whole body of Russian folklore about any close relations between a man and a female bear. It means that the bear in Russian fairy tales is an embodiment of masculinity upgraded to the level of the mythical master of the forest and the ritual specialist.

The fourth remaining plot of the Russian fairy tale with a bear character marked as CIP 650A – "Ivan the Bear's Ear" – is a narrative with a male protagonist born by a woman from a bear. He possesses extraordinary strength and accomplishes unbelievable feats. This plot refers to stories about innocent, persecuted heroes.

Taking four tale genres as examples, we have tracked the realisation of the bear image in the system of Russian folk tale characters. In tall tales, we see as the "impossible formula" that a bear flies through the skies. In Russian folklore, birds usually fly between this world and the other in time and space. Birds are stable metaphors for femininity, so, the tall tale states that, even in a fairy tale, the correlation of the image of a bear with the categories of bird/feminine is impossible. The bear is a masculine image. The cumulative tales strengthen the bear's reputation as a mighty force. The trickster tales usually end with the bear losing the competition with the trickster (a man, a woman, a fox, and so on), thus allowing storytellers and their audience to dream about a victory over a mighty force and the established world order. Becoming skilled in tricks provides a chance of overcoming prevailing forces by cunning. The plots of fairy tales with the bears as donors and antagonists against a mythological background describe a provisional/ritual marriage of the woman and the bear. Such a marriage legitimises the supernatural powers of women. These powers are the appraisal of her skills (spinning, weaving, singing) and they demonstrate her erotic (fertile) capacities as well as the knowledge of the border between life and death. So, the bears in Russian folk tales play various and multiple roles: the embodiment of almost existential horror, the simpleton, the metaphysical master, and the ritual specialist. However, in none of his roles so far is the bear domesticated: he is a physical and metaphysical power that can be conquered by artfulness and deception, but in any other respect is quite unmanageable.

BIBLIOGRAPHY

AARNE/THOMPSON 1964: A. AARNE/S. THOMPSON, The types of the folktale. A classification and bibliography. Antti Aarne's "Verzeichnis der Märchentypen", translated and enlarged by Stith Thompson. Second revision (Helsinki 1964).

ANDREEV/AARNE 1929: N. P. ANDREEV/A. AARNE, Index of fairy tale plots according to the Aarne system (Ukazatel' skazochnykh syuzhetov po sisteme Aarne) (Leningrad 1929).

BATESON 2000: G. BATESON, Steps to an Ecology of Mind: Collected Essays in Anthropology, Psychiatry, Evolution, and Epistemology (Chicago 2000).

BERNSHTAM 1999: T. A. BERNSHTAM, "Slyly-wisely to a needlewoman" (embroidery-sewing in the symbolism of maiden maturity among the Eastern Slavs) ("Khitro-mudro rukodel'itse": [vyshivaniye-shit'ye v simvolizme devich'yego sovershennoletiya u vostochnykh slavyan]). In: Woman and the material world of culture among the peoples of Russia and Europe: collection of articles (Zhenshchina i veshchestvennyy mir kul'tury u narodov Rossii i Yevropy). Compiled by L. S. Lavrentieva/T. B. Shchepanskaya; edited by T. A. Bernshtam (St. Petersburg 1999) 191–250.

BERNSHTAM 2003: T. A. BERNSHTAM, Bird symbolism in the traditional culture of the Eastern Slavs (Ptich'ya simvolika v traditsionnoy kul'ture vostochnykh slavyan). In: Folklore Studies at St. Petersburg University, 2003. http://folk.spbu.ru/Reader/bernshtam1.php?rubr=Reader-articles.

CIP: COMPARATIVE INDEX OF PLOTS of East-Slavic Folk Tales (Sravnitel'nyy ukazatel' syuzhetov "Vostochno-slavyanskaya skazka"). Edited by K. V. Chistov, compiled by L. G. Barag/I. P. Berezovsiy/K. P. Kabashnikov/ N. V. Novikov (Leningrad 1979).

DE CERTEAU 1984: M. DE CERTEAU, The Practice of Everyday Life (Berkeley 1984).

Durakova 2017: K. V. Durakova, Folk tale repertoire of Nina Fyodorovna Klimova (Skazochnyy repertuar Niny Fedorovny Klimovoy). Folklore Archive of St. Petersburg State University. Field Reports Collection (St. Petersburg 2017).

Folklore treasures 1998: Folklore treasures of the Moscow land. Fairy tales and non-fairy tales (Fol'klornyye sokrovishcha Moskovskoy zemli. Skazki i neskazochnaya proza) 140 (Moscow 1998).

Lotman 1992: Yu. M. Lotman, The Origin of Plot in the Light of Typology (Proiskhozhdeniye syuzheta v tipologicheskom osveshchenii). In: Yu. M. Lotman, Selected Articles in three volumes. Vol. 1: Articles on Semiotics and Typology of Culture (Tallinn 1992) 224–242.

Lubok 1984: The Lubok. Russian Folk Pictures in the 17th to 19th Centuries (Leningrad 1984).

Ostrovskiy 1990: A. B. Ostrovskiy, Russian Folk tale "Bear on Lime Leg": from Structural Interpretation to Historical Roots (Russkaya skazka "Medved' na lipovoy noge": ot strukturnoy interpretatsii k istoricheskim kornyam). Paganism of the Eastern Slavs (Yazychestvo vostochnykh slavyan) (Leningrad 1990).

Propp 1928: V. Ya. Propp, The Morphology of the Fairytale (Morfologiya volshebnoy skazki) (Leningrad 1928).

Propp 1976: V. Ya. Propp, The Genre Classification of Russian Folklore (Ghanrovaja klassifikacija russkogo folklora). In: B. N. Putilov (ed.), The Folklore and the Reality (Folklor i deistvitel'nost') (Moscow 1976) 46–82.

Propp 1984: V. Ya. Propp, Russian Fairy Tale (Russkaja skazka). Ed. by K. V. Chistov, V. I. Eremina (Leningrad 1984).

Riabov/De Lazari 2012: O. Riabov/A. De Lazari (eds.), "Russian Bear": History, Semiotics, Politics ("Russkiy medved": Istoriya, semiotika, politika) (Moscow 2012).

Scott 1985: J. Scott, Weapons of the Weak: Everyday Forms of Peasant Resistance (Yale 1985).

Simina 1975: G. Ya. Simina, Pinega fairy tales. Collected and recorded by G. Ya. Simina (Arkhangelsk 1975).

Stepanov [no date]: A. V. Stepanov, How to Wash and Bath in the Russian Stove. https://youtu.be/pUHApERY0G8.

Zhepnikovska 2012: I. Zhepnikovska, A bear in a Russian fairy tale: sketch of the study of the problem (Medved' v russkoy volshebnoy skazke: eskiz issledovaniya problemy). In: Ryabov/De Lazari 2012, 62–71.

Associate Prof. Inna Veselova
Russian Literature Department
Philological Faculty
St. Petersburg State University
St. Petersburg
Russia
veselinna@mail.ru

Bears bring spring: An anthropological view on the role of the bear in middle European winter feasts

By Jet Bakels and Anne Marie Boer

Keywords: Bears, French Pyrenees, fête de l'ours / *bear feasts, fertility, pact with wild animals, animal symbolism*

Abstract: "*Animals are good to think with*", *according to the French anthropologist Claude Levi-Strauss. They provide traits that help us to express our thoughts: Tigers thus symbolise ultimate strength, foxes ultimate cunning, and the bear is known to bring spring as soon as it emerges from its den after a long winter sleep. According to a widespread folk tradition, the bear is therefore associated with fertility. Several villages in the Pyrenees are known for their annual bear feasts with ritual elements: the* fête de l'ours. *During this ceremonial feast, a man dressed as a bear is chased but also cherished and recognised for his fertility-bringing role in the lifecycle. In present-day society we only seem to fear the wild strength of the bear, but we might reconsider our attitude towards wild animals and take an example from the feast of the bear. We know very well how to cope with danger, as we take planes, trains and cars every day. So let us consider the multi-layered and ancient relationship humans had with the bear as an inspiration for a new pact with wild nature!*

Introduction

"Animals are not only good to eat, but also good to think with", said the French anthropologist Claude Lévi-Strauss (Lévi-Strauss 1964, 89 [French original: 1962]). Tigers symbolise ultimate strength, foxes ultimate cunning, and bears represent fertility. Animals thus provide traits that help us to express our own thoughts. They provide a means of ordering the world. Animals are also linked to seasons, especially spring – the explosion of renewed fertility in nature that we see around us at Easter. Nowadays, we do not emphatically dwell on the changing of the seasons, but this is a relatively new, 20th-century development. For the peasant population of Europe, not so long ago, the timely arrival of spring made the difference between icy cold and heat, dark and light, hunger and food, death or survival.

From the Balkans to the Pyrenees, the bear is known to bring spring as soon as it emerges from its den after a long winter's sleep. According to a widespread folk tradition, this happens on the second day of February. Whether or not it stays outside its den is crucial to how the winter progresses. If the bear is frightened by its own shadow – caused by the sun or moon when the sky is clear and therefore cold – it will return underground and sleep for another forty nights before spring begins. If not, the bear, often with cubs born in its winter lair, sets off and spring can begin. In countries such as Hungary and Romania, an army of journalists throngs the bear enclosures of the larger zoos on February 2. What are the bears doing? This is the crucial question that occupies the minds of newspapers and television, just like in America where, also on February 2, the behaviour of the groundhog is attributed this predictive value. The theme was explored in the 1993 film, Groundhog Day, starring Bill Murray as the grumpy weatherman.

A HESITANT RELATIONSHIP

Our relationship with wild nature is changing and shows different tendencies. In Europe, awareness of the loss of biodiversity is increasing, as is demonstrated by recovery efforts such as the reintroduction of large predators: the bear and the wolf. Three bears from the Balkans were introduced into the French Pyrenees in 1996 and another five in 2006 to supplement the minimal local population. Today, 24 bears roam the mountains, much to the chagrin of local sheep farmers, who regularly find killed animals in their herds. The government is also not always in line with policies that protect nature.

This was also evident from the case of Bruno, the bear who roamed the border area between Germany and Austria (FOHRMANN 2006). He had grown used to people. According to the local gendarmerie, this posed a danger, and he was therefore shot in 2006, without having threatened one person. We seem to have forgotten how, not so long ago, we lived with bears in large parts of Europe.

When we look for inspiration for a more equal and spiritual bond with wild nature, we tend to look for it in cultures outside of Europe, where that bond would be more "intact". In doing so, we risk overlooking the fact that in our own past, and in some places in the present, there are inspiring examples of a complex and multi-layered relationship with this impressive "king of the forest", whereby the boundaries between "nature" and "culture" overlap (PASTOREAU 2011). It turns out, however, that in Eurasia we have a wealth of bear myths and ritual festivals in which traditional bear hunts and the fertility and prosperity-bringing powers of this animal are depicted (see different papers, this volume). In short, there is a diversity of stories and performances in which we recognise deeply-rooted cultural values.

R. M. FRANK (2010; 2015; 2016) has done extensive interdisciplinary research on bear stories and rituals. This connects European and Siberian/Indo-American mythologies with bear festivals such as those celebrated in the Pyrenees. Frank considers the bear to be an ancestor of humans (cf. FRANK, this volume). The bear plays his role as a central actor in good-luck visits; these are seasonal festivals in which a group of masked men, including a bear leader and a real or disguised bear, go from house to house receiving food and drink in return for blessings.

This tradition seems to have evolved in, or has blended in with, carnival. R. M. Frank sees the annual spring time bear feasts as connected to fertility, yet mainly understands them as a re-enactment of the traditional bear hunt, where the bear is killed and resurrected. This is a ceremony that is still performed by the Siberian bear-hunting Khanty (cf. RYDVING, this volume). In the bear feasts held today in the Pyrenees, however, it concerns farmers who, perhaps even more than the hunters, depend on the return of the sun and fertility for their cattle and fields.

So, the bear as a fertility symbol and harbinger of spring is still very much alive in the Pyrenees. Bear festivals are held in three villages on the French side of the Pyrenees. The bear festivals are listed on the French National Inventory of Intangible Cultural Heritage. A bear (sometimes more than one, depending on the location), i.e. a fierce imitation of the real animal, storms through the streets and grabs the girls. He fully celebrates his wildness, strength and fertility and by doing so he engages the whole community. The three villages, Arles-sur-Tech, Saint-Laurent-de-Cerdans and Prats-de-Mollo, lie close together in the lowlands on the eastern side of the mountain range (Fig. 1). As Dutch anthropologists interested in the relationship between humans and wild animals, we are used to looking for our field of work in distant areas. The fact that "wildness" appears to be so easily accessible made us decide to travel to Arles-sur-Tech to experience the bear festival (Figs. 2–10).

In this article, we focus on the Arles-sur-Tech bear feast we attended in February 2016 (actually there were two similar feasts; on the first day, it was performed by the youth – a recent addition – and on the second day the "real" feast with adults was performed). What does this bear festival tell us about the place the bear occupies in the world view of the inhabitants of the region? What can we say about the bear's symbolic role? We also take a trip and compare the role of the bear in this ritual with that of other great man-eaters, the European wolf and the Asian tiger. And finally, we ask the question: Are bear feasts doomed to extinction with the disappearance of the bear from Europe, or is an old tradition constantly being given new meaning? What inspiration can we draw from this old connection with the bear and the rituals surrounding it, to reshape our difficult relationship with dangerous animals?

The 21st century and the feast of the bear

Winters have been mild throughout Europe in recent years. Snow only covers the peaks of the Pyrenees as we settle into Mas de Luna, a farmhouse just outside Arles-sur-Tech. On the morning of February 2, in the first pale light, men and women gather in the village café. They are wearing hunter's clothes; sturdy boots and camouflage jackets. They take their first glass of Muscat, the sweet white wine of the region, and begin to paint each other's faces with broad stripes of soot.

One of the hunters wears a curious mix of clothes. This is the trapper, a "hunter from Canada". His leather cowboy hat refers to that distant origin. The hat contrasts strikingly with his shirt, leggings and espadrilles, part of the Catalan costume of the region. The trapper is the hunter who will soon catch the bear. His hat makes him an unknown, unidentifiable person to the bear. Thus, the spirit of the dead bear will never be able to find the trapper to take revenge, the hunters explain.

All morning, the trapper and his gang move from café to café to recruit hunters. The crowd grows and the wine flows. The *poron*, a glass carafe with a long drinking spout, is passed from hand to hand. Robert Bosch (the local specialist and author of "Fêtes de l'ours en Vallespir" [The Bear Feasts of Vallespir]; Bosch 2013) will later cheerfully explain that the comparison with semen is lurking here. In the meantime, the trapper has been joined by his fiancée of the day, Rosetta. Rosetta is also in costume, but her long blonde braids and her breasts are fake: Rosetta is represented by a man.

The couple dances lovingly through the streets of Arles-sur-Tech. They stop at a small square. The crowd assembles in a circle around the hunter and Rosetta. In deadly silence, the trapper gives his first speech, his sermon. He announces that bear tracks have been found around the village and that the vile beast, the *mala bestia*, must be caught. The whole village cheers. Around noon, the group gathers at long tables in one of the village squares. They eat and drink with the other villagers, while the hunters loudly make oblique jokes.

After the meal, there is only one objective: the bear hunt. Accompanied by an orchestra composed of drums, bells and whistles, the mob heads for the river at the edge of the village. The trapper, Rosetta and the hunters cross the river. They comb the bushes on the other side and soon find the bear. The bear is wearing brown woolly overalls, but his head is covered by an impressive hundred-year-old mask, a wooden frame covered with goatskin and fitted with 26 large wooden teeth.

There is a struggle, the bear is chained up and then taken away to the village. But he soon escapes again. The bear is constantly making advances towards Rosetta, who beats him with her handbag. He gets beaten up by the trapper and then grabs the defiant village girls. Screaming, the girls run away, but not too far as being caught is bad, but not being caught seems even worse.

The music starts with the recognisable bear melody and they continue on. Now the bear takes off, slapping the plastic wine baskets that sturdy boys hold over their heads. The baskets are frighteningly painted with eyes and decorated with tufts of broom. The barrel bearers put up a fight, but they keep falling over. Robert Bosch explains: Evil is whisked away and the new spring, the green, is liberated (interview with the authors, 2016).

The bear now makes a spurt. Big and hairy, strong and determined, he chases Rosetta right up on to the French balconies in the narrow streets. When he wants to take her from behind, she seems to appreciate it. But the trapper intervenes. Then the bear runs into the crowd, slings a girl over his shoulder and disappears with her into a den of pine branches. In this den their "marriage" takes place. The crowd cheers, the den shakes dangerously. When the bear comes out, the trapper shoots him.

In deadly silence, the hunters and Rosetta hoist the bear onto a chair. He gets a shaving cap around his head and is shaved by the trapper with a wooden axe and big gestures. Then the orchestra starts up again, the bear frees himself from his mask and takes a gulp of air. The village cheers and dances the *Sardana* together with the bear who, stripped of his wild hair, has now become human. The bear, whose name is Sebastien, turns out to be a valued member of the local rugby team.

Further remarks on the bear feast

It is difficult to estimate how old these "bear festivals" are, and appearances can be deceptive. What seems old, sometimes turns out to be a recent invention. The bear mask of Arles-sur-Tech, the locals say, is about 100 years old, but the bear skin for Saint-Laurent de-Cerdans's suit was recently imported from Canada. The bear claw, which the mayors of the three villages solemnly hand over to each other every year, is also a very recent "tradition". Other aspects may seem relatively new but have deep roots. For example, the hunter's humorous sermon contains medieval terms and concepts. Two themes seem dominant in the present-day performance of the symbolic hunt of the animal: On the one hand, there is its "killing" and resurrection and, on the other hand, there is the waking up of the bear on February 2, its sexual activities and final "marriage" in the constructed green in the village centre.

The theme of the hunt that we mentioned earlier has been linked to Siberian tribal groups that seem to have related customs. The interconnected rituals surrounding the dead bear go back to the prehistoric hunter-gatherer cultures of Eurasia (FRANK 2015). The theme of the bear that brings forth spring, fertility and physical power, however, is, as we would like to stress, also crucial to the agrarian societies in the northern hemisphere, who suffered cold and lack of food at the end of the winter. In the present-day bear feasts of the Pyrenees and the related bear feasts in Europe, such as the one in Comanesti in Romania, as in oral traditions in northern and middle Europe, the role of the bear as a bringer of fertility is central.

In the bear festival of Arles-sur-Tech, the bear is an ambivalent creature. The animal is portrayed as a threat that must literally be "stripped of its wild hair". The association of hair and fur with primal strength is universal and can also be found in the Bible. The bear in the feast that we witnessed is shaved by Rosetta and the hunter and, in this way, tamed and made manageable. Only then can he be accepted into the community. But at the same time, he is a force from the wilderness, penetrating the village and revitalising man and nature.

The theme of the bear kidnapping and "marrying" a woman is widespread in Europe and Eurasia and is certainly very old. According to Robert Bosch, there is a cultural layer recognisable throughout Europe in which fantasies about the power and fertility of the bear are elaborated in stories, rituals and cult objects, a theme we also find discussed in literature (BOSCH 2013; see also LAYOUX 1996; PASTOUREAU 2011; cf. other papers in this volume). An echo of this can be found, for example, in the coats of arms of specific families. Place names are often linked to a story about the defeat by or the marriage to a bear, who in this way passes on its power to a city founder or a lineage group. For example, the *Gesta Danorum* by Saxo Grammaticus, the "Deeds of the Danes", dating to *c.* 1200, tell of the Danish king, Sven II (1047–1076), who is said to be descended from a bear (his great-grandfather was said to be the son of a bear and a woman he abducted) and to have owed his great strength to it (Pastoureau 2011, 79). That this marriage option was taken seriously is shown in a remarkable writing by William of Auvergne, intellectual and bishop of Paris from 1228 to 1249. William relates the story of a woman who was abducted by a bear and had three hybrid sons. After seven years, the woman and her children were freed and reintegrated into society. The sons were even knighted. The only visible sign that their father was from the forest was their "abundant hairiness" (Pastoureau 2011, 77). Remarkably, the bishop acknowledges that the bear's sperm is "almost identical to human sperm, that bear and woman are interfertile, and that the offspring of their union are fully human" (ibid., 78). This is quite a different version than the romantic union between woman and bear in the well-known fairytale versions, from the Grimms' fairytales to Gabrielle-Suzanne Barbot's "The Beauty and the Beast", in which "the beast" always turns out to be an enchanted prince and not a real animal at all.

A marriage and/or family affiliation to impressive, dangerous (often man-eating) animals is not restricted to the bear. Other dangerous animals, such as the wolf and tiger, have their own characteristics but share their impressive power and their ability to kill. Can we understand the role of the bear in the European and Eurasian past and present better by a short detour that will allow us to see how other dangerous animals have been conceptualised?

Pacts with dangerous animals – tigers, wolves, bears

Because the bear is, in the European context, the animal that resembles humans the most, a marriage is perhaps imaginable. But marriages are also mentioned in myths concerning tigers and wolves. A Turkish legend, for example, tells of a young boy who survived a raid on his village. A she-wolf finds the injured child and nurses him back to health. He subsequently impregnates the wolf who then gives birth to ten half-wolf, half-human cubs. One of these, Ashina, becomes their leader and establishes the Ashina clan, which ruled the Göktürks (T'u-chueh) and other Turkish nomadic empires (Findley 2005, 38).

Further to the east, on the Indonesian island of Sumatra, the tiger is the central animal in mythology. It is also the most feared predator, entering villages and preying on cattle and sometimes people. This has not led to tiger hunts; on the contrary, until recently, local populations have declined to kill tigers for money. The reason is that the tiger is still revered as an ancestral being, an incarnation of an important ancestor and related to humans by marriage. In one myth, a Sumatran farmer's son marries a princess of the tiger clan. The members of this clan are living in a village in the jungle in the form of people, but outside of the village they turn into tigers. The marriage results in a separation, but also establishes a pact between the tiger people and humans. In this pact they agree to respect each other as family and not to bother each other "leaving the village to the people and the forest to the tigers" (Bakels 2004). People still refer to this pact when explaining why they do not fear tigers. There is, however, one exception to this non-violence pact. That is when a person trespasses the rules of behaviour. Only then will the tiger attack this person or their cattle. The opposite scenario is also important: Only a tiger that enters a village and kills can be killed, and it will then be buried with ritual ceremony.

Such a symbolic pact has many advantages. It transforms an uncontrollable natural force (a tiger attack) into controllable social behaviour (behaving according to the law). This can be advantageous to a society as a whole, where disruptive anti-social behaviour is sometimes difficult to correct. Last but not least, this pact protects the tiger in Indonesia to a certain extent. Interestingly enough, such pacts between man and dangerous animals also existed in Europe, probably on a much wider scale than we realise. A famous example of such a pact is that between St Francis of Assisi and the wolf. According to the story, a wolf terrorised the people of Gubbio. Francis made a pact with the wolf, in which the wolf promised to refrain from eating cattle and humans, and the people of Gubbio promised to feed the animal daily. The scene was captured by the painter Sassetta around 1440; we see Francis writing down the appointments and the wolf watching patiently (VAN OS et al. 2016).

But back to the bear! Interestingly, there are also some reminiscences of such a pact with the bear, where the animal seems to play a role as corrector of anti-social behaviour. For example, in the earlier-mentioned Khanty mythology and rituals connected to the dead bear, there is the idea that the soul of the slain bear "reports" to the bears in heaven about whether it has been treated properly. If so, bears will be reborn in the spring. R. M. Frank mentions a myth that speaks of a kind of pact in which a bear cub comes down from the sky, after agreeing not to harm people, or eat their food supplies. The cub is to teach humans how to carry out the bear feast. That pact, however, is between the bear cub and its bear mother or father dwelling in the sky (FRANK 2015, and personal communication): It is not directly between bears and humans, although the cub is not summoned to hurt people, which is the key notion in the pact with the tiger and the wolf. One wonders what more rules of behaviour were possibly connected to this reporting to the bear in heaven. Could it be that, as came to the fore in the pact with the tiger, man and bear promised to respect each other's domains? Was there also once in Europe a "pact with the bear" – a tendency to kill only those bears who killed sheep or people? There are references to "unjustly killed" bears (PASTOUREAU 2011, 82), but more research is clearly needed.

A NEW NARRATIVE OF PEOPLE AND BEARS

Meanwhile, our relationship with "the wild" has not become any easier. While bears have been reintroduced into the Pyrenees, and also into northern Italy in 1999, there has yet to be a well-understood "pact with the bear" in our contemporary relationship with these creatures. In 2006, it was the already-mentioned bear Bruno who, straying into Germany, was deemed too dangerous and therefore shot. In 2013, an "at risk" bear named M13 was shot in the Swiss canton of Graubünden. Although he was clumsy and not very shy, he had not yet done any real harm to anyone.

This raises questions, including our ability in Europe to coexist with such a potentially dangerous animal. For one thing, there seems to be an essential difference between the way Canadians and Americans react to "bears around the house" and the way Europeans react. The over-excited reporting of the bear Bruno's whereabouts, of what is in fact "business as usual" in the US and Canada, where people have frequently lived with bears and grown up with them, illustrates how we in Europe have lost contact with bears, and with the idea of co-existing with a dangerous animal.

In most European countries, the bear disappeared generations ago or has become very rare (cf. ZEDROSSER/SVENSSON, this volume). The bear is a stranger, we are no longer used to him, he is not part of our lives and we find even the smallest chance of unpleasant confrontations unacceptable. The question is whether we can change that basic attitude. Can we form a new relationship, a new pact with the bear? It is understandable but strange that other potential dangers for Europeans, such as an attack by a dog or a collision by a car, are acceptable to us, but a bear in our path is not.

Could we build a moral circle, a "circle of trust", a concept originally meant to indicate the group of people that you trust, but one that could be extended to animals, including the bear (cf. WENZ 1988)? How do we get the bear into this circle? Perhaps the old multi-layered relationships we had with the bear can inspire us to accept him once again as a harbinger and bringer of fertility, strength and happiness.

And the bear feasts? This kind of traditional feast and ritual performance have a layered and poetic power of expression. The feelings of belonging that the bear festival evokes is obvious, even to us outsiders, and it seems to be the main reason the feast is performed. Young men and women come back to their home villages for the weekend, and they look forward to next year's festival as soon as it is over: It has become an identity symbol. Perhaps the bear festival is even more popular than ever. As the world grows larger, local – in this case Catalan – distinctiveness is more cherished.

Bibliography

Bakels 2004: J. Bakels, Farming the forest edge: perceptions of wildlife among the Kerinci of Sumatra. In J. Knight (ed.), Wildlife in Asia. Cultural Perspectives (London, New York 2004).

Bosch 2013: R. Bosch, Fêtes de l'ours de Vallespir (Perpignan 2013).

Findley 2005: C. V. Findley, The Turks in World History (Oxford 2005).

Fohrmann 2006: P. Fohrmann, Bruno alias JJ1: Reisetagebuch eines Bären (Berlin 2006).

Frank 2010: R. M. Frank, Hunting the European Sky Bears: German "Straw-bears" and their relatives as transformers. In: M. Rappenglück/B. Rappenglück (eds.), Symbole der Wandlung – Wandel der Symbole. Proceedings of the Gesellschaft für wissenschaftliche Symbolforschung / Society for the Scientific Study of Symbols, May 21–23, 2004, Kassel, Germany (München 2010) 141–166.

Frank 2015: R. M Frank, Bear Ceremonialism in relation to three ritual healers: The Basque salutariyua, the French marcou and the Italian maramao. In: E. Comba/D. Ormezzano (eds.), Uomini e Orsi: Morfologia del Selvaggio (Turin 2015) 41–122.

Frank 2016: R. M. Frank, Sky Bear research: Implications for "Cultural Astronomy." In: J. P. Hernández/C. González/G. Magli/D. Nadali/A. Polcaro/L. Verderame (eds.), Proceedings of SEAC 2015 Conference: Astronomy in Past and Present Cultures, Rome, 9–13 November 2015, Rome University "La Sapienza". Mediterranean Archaeology & Archaeometry 16(4), 2016, 343–350.

Layoux 1996: J.-D. Lajoux, L'Homme et l'ours (Grenoble 1996).

Lévi-Strauss 1962: C. Lévi-Strauss, Le totémisme aujourd' hui (Paris 1962).

Lévi-Strauss 1964: C. Lévi-Strauss, Totemism. Transl. R. Needham (London 1964).

Van Os et al. 2016: H. van Os/G. P. Freeman/M. Brüggen Israëls/B. Roest/K. van Dooren/E. Verheggen/ F. Bosmans, Franciscus Van Assisi (Zwolle 2016).

Pastoreau 2011: M. Pastoureau, The Bear. History of a Fallen King (Harvard 2011).

Wenz 1988: P. S. Wenz, Environmental Justice (Albany 1988).

Dr. Jet Bakels
Dutch Centre for Intangible Cultural Heritage
Arnhem
The Netherlands
jet.bakels@planet.nl

Dr. Anne Marie Boer
Retired curator of Museon
Den Haag
The Netherlands
amboer@xs4all.nl

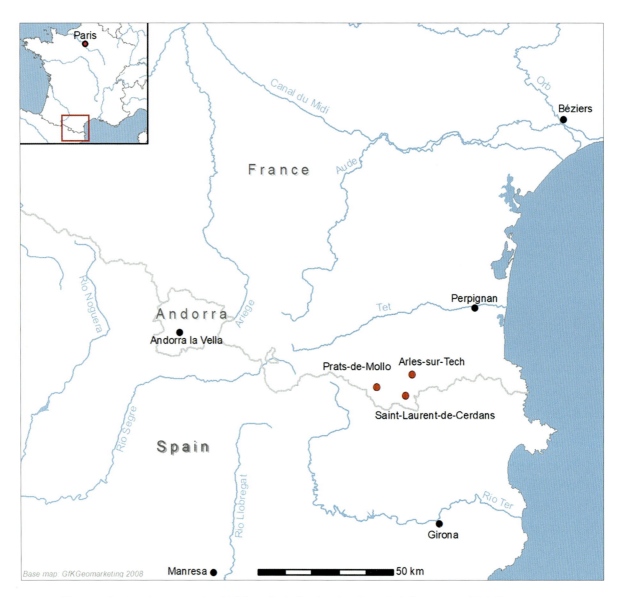

Fig. 1. Villages in the French Pyrenees in which bear festivals take place (map GIS department, ZBSA).

Fig. 2. Bear festival in 2016, Arles-sur-Tech, French Pyrenees: The bear is discovered in the forest (youth performance; photo J. Bakels/A. M. Boer).

Fig. 3. Bear festival in 2016, Arles-sur-Tech, French Pyrenees: The bear is taken to the village (youth performance; photo J. Bakels/A. M. Boer).

Fig. 4. Bear festival in 2016, Arles-sur-Tech, French Pyrenees: The bear, the hunter and Rosetta on a balcony (youth performance; photo J. Bakels/A. M. Boer).

Fig. 5. Bear festival in 2016, Arles-sur-Tech, French Pyrenees: A recurring fight between the bear, the hunter and Rosetta (photo J. Bakels/A. M. Boer).

Fig. 6. Bear festival in 2016, Arles-sur-Tech, French Pyrenees: A recurring fight between the bear, the hunter and Rosetta (photo J. Bakels/A. M. Boer).

Fig. 7. Bear festival in 2016, Arles-sur-Tech, French Pyrenees: The bear in his costume (photo J. Bakels/A. M. Boer).

Fig. 8. Bear festival in 2016, Arles-sur-Tech, French Pyrenees: The bear and green branches symbolising spring (photo J. Bakels/A. M. Boer).

Fig. 9. Bear festival in 2016, Arles-sur-Tech, French Pyrenees: The bear sits to be shaved (photo J. Bakels/A. M. Boer).

Fig. 10. Bear festival in 2016, Arles-sur-Tech, French Pyrenees: The bear and the hunter after the performance (photo J. Bakels/A. M. Boer).

What are those bears doing there?
On a painting from early Italian art

For Penny

By Henk van Os

Keywords: Early Italian art (1250–1400), Dittico di Santa Chiara, *St Francis, stigmatisation, bears*

Abstract: The monumental Dittico di Santa Chiara, *which belongs to early Italian art (1250–1400), has seen a lot of discussions about its date and attribution. One part of it shows the culmination of the life of St Francis – his stigmatisation. Until now, however, one meaningful detail in this painting has been overlooked – two adult bears and a cub that is, as yet, a shapeless lump in the grotto. It was first mentioned by Ovid that the mother bear licks its young into bear-shape. In medieval times, this licking by the mother symbolised how Christ converted the gentiles.*

Over 50 years ago, the Pinacoteca Nazionale in Siena was the principal location for my research on early Italian art (1250–1400). It is the only museum that possesses a complete overview of two centuries of medieval painting. During this period, all of Europe spoke, according to Bruno Santi, of Sienese painting as *quella fascinosa miscela – insomma – di semplicità e di grandeur –* that fascinating mixture of simplicity and grandeur.

With the passage of time, very little has changed in the Pinacoteca, and it is very quiet there. The day-trippers come to see the town's Duomo and gather on the Campo, the scallop-shaped square that is still the focus of the city's civic pride. They rarely venture any further than that.

Almost every time I revisit the tranquil Pinacoteca, I discover something new in those paintings that I have known for so long. Five years ago, I had an experience like this almost at the moment I walked in the door. The second room contains a monumental diptych of St Clare of Assisi (the *Dittico di Santa Chiara*; Fig. 1) measuring about 120 x 170 cm, including a painting of the high point of the life of St Francis – his stigmatisation. Through the intensity of his prayers to the suffering Christ, he received the wounds of the crucified Saviour on his own body. Francis is depicted in a kneeling pose but, at the same time, he is "reaching up to the eternal Godhead", as mystics would put it.

There has been a great deal of art-historical discussion about the date and attribution of this work. Is it an early painting by the first great Sienese painter, Guido da Siena? Or was it painted by an assistant? As a result of this discussion, hardly anyone has ever given a thought to the two massive bears and the grotto containing a shapeless cub. The bears have taken the rugged forests of Mount La Verna in Tuscany as their corner of paradise. I had paid far too much attention to Francis and his prayers to realise just how remarkable those bears actually were. Realistic wild animals like this are nowhere to be seen in 13th-century Italian paintings. You might come across some little creature now and then, but these two stand out. Why are they there? I asked myself that question, but I found no satisfactory answer in art-historical literature.

However, medieval books on animals, bestiaries, give detailed descriptions of bears. They are portrayed as a terrifying menace, but also as lovable creatures. It is related there, too, that a new-born bear cub starts out as a shapeless lump. Only after his mother has licked it repeatedly does it assume the shape of a bear; the mother appears as a sculptor. This story was first told by Ovid in his Metamorphoses (chapter 15: 379–381). The authors of the medieval bestiaries often added the remark that this creative licking on the part of the mother symbolises how Christ converted the gentiles. After all, in the scholastic period everything in nature only assumed meaning after it had been Christianised.

It is very likely that this tale of the licking bear is being portrayed in the painting. Thus the symbolism of the bears would represent an iconographic theme that I, because they were "just bears", had tended to overlook – as had perhaps other spectators as well. Three bears are depicted: One of the adults is nibbling on the berries of a shrub, while the other fully-grown one witnesses the miracle of the stigmata, the most wondrous metamorphosis in Christendom. The third bear is that lump in the grotto. The mother still has a lot of work to do on it.

The most detailed description of the bears at Francis' stigmatisation dates, like the painting, from the end of the 13th century. Its title is "Reflections on the Stigmata" (anonymous). It is a sort of compilation of old stories about the miracle that Francis experienced. The first of the five reflections tells how Francis came to be on that high mountain.

At a reception in 1224, Francis met the immensely wealthy owner of Mount La Verna, Roland of Chiusi and Casentino. This man had heard of the remarkable hermit and offered him the mountain as a perfect location for prayer. It was too rugged for hunting, but for hermits it was ideal. Francis and two of his pious brothers would live in such proximity to the Creator that they could co-exist in peace with all earth's creatures. Francis accepted Roland's offer, and Roland then gave him and his two disciples 50 armed men to scout out a suitable location for their life of prayer. Fifty men! That shows how dangerous the wild animals were. And the most prominent of them all was, of course, the bear. But even if there were bears, snakes or wolves, that did not deter Francis and his companions from their intense devotion. During his prayers, he often gestured towards the heavens with his arms. He brought the sign of the cross to Jesus and prayed that he might be able to share Jesus' cross and suffering. One night it happened: He was united in prayer with Jesus. And the bears watched it happen.

FURTHER READING

Saint François D'Assise. Documents ecrits et premières biographies. Eds. T. Desbonnets/D. Vorreux (Paris 1968).

"Bär". In: Lexikon der Christlichen Ikonographie, vol. 1. Ed. E. Kirschbaum (1968) col. 242–244.

Editing: Pauline van der Hoeven
Translation: David Barick

Dr. Henk van Os
Former director of the Rijksmuseum
Amsterdam
The Netherlands

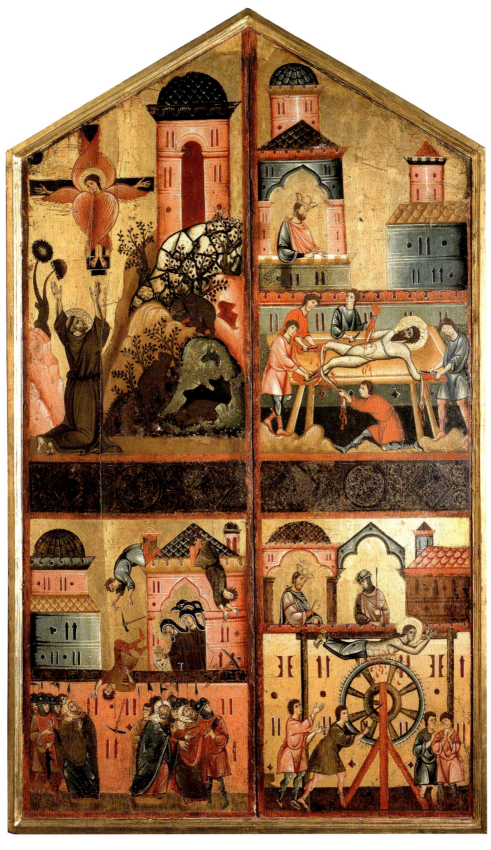

Fig. 1a. The Dittico di Santa Chiara. *Overall view (Pinacoteca Nazionale in Siena, Italy, © Bridgeman Images).*

Fig. 1b. The Dittico di Santa Chiara. *Detail: the stigmatisation of St Francis, with three bears being present (Pinacoteca Nazionale in Siena, Italy, © Bridgeman Images).*

Bear-human interactions: Archaeological and ethnographic investigations in North American indigenous cultures

By Kerry Hull

Keywords: Native Americans, bears, ceremonialism, bear costuming, shamanism

Abstract: Bears have held a special position in Native American societies for millennia. Archaeological and ethnographic data illuminate deep ideological connections to bears from at least 8000 BC. This study investigates early indigenous groups of North America such as the Adena, Hopewell, and Mississippian cultures (c. 800 BC – AD 1600) to assess the roles of bears in myth, ritual, and as markers of clan identity. Post-Conquest ethnographic sources offer a wealth of data showing the status of bears. In this study, bear paraphernalia such as claws, teeth, paws, bear skins, and bear headdresses are examined in various ritual contexts such as healing, dance, and shamanism. The ceremonialism associated with bear hunts, especially those involving bear costuming and tobacco rites, are also discussed as a means of understanding the complex mythological relationship between bears and humans.

Introduction

Various species of bears inhabit parts of North America, most commonly in Alaska and Canada. While regular bear-human contact is not common in most parts of the United States, their influence on indigenous cultures, past and present, has been profound. In this paper, I investigate the roles of bears throughout known history in the lives of Native American groups in North America. I draw upon archaeological and ethnographic data to contextualize the importance of different species of bears among numerous North American indigenous groups. I also look at bears in ceremonial contexts based on archaeological evidence, primarily in the form of bear skull rituals, bear costuming, and bear paraphernalia, such as bear teeth, claws, and paws. I also provide ample ethnographic data as possible lenses through which to interpret the use of these items in pre-Columbian societies of North America. Furthermore, I detail the unique status bears hold among Native American groups as semi-human, healers, and warriors.

In the context of the present volume, which focuses on bears in northern Europe, it is hoped that this study of bears in North America will provide a comparative discussion for readers since many of the traditions surrounding bears in Europe have close correlates in North America.

The bears of North America

Three bear species are found in North America – the American black bear (*Ursus americanus*), the grizzly or brown bear (*Ursus arctos*), and in the Artic regions the polar bear (*Ursus maritimus*).[1] During the 19th century, due to negative attitudes toward bears and hunting, bear populations of all species precipitously declined, resulting in a steep drop in the number of grizzlies in most of the United States as well as a decrease in black bears in the southern states (MILLER 1990, 357). Today bear management efforts have led to a stability in bear populations (JORGENSEN et al. 1978; WILL 1980), principally between the 1980s and 2000s (MILLER et al. 2011; SCHEICK/MCCOWN 2014, 24). Current estimates place the number of grizzly bears in North America at between 52,000 and 63,000 (PEEK et al. 1987, 160).

Before discussing the impact bears have had on traditional cultures in North America, it is important to understand that the historical range of North American bears was substantially larger than our current situation. Black bears, for example, traditionally covered most forested areas of North America and Mexico (PELTON et al. 1999; LAVIVIÈRE 2001; PELTON 2003), but have lost 41 % of their historical range since the arrival of the Europeans (LALIBERTE/RIPPLE 2004, 126). Grizzly bears' populations also historically covered much larger areas of North America, from Canada into Mexico, but today have suffered a 55 % decrease in historical range since European contact (LALIBERTE/ RIPPLE 2004, 126), primarily due to human interactions (MATTSON/MERRILL 2002).[2] The broad extent of bear populations throughout North America ensured human encounters and a place in human mythology for the bear. HALLOWELL's (1926) classic study *Bear Ceremonialism in the Northern Hemisphere* expertly described the ethnographic history of bear reverence throughout Eurasia and North America, even pointing out the surprising commonalities between these distant regions.[3]

Indigenous cultures and bears

Indigenous populations in North America have developed complex cultural connections to the bears. Early archaeological evidence indicates that some form of bear ceremonialism was in place during the early Archaic Period (8000–500 BC; Fig. 1; Table 1). One of the earlier bear artifacts found in North America in California was incorporated into that state's flag by 1846. In 1985, a graduate student excavating at the Allan O. Kelly Archaeological Dig (CA-SDI-9649) in San Diego County, California, found a chipped bear-shaped stone crescentic made of meta-volcanic rock, measuring 6.5 cm long x 3.1 cm wide (Fig. 2). It was associated with artifacts in an Dieguito EQ-early La Jollan transition period midden, which dates it between 7000–8000 years old (KOERPER/FARMER 1987, 283) and suggests some form of bear reverence during the Archaic period.

Among the numerous archaeological cultures in North America (here defined as the United States [including Alaska] and Canada), three major cultures existed in the Eastern Woodlands and Southeastern cultural areas of what is now the eastern United States: 1) the Adena culture (1000–200 BC; Early Woodland Period), consisting of several hundred sites in the Ohio Valley and others in neighboring states; 2) the Hopewell culture (100 BC to AD 500; Middle Woodland Period), comprised of

---

1   This study will focus on the American black bear (*Ursus americanus*) and the grizzly or brown bear (*Ursus arctos*).
2   In 1975, grizzly bears were included as an endangered species in the Endangered Species Act (ESA; 16 U.S.C. 1531–1544).
3   Many have noted the similarities between Eurasian and North American traditions related to bears. For example, FRAZER (1912, 224) wrote: "The reverence of hunters for the bear which they regularly kill and eat may thus be traced all along the northern region of the Old World, from Bering Strait to Lappland. It reappears in similar form in North America". For more on such areas of convergence, see HALLOWELL 1926, 21–22, 29–31. See also LABBÉ 1903, 231.

hundreds of autonomous villages, which extended from Lake Ontario down to Florida, with one of the larger populations in the Ohio region; and 3) the Mississippian culture (AD 800–1600), primarily located in the midwestern, eastern, and southeastern United States. All three cultures are commonly referred to as the "Mound Builders". Archaeological data from these cultures, particularly among surviving Hopewell artifacts, show that bears held a position of status in their society.

The artistic skill of these early cultures is on full display in many of the bear-related artifacts that have come down to us. For example, in 1915, an exceptionally beautiful mica (a silicate mineral from the southern Appalachian region) bear was discovered measuring 11.25 cm high and 16.06 cm wide (Fig. 3). The item belongs to the Hopewell culture, dating to 100 BC – AD 500, and was found during excavation in Tremper Mound in Rush Township, Scioto County, Ohio. Another highly artistic creation is a stone bear effigy discovered at the Middle Woodland (250 BC – AD 500) Hopewell site of White Oak Point in Minnesota that boasts copper inlays in its eyes (Fig. 4; cf. MATHER 2019, 201–202 fig. 51). Bear effigies and bear effigy tools are also attested in Massachusetts, New Hampshire, and Connecticut (WILLOUGHBY 1935, 37, 166). However, it was the use of bear incisors and claws, both real and artificial, where creative expression truly met cultural saliency.

BEAR TEETH AND CLAWS

Without a doubt, the most commonly encountered bear artifacts at North American indigenous sites are teeth and claws. The ornamental use of bear canine teeth dates back to Late Archaic (3000–1000 BC) times and continued through Native American indigenous cultures (BERRES et al. 2004, 17). Excavations at the Hopewell Mound Group of Ohio produced an abundance of grizzly and brown bear teeth in Hopewell burials. Among these bear teeth some were plated with copper (Mound 20; MOOREHEAD 1922, 95) or inlaid with pearls[4] (Mound 25; ibid., 111), and various faux bear teeth were made out of copper (ibid., 126), wood (ibid., 128), deer antler (ibid., 159), and one possibly made from limestone (ibid., 156; Fig. 5a–d). Bear teeth are found at various sites throughout the Hopewell region (GREBER 1979, 53; TOMAK 1994, 38), with more than 1,000 known from Hopewell sites by the 1970s (SEEMAN 1979, 372). In some cases, meteoric iron is set into the canine tooth (e.g. Seip Mound 2; PRUFER 1961, 342). Bear teeth were often made into necklaces (Fig. 6; PHELPS/BURGESS 1964, 199; IRETON 1982; BERTINO 1994). Further north, the Central Esquimo used bear teeth as fishing lures (see WILLOUGHBY 1935, 49). Bear teeth and claws could also be indicators of clan society membership.

Bear claws, both artificial and natural, were included in numerous burials in the Hopewell culture. A bear effigy claw made of copper was discovered in the Hopewell Mound Group of Ohio (Fig. 7; MOOREHEAD 1922, pl. LXIX). A number of copper claw replicas,[5] likely those of a bear, were found in Hopewell Mound City Group earthworks in southern Ohio dating to the Middle Woodland period (200 BC – AD 500; Fig. 8). Bear claw replicas were also produced from canal coal and mica. Abundant ethnographic data show that bear claws were often kept by hunters who killed a bear, and they were worn as ornamental jewelry for status and power. Among the East Cree, SKINNER (1911, 69) notes: "If an Eastmain hunter is alone when he kills a bear, he cuts off the middle toe and claw of the right forefoot. Upon returning to camp he gives this token to the person who is to carry the body from the woods, usually his wife, if the hunter is married. This individual takes a companion and fetches the carcass." The claw is "wrapped in cloth, beaded or painted, or both, and kept as a memento of the occasion" (SKINNER 1911, 69). Bear claws were commonly formed into necklaces,

---

4   Grizzly teeth inlaid with pearls may be import items from the Rockies (PRUFER 1964, 93).
5   Ohio History Connection museums: HOCU-2781 and HOCU-2728.

which are often found around the chest region of Hopewell burials. Indeed, one of the most common uses of bear claws in North American indigenous traditions is making necklaces (cf. SPARKMAN 1908, 199; Fig. 9).

Bear claws could have pedestrian functions such as among the Chipewyan, where bear claws (*maskukásiyǔ*) were used as a form of dice play (CURTIS 1928, 69), and among the Lenepe, where bear teeth were used for gaming and conjuring.[6] However, the dominant use was for ritual or spiritual purpose. Among various Northwest Coast groups bear claws could be worn as a type of crown by shamans (SHEARAR 2008, 20). In some cases, a single bear claw was woven into the hair as a token of power, such as Cheyenne dancers in the Medicine Lodge or Sun Dance ceremony (GRINNELL 1923, 280). In the early 20th century in Montana, a famous Gros Ventre (also known as Atsina or A'aninin) chief named A-kich-kot-ope once killed a large number of Blackfeet. The Blackfeet were able to kill all of the Gros Ventre, except their chief, who disappeared into a bush, where he made grizzly bear noises. He taunted the Blackfeet by saying "My power is very great". When they finally killed him, they discovered he was a medicine man whose "power came from the grizzly bear". He was wearing a grizzly skin and a necklace of grizzly claws when they killed him (MCCLINTOCK 1923, 107–108). In addition, they also found a grizzly claw in his hair – the source of his power. So afraid were they that A-kich-kot-ope's power might escape that they burned his body on the spot (MCCLINTOCK 1910, 56).

The power of bears is often linked to their erratic and frenzied characteristics, their unpredictability. Shamans transforming into grizzlies "often become uncontrollable and frenzied while dancing, and are said to have short lives unless they cleanse themselves after such a seizure" (WHERRY 1964, 80).[7] Bears are often associated with warriors because of their fierceness (SHEARAR 2008, 20), and because of their power and volatility.[8] Consequently, bear paraphernalia were often worn in battle. The Cheyenne would paint grizzly images on their war shields precisely because bears are courageous and hard to kill (TEIT 1900, 249), and anyone who bore such a shield would share in those qualities (GRINNELL 1923, 192). Bear paws painted on one's face symbolized strength among the Eastern Cree (SKINNER 1911, 21–22). Bear claws, skins, teeth, and other body parts were worn in battle, and battle armor often depicted bears (KENNEDY/STEVENS 1972, 96). And when in battle, bear paraphernalia or icons also had protective powers (ROCKWELL 1991, 94).

BEAR PAWS

Like the individual claws, bear paws were also considered items of great potency and so were used in ceremonial contexts (SPARKMAN 1908, 210; KINIETZ/RAUDOT 1940, 159–160). Bear paws are attested as paraphernalia of North American shamans (HALLOWELL 1926, 78). Pomo shamans of San Felipe, California, who have as their guardian Bear, used bear paws to empower themselves to cure someone afflicted by witchcraft (WHITE 1932, 47). The Navajo usually only hunt bear in order to get a paw to use in ceremonies, since the right and left front paw of a bear were required for the Navajo Male and Female Mountain Way Chant, a ceremony performed to help individuals with anxiety or antisocial issues (HILL 1938). In the early 20th century, during initiation ceremonies among Keres in New Mexico, the shamans gathered together holding the skin and claws from a bear's foreleg, and one by

---

6 For an example, see the Smithsonian Institute collection, USNM Number E6881-0.
7 Stories abound of individuals able to change themselves into bears (MCCLINTOCK 1923, 230).
8 Indeed, the English word "berserk" comes from the Norse term *berserk* < Old Norse *berserkr* "raging warrior of superhuman strength", possibly further deriving from **ber-* "bear" and *serkr* "shirt". For a fuller discussion on the controversies surrounding this etymology, see SUNDQVIST, this volume.

one they arose to strike the young man with bear paws to help empower the youth (Curtis 1926, 91).⁹ There is sometimes a prohibition against eating bear paws, especially for certain members of society. For instance, the Plains-Cree believe that children should never eat bear paws lest they become savage like the bear during their impressionable years (Skinner 1914, 510). In addition, women were not allowed to eat bear paws among the Eastern Cree (Skinner 1911, 69). Furthermore, bear paws could have an ornamental function as they did with the Pomo (Gifford/Kroeber 1937, 166). Among the Blackfoot of Montana, men who achieved a certain level of status in a tribe would wear a grizzly bear paw on a necklace – an object of considerable value in their culture. Indeed, the separation of the Kickapoo and the Shawnee of the Ohio Valley region was said to have been caused by a disagreement over a bear foot (Trowbridge 1938, 69).

Bear teeth, claws, and paws are, on the one hand, relics, mementos from a successful bear hunt. However, they also fit prominently into social systems of exchange, shamanistic ritual (including healing), and societal status. It is crucial to note that the teeth, claws, and paws are the most dangerous parts of the bear, which also helps to explain their value as objects of inherent power. In essence, their removal from the bear divests the bear of that power and imbues humans with it.

Bear masks and skull rites

Another form of appropriating bear attributes was through costuming (Fig. 10). Bear skins and bear masks were often used to imitate bears, usually in ritual contexts. Early examples of this type of behavior are attested in part in the archaeological record, but much more fully in ethnographic studies. Bear headdresses were used by the Hopewell, but no actual bear headgear has survived the ravages of time. Copper headdresses, however, have been found, such as one discovered in Hopewell Burial 11 in Hopewell Mound 25, which was first interpreted as avian by Shetrone (1926, 68–72; see Giles 2013, fig. 2.1). A more recent treatment of the object identified it as representing a two-headed raptor (Giles 2013). An even more elaborate copper bear headdress was unearthed in Mound 13 of the Mound City group made from copper sheets secured by rivets (Fig. 11; see Converse 1983, 21). Also, another copper headdress from Mound 25, Burial 4 of the Hopewell Group, Ross County, Ohio (archaeological site number 33-Ro-27), currently in the Ohio State Museum, includes a bear claw design (#13885; Fig. 12). And in some cases, actual bear skulls were used in ritual contexts. In Williams Cemetery, Wood County, Ohio, an extraordinary artifact was discovered – a bear skull that was modified to be used as a human mask. There were three pairs of holes drilled into the skull so that the bear skin could be attached (Berres et al. 2004, 15).

Bear costuming is a practice that is well attested in the archaeological record. For instance, a figurine fragment from the Hopewell Wenger #4 site in Illinois (150 BC – AD 350) depicts a man wearing what appears to be a full bear costume with his head covered by a bear headdress (Fig. 13; Koldehoff 2006). The Wray figurine, found in the Newark Earthworks, Licking Col, Ohio, in 1881, is an example of a complete individual wearing a bear costume (Fig. 14; Dragoo/Wray 1964). This figure is of a man, possibly a shaman, holding a detached human head (possibly a "trophy") in his right hand. The man is wearing a bear skin with an attached bear headdress and bear claws hanging by his hands. Although we lack context, it is possible that this figurine represents a bear shaman involved in a transformation or some other type of ritual (cf. Romain 2009, 39–42).

---

9 According to many North American traditions, bear paws also have secretive powers for bears themselves as it is thought that bears sustain themselves during hibernation by sucking on their paws (Hallowell 1926, 27–29).

Ethnographic data also point to bear costuming as a prominent part of North American tribe ceremonialism. Blackfoot dancers in Montana would don a bear headdress that had two bear claws on it like horns (McClintock 1923, 288). The Brave Dog clan members associated with bears would also wear a bear headdress, patches of bear skin on their arms with bear claws attached by the hands, the faces painted to resemble that of a bear, the lower part of their bodies being naked (Fig. 15; McClintock 1923, 285). A bear skin costume and bear mask similar to the Wray figurine were also used by the Lenape in performances in which one was said to become "imbued with the spirit" of the bear once the mask was put on (Harrington 1913, 45).

Bear costuming and masking were likewise a key component of bear hunts. Before setting out, it was common among many North American tribes to participate in a bear dance in order to assure a successful hunt. In the mid-1800s, Domenech (1862, 445–446) described the pre-hunt bear dance among the Natchez as an appeal to "le génie des ours" that the hunt would be productive. The Medicine Man would lead the ceremony while dressed entirely in a bear skin and wearing the head of a bear to mask his face. All the dancers are similarly dressed as they imitate the movements of a bear while dancing in a circle.

While bear heads covered with natural fur were commonly in pre-hunt masking ceremonies, bear mandibles and full skulls were also powerful ritual items post-hunt. There is evidence of bear skulls as trophies at Tremper Mound, one of the earliest Hopewell sites in Ohio (Seeman 1977; 1979, 570). After a successful bear hunt, bear skulls were cleaned, dried, and often painted before being set in a specific place, commonly hung in a tree (Skinner 1911, 70). The Montagnais (Innu) of Canada would place the head of the killed bear on a pole in the very spot where it was killed as a "trophée" to showcase their successful hunt (Perrot 1864, 200–201). According to Hallowell (1926, 135): "In by far the majority of the tribes studied in northern North America and Siberia as well, a special emphasis is placed upon the preservation of the bear's skull, which is usually placed upon the branch of a tree in the woods, on a pole in some instances, or deposited in an ostensibly sacred place in the forest, sometimes along with the skulls of other animals." After killing a bear, the Timagami Ojibwe would hold a bear feast. The skull of the bear which has been painted with black stripes is presented to the most venerable in the group. The skull is then attached to the lower jawbone and placed on a tree branch facing a prominent point in the landscape, thereby marking it as a place where a bear feast had taken place (Speck 1915, 25–26).[10] In other cases, bear skulls were placed on wooden poles[11] "to appease the spirit of this important animal"[12] (Strong 1930, 5) among the Labrador. Among the Cree, decorated bear skulls were attached to upright, 4.5 m high poles in a line, all facing the rising sun (Flannery/Chambers 1985, 10). In some cases, the display of a bear skull could also have an apotropaic function in society; the Kutenai would place a bear skull on an altar and offer up prayers to stop grizzly attacks on their people (Curtis 1911, 140–1).

One particular ritual involving bear skulls is worthy of a more detailed examination, that of tobacco offerings that are literally "introduced" into recently killed bears using pipes.

---

10 The Liuseno would instead erect a stone where a bear was killed (Sparkman 1908, 199).
11 Similar rites are found among the Ainu of Japan. On the first day of the bear ceremony, a bear cub was killed. On the second day, the bear skull was cleaned and decorated before being placed on a forked pole, whereupon prayers were offered. It was then turned toward the east, the direction of its destination, *Kamui-moshir*, the supernatural place where gods and the dead live (Yamada 2018, 39–41). For related practices in Eurasia, see Germonpré/Hämäläinen 2007.
12 Likewise after the dance and successfully killing a bear, the Natchez "prie le génie de l'animal de ne pas se fâcher et de ne pas lui être contrair dans une autre expédition" ("pray to the spirit of the animal to not become angry and to not work against him in another expedition"; Domenech 1862, 445–446).

In order to properly contextualize the use of pipes in bear-related ceremonies, a brief discussion of tobacco pipes in ancient North American indigenous traditions is necessary. Adena, Hopewell, and Mississippian cultures all participated in the production of effigy pipes. Pipes from the Adena culture have been found at the Eva Site that date to *c.* 2000 BC (Lewis/Lewis 1961, 66). Among the Hopewell, the two primary sites where effigy pipes have been recovered are Mound City, Ohio, where a large number of stone pipes were found in Mound 8 during the mid-1800s, and Tremper Mound, Ohio, where 136 pipes of various types were discovered in 1915. These stone pipes were commonly made in the shape of humans or various animals, such as raccoons, dogs, porcupines, minks, mountain lions, numerous types of birds, and beavers (Townsend 1954). A number of bear effigy pipes have also been discovered, such as one found in Tremper Mound (A 125/000028; Fig. 16). Effigy pipes were also ubiquitous throughout the later Mississippian culture (see Steponaitis et al. 2019). Pipes resembling those of the Mississippian culture were still being made by the Cherokee into the 19th century (Brown 1926, fig. 231).

Adena, Hopewell, and Mississippian cultures' pipes were often made of pipestone. Pipestone, which was commonly extracted from quarries near the Scioto Valley, Ohio,[13] refers to "claystones and massive, fine-grained, carvable metamorphic rocks" (Hughes et al. 1998, 711). Catlin (1841, 234) notes from his extensive experience in the early 1800s among Native American peoples that some of their pipes were made of "pipe-stone". He states that the quarry from which the stone was taken was "a place of such vast importance to the Indians – as given to them by the Great Spirit, for their pipes" and that this stone was "strictly forbidden to be used for anything else". Catlin also illustrated a wooden pipe with a bear carved near the bowl (Fig. 17; Catlin 1841).

Stone pipes seem to be ritual items that were not made for daily use.[14] A ceremonial function for pipes may be evident from an early example of a bear feast at the early Late Woodland (AD 1000–1300) Carpenter Brook site in Onondaga County, New York, which was part of the Mississippian culture.[15] Excavations by Ritchie in the 1940s produced bones from at least seven individual black bears together with a smoking pipe and a clay phallic effigy pipe (Ritchie 1947; 1965). The presence of the pipes buried with the bear remains suggests the pipes served a purpose in the bear feast ritual, something that can be confirmed through numerous North American tribal traditions. For example, during a bear feast following a successful hunt, Hallowell (1926, 64) notes this tradition of the Cree of Labrador: "In cases where there are several hunters together and a bear is killed, it is customary for them all to sit down around the carcass after the beast has been slain. The oldest man in the group then makes a bark pipe, smokes a while and blows a few puffs into the air before they proceed to eat the bear. The improvised pipe is then thrown away." Among the Eastern Cree, after a bear has been killed, the hunters place tobacco in the bear's mouth and then they smoke pipes over the bear (Skinner 1911, 69). The Cree of Moose Fort have a different procedure. They cut off the bear's head, and then a "large stone pipe was laid beside the head and a plug of tobacco placed upon it". The men around the bear then smoked the pipe, each with a single puff, before all taking one bite out of the head meat (Skinner 1911, 71).

Similar to the East Cree and the Cree of Moose Fort, the Plains-Cree perform a related tobacco ceremony. After killing a bear, the Plains-Cree would choose a worthy elderly man and give him a

---

13 Emerson et al. (2013) have recently shown that Hopewell Tremper Mound pipe raw-material came from Illinois and Minnesota, but that later Mound City cache pipes came almost exclusively from local limestone and pipestone.
14 Hopewell pipes sometimes show evidence of deliberate breaking, presumably to ritually "kill" the spirit contained in them (Gehlbach 2006).
15 Berres et al. (2004, 11) date this assemblage to *c.* AD 1000–1300.

large stone pipe with tobacco. He would smoke it and say a prayer to the gods that the bear was not killed for sport but for food (Skinner 1914, 510). In other cases, a pipe is given directly to the bear while invoking a prayer of good will for the slain bear (Skinner 1914, 514). A Cree hunter would sit by the recently killed bear and say: "Black Food, do not be angry. Do not let the other bear spirit be angry… When you go back to *Memekwesiw*,[16] tell him how I have treated you" (Rockwell 1991, 36). Among the neighboring Ungava of the Labrador region, hunters would also stick the pipe into the mouth of the bear and say: "My grandfather, I will light your pipe" (Hallowell 1926, 64). Carr et al. (2008, 516) note a related tradition in which pipes could be used to "blow into the nostrils of a killed bear to appease it". In the 1760s, Alexander Henry witnessed a similar use of pipes after killing a grizzly: "The pipes were now lit; and Wawatam blew tobacco smoke into the nostrils of the bear, telling me to do the same, and thus appease the anger of the bear on account of my having killed her" (Quaife 1921, 139–140). In this context, pipes were a mechanism for making a final offering[17] of tobacco to (and *into*) the bear so that other bears would not hold a grudge against the hunter.

During Vicomte de Chateaubriand's travels to America in 1791, he witnessed a bear hunt among the Natchez. He records that after killing the bear, the hunter "lights his pipe, puts the bowl [of the pipe] into the bear's mouth, and, blowing down the tube, fills the throat of the animal with smoke".[18] The hunter then implores the spirit of the bear not to try to thwart any future hunting expeditions (Vicomte de Chateaubriand 1928).[19] The use of stone pipes in close connection with bear rituals and bear feasts from these ethnographic sources strongly suggests a direct link to the pre-Columbian assemblage of bear bones and ceramic pipes at the Carpenter Brook site discussed above (cf. Ritchie 1947; 1965).

The connection between tobacco offerings and pipes to bears in North American indigenous traditions is multi-layered, beyond bear hunting rituals. Wooden pipes are also imbued with power in other bear-related contexts. Among the Blackfeet, medicine pipes were particularly potent objects. During a ceremony to transfer the Medicine Pipe from one person to another,[20] members of the Society don bear skins, chant bear songs, hold up bear paws, make lumbering movements like a bear, and carefully shake the pipe "in imitation of a bear" (McClintock 1910, 262–264). "Of all the Blackfeet medicines," states McClintock (1910, 267), "the pipe is believed to have the greatest power, but it also brings the greatest burden". In the pipe itself resided the power of the bear (McClintock 1923, 290). He also notes that the "word 'bear' must never be named before the Pipe, lest it cause bad dreams and bring sickness upon your family – the word 'badger' should always be used instead. The Evil Power in such a violation may be averted by burning sweet pine as incense" (McClintock 1910, 267–268; 1923, 296). The pipe could only be brought into the presence of a bear skin with great care, being cautious to use the word "badger" instead of "bear" (McClintock 1923, 423), always being held with two hands "just as the bear does" (McClintock 1910, 269). During the dance that preceded the transference of the Medicine Pipe, the chief, while dressed in bear costuming, would imitate a bear in dance, moving back and forth, breathing hard, and digging in the ground as if he were looking for insects (McClintock 1923, 295).

---

16 *Meme:kwe:ši:w* is the being who presides over "clawed" animals on earth. See Flannery/Chambers 1985.
17 If an offering is not made to the killed bear, the Ojibwa say that the hunter "will be punished supernaturally for he will meet a bear who will maul him or kill him in revenge" (Landes 1937, 137).
18 In a later retelling of the event in French, Vicomte de Chateaubriand simply says "il allume sa pipe, la met dans la geule de l'ours" ("he lights his pipe, he places it in the mouth of the bear") without a mention of the bowl of the pipe (Vicomte de Chateaubriand 1857, 148).
19 He then cuts out the string of the bear's tongue, which will be burned back in the village, and depending on how it crackles, it indicates whether the spirit of the bear is appeased or not (Vicomte de Chateaubriand 1928).
20 The Blackfoot Medicine Pipe is said to represent the authority to conduct all such rites (McClintock 1910, 160).

It is clear, therefore, that bears and tobacco/pipes have a deep and abiding connection in North American indigenous societies. In both hunting and non-hunting contexts tobacco was a ritual offering par excellence for bears, and pipes were an essential ritual instrument used to carry that offering to the bear.

Bear-human: A near approximation

"Bears are like people except they can't make fire," state the Yavapai of Arizona (Gifford 1933, 241). The belief that bears and humans are closely related pervades North American indigenous traditions. They are omnivorous, cohabit areas with humans, and stand up on their two hind legs like humans. The Cree of eastern California summarize the complex nature of bears by saying a bear is "a furry person, a relative, that goes underground when the earth sleeps and emerges when it awakes" (Wright 2013, 55). Landes (1968, 27) similarly notes that among the Chippewa bears were thought to be "quasi-human, in anatomy, erect carriage, cradling of young with the forearms, enjoyment of sweets and liquors, manner of drinking liquid, shows of intelligence, [and] inclination to moderate behavior despite great physical strength". Bears, especially when skinned, are said to strongly resemble human beings. Sometimes humans are said to have been born of bears. Franz Boas noted that the Nlaka'pamux of south central British Columbia believe that twins are the children of grizzly bears (Rohner 1969, 203). There are documented cases of women breastfeeding bear cubs as if they were their own (Schaeffer 1966, 16). Indeed, bears are often referred to with human kinship terms throughout North America and Eurasia, especially during bear hunts.[21] A common label of affinity is "great-grandfather",[22] "grandmother"/"grandfather",[23] or "old man".[24] In the mid-1700s, Alexander Henry experienced the heartfelt reaction of an Ojibwa named Wawatam Ojibwa to his killing of a grizzly bear. "The bear being dead, all my assistants approached, and all [...] took her head in their hands, stroking and kissing it several times; begging a thousand pardons for taking away her life: calling her their relation and grandmother; and requesting her not to lay the fault upon them, since it was truly an Englishman that had put her to death" (Quaife 1921, 139). In these traditions, bears are viewed relatives to humans, and their killing, while sometimes deemed necessary, is always a solemn experience.

"Bear doctors": Shamanistic traditions

One particular class of shamans, especially in California, is known as "bear doctors".[25] Of these bear shamans Kroeber (1907, 331) explained: "A special class of shamans found to a greater or less extent among probably all the Central tribes, though they are wanting both in the Northwest and the South, are the so-called bear doctors, shamans who have received power from grizzly bears, often by being taken into the abode of these animals – which appear there in human form –, and who after their

---

21 The Eastern Cree say that bears can understand everything said to them (Skinner 1911, 73), so using alternate names for them is a hunting strategy. For a full discussion of this phenomenon both in Eurasia and North America, see Hallowell 1926.
22 For this term among the Luiseno, see Gifford 1916, 209.
23 Such as among the Montagnuis-Naskapi. See Hallowell 1926, 44. Also, to a rattlesnake or a bear, a Pomo shaman will say: "Grandfather, I am not going to bother you. Let me go by safely." See Gifford/Kroeber (1937, 202).
24 For the use of this term among the Sauk, see Skinner 1923, 21.
25 The term *Gauk burakal* among the Pomo of California means "bear doctors", but translates literally as "human bear" (Lyon 1996, 86–87).

return to mankind possess many of the qualities of the grizzly bear, especially his apparent invulnerability to fatal attack. The bear shamans can not only assume the form of bears, as they do in order to inflict vengeance on their enemies, but it is believed that they can be killed an indefinite number of times when in this form and each time return to life."

KROEBER (1925, 259) notes that "bear doctors" do not attribute their magical powers to having the bear as one's guardian spirit,[26] but rather by the sheer possession of the bear paraphernalia. "In short," writes KROEBER (1925, 259), "he was the possessor of a fetish that increased his strength and endurance, and not a shaman at all, if the native information available may be relied on [...] There can thus be no doubt that the basis of the belief throughout California is shamanistic, and that the bear doctor falls into a class with the malignant shaman or evil witch."[27] Bear doctors can be male or female, but they must have a female assistant who aids with details such as sewing the doctor's all-important bear suit, known as a *gawī*, from a grizzly-bear skin, which imbued the wearer "with rapidity of motion and great endurance" (BARRETT 1917, 452–456, 464). Pomo bear doctors had secret hiding places, such as a cave, where they were instructed in the ritual songs and how to use ritual paraphernalia (LYON 1996, 86–87). They also work only at night under the cover of darkness, since they are then safe from hunters, and if any light from the moon appeared, they would immediately cease working (BARRETT 1917, note 6).

The Pomo also believe that bears can appear in dreams to instruct someone (often a bear doctor) how to commit murder. The person appearing in the dream has a body part of a bear, which gives him power "to make [the person] a bear", so that they can kill another – a process known as "bearwalking" (DORSON 1952, 34). DE LAGUNA (1987, 88) notes that this type of "Bear-walking" was practiced by "bad doctors" (i.e. shamans).

While Pomo and Yuki bear doctors do not heal, but rather are often menacing, in other North American indigenous groups bear doctors are expressly linked to curation,[28] as are certain bear societies.[29] Healing bear doctors dressed in bear skins would wear bear claw accoutrements, and often live and act like bears in order to solidify their power through mimicry. Early explorers often expressed amazement (and, at times, ridicule) at the bear doctors' costuming. George Catlin, an American artist who documented various Native American cultures, illustrated (Fig. 18) and wrote a description of a bear shaman summoned to perform a healing rite (CATLIN 1841, 38): "His entrée and his garb were somewhat thus [...] he approached the ring with his body in a crouching position, with a slow and tilting step; his body and head were entirely covered with the skin of a yellow bear, the head of which (his own head being inside of it) served as a mask; the huge claws of which also, were dangling on his wrists and ancles; in one hand he shook a frightful rattle, and in the other brandished his medicine spear or magic wand; to the rattling din and discord of all of which, he added the wild and startling jumps and yelps of the Indian, and the horrid and appalling grunts, and snarls, and growls of the grizzly bear, in ejaculatory and guttural incantations to the Good and Bad Spirits, in behalf of his patient; who was rolling and groaning in the agonies of death, whilst he was dancing around him, jumping over him, and pawing him about, and rolling him in every direction."

---

26 Others beyond bear doctors can have bears as their guardian spirit. Women who have the black or brown bear as their guardian spirit become industrious, good mothers, and skilled housekeepers, whereas men become expert at hunting and show great endurance (WHERRY 1964, 80).

27 RATZEL (1888, 99) notes that in California there were "Algunos impios wintunes se convertieron en osos grises, animales à quienes se tiene por encarnación de todo lo mal" ("Some ungodly Wintuns turned into gray bears, animals who are considered to be the incarnation of all evil").

28 In Tekwa, the word *kieh* means both "bear" and "doctor," showing the intimate relationship between them (WRIGHT 2013).

29 Among the Seneca, the Bear Society (*Nia'gwai'' oä''no'*) would meet to cure one of their members who was afflicted with an illness, such as a fever or rheumatism. The ceremony would begin by offering tobacco to the spirits of the bears, followed by various songs and a curative bear dance (PARKER 1909, 177).

The reason for the elaborate costuming and behavior imitation of bears was meant to link the individual to the curative abilities of bears. Bears were commonly viewed as having secretive powers related to healing, such as the Zuni who saw bears as the preeminent curing animal (STEVENSON 1904, 23). Likewise, for the Cheyenne, the bear is "a great medicine animal," who "possesses power – spiritual power," so that he can heal himself, and can heal other bears (GRINNELL 1923, 105). Also, in the Great Plains region, prevailing mythology viewed bears as healers par excellence. The Lakota in North and South Dakota have bear doctors known as *mato wapiye*, who live and eat like bears. They usually perform healing ceremonies in darkened rooms. If medicinal herbs are needed for the ceremony, they dig them up as bears would, using bear claws (LYON 1996, 166). Among the Teton Sioux, the elder Śiya´ka stated: "We consider the bear as chief of all the animals in regard to herb medicine, and therefore it is understood that if a man dreams of a bear he will be expert in the use of herbs for curing illness. The bear is regarded as an animal well acquainted with herbs because no other animal has such good claws for digging roots" (DENSMORE 1992, 195). In fact, ROCKWELL (1991, 76) suggests it is probably the fact that bears eat and gather plants and roots – the primary elements of medicine – that facilitated this association between bears and healing.

CONCLUSION

As semi-human, bears already hold a unique place among animals in indigenous North American thought. This study has described the ways in which bear ideology informs ritual, myth, and identity in these societies. Archaeological evidence indicates that some degree of bear ceremonialism, present from at least *c.* 8000 BC, and possibly several millennia earlier, existed in North America. Adena, Hopewell, and Mississippian cultures (*c.* 800 BC – AD 1600) incorporated bear paraphernalia and ideology into Clan divisions, healing rites, and mimetic dances. Post-Conquest ethnographic data provide a wealth of contextualization for many of the pre-Columbian artifacts related to bears. What becomes abundantly clear through all of these lines of evidence is that bears were revered for their near-humanness, their intimate knowledge and abilities in curing illness, and, perhaps most obviously, their power as apex land predator in North America.

BIBLIOGRAPHY

BARRETT 1917: S. A. BARRETT, Pomo Bear Doctors. Publications in American archaeology and ethnology 12, No. 11 (Berkeley 1917).

BERRES et al. 2004: T. E. BERRES/D. M. STOTHERS/ D. MATHER, Bear imagery and ritual in northeast North America: an update and assessment of A. Irving Hallowell's work. Midcontinental Journal of Archaeology 29(1), 2004, 5–42.

BERTINO 1994: L. BERTINO, The significance of bear canine artifacts in Hopewell context. Master's Thesis, Ball State University, Department of Anthropology (Muncie 1994).

BROWN 1926: C. S. BROWN, Archeology of Mississippi. Mississippi Geological Survey (Mississippi 1926).

CARR et al. 2008: C. CARR/R. WEEKS/M. BAHTI, The functions and meanings of Ohio Hopewell ceremonial artifacts in ethnohistorical perspective. In: D. T. CASE/ C. CARR, The Scioto Hopewell and Their Neighbors (New York 2008) 501–521.

CATLIN 1841: G. CATLIN, Letters and Notes on the Manners, Customs, and Condition of the North American Indians: Written During Eight Years' Travel Amongst the Wildest Tribes of Indians in North America (London 1841).

CONVERSE 1983: R. N. CONVERSE, A Hopewell Copper Headdress. Ohio Archaeologist 33(3), 1983, 21.

CURTIS 1911: E. S. CURTIS, The North American Indian VII. The Yakima. The Klickitat. Salishan tribes of the interior. The Kutenai (Seattle 1911).

CURTIS 1926: E. S. CURTIS, The North American Indian 16. The Tiwa, the Keres (Cambridge 1926).

CURTIS 1928: E. S. CURTIS, The North American Indian (1907–1930) 18. The Chipewyan. The Western woods Cree. The Sarsi (Cambridge 1928).

DE LAGUNA 1987: F. DE LAGUNA. Atna and Tlingit Shamanism: Witchcraft on the Northwest Coast. Arctic Anthropology 24(1), 1987, 84–100.

Densmore 1992: F. Densmore, Teton Sioux Music & Culture. Reprint edition (Lincoln 1992; originally published 1918).

Domenech 1862: E. H. D. Domenech, Voyage pittoresque dans les grands déserts du Nouveau Monde (Paris 1862).

Dorson 1952: R. M. Dorson, Bloodstoppers & Bearwalkers: Folk Traditions of the Upper Peninsula (Cambridge 1952).

Dragoo/Wray 1964: D. Dragoo/C. Wray, Hopewell Figurine Rediscovered. American Antiquity 30(2), 1964, 195–199.

Emerson et al. 2013: T. Emerson/K. Farnsworth/S. Wisseman/R. Hughes, The allure of the exotic: Reexamining the use of local and distant pipestone quarries in Ohio Hopewell pipe caches. American Antiquity 78(1), 2013, 48–67.

Flannery/Chambers 1985: R. Flannery/M. Chambers, Each Man Has His Own Friends: The Role of Dream Visitors in Traditional East Cree Belief and Practice. Arctic Anthropology 22(1), 1985, 1–22.

Frazer 1912: J. G. Frazer, Spirits of the corn and of the wild (London ³1912).

Gehlbach 2006: D. Gehlbach, D. Hopewell platform pipes, tools of magic and mystery. Central States Archaeological Journal 53(3), 2006, 130–133.

Germonpré/Hämäläinen 2007: M. Germonpré/R. Hämäläinen, Fossil Bear Bones in the Belgian Upper Paleolithic: The Possibility of a Proto Bear-Ceremonialism. Arctic Anthropology 44(2), 2007, 17–18.

Gifford 1916: E. W. Gifford, Clans and Moieties in Southern California 14 (Berkeley 1916).

Gifford 1933: E. W. Gifford, The Southeastern Yavapai (Berkeley 1933).

Gifford 1940: E. W. Gifford, CED: XII, Apache-Pueblo. University of California, University Records 4 (Berkeley 1940).

Gifford/Kroeber 1937: E. W. Gifford/A. L. Kroeber, Culture Element Distributions: IV Pomo. University of California Publications in American Archaeology and Ethnology 37(4), 1937, 117–254.

Giles 2013: B. Giles, A contextual and iconographic reassessment of the headdress on Burial 11 from Hopewell Mound 25. American Antiquity 78(3), 2013, 502–519.

Greber 1979: N. Greber, Variations in social structure of Ohio Hopewell peoples. Midcontinental Journal of Archaeology 4(1), 1979, 35–78.

Grinnell 1923: G. B. Grinnell, The Cheyenne Indians: their history and ways of life 1 (New Haven 1923).

Hallowell 1926: A. I. Hallowell, Bear Ceremonialism in the Northern Hemisphere. American Anthropologist, New Series, 28(1), 1926, 1–175.

Harrington 1913: M. R. Harrington, A Preliminary Sketch of Lenape Culture. American Anthropologist 15(2), 1913, 208–235.

Hill 1938: W. W. Hill, The Agricultural and Hunting Methods of the Navaho Indians. Yale University Publications in Anthropology, Number 18 (New Haven 1938).

Hughes et al. 1998: R. E. Hughes/D. M. Moore/T. E. Berres/K. B. Farnsworth, Revision of Hopewellian trading patterns in Midwestern North America based on mineralogical sourcing. Geoarchaeology: An International Journal 13(7), 1998, 709–729.

Ireton 1982: J. Ireton, The Bils Site. Ohio Archaeologist 32(3), 1982, 18.

Jorgensen et al. 1978: C. J. Jorgensen/R. H. Conley/R. J. Hamilton/O. T. Sanders, Management of black bear depredation problems. In: R. D. Hugie (ed.), Fourth Eastern Black Bear Workshop (Greenville 1978), 297–319.

Kennedy/Stevens 2013: D. Kennedy/J. R. Stevens, Recollections of an Assiniboine Chief (Toronto 2013).

Kinietz/Raudot 1940: W. V Kinietz/A. D. Raudot, The Indians of the Western Great Lakes: 1615–1760 (Ann Arbor 1940).

Koerper/Farmer 1987: H. C. Koerper/M. F. Farmer, A Bear-shaped Crescentic from Northern San Diego County, California. Journal of California and Great Basin Anthropology 9(2), 1987, 282–288.

Koldehoff 2006: B. Koldehoff, Hopewellian Figurines from the Southern American Bottom. Illinois Archaeology: Journal of the Illinois Archaeological Survey 18, 2006, 185–193.

Kroeber 1907: A. L. Kroeber, Religion of the Indians of California. University of California Publications in American Archaeology and Ethnology IV (Berkeley 1907).

Kroeber 1925: A. L. Kroeber, Handbook of the Indians of California. Bureau of American Ethnology, Bulletin 78 (Washington D.C. 1925).

Labbé 1903: P. B. R. Labbé, L'île de Sakhaline (Paris 1903).

Laliberte/Ripple 2004: A. S. Laliberte/W. J. Ripple, Range contractions of North American carnivores and ungulates. Bioscience 54(2), 2004, 123–138.

Landes 1937: R. Landes, Ojibwa Sociology (New York 1937).

Landes 1968: R. Landes, Ojibwa Religion and the Midewiwin (Madison 1968).

Lavivière 2001: S. Lavivière, Ursus americanus. Mammalian Species 647, 2001, 1–11.

Lewis/Lewis 1961: T. M. N. Lewis/M. D. K. Lewis, Eva: An Archaic Site (Knoxville 1961).

Lyon 1996: W. S. Lyon, Encyclopedia of Native American Healing (New York 1996).

Mather 2019: D. Mather, The Archaeobiology of Bears and Bear Ceremonialism in Minnesota. Unpubl. Ph.D. dissertation, University of Minnesota (Minneapolis 2019).

Mattson/Merrill 2002: D. J. Mattson/T. Merrill, Extirpations of grizzly bears in the contiguous United States, 1850–2000. Conservation Biology 16, 2002, 1123–1136.

McClintock 1910: W. McClintock, The Old North Trail: Or, Life, Legends and Religion of the Blackfeet Indians (London 1910).

McClintock 1923: W. McClintock, Legend of the Bear Spear by Onesta (Blackfoot) from Old Indian Trails (New York 1923).

Miller 1990: S. Miller, Population Management of Bears in North America. Bears: Their Biology and Management 8, 1990, 357–373.

Miller et al. 2011: S. Miller/J. W. Schoen/J. Faro/D. Klein, Trends in Intensive Management of Alaska's Grizzly Bears, 1980-2010. The Journal of Wildlife Management 75(6), 2011, 1243–1252.

Mills 1922: W. C. Mills, Explorations of the Mound City Group. Ross County, Ohio. American Anthropologist, New series, 24(4), 1922, 397–431.

Moorehead 1922: W. Moorehead, The Hopewell Mound Group of Ohio. Publications of the Field Columbian Museum. Anthropological Series 6(5), 1922, 75–185.

Parker 1909: A. C. Parker, Secret medicine societies of the Seneca. American Anthropologist 11(2), 1909, 161–185.

Peek et al. 1987: J. Peek/M. Pelton/H. Picton/J. Schoen/ P. Zager, Grizzly Bear Conservation and Management: A Review. Wildlife Society Bulletin (1973–2006) 15(2), 1987, 160–169.

Pelton 2003: M. R. Pelton, Black bear. In: G. A. Feldhamer/B. C. Thompson/J. A. Chapman (eds.), Wild mammals of North America: biology, management, and conservation (Baltimore 2003) 547–555.

Pelton et al. 1999: M. R. Pelton/A. B. Coley/T. H. Eason/D. L. Doan-Martinez/J. A. Pederson/F. T. van Manen/K. M. Weaver, American black bear conservation action plan. In: C. Servheen/S. Herrero/B. Peyton (eds.), Bears, status survey and conservation action plan. International Union for Conservation of Nature/ Species Survival Commission Bear Specialist Group – Polar Bear Specialist Group, International Union for Conservation of Nature Conservation Library (Cambridge [UK] 1999) 144–156.

Perrot 1864: N. Perrot, Memoire sur les moeurs, constumes et religion des sauvages del, Amerique septentrionale (Montreal 1864).

Phelps/Burgess 1964: D. S. Phelps/R. Burgess, A possible case of cannibalism in the Early Woodland period of eastern Georgia. American Antiquity 30(2), 1964, 199–202.

Prufer 1961: O. H. Prufer, Prehistoric Hopewell meteorite collecting: Context and implications. The Ohio Journal of Science 61(6), 1961, 341.

Prufer 1964: O. H. Prufer, The Hopewell cult. Scientific American 211(6), 1964, 90–105.

Quaife 1921: M. M. Quaife (ed.), Alexander Henry's Travels and Adventures in the Years 1760–1776, edited with Historical Introduction and Notes (Chicago 1921).

Ratzel 1888: F. Ratzel, Las razas humanas (Spain 1888).

Ritchie 1947: W. A. Ritchie, Archaeological Evidence for Ceremonialism in the Owasco Culture. Researches and Transactions of the New York State Archaeological Association, Lewis H. Morgan Chapter XI,2 (New York 1947).

Ritchie 1965: W. A. Ritchie, The Archaeology of New York State (Garden City 1965).

Rockwell 1991: D. Rockwell, Giving Voice to Bear: North American Indian Myths, Rituals and Images of the Bear (Niwot 1991).

Rohner 1969: R. Rohner (ed.), The Ethnography of Franz Boas: Letters and Diaries of Franz Boas Written on the Northwest Coast from 1886 to 1931 (Chicago 1969).

Romain 2009: W. F. Romain, Shamans of the lost world: A cognitive approach to the prehistoric religion of the Ohio Hopewell (Lanham 2009).

Schaeffer 1966: C. E. Schaeffer, Bear Ceremonialism of the Kutenai Indians. Studies in Plains Anthropology and History 4 (Browning 1966).

Scheick/McCown 2014: B. Scheick/W. McCown, Geographic distribution of American black bears in North America. Ursus 25(1), 2014, 24–33.

Seeman 1977: M. F. Seeman, Stylistic Variation in Middle Woodland Pipe Styles: The Chronological Implications. Midcontinental Journal of Archaeology 2, 1977, 47–66.

Seeman 1979: M. F. Seeman, The Hopewell Interaction Sphere: The Evidence for Interregional Trade and Structural Complexity. Prehistory Research Series 5.2 (Indianapolis 1978).

Shearar 2008: C. Shearar, Understanding Northwest Coast Art: A Guide to Crests, Beings and Symbols (Seattle 2008).

Shetrone 1926. H. C. Shetrone, Explorations of the Hopewell Group of Prehistoric Earthworks. Ohio Archaeological and Historical Quarterly 34, 1926, 1–227.

Skinner 1911: A. Skinner, Notes on the Eastern Cree and Northern Saulteaux. Anthropological Papers of the American Museum of Natural History IX.I (New York 1911).

Skinner 1914: A. Skinner, Political organizations, cults, and ceremonies of the Plains-Ojibway and Plains-Cree Indians. Anthropological papers of the American Museum of Natural History 11.6 (New York 1914).

Skinner 1923: A. Skinner, Observations on the Ethnology of the Sauk Indians. In: Bulletin of the Public Museum of the City of Milwaukee 5.1 (Milwaukee 1923) 1–57.

Sparkman 1908: P. S. Sparkman, The culture of the Luiseno Indians. University of California Publications in American Archaeology and Ethnology 8(4), 1908, 187–234.

Speck 1915: F. G. Speck, Myths and Folk-lore of the Timiskaming Algonquin and Timagami Ojibwa. Canada Department of Mines, Geological Survey Memoir 71, No. 9, Anthropological Series (Ottawa 1915).

Steponaitis et al. 2019: V. P. Steponaitis/V. J. Knight Jr./ G. E. Lankford, Effigy pipes of the Lower Mississippi Valley: Iconography, style, and function. Journal of Anthropological Archaeology 55, 2019. https://doi.org/ 10.1016/j.jaa.2019.101070.

Stevenson 1904: M. C. Stevenson, The Zuñi Indians: Their Mythology, Esoteric Fraternities, and Ceremonies. US Bureau of American Ethnology (Washington 1904).

Strong 1930: W. D. Strong, 1930 Notes on Mammals of the Labrador Interior. Journal of Mammalogy 11(1), 1930, 1–10.

Teit 1900: J. A. Teit, The Thompson Indians of British Columbia. Publications of the Jesup North Pacific Expedition, Vol. 1, Part 4. Memoirs of the American Museum of Natural History 2 (New York 1900).

Tomak 1994: C. H. Tomak, The Mount Vernon site: A remarkable Hopewell mound in Posey County, Indiana. Archaeology of Eastern North America 22, 1994, 1–46.

Townsend 1954: E. Townsend, Hopewell or Adena Effigy Pipes. Central States Archaeological Journal 1(2), 1954, 45–48.

Trowbridge 1938: Ch. C. Trowbridge, Shawnese Traditions: C. C. Trowbridge's Account. Edited by V. Kinietz and E. W. Voegelin. Museum of Anthropology, University of Michigan, Occasional Contributions 9 (Ann Arbor 1938).

Vicomte de Chateaubriand 1828: F. Vicomte de Chateaubriand, Travels in America and Italy (London 1828).

Vicomte de Chateaubriand 1857: F. Vicomte de Chateaubriand, Voyage en Amérique (Paris 1857).

Wherry 1964: J. H. Wherry, The totem pole Indians (New York 1964).

White 1932: L. A. White, The Pueblo of San Felipe. American Anthropological Association 38 (Menasha 1932).

Will 1980: G. B. Will, Black bear-human conflicts and management considerations to minimize and correct these problems. In: J. M. Collins/A. E. Ammons (eds.), Fifth eastern workshop on black bear management and research (Wrightsville Beach 1980) 75–88.

Willoughby 1935: C. C. Willoughby, Antiquities of the New England Indians: with notes on the ancient cultures of the adjacent territory 17 (Cambridge 1935).

Wright 2013: L. Wright, The Bear Book: Readings in the History and Evolution of a Gay Male Subculture (Binghamton [NY] 2013).

Yamada 2018: T. Yamada, The Ainu Bear Ceremony and the Logic behind Hunting the Deified Bear. Journal of Northern Studies 12(1), 2018, 35–51.

Prof. Kerry Hull
Brigham Young University
USA
kerry_hull@byu.edu

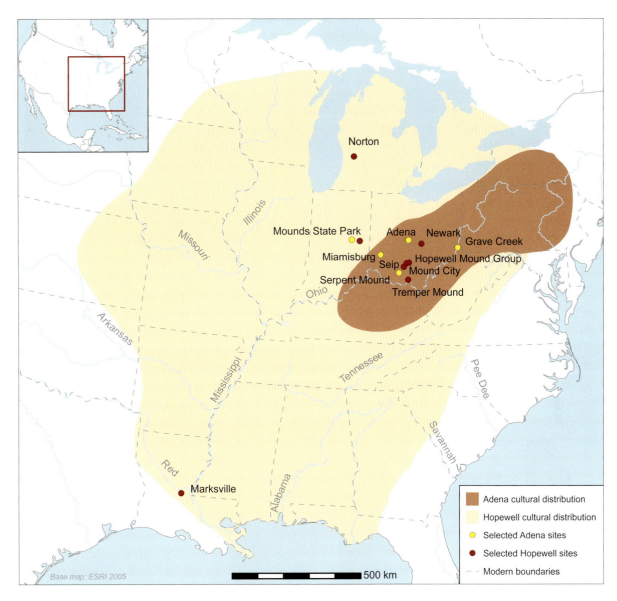

Fig. 1. Selected sites mentioned in the text (map GIS department, ZBSA).

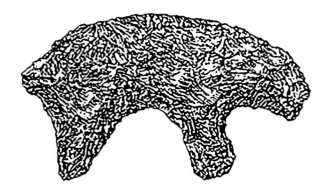

Fig. 2. Chipped bear-shaped stone crescentic, discovered at the Allan O. Kelly Archaeological Dig in San Diego County, California, dating to c. 8000–7000 BC (#CA-SDI-9649; after photo in KOERPER/FARMER 1987, fig. 1).

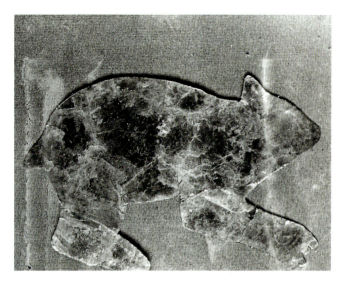

Fig. 3. Bear effigy made of mica, Hopewell Tremper Mound, Ohio (Ohio History Connection Archaeology Photograph Collection).

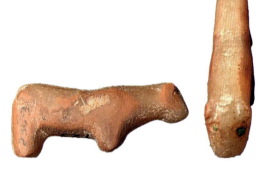

Fig. 4. Bear effigy discovered at the Hopewell Middle Woodlands (200 BC – AD 500) White Oaks Point site in Minnesota with copper inlays in its eyes (after MATHER 2019, 201–202, fig. 51). Not to scale.

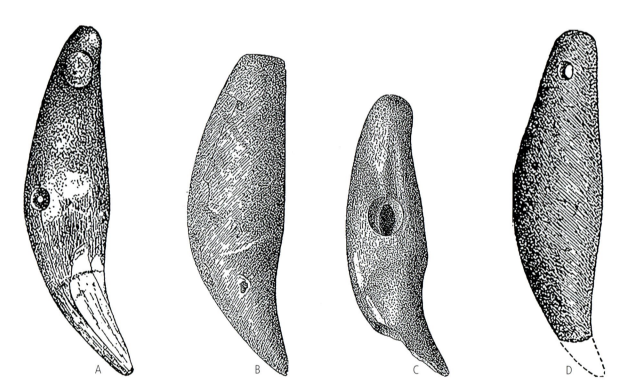

Fig. 5. Modified or artificial bear teeth from Mound 25 of Hopewell Mount, Ohio. a: Bear tooth inlaid with pearls (after MOOREHEAD 1922, fig. 49, right); b: Faux bear tooth made of wood and covered in copper (after MOOREHEAD 1922, fig. 57); c: Faux bear tooth made of deer antler (after MOOREHEAD 1922, fig. 58); d: Faux bear tooth possibly made of limestone (after MOOREHEAD 1922, fig. 58). Not to scale.

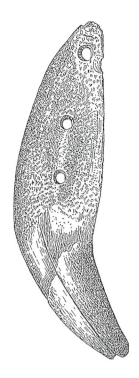 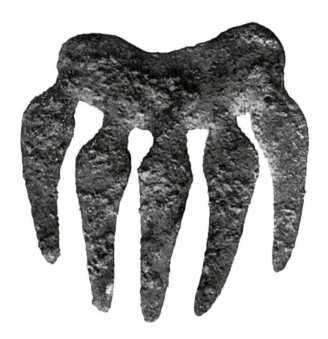

Fig. 6. Bear tooth likely used as a pendant during the Hopewell Middle Woodlands period (200 BC – AD 500; after MOORE-HEAD 1922, fig. 48).

Fig. 7. A bear effigy paw made of copper, discovered in the Hopewell Mound Group of Ohio (after MOOREHEAD 1922, pl. LXIX).

Fig. 8. Copper claw replicas, likely of a bear, found in the Hopewell Mound City Group earthworks in southern Ohio, dating to the Middle Woodland period (200 BC – AD 500; photo courtesy of Ohio History Connection, HOCU-2781, HOCU-2728, HOCU-2734, HOCU-2736).

Fig. 9. Bear claw necklace of the Nez Perce of northern central Idaho (c. 1900; photo courtesy of Nez Perce National Historical Park).

Fig. 10. A Qagyuhl man in full bear costuming (photo E. S. Curtis, 1914, courtesy of Charles Deering McCormick Library of Special Collections, Northwestern University Libraries. "Náne – Qágyuhl", Edward S. Curtis's The North American Indian. Accessed Wed Nov 02, 2022. https://dc.library.northwestern.edu/items/13038d9c-1fff-40c5-9502-3a847a7d2e42).

Fig. 11. Hopewell Copper bear effigy headdress from Burial 3, Mound 13, at the Mound City Group, Ohio, dating to 100 BC – AD 500 (after MILLS 1922, 413, fig. 38; graphically reworked by M. Bolte, ZBSA).

Fig. 12. Copper headdress from Mound 25, burial 4, of the Hopewell Group, Ross County, Ohio, with bear-claw design (Ohio State Museum, #33-Ro-27; photo K. Hull; graphically reworked by M. Bolte, ZBSA).

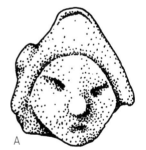 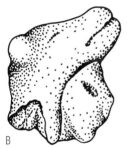

Fig. 13a–b. Fragment of a figurine from the Hopewell (100 BC – AD 500) Wenger #4 site in Illinois, wearing a full bear costume with the head covered by a bear headdress (after KOLDEHOFF 2006, fig. 2, used with permission from author).

Fig. 14. a: The Wray figurine showing a possible shaman wearing a bear costume, found in the Hopewell (100 BC – AD 500) Newark Earthworks, Licking Col, Ohio; b: Upside down view of severed human head in lap of Wray figurine (drawing A. C. Lange, after a template).

Fig. 15. Blackfoot Brave Dog clan members, one with a bear claw necklace (after McClintock 1923, 285).

Fig. 16. Bear effigy pipe discovered in Hopewell Tremper Mound in Ohio (100 BC – AD 500; A 125/000028; courtesy of the Ohio History Connection).

Fig. 17. A 19th-century pipe, possibly Sioux, with a bear and human carved near the bowl, that portion being carved from pipestone (graphically reworked by L. F. Thomsen after CATLIN 1841, pl. 98).

Fig. 18. Illustrations of a 19th-century Blackfoot bear shaman during a healing rite (after CATLIN 1841, pl. 19).

Table 1. Major cultural periods in North America.

Paleoindian	16000–8000 BC
Archaic period	8000–500 BC
Woodland period	500 BC – AD 1100
Mississippian period	1100–1541
Proto-historic period	1450–1750
Historic period	1542 – present

Bears in the starry sky

By Ernst Künzl

Keywords: Circumpolar constellations, the Great Bear, the Little Bear, wain, Mesopotamia, Homer, Greece, Germanic peoples

Abstract: Our names for the constellations of both hemispheres largely stem from ancient Greece. The Great Bear and the Little Bear belong to the circumpolar constellations that never set. The Great Bear's alternative appellation, the Wain, came from Mesopotamia and was known to the Greeks but never depicted by them. The prolific ancient literature is divided into "astrothetical" works (star catalogues) and "mythographical" works (the connection of the constellations to Greek myth). The two she-bears were associated with varying mythological conceptions, while transformation myths formed the background to many constellation names: Heroes, humans, and animals were transferred to the sky as stars by the gods (catasterism). The two she-bears were thus perceived as two wet nurses of Zeus, the king of the gods, on the island of Crete; other authors attempted to demonstrate a connection with Arcadia. The Germanic peoples again saw a wain in each of the two constellations, as it had formerly been the case in Mesopotamia.

In the far north of the northern hemisphere's starry sky stand the circumpolar constellations that never set, including the Great Bear (Ursa Major) and the Little Bear (Ursa Minor). These constellation names of ours can be directly traced back to those of Greek antiquity (Künzl 2018). In Greece and along the Greek Aegean coast of western Asia Minor, the Great Bear was known since Homer's Iliad, i.e. from *c.* 800 BC onwards.

Did Homer in his day come to know bears in western Asia Minor? Where around the Mediterranean Sea did bears live at the time? The answer is simple: almost everywhere. In our present-day sense, the brown bear (*Ursus arctos* L.) is perceived as an animal of the north, due to its occurrence in northern parts of Scandinavia and Russia. However, the bears in the mountains of south and southeast Europe remind us not to lose sight of Mediterranean geography: bears still live in the Carpathians, the Balkans, in Abruzzo, the Pyrenees, and in northwest Spain. In antiquity, conditions were yet different, with bears existing in the entire Mediterranean region, in Spain, Italy, Greece, and Asia Minor, but also in North Africa outside of Egypt (Toynbee 1983, 83–91; Bevan 1986; 1987; Hünemörder 2002). In the Roman arenas, bears had to fight with each other and against humans. Bear hunts were a royal topic of interest. Emperor Hadrian (AD 117–138) hunted bears in Greece (Toynbee 1983, 83) and Asia Minor; he killed a she-bear in Mysia in Anatolia, and on this site founded the city of Hadrianotherai ("Hadrian's kill"; *Historia Augusta*, Hadrian 20,13; Toynbee 1983, 83).

Depictions of bears are found everywhere in the Mediterranean area, albeit more from the Roman imperial period in the centuries following the birth of Christ than from the time before. The excellent bronze she-bear in Aachen cathedral came from Rome, where it had served as a fountain figure.

It was originally part of a Greek sculpture group of a hunting party of the 3rd century BC, and had probably come to Italy as part of the gigantic Roman art theft activities from the 3rd to the 1st centuries BC (Fig. 1; cf. Künzl 2003). Subsequently the bear goddess Artio from Muri near Bern, with her brown bear, stands out among the ancient bear depictions in a Roman work of the 2nd century AD (Fig. 2; cf. Bachofen 1863; 2020; Kaufmann-Heinimann 2002). In the Roman army of the imperial period, bears played an enduringly visible role; the standard bearers and signallers of the Roman legions (*signiferi*) primarily wore bear skins for deterrence, since their hands were not free to carry weapons (Künzl 2008, 23 figs. 31–32).

The Olympian gods of the Greeks had their corresponding animals, such as Zeus his eagle or Apollo a griffin. The animal goddess par excellence was Artemis, the Roman Diana. Her preferred, though not exclusive animal was the hind. One measure for the animal-deity relationships are the consecration figurines and sacrificial leftovers in Greek sanctuaries. Here, a differentiated picture emerges with respect to the bear. In the pre-Christian centuries, bears lived in Greece and Asia Minor, though they were still rare as a subject in religious ritual. Consecrations of bear depictions in Greek sanctuaries are scarce and limited to the goddess Artemis. Known as the Mistress of Animals (*Potnia Theron*), Artemis received such votive offerings in her temples on the island of Thasos, in Athens, and in the Peloponnese in particular (Argos, Tegea, Sparta; Bevan 1986, 338, 346; 1987). Arcadia in the Peloponnese is classic Greek bear country, which we will come back to with regard to the constellations. In the sanctuary of Artemis Brauronia in Brauron, Attica, the child-like ritual attendants were called *Arktoi* ("she-bears"; Bevan 1986, 18–19; Lohmann 1997).

Now these analyses concern a period partly far later than the first appearance of the she-bear as a constellation in the Homeric Iliad. What do we thus know about the two she-bears in the northern sky? Our knowledge rests on both archaeological evidence and ample ancient written sources. In Greek and Roman literature, two currents can be distinguished. The ancient scientific literature on the starry sky ("astrothetical" literature) often only casually takes note of the mythological explanation for the constellation names, focussing instead on star configurations. This literature encompasses the ancient fixed star catalogues: Eudoxus (4th century BC), Hipparchus (2nd century BC), the third book on astronomy by Hyginus (1st century BC – 1st century AD), and Claudius Ptolemy (2nd century AD).

The mythological tradition of ancient literature, on the other hand, primarily concerns itself with the explanation of the individual star sagas and constellations. This series in the first place includes Aratus (3rd century BC), followed by his Latin versions composed by Cicero (1st century BC), Germanicus (1st century AD), and Avienus (4th century AD). This is, however, only one branch within the field of mythological star sagas. Beside it, many Greek writers were involved with the composition of the mythological starry sky.

The ancient explanations of individual constellations could vary considerably, since different local traditions existed. Our naming sequence follows that of Claudius Ptolemy from the 2nd century, which had developed since late Hellenism and subsequently established itself in Ptolemy's version (Fig. 3; cf. Küentzle 1897; Gundel 1922; Boll/Gundel 1924–1937; Künzl 2005, 60–77; 2018). *Ptolemy* (*Geogr.* 1.22) calls a constellation an asterism, a "grouping of stars", signifying the charting and naming of the stars, i.e. an active process that ends with the appearance of a complete constellation. The term "catasterismi", introduced by Eratosthenes, designated the "placings among the stars", i.e. the placement of a being or object among the stars.

What do the stars in the sky say? There is the basic explanation that the stars represented an apotheosis, and that heroes such as Perseus, Heracles, and Orion lived on in the world of the stars as immortal beings, with the result, of course, that a codification of the myth, as it were, must be assumed: Perseus forever and ever fights the sea monster, and Andromeda forever and ever is chained to the rock. The other possibility is that the constellation names signified image denotations with a mythological or different background, which were then treated in an undogmatic manner.

Transformation (metamorphosis) and star apotheosis (placings among the stars; catasterism) were the categories concerning celestial processes when the constellations were named. Metamorphoses could be gruesome, as in the transformation of Princess Skylla into a multi-limbed monster while she was bathing in the sea (*Ovid, Metam.* 14,18–70). A divine metamorphosis could, however, also impart immortality in the sky, when the gods transferred a hero and his deed to the heavens, such as in the case of Perseus and Andromeda. These notions should not be interpreted too pedantically, since the monster (Cetus; Ketos) killed by Perseus likewise appears in the sky, and strictly speaking thus also became immortal (details in BOLL/GUNDEL 1924–1937; KÜNZL 2018).

An important element of the Greek view of the sky was the indicated relationship and hence movement of the constellations with respect to each other (HÜBNER 2004). Orion seems to follow the Pleiades, and Boötes the two she-bears, while the sea monster Cetus looks up towards Andromeda in order to devour her (HÜBNER 2004, 155).

The Greeks closely observed the movement of the stars, since they were hoping for calendrical help with agriculture and seafaring. The orientation of sailors towards specific constellations is already described in Homer's Odyssey, while the connection between constellations and seafaring probably was a Minoan legacy (WIESNER 1968, 33).

With the Greeks, we come across constellations and star names already in their great early poetry, the epics of Homer and the didactic poems of Hesiod. The shield of Achilles, with its rich sculptural decoration, is described by Homer, including the view of the heavens (*Homer, Iliad* 18, 483–489):

> *Therein he wrought the earth, therein the heavens, therein the sea,*
> *and the unwearied sun, and the moon at the full,*
> *and therein all the constellations wherewith heaven is crowned —*
> *the Pleiades, and the Hyades and the mighty Orion,*
> *and the Bear, that men call also the Wain,*
> *that circleth ever in her place, and watcheth Orion,*
> *and alone hath no part in the baths of Ocean.*

The rim of the shield is entwined by the ocean, the centre features the star chart of the northern sky, which almost appears like a planisphere of later types; the image of the she-bear (Ursa Major; *Arktos Megale* in Greek), dominating the northern sky, as well as the two constellations of Taurus (Bull, represented by the Hyades and the Pleiades) and Orion (Fig. 4), are mentioned. Moreover, Homer is familiar with Boötes ("he who ploughs with oxen"; also called Arctophylax, "bear-keeper"; cf. HÄBLER 1897; KNAACK 1897) and its brightest star, Arcturus. Homer has Calypso advise Odysseus to orient himself by Boötes, amongst others, during his voyage eastwards. Boötes thus already was an important seafaring mark at an early date (WENSKUS 1990). Odysseus sails away from Calypso's island to the east, and is guided by the Pleiades, Boötes, and Ursa Major (*Homer, Odyssey* 5, 271–277):

> *(...) nor did sleep fall upon his eyelids,*
> *as he watched the Pleiads, and late-setting Bootes,*
> *and the Bear, which men also call the Wain,*
> *which ever circles where it is and watches Orion,*
> *and alone has no part in the baths of Ocean.*
> *For this star Calypso, the beautiful goddess,*
> *had bidden him to keep on the left hand as he sailed over the sea.*

The northern hemisphere, as well as correspondingly the southern one, has a zone of stars around its respective pole, which always remain visible. In the north, these so-called circumpolar constellations that never set encompass the constellations Cassiopeia and Cepheus, as well as Draco (Dragon), the Little Bear (Ursa Minor), and the Great Bear (Ursa Major; Fig. 5). On the flat star charts, the planispheres, these circumpolar constellations occupy the centre of the depiction, which is arranged around the stellar north pole (Fig. 6). It is these constellations of the north that Homer refers to in his phrasing that the she-bear does not bathe in the ocean (*Homer, Odyssey* 5, 273–275), meaning that it never sets.

From among the circumpolar constellations, Homer was indeed only familiar with the Great Bear; he simply calls it the she-bear (*Arktos*). The Little Bear (Ursa Minor) was named only later. In antiquity, the definition of Ursa Minor was attributed to Thales of Miletus (6th century BC; BOLL/ GUNDEL 1924–1937, col. 869–870). With the appearance of the Little Bear, the Great Bear received its epithet ("great"). The alternative name for this constellation, the Wain, was already used by Homer, and his reference is incidentally the earliest instance for different appellations of one constellation. Another name, the Seven Sisters, was added, usually denoting the Pleiades but sometimes also the Great Bear (SCHERER 1953, 138–139). Though the Great Bear (Ursa Major; BOLL/GUNDEL 1924–1937, col. 873–881) was also called a wain in literary terms, the wain does not appear in Roman depictions (and therefore presumably also did not in the lost Greek illustrations).

The Wain as a constellation was a creation of the Mesopotamian astronomers in Babylon (JEREMIAS 1913, 33, 126, 128, 292; on the constellations of the Babylonians in general, see JEREMIAS 1909). The literary sources, mainly from the time of the Assyrian kings Assurnasirpal II (884–859 BC) and Assurbanipal (669–627 BC; cf. JEREMIAS 1913, 33), are chronologically so close to Homer's epics that a direct influence of the Babylonian constellation of the Wain on Homer can be assumed: "The she-bear, which some also call the Wain" thus would mean that the Babylonian astronomers named it the Wain. The beginnings of astronomy in Mesopotamia date back to the 2nd millennium BC (HUNGER 1997).

From Mesopotamia came the conception of the zodiac, an approximately 50°-wide belt in the sky on which the seven planets – as known in antiquity – moved, including Mars, Mercury, Jupiter, Venus, Saturn, as well as the sun and the moon. The Greeks subsequently adopted the Mesopotamian zodiac, together with some constellations such as Scorpio, Sagittarius, and Capricorn. Others were invented by the Greeks in a new form (GUNDEL 1992, 12, 36, 153; KÜNZL 2005, 42–43).

The constellation of the Wain must have been popular in the Orient; in Assur, a building inscription of King Sennacherib (705–681 BC) calls one of the north-facing gates the "Gate of the Wain Constellation" (JEREMIAS 1913, 126). A depiction of the constellation of the Wain was recognised on a cylinder seal (JEREMIAS 1913, 292–293 fig. 193). The Wain as a constellation was also known in China. A Chinese relief from AD 147 shows a star deity seated in a wain (JEREMIAS 1909, col. 1489–1490 fig. 50; 1913, 128 fig. 107). The Plough or Big Dipper was, just as Orion, and Ursa Minor or the Little Dipper, perceived as a cluster of seven stars in Babylon, with Ursa Minor or the Little Dipper also being defined as a plough (JEREMIAS 1913, 128). The bear, however, remained unknown in the Mesopotamian sky.

The Romans were initially unfamiliar with either the bear or the wain, and instead called the Great Bear *Septentriones*, i.e. the Seven Threshing Oxen (SCHERER 1953, 134–137) in the early period. When the Little Bear was subsequently recognised as its own constellation, this appellation was also transferred to it, and it became known as the Seven Little Threshing Oxen (*Septentriones Minores*; according to Varro; cf. BOLL/GUNDEL 1924–1937, col. 870). The Little Bear (Ursa Minor; BOLL/GUNDEL 1924–1937, col. 869–871) was called "dog's tail", too (*Canis cauda* [*Kynosura* in Greek]). The melange of terms for these two main constellations around the north pole becomes even more complex considering that the appellation of "wain" (*hamaxa* in Greek) was used for both the Great Bear and the Little Bear. Aratus (3rd century BC) calls them two she-bears, which frame the pole and jointly circle it, and therefore, they would also be called "wain" (*Aratus* 26–27). That the two she-bears are running after each other while circling the pole is also shown in ancient and medieval depictions (cf. Figs. 5–6).

Celestial globes are among the most important archaeological documents. Some of them only depict the zodiac and therefore have limited informative value (Gundel 1992 *passim*; Künzl 2005 *passim*). Complete depictions of the constellations in both hemispheres are, by contrast, rare. The large celestial globe on the Farnese Atlas in Naples (marble, early imperial period; Thiele 1898) and the small celestial globe in Mainz (brass, AD 150–220; cf. Künzl 2005, 60–70) are outstanding examples (cf. Fig. 3).

The inverted depiction of the constellations on the celestial globes of antiquity is not an error; this was a specific view of the ancient conception of the globe (Künzl 2005, 65). The Roman-era celestial globes show the starry sky with the eye of an observer who was imagined to be at the centre of the globe's interior. The correct view of the vaulted sky was thought to be from the centre of the globe, as if the imaginary observer was a tiny being in the middle of the sphere and the starry sky ought to orient itself towards this viewer. Therefore, the constellations on the ancient globes appear inverted; on them, the Great Bear and Leo (Lion), for example, look to the left (cf. Fig. 4), while modern star charts correctly show them looking to the right (Fig. 7). The ancient celestial globes are viewed by us as if from the outside of the vaulted sky and onto the imaginary external skin of the starry sky. Nevertheless, modern decorative celestial globes are illogical on their part by applying the correctly orientated star charts from the outside on the spherical shape of a globe. The ancient idea of a celestial globe with a (theoretically) tiny observer in its interior was only realised by the modern planetarium in its own way (developed by Walther Bauersfeld, Jena, 1919–1926); in a projection planetarium, we can let a moving and correctly orientated vaulted sky pass by in front of our eyes.

Hyginus (2.2) comments on the Little Bear in as much detail as on the Great Bear. The women Helice (Gundel 1912) and Cynosura (Gundel 1924) were wet nurses of Zeus as a child on Crete, and were transferred to the sky as she-bears; Helice ("the revolving one") as the Great Bear and Cynosura ("dog's tail"; Stoll 1890–1897) as the Little Bear. These interpretations go back to Aratus and the historian Aglaosthenes of Naxos, while it was Aratus who drew up the subsequently definitive connection to Crete (Gundel 1912). Following a few discussions on the term of the wain, Hyginus also mentions the name Phoenice (i.e. the lady from Phoenicia) for the Little Bear, the purported reason being that the Phoenician seafarers liked to orient themselves by the constellation of Ursa Minor (*Hyginus* 2.2). For him, the wain-based variants were a purely inner-Greek discussion. He was already unfamiliar with the wain constellation's Babylonian predecessor.

Arcadia in the Peloponnese was a land of wild animals and gruesome myths. The versions of Ursa Major, Ursa Minor, Boötes, and Hercules (*Engonasin*) that are linked to Arcadia instead of Phoenicia and Crete, demonstrate the endeavour to create a local complex of myths with Arcadia as the centre in the sky (Fig. 8). The author of this version was an unknown Greek poet or scholar of Hellenism. This Arcadian version did not prevail in the long run, according to the evidence in the ancient written sources (Künzl 2018, 52–53). One of the main subjects here was the transformation of Princess Callisto of Arcadia, who was one of the many lovers of Zeus, into a she-bear (Boll/Gundel 1924–1937, col. 874–875).

Hyginus enumerates a series of mythical variants for his readers in the early Roman imperial period, which are to be linked with the two she-bears. The common denominator was the mountainous region of Arcadia in the Peloponnese in southern Greece. Werewolf stories likewise originated in the diverse myths surrounding King Lycaon of Arcadia; behind the name Lycaon is *lykos* (for "wolf" in Greek). According to *Hyginus* (2.1), the cruel King Lycaon had a daughter, Callisto, who loved to hunt and was loved by the goddess Artemis. Callisto was supposed to, and wanted to, remain chaste for the sake of Artemis but was impregnated by Zeus and, as a punishment, transformed into a she-bear by Artemis. As a she-bear, she gave birth to a son, Arcas. Zeus (in other sources Artemis) transferred Callisto to the sky as the Great Bear, and likewise her son Arcas as Boötes (Arctophylax; for the Arcadian transformation myths in more detail, see also *Ovid, Metam.* 2, 405–507).

Other mythical versions include Zeus' jealous wife, Hera, in the events. Even for the reason why Ursa Major never set in the sky a story of jealousy among gods was invented: Tethys, goddess of the sea and wife of Oceanus, did not want to welcome the she-bear in her realm, meaning that it is not allowed to dive into the waters of the ocean and bathe therein: "[and the Bear, which] alone has no part in the baths of Ocean" (*Homer, Odyssey* 5, 274–275), the reason being that Tethys was Hera's wet nurse, and Hera had a grudge against Callisto as the lover of Zeus. This scholarly version, however, was presumably invented only in the Hellenistic period (WAGNER 1895).

According to the historian Araithos of Tegea, Ursa Major was not Callisto ("the most beautiful") but Megisto ("the largest"), and likewise an Arcadian and a daughter of Ceteus, the son of King Lycaon. Ceteus was then purportedly transferred to the sky as the constellation of Engonasin (which is otherwise interpreted as Hercules). For Ursa Minor, the Little Bear, Aratus' interpretation led the way, who oriented himself by Aglaosthenes of Naxos (*Aratus* 35–37). He connects the two she-bears with the myth of Zeus on Crete; the two she-bears Helice ("the revolving one"; Ursa Major) and Cynosura ("dog's tail"; Ursa Minor; STOLL 1890–1897) took care of Zeus as a child on Crete as wet nurses in the cave at Mount Ida. Another version of the myth declared Ursa Minor the hunting dog of Callisto of Arcadia, which was transferred to the sky as a constellation together with its mistress. In the case of the constellation of Boötes ("ox-driver"), and under the name of Arctophylax ("bear-keeper"), the references to the she-bears (*Arktoi*; *Ursae*) had the most success in antiquity; this had to do with the name Arcturus, the brightest star in the constellation, and referred to the landscape of Arcadia in southern Greece and the myth of the Arcadian clan hero Arcas and his mother Callisto.

For the Greeks, the names for the constellation of Boötes alternated between the wain and the bear (cf. Fig. 4, above right). The older name Boötes ("ox-driver") still contains the reference to the oxen of the wain. Since it was the she-bears that prevailed as a definition in the northern sky, the name of Boötes has changed to Arctophylax (bear-keeper; *Aratus* 92) since the 3rd century BC, which then remained predominant in the Roman writings (BOLL/GUNDEL 1924–1937, col. 886–887). Both Boötes and Arctophylax describe the relationships to the circumpolar constellations of the wain and the she-bear, respectively, in terms of views and gestures.

The starry sky of the Greeks dominated the ancient Roman period, too, and remains valid for us to this day. Other peoples, however, had their own astral systems. Unfortunately, nothing is known about the starry sky of the Celts, for example. This has to do with the fate of the druids. There was great respect for their intellectual achievements; they explored the stars and their movements, the size of the universe, natural science, and the power of the immortal gods, sharing their knowledge with the youth (*Caesar, Gallic War* VI 14). As the druids in Caesar's time, i.e. in the 2nd to 1st century BC, knew Greek and thus were capable of gaining access to Greek astronomy, specifics on the Celtic starry sky were to be expected. The Romans certainly ensured that in the hundred years between Caesar and emperor Nero, the caste of the druids as the intellectual elite of the Celts was eradicated; following the mass murder of druids by the Romans on the island of Mona (Anglesey) in the Irish Sea in the year 61 (*Tacitus, Ann.* XIV 29–30) there were no druids anymore. Their astronomical knowledge has since been lost.

It is likewise difficult to visualise the starry sky of the Germanic peoples. The ancient writers, including Tacitus with his *Germania* in particular, were either moralising ethnographers or primarily interested in military history. Germanic star names from the Migration Period and the time before Christianisation (in the north), therefore need to be retrojected from medieval sources, including all the uncertainties about the emergence of these individual names (REUTER 1934, celestial map fig. 83; MÜLLER 1970, 133–134 fig. 72; on the Germanic star names, see DREWS 1923, 89–126; REUTER 1934,

219–285; Scherer 1953, 129–131; Künzl 2018, 115–120). Any kind of "astrothetical" literature from the Germanic world is missing.

The Germanic pictorial records from the 3rd to the 8th century present a further problem. Many depictions are not easy to decipher, and as far as they can be interpreted, references to stars and constellations seem to be missing; this applies to the bracteates of the Migration Period as well as to the decorated weapons and the Gotlandic picture stones, with the exception of the two golden Gallehus horns from Jutland (Oxenstierna 1956; Brøndsted 1963, 324; Sommerfeld 2004, 86–87; Andersson 2008, 54), probably works of the Migration Period from the early 5th century. These contain reliefs of a pandemonium comprising warriors, combat scenes, animals, and mythical creatures. What is striking is the large number of stars as a filler motif in the background on one of the two golden horns. The obvious assumption has been that the animals and characters were linked to the Germanic starry sky of the 5th century (e.g. Betz 1927, 642–643), which cannot, however, be corroborated without utilisable written records (for a thesis on references to the solar eclipse of AD 413 and the lunar eclipse of AD 412, see Hartner 1969).

The most important sources continue to be written records. The methodological drawback is that these sources are spread across more than half a millennium, from the 8th to the 13th century, while some records date from an even later period. The published proposals should therefore be taken with a pinch of salt, and this also applies to the hypothetical drawing of the Germanic starry sky (Reuter 1934, fig. 83, and later Müller 1970, 134 fig. 71; critical of the hypotheses of Reuter 1934: Zinner 1935 *passim*; Bauer 1937, 47).

Examining some prominent constellations, we can already anticipate the outcome – bears are not to be found in them. The head of the bull, together with the Hyades and Aldebaran, in Taurus, was called *ulfs keptr* ("wolf's jaws"), according to an Old Icelandic source from the 13th century (gloss in the *Cod. 1812*, Reuter 1934, 274; Scherer 1953, 149). In Old Saxon glosses of the 8th and 9th centuries, the jostling boars appear as Orion's stars, which might have been a mix-up with the Pleiades (Steinmeyer/Sievers 1879, 496; 1882, 341; Reuter 1934, 273, 280; Scherer 1953, 148). On the other hand, the Pleiades are called the Seven Sisters; in Old High German glosses of the 9th century they appear under the names of *sibunstirri* and *sipunstirni* (Steinmeyer/Sievers 1882, 8; 1895, 606–607; Scherer 1953, 145).

The term "women's wain" (*kvennavagn*) for the Little Bear is rare and appears only once in an Old Icelandic source (*Cod. 1812*) in around 1250: "the bears that we call wain and women's wain" (translation after Reuter 1934, 249; Scherer 1953, 140). More numerous are the records for the Wain as the Great Bear (Reuter 1934, 183–184, 250–254; Drews 1923, 94). For Ursa Major, which already Homer knew as the Wain, the Wain dominates the scene, either simply as a wain or as *karlvagn* ("men's wain", Odin's or Thor's wain); apart from the stated Old Icelandic source, the records are, however, from a later period (cf. Scherer 1953, 140).

The Wain (the Great Bear of the south), the Hyades, and the Pleiades, thus were represented as constellation names with the Germanic peoples as well as Homer. The brightest star in the sky with the exception of the planets, Sirius in Canis Major ("greater dog") was associated with Loki, the fire-being and shapeshifter: *Loka brenna* ("Loki's fire") denotes Sirius in Icelandic; the appellation is, however, not yet verified in Old Norse (Reuter 1934, 280).

The names of other gods, too, are found in the sky. Highly poetic is the myth of the two eyes of the giant Thiassi (Thiazi), which Thor threw onto the sky (*Poetic Edda*, Songs of the Gods 7, Harbard's Song, verse 19). That Thiassi's eyes are identical with Castor and Pollux in Gemini (Reuter 1934, 282–283) may seem obvious, but is not literally stated in the texts (Zinner 1935, col. 695 calls the interpretation "arbitrary"). The catasterism of Thiassi's eyes was also attributed to Odin and not to Thor (*Prose Edda*, 2. Skáldsk. 56; Reuter 1934, 282). Out of the impressive constellation of Orion, which the Greeks (and we, too, along with them) always saw as a unit, many peoples picked out the three prominent belt stars, giving them special names, with the predominant terms being beam, pole, stick, and spindle. In the Nordic tradition, "Frigg's distaff" denoted the three belt stars of Orion. The designation is, however, unverified in Old Norse; the name primarily appears in Swedish records (cf. REUTER 1934, 272–273).

A very different and remarkably independent area of the Germanic view of the sky was the connection of the Milky Way (*Via lactea*) with persons of contemporary history. This began in Belgium and northern France with names such as Huldenstraat, Vroneldenstraat, or Brunelstraat for the Milky Way, which denoted the Merovingian queen Brunhilda (Brunichild, Visigothic by birth, † 613; BOLL/GUNDEL 1924–1937, col. 1026–1027; REUTER 1934, 283–284).

The Germanic connection to contemporary history reached a climax in the appellation "Iring's Way" for the Milky Way (REUTER 1934, 283–284; KÜNZL 2018, 119–120). The Thuringian defeat by the Franks and the demise of their kingdom in the early 6th century AD came about due to an inheritance dispute and Iring's betrayal (on Iring in general, see WEDDIGE 1989). The main source is the Saxon aristocrat Widukind of Corvey († 973), writing long after the events themselves, with his three-volume Saxon chronicle composed in Latin. Widukind confirms that in his day, the Milky Way was still named after the Thuringian Iring (*Widukind, Rerum Gestarum Saxonicarum* I, c. XIII; WEDDIGE 1989, 30, 63–69, 176).

During Widukind's time, i.e. in the 10th century, "Via lactea" had otherwise become the common name for the Milky Way; this was ensured by the illustrated codices with Aratus's/Germanicus's texts, which had by then become more widely distributed. There is no good manuscript of Iring's Song, and its contents have to be reconstructed from miscellaneous sources (on the personage of Iring in the various heroic epics, see WEDDIGE 1989, 99–118). That Widukind heard the Milky Way still being called Iring's Way 400 years after the demise of the Thuringian kingdom provides a glimpse of how impressive the epic must have been. But the astronomical literature of the European Middle Ages was focussed on the editions of the ancient texts. Iring could not find entrance into these, and was therefore again forgotten. As a reflection of partly historical events and partly of a Migration Period heroic song, the appellation Iring's Way for the Milky Way remains a fascinating isolated phenomenon.

Apart from the hero Iring, Odin, Frigg, Thor, and Loki, as prominent exponents of the Germanic realm of the gods, are also represented as constellations (BAUER 1937, 47). Bears are, however, unfortunately again not to be found therein.

BIBLIOGRAPHY

Primary sources

Aratus: Aratus, Phaenomena. Edited with introduction, translation and commentary by D. Kidd (Cambridge 1997).

Caesar, Gallic War: C. Ivlii Caesaris Commentarii rervm gestarvm, vol. 1. Edited by W. Hering (Berlin 1997).

Cod. 1812: see REUTER 1934; SCHERER 1953.

Historia Augusta: The Scriptores Historiae Augustae in three volumes with an English translation by D. Magie (Cambrige 1921).

Homer, Iliad: www.theoi.com/Text/HomerIliad18.html, translation by A. T. Murray.

Homer, Odyssey: www.theoi.com/Text/HomerOdyssey5.html, translation by A. T. Murray.

Hyginus: Hyginvs, Fabvlae. Edited by P. K. Marshall (Monachii 2002).

Ovid, Metam.: P. Ovidii Nasonis Metamorphoses. Edited by W. S. Anderson (Leipzig ^{3}1985).

Poetic Edda: U. Dronke, The Poetic Edda (Oxford 1969).

Prose Edda: Heimir Pálsson/A. Faulkes (eds.), Snorri Sturluson: The Uppsala Edda, DG 11 4to (PDF) – London: The Viking Society for Northern Research. Norse with English translation: Uppsala Edda.pdf. vsnrweb-publications.org.uk; A. Faulkes (ed.), Edda: Skáldskaparmál, vol. 1: Introduction, Text and Notes, vol. 2: Glossary and Index of Names (London 1998; 2007).

Ptolemy, Geogr.: Klaudios Ptolemaios, Handbuch der Geographie: griechisch – deutsch. Einleitung, Text und Übersetzung, Index. Edited by A. Stückelberger/G. Graßhoff, with F. Mittenhuber (Basel 2017).

Tacitus, Ann.: P. Cornelii Taciti, Libri qui supersunt. Ed. E. Kostermann. Teil 1: Ab excessv Divi Avgvsti (Leipzig ³1971).

Widukind, Rerum Gestarum Saxonicarum: B. S. Bachrach/ D. S. Bachrach, Widukind of Corvey: Deeds of the Saxons. Medieval texts in translation (Washington 2014).

Secondary sources

ANDERSSON 2008: K. ANDERSSON, Gold des Nordens. Skandinavische Schätze – von der Bronzezeit bis zu den Wikingern (Stuttgart 2008).

BACHOFEN 1863: J. J. BACHOFEN, Der Baer in den Religionen des Alterthums (Basel 1863).

BACHOFEN 2020: J. J. BACHOFEN, Der Bär in den Religionen des Altertums. In: A. Bollinger/A. Kaufmann-Heinimann/U. Breitenstein (eds.), Johann Jakob Bachofens gesammelte Werke V. Archäologische Schriften (Basel 2020) 117–186.

BAUER 1937: K.-G. BAUER, Sternkunde und Sterndeutung der Deutschen im 9.–14. Jahrhundert unter Ausschluß der reinen Fachwissenschaft. Germanische Studien 186 (Berlin 1937).

BETZ 1927: K. BETZ, Astralmythologie. In: E. Hoffmann-Krayer/H. Bächtold-Stäubli (eds.), Handwörterbuch des deutschen Aberglaubens 1 (Berlin 1927) col. 632–645.

BEVAN 1986: E. BEVAN, Representations of Animals in Sanctuaries of Artemis and other Olympian Deities. British Archaeological Reports, Intern. Ser. 315 (Oxford 1986).

BEVAN 1987: E. BEVAN, The Goddess Artemis, and the Dedication of Bears in Sanctuaries. Annual of the British School at Athens 82, 1987, 17–21.

BOLL 1903: F. BOLL, Sphaera. Neue griechische Texte und Untersuchungen zur Geschichte der Sternbilder (Leipzig 1903).

BOLL/GUNDEL 1924–1937: F. BOLL †/W. GUNDEL, Sternbilder, Sternglaube und Sternsymbolik bei Griechen und Römern. In: Roscher – Ausführliches Lexikon der griechischen und römischen Mythologie VI, supplements (Leipzig, u.a. 1924–1937) col. 867–1071.

BRØNDSTED 1963: J. BRØNDSTED, Nordische Vorzeit 3. Eisenzeit in Dänemark (Neumünster 1963).

DREWS 1923: A. DREWS, Der Sternhimmel in der Dichtung und Religion der alten Völker und des Christentums. Eine Einführung in die Astralmythologie (Jena 1923).

GUNDEL 1912: W. GUNDEL, Helike. In: Realencyclopädie der classischen Altertumswissenschaft (RE) VII (Stuttgart 1912) col. 2858–2862.

GUNDEL 1922: W. GUNDEL, Sterne und Sternbilder im Glauben des Altertums und der Neuzeit (Bonn 1922).

GUNDEL 1924: W. GUNDEL, Kynosura 6. In: Realencyclopädie der classischen Altertumswissenschaft (RE) XII/1 (Stuttgart 1924) col. 37–41.

GUNDEL 1992: W. GUNDEL, Zodiakos. Tierkreisbilder im Altertum. Kosmische Bezüge und Jenseitsvorstellungen am antiken Alltagsleben. Kulturgeschichte der antiken Welt 54 (Mainz 1992).

HÄBLER 1897: A. HÄBLER, Bootes 1. In: Paulys Realencyclopädie der classischen Altertumswissenschaft (RE) III,1 (Stuttgart 1897) col. 717–718.

HARTNER 1969: W. HARTNER, Die Goldhörner von Gallehus. Die Inschriften. Die ikonographischen und literarischen Beziehungen. Das Entstehungsdatum (Wiesbaden 1969).

HÜBNER 2004: W. HÜBNER, Zur Ikonographie des Sternenhimmels. In: W. Hübner/K. Stähler (eds.), Ikonographie und Ikonologie. Interdisziplinäres Kolloquium Münster 2001. Eikon – Beiträge zur antiken Bildersprache 8 (Münster 2004) 147–168.

HÜNEMÖRDER 2002: Ch. HÜNEMÖRDER, Bär. In: Der Neue Pauly 12/2 (Stuttgart 2002) col. 912.

HUNGER 1997: H. HUNGER, Astronomie. In: Der Neue Pauly 2 (Stuttgart 1997) col. 127–130.

JEREMIAS 1909: A. JEREMIAS, Sterne (bei den Babyloniern). In: Roscher – Ausführliches Lexikon der griechischen und römischen Mythologie (Leipzig 1909) col. 1427–1500.

JEREMIAS 1913: A. JEREMIAS, Handbuch der altorientalischen Geisteskultur (Leipzig 1913).

KAUFMANN-HEINIMANN 2002: A. KAUFMANN-HEINIMANN, Dea Artio, die Bärengöttin von Muri. Römische Bronzestatuetten aus einem ländlichen Heiligtum. Glanzlichter aus dem Bernischen Historischen Museum 9 (Bern 2002).

KNAACK 1897: G. KNAACK, Bootes 2. In: Paulys Realencyclopädie der classischen Altertumswissenschaft (RE) III,1 (Stuttgart 1897) col. 718–719.

KÜENTZLE 1897: H. KÜENTZLE, Über die Sternsagen der Griechen I (Karlsruhe 1897).

KÜNZL 2003: E. KÜNZL, Die Bärin im Dom zu Aachen. Jahrbuch RGZM 49, 2002 (2003), 1–39.

KÜNZL 2005: E. KÜNZL, Himmelsgloben und Sternkarten. Astronomie und Astrologie in Vorzeit und Altertum (Stuttgart 2005).

KÜNZL 2008: E. KÜNZL, Unter den goldenen Adlern. Der Waffenschmuck des römischen Imperiums (Regensburg 2008).

KÜNZL 2018: E. KÜNZL, Helden am Himmel. Astralmythen und Sternbilder des Altertums (Mainz 2018).

LOHMANN 1997: H. LOHMANN, Brauron. In: Der neue Pauly 2 (Stuttgart 1997) 762–763.

MÜLLER 1970: R. MÜLLER, Der Himmel über dem Menschen der Steinzeit. Astronomie und Mathematik in den Bauten der Megalithkulturen. Verständliche Wissenschaft 106 (Berlin 1970).

OXENSTIERNA 1956: E. Graf OXENSTIERNA, Die Goldhörner von Gallehus (Lidingö 1956).

Reuter 1934: O. S. Reuter, Germanische Himmelskunde. Untersuchungen zur Geschichte des Geistes (München 1934).

Scherer 1953: A. Scherer, Gestirnnamen bei den indogermanischen Völkern. Indogermanische Bibliothek, Dritte Reihe: Untersuchungen. Forschungen zum Wortschatz der indogermanischen Sprachen 1 (Heidelberg 1953).

Sommerfeld 2004: C. Sommerfeld, Mythische Geschichten aus der Bronzezeit – Ein phantastischer Ausblick. In: H. Meller (ed.), Der geschmiedete Himmel. Die weite Welt im Herzen Europas vor 3600 Jahren (Stuttgart 2004) 82–87.

Steinmeyer/Sievers 1879: E. Steinmeyer/E. Sievers, Die althochdeutschen Glossen 1 (Berlin 1879).

Steinmeyer/Sievers 1882: E. Steinmeyer/E. Sievers, Die althochdeutschen Glossen 2 (Berlin 1882).

Steinmeyer/Sievers 1895: E. Steinmeyer/E. Sievers, Die althochdeutschen Glossen 3 (Berlin 1895).

Stoll 1890–1897: H. W. Stoll, Kynosura. In: Roscher – Ausführliches Lexikon der griechischen und römischen Mythologie II (Leipzig 1890–1897) col. 1706.

Thiele 1898: G. Thiele, Antike Himmelsbilder. Mit Forschungen zu Hipparchos, Aratos und seinen Fortsetzern und Beiträgen zur Kunstgeschichte des Sternenhimmels (Berlin 1898).

Toynbee 1983: J. M. Toynbee, Tierwelt der Antike. Kulturgeschichte der antiken Welt 17 (Mainz 1983).

Wagner 1895: R. Wagner, Arktos 2. In: Paulys Realencyclopädie der classischen Altertumswissenschaft (RE) II,1 (Stuttgart 1895) col. 1172–1173.

Weddige 1989: H. Weddige, Heldensage und Stammessage. Iring und der Untergang des Thüringerreiches in Historiographie und heroischer Dichtung. Hermaea N. F. 61 (Tübingen 1989).

Wenskus 1990: O. Wenskus, Astronomische Zeitangaben von Homer bis Theophrast. Hermes Einzelschriften 55 (Stuttgart 1990).

Wiesner 1968: J. Wiesner, Griechische Sternbilder der Frühzeit. Raggi 8(2), 1968, 29–43.

Zinner 1935: E. Zinner, Rezension von Reuter 1934. Deutsche Literaturzeitung 56, 1935, col. 692–698.

Dr. Ernst Künzl
Eckental
Germany
info@archaeologie-sachbuch.de

Fig. 1. She-bear sculpture. Original: Greek work from the 3rd century BC, bronze, height 84 cm, Aachen, cathedral porch. Depicted is a copy. Römisch-Germanisches Zentralmuseum (RGZM), Mainz, Germany (photo E. Künzl).

Fig. 2. The she-bear goddess Artio from Muri, Canton of Bern, Switzerland. Roman bronze figurine, late 2nd century AD, length of base 28.6 cm. Bernisches Historisches Museum, Bern, Switzerland (photo St. Rebsamen).

1195

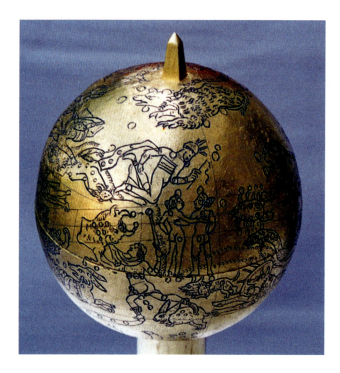

Fig. 3. The Great Bear (on top) on a Roman celestial globe with 48 constellations. Original: AD 150–220, brass, diameter 11 cm. Depicted is a galvanoplastic copy with dark line colouring. Römisch-Germanisches Zentralmuseum (RGZM), Mainz, Germany (photo E. Künzl).

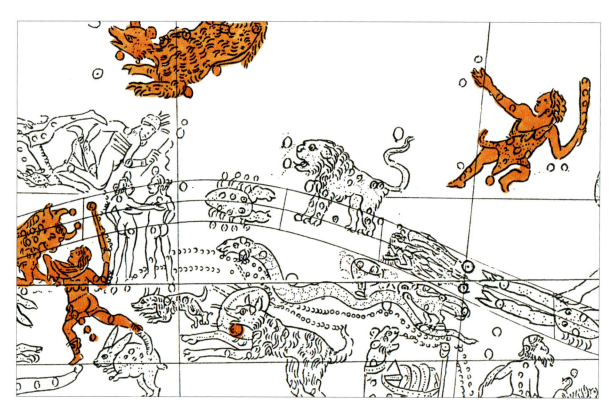

Fig. 4. The stars of Homer. From below in clockwise direction, coloured: Sirius in Canis Major (the Greater Dog), Orion, the Pleiades in Taurus (the Bull), Ursa Major (the Great Bear), and Boötes (the Ox-driver). Depiction on Roman celestial globe, AD 150–220, brass. Römisch-Germanisches Zentralmuseum (RGZM), Mainz, Germany (drawing courtesy of RGZM, Mainz; colours by E. Künzl).

Fig. 5. Circumpolar constellations: the Great Bear (Ursa Major), the Little Bear (Ursa Minor), and Draco. Depiction on Roman celestial globe, AD 150–220, brass, diameter 11 cm. Römisch-Germanisches Zentralmuseum (RGZM), Mainz, Germany (drawing courtesy of RGZM, Mainz; colours by E. Künzl).

Fig. 6. Circumpolar constellations (Ursa Major, Ursa Minor, Draco) on a star chart (planisphere) from the 15th century, following late antique source materials. Codex Vaticanus graecus 1087-310v. Vatican, Bibliotheca Apostolica Vaticana, Rome, Italy (after BOLL 1903, plate 1 [detail]).

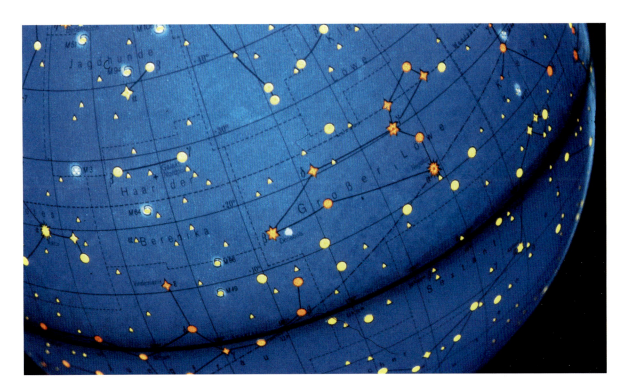

Fig. 7. Modern decorative celestial globe. Coma Berenices (Berenice's Hair) and Leo (Lion) are in the centre, Leo correctly faces to the right (photo E. Künzl).

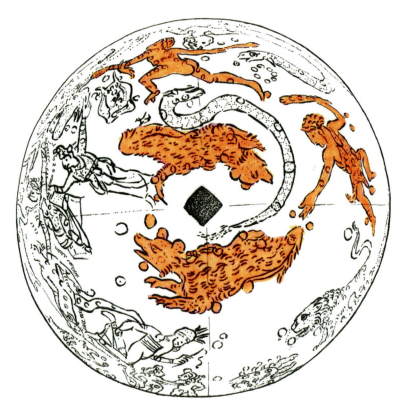

Fig. 8. References to Arcadia. From below in clockwise direction, coloured: Ursa Major (the Great Bear), Ursa Minor (the Little Bear), Engonasin (the Kneeler; Hercules), and Arctophylax (the Bear-keeper; Boötes) were associated with Callisto, Cynosura, Arcas, and Ceteus of Arcadia. Depiction on Roman celestial globe, AD 150–220, brass, diameter 11 cm, Römisch-Germanisches Zentralmuseum (RGZM), Mainz, Germany (drawing courtesy of RGZM, colours by E. Künzl).